ENGINEERING ANALYSIS OF SMART MATERIAL SYSTEMS

ENGINEERING ANALYSIS OF SMART MATERIAL SYSTEMS

Donald J. Leo

Department of Mechanical Engineering
Virginia Polytechnic Institute and State University
Blacksburg, Virginia

JOHN WILEY & SONS, INC.

This book is printed on acid-free paper. ♾

Published by John Wiley & Sons, Inc., Hoboken, New Jersey.
Published simultaneously in Canada.

Wiley Bicentennial Logo: Richard J. Pacifico

For general information on our other products and services, please contact our Customer Care Department within the United States at (800) 762-2974, outside the United States at (317) 572-3993 or fax (317) 572-4002.

Wiley also publishes its books in a variety of electronic formats. Some content that appears in print may not be available in electronic formats. For more information about Wiley products, visit our web site at www.wiley.com

Library of Congress Cataloging-in-Publication Data:

Leo, Donald J.
Engineering analysis of smart material systems/by Donald J. Leo.
p. ; cm.
Includes bibliographical references.
ISBN 978-0-471-68477-0
1. Smart materials. 2. Electronic apparatus and appliances. I. Title.
TA418.9.S62L46 2007
620.1′.1–dc22

2006103082

Printed in the United States of America

10 9 8 7 6 5 4 3 2 1

To my wife, Jeannine Alexander, and sons, Jonathan and Matthew Leo

CONTENTS

PREFACE

The field of smart materials has grown considerably in the past ten to fifteen years as topics have transitioned from basic research to commercial applications. In the beginning stages of my career it was still rather exotic for a device or structure to be called "smart." It was the early 1990s, and while the integration of electronics into everyday devices was underway, the revolutions that would become wireless communication and global networking were only beginning. Today it is somewhat trite to call a device "smart" since almost all engineering systems have some measure of networked "intelligence" to them through the integration of sensors, electronics, and actuators. Hopefully this book will provide a basis of understanding for many of the materials and material systems that underlie the analysis and design of "smart" devices.

This book grew out of a series of courses taught by myself to graduate and undergraduate students at Virginia Tech in the late 1990s and early 2000s. The primary purpose of this book is to provide educators a text that can be used to teach the topic of smart materials and smart material systems. For this reason the book is organized into two introductory chapters, four chapters that teach the basic properties of several types of smart materials, and a final four chapters that apply the material concepts to engineering application areas. Each chapter contains worked examples and solved homework problems that reinforce the mathematical concepts introduced in the text. A secondary purpose of the book is to provide practicing engineers a text that highlights the basic concepts in several types of applications areas for smart materials, areas such as vibration damping and control, motion control, and the power considerations associated with smart material systems.

As with any project of this size there are a number of people to thank. Early motivation for writing a book came from discussions with my mentor, colleague, and friend Dr. Daniel Inman. He told me on a number of occasions that writing a book is a rewarding experience and in a couple of years I will most certainly agree with him. I also want to thank all of the graduate students that I have worked with over the years. In particular, one of my Masters students, Miles Buechler, now at Los Alamos National Laboratory, and one of Dan's students, Pablo Tarazoga, gave me valuable corrections on early versions of the manuscript. Dr. Kenneth Newbury, Dr. Matthew Bennett, Dr. Barbar Akle, Dr. Curt Kothera, and Dr. Kevin Farinholt also deserve special mention because much of the work I have written about on polymer actuators has been performed in collaboration with them. I would also like to thank Dr. Zoubeida Ounaies and Dr. Brian Sanders for using early versions of this manuscript in their courses and providing me feedback on the topics. Finally, my editors at Wiley, Bob Argentieri and Bob Hilbert, require special thanks for their patience during my three job changes that occurred over the course of writing this text.

The most important thanks, of course, goes to my wife Jeannine and two sons, Jonathan and Matthew, for supporting me throughout this project. To them this book is dedicated.

ENGINEERING ANALYSIS OF SMART MATERIAL SYSTEMS

1

INTRODUCTION TO SMART MATERIAL SYSTEMS

The purpose of this book is to present a general framework for the analysis and design of engineering systems that incorporate *smart materials*. Smart materials, as defined in this book, are those that exhibit coupling between multiple physical domains. Common examples of these materials include those that can convert electrical signals into mechanical deformation and can convert mechanical deformation into an electrical output. Others that we will learn about are materials that convert thermal energy to mechanical strain, and even those that couple the motion of chemical species within the material to mechanical output or electrical signals. We focus on developing an understanding of the basic physical properties of different types of materials. Based on this understanding we develop mathematical models of these smart materials and then incorporate these models into the analysis of engineering systems. Through a basic understanding of smart material properties and how they are integrated into engineering systems, we will gain an understanding of engineering attributes such as range of motion, ability to generate force, and the speed of response of the materials.

Central to the book is the development of methods for analyzing and designing systems that incorporate smart materials. We define a *smart material system* as an engineering system that utilizes the coupling properties of smart materials to provide functionality. This is a very broad term, but it is suitable in relation to the organization of this book. Smart material systems that fit this definition include a machine that utilizes an electromechanical transducer as a means of real-time monitoring of its "health" and a semiconductor wafer manufacturing system that uses a smart material to control the motion of its positioning stage with nanometer-level accuracy. Other examples include the use of smart material ceramics as a means to control the vibrations of jet fighters or to reduce the vibrations of sensitive optical equipment. Under the broad definition that we have proposed, smart material systems also include the use of pseudoelastic wires for low-force interconnects or as a means of controlling the expansion of stents, biomedical devices that are useful in the treatment of cardiovascular disease.

All of these applications of smart material systems require a knowledge of the basic properties of various types of smart materials, methods for modeling the coupling mechanisms within these materials, and mathematical approaches to incorporating material models into models of engineering systems. Not surprisingly, then, these three topics are the central themes of this book. In addition to these central themes, we also study specific topics in the use of smart materials: the use of smart material systems for motion control, vibration, suppression, and power considerations associated with smart material systems.

1.1 TYPES OF SMART MATERIALS

The study of smart materials and smart material systems is a diverse discipline. Over the past 10 to 20 years, a number of materials have been given the term *smart* based on their interesting material properties. Some of these materials exhibit a volume change when subjected to an external stimulus such as an electric potential; others shrink, expand, or move when heated or cooled. Still other types of smart materials produce electrical signals when bent or stretched. Other names for these types of materials are *intelligent materials, adaptive materials*, and even *structronic materials*.

As noted earlier, we define smart materials as those that convert energy between multiple physical domains. A *domain* is any physical quantity that we can describe by a set of two state variables. A more mathematical definition of state variables is provided in Chapter 2, but for now a *state variable pair* can be thought of as a means of defining size or location within a physical domain. An example of a physical domain that we study at length is the *mechanical domain*, whose state variables are the states of stress and strain within a material. Another example of a physical domain is the *electrical domain*, whose state variables are the electric field and electric displacement of a material. Other examples are the *thermal, magnetic*, and *chemical domains* (Figure 1.1).

Defining physical domains and associated state variables allows us to be more precise in our definition of the term *coupling*. Coupling occurs when a change in the state variable in one physical domain causes a change in the state variable of a separate physical domain. Coupling is generally denoted by a term that is a combination of the names associated with the two physical domains. For example, changing the temperature of a material, which is a state variable in the thermal domain, can cause a change in the state of strain, which is a mechanical state variable. This type of

Mechanical	Electrical	Thermal	Magnetic	Chemical
Stress	Electric field	Temperature	Magnetic field	Concentration
Strain	Electric displacement	Entropy	Magnetic flux	Volumetric flux

Figure 1.1 Examples of physical domains and associated state variables.

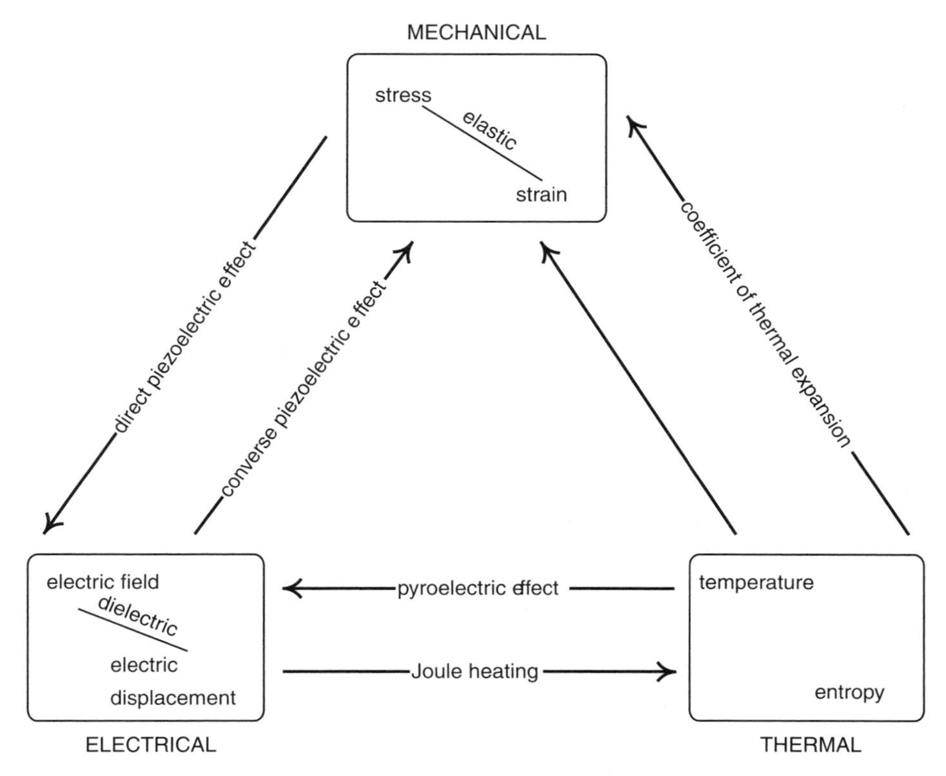

Figure 1.2 Visual representation of coupling between physical domains.

coupling is called *thermomechanical coupling* because the coupling occurs between the thermal and mechanical physical domains.

A visual representation of the notion of physical domains and the coupling between them is shown in Figure 1.2. Each rectangle represents a single physical domain, either mechanical, electrical, or thermal. Listed in each rectangle are the state variables associated with the domain. The bridge within the rectangle is the physical property that relates to the state variables. The elastic properties of a material relate the states of stress and strain in the material, and the dielectric properties relate the electrical state variables. Coupling between the physical domains is represented by the arrows that connect the rectangles. For example, the electrical output produced by a thermal stimulus is termed the *pyroelectric effect*. Similarly, the variation in mechanical stress and strain due to a thermal stimulus is termed *thermal expansion*.

In this book we concentrate on materials that exhibit one of two types of coupling: electromechanical or thermomechanical. Electromechanical materials are characterized by their ability to convert an electrical signal into a mechanical response, and in a reciprocal manner, to convert a mechanical stimulus into an electrical response. The fact that a reciprocal relationship exists between electrical and mechanical coupling

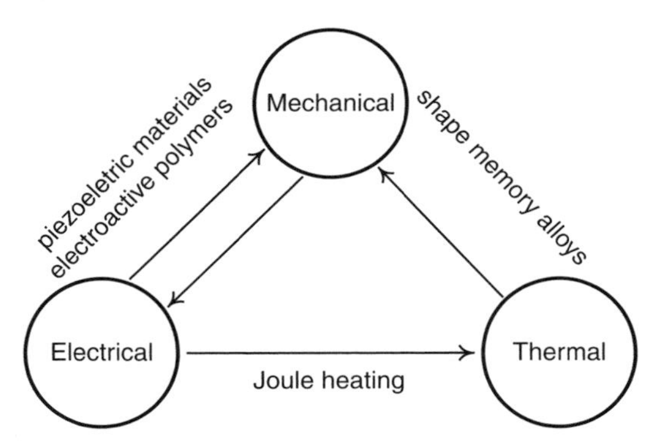

Figure 1.3 Various types of coupling exhibited by the materials studied in this book.

motivates us to define these materials as exhibiting *two-way coupling*. The thermomechanical materials studied in this book are an example of material that exhibits what we call *one-way coupling*. The thermomechanical materials discussed produce mechanical deformation when heated, but unlike two-way electromechanical materials, they do not produce a measurable temperature rise due to the mechanical deformation.

We focus on understanding the coupling properties of three types of smart materials. *Piezoelectric materials*, the first set of materials studied, convert energy between the mechanical and electric domains. *Shape memory alloys*, the second set of materials studied, are thermomechanical materials that deform when heated and cooled. The third class of materials that we study form a subset of *electroactive polymers* that exhibit electromechanical coupling. The electroactive polymer materials we study are functionally similar to piezoelectric materials but exhibit much different electromechanical response characteristics.

The relationship between electrical, thermal, and mechanical domains is shown concisely in Figure 1.3 for the materials studied in this book. The three vertices of the triangle represent the physical domains and the interconnections between the vertices are materials studied in this book that exhibit coupling behavior. We introduce methods to study the three broad classes of materials described above. The electromechanical coupling in piezoelectric materials is studied in depth. This study requires us to define the fundamental physical properties of piezoelectric materials and mathematical representations of electromechanical coupling in these materials. The thermomechanical behavior of shape memory alloys is also studied. As with piezoelectric materials, shape memory alloys are discussed in relation to their fundamental material properties and in the context of engineering models for these materials. The third broad class of materials, electroactive polymers are a class of materials that exhibit electromechanical coupling. These materials are functionally similar to piezoelectric devices but have a number of interesting properties that makes them useful for applications in which piezoelectric devices are not appropriate. Examples of these three types of materials are shown in Figure 1.4.

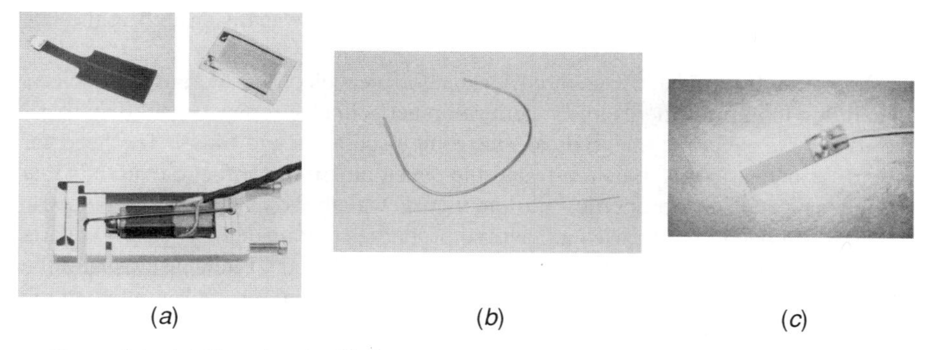

(*a*) (*b*) (*c*)

Figure 1.4 (*a*) Piezoelectric, (*b*) shape memory, and (*c*) electroactive polymer materials.

1.2 HISTORICAL OVERVIEW OF PIEZOELECTRIC MATERIALS, SHAPE MEMORY ALLOYS, AND ELECTROACTIVE POLYMERS

The story of the materials covered in this book starts in the late nineteenth century, when the Curie brothers discovered that several natural materials, including quartz and Rochelle salt, exhibited an interesting property. The Curies demonstrated that an electrical output was produced when they imposed a mechanical strain on the materials. They demonstrated this coupling by measuring the charge induced across electrodes placed on the material when it underwent an imposed deformation. They denoted this effect the *piezoelectric effect*. Several years later it was demonstrated that piezoelectric materials also exhibited the reciprocal property: namely, that a mechanical strain was induced when an electric signal was applied to the material.

The electromechanical coupling that was discovered by the Curies was interesting but unfortunately, not very useful, due to the fact that the amount of electrical signal produced by the mechanical deformation (and the amount of mechanical deformation produced by an electrical input) was "small." In modern-day terms, we would call the coupling *weak* compared to that of other materials. The utility of this new material was also limited by the fact that precise instrumentation for measuring the electrical or mechanical output of the material did not exist. It would be a number of years before precision instrumentation was available for measuring and applying signals to piezoelectric materials.

Interest in piezoelectric materials increased in the early twentieth century, due to the onset of World War I and the development of new means of warfare. One weapon that gained prominence during the war was the submarine, which was used very effectively against Great Britain in an attempt to destroy the trade routes that supplied the nation. To combat the submarine threat, a Frenchman named Langevin developed an underwater device, a *transducer*, that utilized a piezoelectric crystal to produce a mechanical signal and measure its electrical response as a means of locating submarines. This work was the basis of sonar and became one of the first engineering applications for the piezoelectric effect that was discovered in the late nineteenth century.

World War II stimulated even more advances in piezoelectric materials and devices. In addition to improvements in sonar, developments in electronics began to motivate the use of piezoelectric materials as electronic oscillators and filters. The need for better piezoelectric materials motivated the development of synthetic materials that exhibited piezoelectric properties. Barium titanate was an early synthetic piezoelectric material that had piezoelectric and thermal properties that made it superior to quartz crystals. Advances in synthetic piezoelectric materials led to their use as ceramic filters for communications and radio, as well as other applications, such as phonograph cartridges for record players.

At the same time that advances were being made in piezoelectric devices, fundamental research in shape memory alloys and electroactive polymers was being performed. Although the shape memory effect was known to exist in certain materials, the first development of materials that exhibited a significant strong shape memory effect occurred in 1965 at the Naval Ordnance Laboratory in the United States. A group of researchers demonstrated that an alloy of nickel and titanium exhibited significant shape memory properties when heated. This alloy, which they named *Nitinol* (for Nickel–titanium–Naval Ordnance Lab), has become one of the most useful variants of shape memory alloys, due to its mechanical properties and its capability for large strain recovery. At approximately the same time, work by Kuhn and Katchalsky demonstrated that polymeric materials would exhibit a change in volume when placed in solutions of different pH values. This was the seminal work for electroactive polymer materials, since it demonstrated that chemical stimulation could induce mechanical strain in a polymer material. This work, which occurred in the late 1940s and early 1950s, was followed by work approximately 20 years later by Grodzinsky in the use of collagen fibers, a naturally occurring polymer, as electromechanical sensors and actuators.

Work continued on the development of improved piezoelectric materials in parallel with the seminal developments in shape memory materials and electroactive polymers. The discovery of barium titanate led researchers to study other material compositions. This work led to the development of lead–zirconate–titanate (PZT) in the 1950s and 1960s. PZT exhibited piezoelectric properties superior to those of barium titanate and continues to be the most widely used piezoelectric material.

1.3 RECENT APPLICATIONS OF SMART MATERIALS AND SMART MATERIAL SYSTEMS

By the late 1970, fundamental developments in piezoelectric materials, shape memory alloys, and electroactive polymers had been made. The last 20 to 25 years have seen an increasing number of engineering systems being developed that utilize these three types of smart materials. For piezoelectric materials, one of the most extensive commercial applications has been the use of motion and force sensors. Piezoelectric

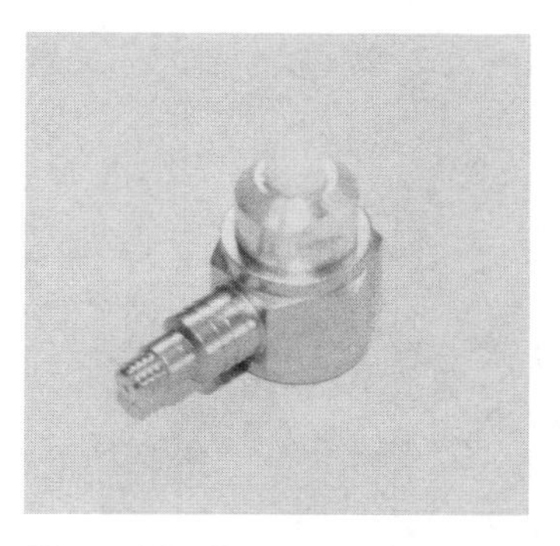

Figure 1.5 Representative commercial applications of piezoelectric materials: (*a*) piezoelectric accelerometer; (*b*) piezoelectric actuator.

crystals are the transducer element for accelerometers, activators, dynamic pressure sensors, and load cells (Figure 1.5). Their advantages for these applications are high mechanical stiffness and low mass, which leads to a fast sensing response. Piezoelectric materials have also been used in microscale devices, also known as *microelectromechanical systems* (MEMs), as the transducer element for sensing elements and as actuating elements for miniature pumps. Another common application of piezoelectric materials has been in MEMs cantilevers for atomic force microscopes (AFMs), which have revolutionized microscopy due to their ability to image and manipulate nanoscale features. A primary component of AFM systems is a piezoelectric cantilever that vibrates at a very high frequency (on the order of hundreds of kilohertz) for the purpose of measuring surface topography or interaction forces between the cantilever and the surface. The high precision of the AFMs is related directly to the high oscillation frequency of the cantilever, which, in turn, is related directly to the material properties of the piezoelectric material. Piezoelectric devices are also central to the positioning stage of an AFM for their ability to position the sample under test with nanometer precision. This high precision is also utilized in the semiconductor manufacturing industry for nanometer-scale positioning of wafers for microprocessor fabrication.

The actuation properties of piezoelectric devices have also enabled new types of electric motors. One of the most studied types of piezoelectric motor is the *inchworm motor*, an example of which is shown in Figure 1.6*a*. Motors based on the inchworm concept are available commercially in a range of speeds and loads. Once again, the ability of the piezoelectric material to oscillate at high-frequencies is central to their use as inchworm motors. Recently, piezoelectric materials have been used as motors for hydraulic pumps for flow control. High-frequency operation is also useful as a sensing modality. Devices known as *surface acoustic wave sensors* have been utilized for measuring small changes in mass on a surface. This sensing modality has been useful in the detection of chemical and biological agents.

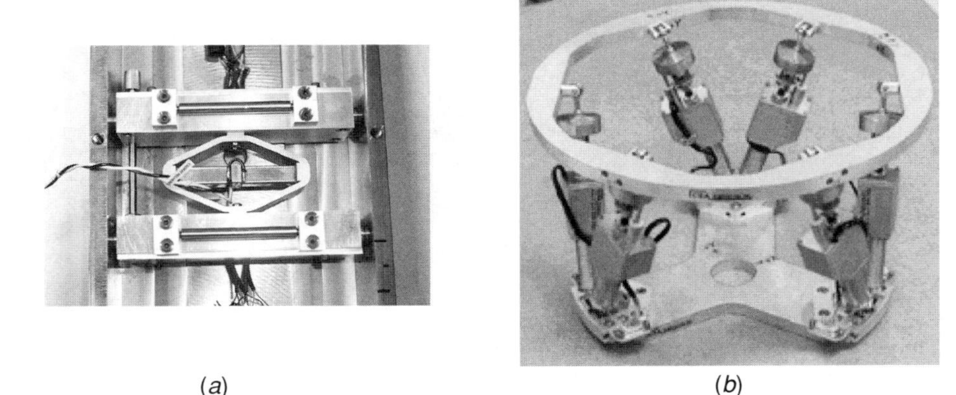

Figure 1.6 (*a*) Inchworm actuator using piezoelectric materials; (*b*) a vibration isolation platform that utilizes piezoelectric actuators for precision positioning (Courtesy CSA Engineering).

Systems that utilize piezoelectric materials have also been developed, in many cases for the purpose of controlling noise and vibration. Research into the use of piezoelectric materials to control the vibration of the tail section of fighter jets has been conducted, as well as motion and vibration control on high-precision space platforms (Figure 1.6(*b*). Less esoteric applications of noise and vibration control has been the use of piezoelectric materials to control noise transmission through acoustic panels, or in their use as devices that tune their frequency automatically to maximize vibration suppression. Applications in noise and vibration control have led to the development of methods for using the same wafer of piezoelectric material simultaneously, as both a sensor *and* an actuator, thus simplifying the design of vibration control systems.

Applications of shape memory alloys have encompassed some of the same engineering systems as those of piezoelectric materials. For example, shape memory

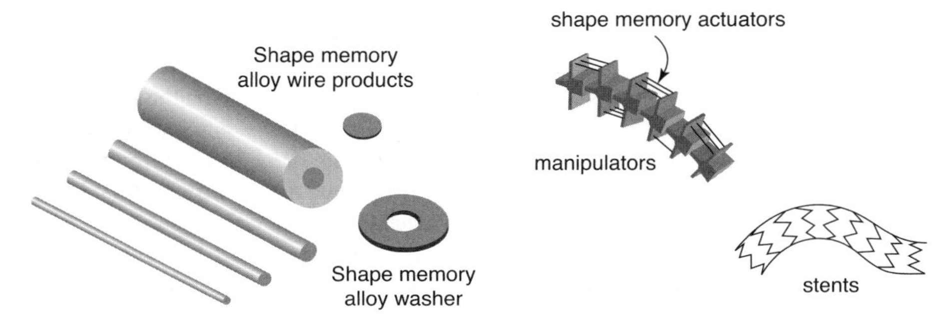

Figure 1.7 Applications of shape memory alloys.

alloy materials have been utilized as motors and actuators for vibration suppression and position control (Figure 1.7). They are particularly well suited to applications that require a large ammount of deformation since shape memory alloys can produce much larger strains than can conventional piezoelectric devices. The disadvantage of shape memory alloys is time response; therefore, the use of shape memory alloys for suppression of high-frequency vibrations is limited by the slow response time of the material. Shape memory materials have also been used for active–passive vibration suppression systems and as positioning devices for systems such as biomimetic hydrofoils (Figure 1.7). The use of shape memory alloys for positioning devices is particularly useful for robotic applications, due to their large range of motion and silent operation.

Some of the most common uses of shape memory alloys, though, is based on their interesting stress–strain behavior. In Chapter 7 we introduce the concept of *pseudoelasticity*, which is a nonlinear relationship between stress and strain that characterizes shape memory alloys. This nonlinear stress–strain behavior enables the use of shape memory alloys in applications such as eyewear and undergarments. The advantage of shape memory alloys for these applications is their ability to undergo large deformations without suffering from plastic deformation. These properties, along with biocompatibility, allow shape memory alloys to be used in medical applications to combat cardiovascular disease. *Stents*, which consist of cylindrical memory alloy mesh that expands when placed in an artery or vein, open the blood vessel and restore blood flow.

The need for large deformation is a motivation for the use of electroactive polymers for applications in motion control. Piezoelectric ceramics and shape memory alloys are hard materials. Electroactive polymers are soft materials that produce a large amount of deformation upon application of an electric potential. To date, commercial uses for electroactive polymers have been limited by the lack of suppliers for most of these materials. This is changing as this book is being written. At least one company that produces and sells standard dielectric elastomer actuators has been formed. Additional companies that sell conducting polymers and ionomeric polymers have also been formed. Electroactive polymers, called *artificial muscles* by some people, are designed to fill a niche in applications that require a large range of motion but generally much smaller forces than those of their ceramic and metal counterparts. Their use as sensors is also being studied. An example is the use of an ionomeric sensor as a novel shear sensor for underwater applications (Figure 1.8).

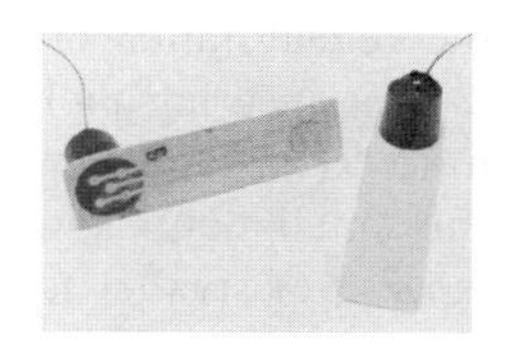
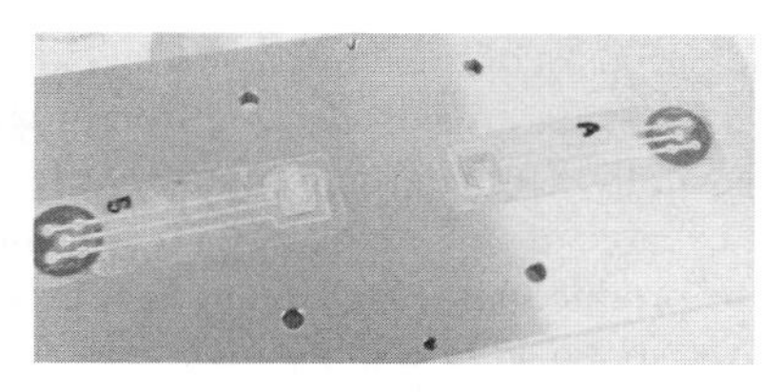

Figure 1.8 Representative application of electroactive polymers (Courtesy Discovery Technologies).

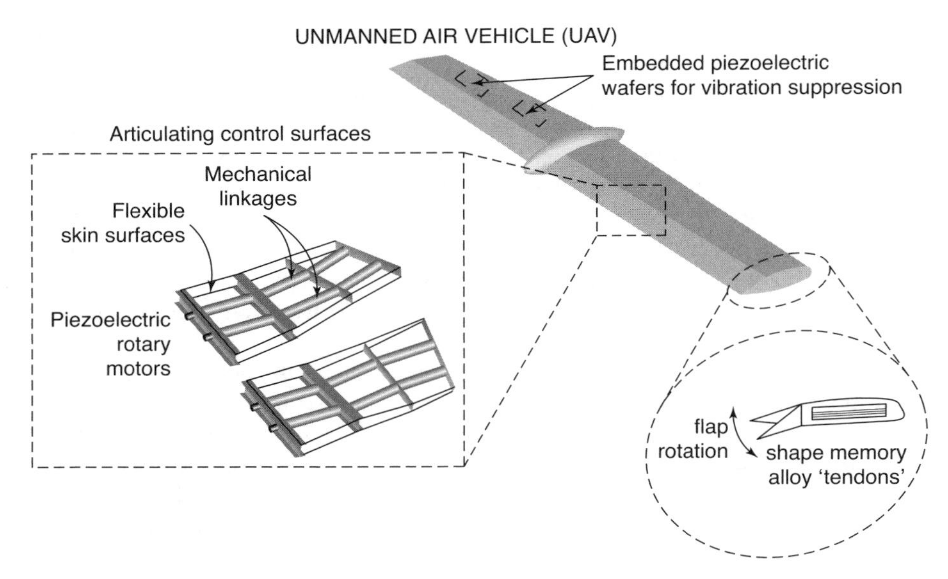

Figure 1.9 Smart wing concept utilizing piezoelectric and shape memory alloy actuators for control of a fixed-wing control surface.

The development of smart materials as actuators and sensors has enabled their integration into larger engineering systems. A concerted effort in the 1990s to develop new types of control surfaces for fixed-wing and rotary aircraft has been one of the most successful efforts to use the unique properties of smart materials for vibration suppression and motion control. Figure 1.9 illustrates the use of piezoelectric and shape memory alloy materials as actuation elements for a deformable aircraft control surface. Shape memory alloy actuators were utilized for large-deflection, low-frequency shape control, while ultrasonic piezoelectric motors were used to control flexible surfaces on the trailing edge of the wing.

The use of piezoelectric materials and shape memory alloys for control of a rotary aircraft has also been studied in depth (Figure 1.10). Piezoelectric materials and shape memory alloys were used as actuation elements to twist the rotor blade to enable higher authority flight control. Noise suppression through control of the flaps has also been studied quite extensively using high-frequency operation of piezoelectric actuators.

One of the purposes of this book is to develop a theoretical framework for understanding the use of smart materials for different types of engineering applications. As alluded to above, the engineering properties of smart materials differ greatly. Some materials are capable of producing large forces but only small motions, whereas others are able to produce large deformations at the expense of smaller forces. Some materials can respond very quickly, whereas the response time of other materials is much slower. We provide a thorough comparison of these materials in terms of

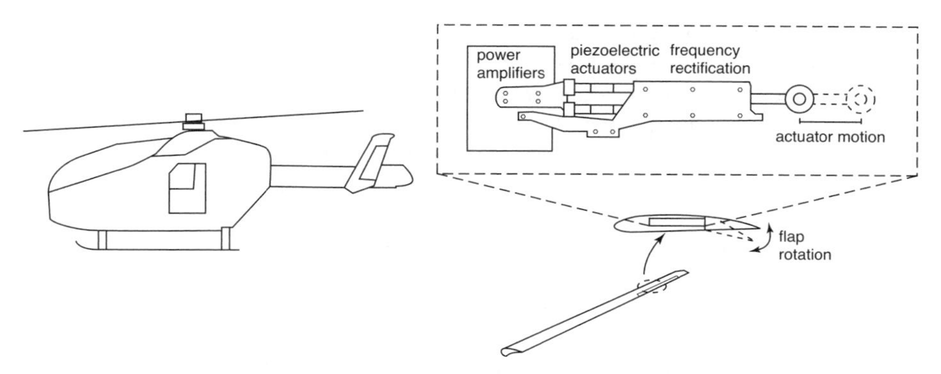

Figure 1.10 Active rotor blade concepts.

important engineering parameters and introduce consistent methods of analyzing smart materials and smart material systems in engineering designs.

1.4 ADDITIONAL TYPES OF SMART MATERIALS

People familiar with the field of smart materials will immediately recognize that the set of material types studied in this book represents only a subset of the complete set of materials that have been labeled *smart*, *active*, or *intelligent*. For example, we do not study a class of materials that have been investigated extensively for smart material applications as well as for innumerable applications in communications systems. These materials, collectively known as fiber optics, have received considerable attention in the smart materials literature for their use as embedded motion sensors for applications such as structural health monitoring. In addition to their sensing properties, fiber optics are most commonly used to replace copper wires in communications networks.

Another set of materials that receive no attention in this book is the broad class of materials that couple a magnetic field to mechanical motion. These materials, known collectively as *magnetostrictive materials*, have been studied extensively and represent a very interesting class of smart materials. They are useful for motion control applications and as elements of sensing systems for nondestructive damage evaluation. Another class of smart materials that exhibit optical coupling properties are the *electrochomic materials*, which have very interesting applications for display devices and for systems that incorporate controlled color changes.

Why so many omissions? First, it was decided early in the development of this book to limit the scope of the discussion so that the three material classes could be treated in depth. It is the opinion of the author that there are a number of excellent research monographs and encyclopedias that address the basic properties of a wide range of materials, leaving a need for a textbook that examines a smaller subset of materials more thoroughly. Thus, we forgo the treatment of magnetostrictive materials in the interest of an in-depth treatment of piezoelectric material systems and a more complete discussion of electroactive polymers.

More important to the development of a textbook, though, is the fact that limiting the scope of the treatment allows us to introduce topics within a general framework that is based on an analysis of constitutive properties and fundamental thermodynamic principles. As an example, we will see in future chapters that we can discuss piezoelectric materials and piezoelectric material systems within the common framework of constitutive properties that are based on fundamental thermodynamic laws. This also allows us to relate these fundamental principles to the material model for shape memory alloys, thus enabling direct comparison between the coupling properties of these two materials.

An additional advantage of limiting the scope to systems that couple mechanical, electrical, and thermal domains is the fact that we can present a pedagogical approach to the subject matter. One of the primary motivations for this book was the perceived need for a treatment that first discussed the basic theory, then moved on to a discussion of various material types based on the theory, and then presented applications of the material models to broad classes of engineering systems. This perspective was facilitated by limiting the scope to the basic theory of mechanics and electrostatics and then applying these principles to the analysis of piezoelectric materials, shape memory alloys, and electroactive polymers.

The pedagogical approach developed follows traditional textbooks more closely than it follows research monographs or edited volumes on the subject of smart materials, one result being the inclusion of numerous worked examples. Their purpose is twofold: They serve as a means of illustrating the theory introduced in a chapter, and perhaps more important, they are designed to illustrate analysis and design parameters that are relevant to engineering applications. For example, in Chapter 4 the examples are written to illustrate the fact that piezoelectric materials are limited in their strain output, which is a major limitation in their use in engineering systems. Although it would be possible for the reader to come to this conclusion by working the equations, it is felt that "tuning" the properties of the examples to illustrate these analysis and design principles will increase the utility of the book.

Another consequence of the pedagogical approach is the inclusion of numerous homework problems at the close of each chapter. The problems, along with their worked solutions, also serve to reinforce the analysis and design principles introduced in the text and worked examples.

In summary, it is hoped that although the range of material types studied in this book is by no means all-inclusive, the depth of the treatment and the associated pedagogical approach will be useful for people interested in the field of smart materials.

1.5 SMART MATERIAL PROPERTIES

Smart materials are, first and foremost, materials, and it is useful to step back and examine the properties of smart materials compared to those of conventional materials.

Two properties that are commonly used to compare engineering materials are density and elastic modulus. The *density* of a material is the mass normalized to the volume, and in SI units it is defined in terms of kg/m^3. The *elastic modulus* of a material is a material property that relates the applied loads on a solid material to the resulting deformation. Elastic modulus is defined formally in Chapter 2, but for now it is only important to note that for the same applied load, materials with a higher elastic modulus will undergo *less* deformation than will materials with a lower elastic modulus. Thus, materials with a higher elastic modulus will be stiffer than "soft" materials that have a low elastic modulus.

The universe of materials spans a wide range of density and modulus values. The density of all materials generally varies over approximately three orders of magnitude. Low values for materials are approximately 0.01 kg/m^3 for foams, and high values can reach $\approx$ 20 kg/m^3 for some metals and ceramics. In contrast, the variation in elastic modulus for materials spans approximately seven orders of magnitude, from approximately 1 kPa for soft foams and elastomeric materials to almost 1000 GPa for certain ceramics.

The smart materials discussed in this book generally fall in the middle of the range of density and modulus values. Piezoelectric materials and shape memory alloys have modulus values on the order of 10 to 100 GPa with a density that is typically in the range 7000 to 8000 kg/m^3. Piezoelectric polymers are softer materials whose elastic modulus is on the order of 1 to 3 GPa, with a density of approximately 1000 to 2000 kg/m^3. Electroactive polymers are generally the softest and least dense materials discussed in this book. Electroactive polymers have moduli that span a wide range—from approximately 1 MPa to greater than 500 MPa—and density values that range from 1000 to 3000 kg/m^3. A summary of these values is shown in Figure 1.11*a*.

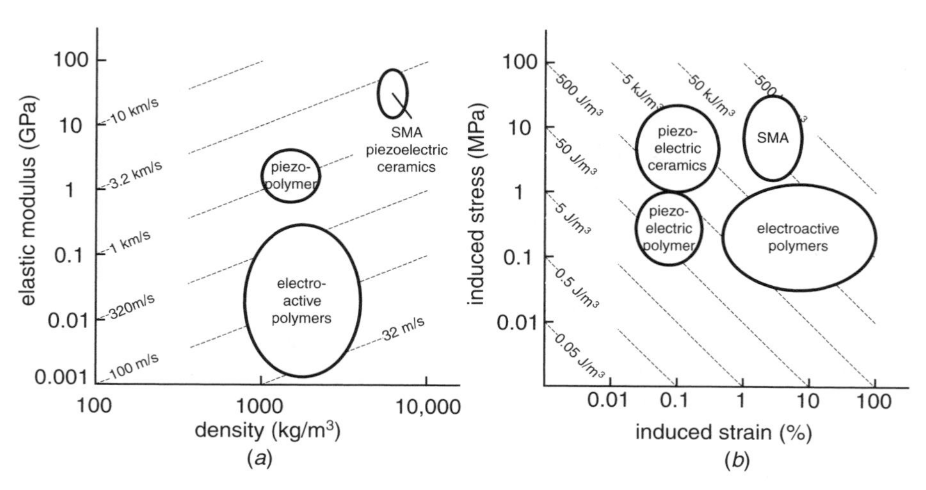

Figure 1.11 Comparison of induced stress and induced strain for the actuator materials studied in this book.

Engineering design often requires materials that have a high modulus and are lightweight. High-modulus lightweight materials would lie in the upper left portion of Figure 1.11*a*. A material property that relates to the modulus and density is the *wave speed*, defined as the square root of the modulus normalized to the density. The dashed lines in Figure 1.11*a* represent lines of constant wave speed, and a higher wave speed indicates that the material has a higher ratio of modulus to density. As we shall see in upcoming chapters, the wave speed also relates to the dynamic properties of a material since the fundamental vibration modes of a structure are proportional to the material wave speed. From Figure 1.11*a* we note that the wave speed of the materials covered in this book generally span values from 10 m/s to over 1 km/s.

The applications discussed in Section 1.4 highlight the use of smart materials as actuators and motors. For this reason a large percentage of this book deals with the analysis and design of systems that incorporate smart material actuators for applications such as motion control, damping, and vibration suppression. Actuator materials are often compared in terms of the force and the motion that they can generate under an applied stimulus. Another important metric for actuator materials is the speed with which they respond to a command stimulus. Force and displacement are examples of *extrinsic properties* (i.e., those that are a function of the geometry of the material or device). As we shall see, it is often useful to compare materials by certain *intrinsic properties*: properties that do not depend on geometry. Intrinsic material properties that are important for actuator comparisons are the stress and strain that are produced by the applied stimulus. *Stress* is defined as the force applied per unit area (note that it is an extrinsic property normalized to geometry), and *strain* is defined as a change in a dimension over the original size of the dimension.

The three material types that we focus on—piezoelectric materials, shape memory alloys, and electroactive polymers—represent three classes of materials that have a range of actuation properties. In general, piezoelectric materials are a class of material that produces small strains, typically only 1 part in 1000. Strains of this magnitude are generally specified as a percentage; therefore, strains of 1 part in 1000 would be specified as 0.001 or 0.1%. Shape memory alloys are materials that produce large strains, typically on the order of 4 to 8%. There are several classes of electroactive polymers, and the strains produced by the various classes of these materials can range from 1% to greater than 100%, depending on the material type.

The stress produced by these three classes of materials also spans a wide range. Hard piezoelectric ceramics can produce actuation stress on the order of tens of megapascal. A *megapascal* is defined as 1×10^6 N/m^2. One way to visualize 1 MPa would be to envision a 1-N force applied over an area with dimensions 1 mm $\times$ 1 mm; this would be equivalent to 1 MPa of stress. Thus, piezoelectric ceramics can produce tens of newtons of force over a 1-mm^2 area. There is a class of piezoelectric polymer materials that are generally much softer than piezoelectric ceramics. Piezoelectric polymer materials produce only 1/10 to 1/100 of the stress produced by a piezoelectric ceramic. Shape memory alloys, on the other hand, produce stress in the same range as that of piezoelectric ceramics (tens of megapascals) while producing strains on the order of 4 to 8%. Electroactive polymer materials are generally soft materials, but they can also produce stress in the range 1 to 10 MPa, due to their large strain capability.

A general comparison of piezoelectric materials, shape memory alloys, and electroactive polymers is shown in Figure 1.11*b* as a plot of the maximum stress produced by the material on the vertical axis versus the maximum strain produced on the horizontal axis. Piezoelectric ceramics generally occupy the upper left portion of the diagram because they produce small strain and large stress, whereas electroactive polymer materials generally occupy the lower right part of the diagram because they are large strain–small stress materials. Shape memory alloys are the materials that push farthest into the upper right part of the diagram, due to the fact they can produce large stress and large strain. In some applications, the most important parameters for the material to possess are the ability to produce stress and strain. In these applications we see that materials that lie in the upper right portion of the diagram will be most desirable.

An understanding of the stress and strain characteristics of various materials allows us to define a related parameter that also serves as a good means of comparing actuator materials. The product of the stress and strain produced by a material is defined as the volumetric energy density. *Energy* is defined as the capacity to do work, and *volumetric energy density* is defined as the capacity to do work per unit volume. Thus, a material with a higher energy density will have a larger capacity to do work per unit volume. The energy density of a material can be as visualized in Figure 1.11*b*: a straight line drawn from the upper left to the lower right of the plot. Each line represents a line of constant energy density. Once again, materials in the upper right portion of the plot are materials that have higher energy density and thus have a larger work capacity per unit volume.

The comparision shown Figure 1.11*b* illustrates that high energy density can be achieved with both stiff and soft smart materials. A stiff material such as a piezoelectric ceramic has an energy density on the order of 10 to 100 kJ/m^3 because it can generate high induced stress (tens of megapascal) even though the maximum strain may be limited to 0.1%. Conversely, a soft material such as an electroactive polymer has an energy density that is on the same order as that of piezoelectric ceramics (larger, in certain cases) because it can produce very large strains (10 to 100%) even though the elastic modulus is much smaller than that of a piezoelectric ceramic.

The speed of response of an actuation material is also important in a large number of engineering applications. There is no consistent definition of speed of response, but this metric is usually thought of as the rate of change of the strain, displacement, or force upon the application of a step change in the applied stimulus. Piezoelectric materials generally have the largest response speed of the materials studied in this book, as shown in Figure 1.11*b*. As we will find in upcoming chapters, the response speed of piezoelectric materials is governed by small changes in the molecular structure. Because these molecular changes occur on very small length scales, piezoelectric materials can respond very fast to changes in the stimulus applied. For example, it is possible to design a piezoelectric material that will change dimensions in a time scale on the order of 1×10^{-6} seconds, or microseconds. The time response of shape memory alloys, on the other hand, is limited by the speed at which the stimulus can cause changes in the molecular structure of the alloy. For certain materials that respond to changes in temperature, the response speed is generally limited to a time scale on the order of 10 to 100 milliseconds, or three to four orders of magnitude smaller than that of piezoelectric material. Electroactive polymers span the range of response speeds. Certain types of electroactive polymer materials can respond on the millisecond time scale, whereas others require 10 to 100 ms to respond. The reasons for these differences are discussed in upcoming chapters.

Figures of merit for sensing applications are also of use for comparing different types of materials. Sensor materials are generally compared in terms of a range of extrinsic properties related to the sensitivity of material, the linearity of response, and the resolution. As an example, we will see that piezoelectric materials are often used as sensors that convert displacement or force to an electrical signal. The sensitivity of the sensor is defined as the electrical output per unit force or displacement, and the resolution of the sensor is the smallest value of force or displacement that can be measured.

An important metric for sensing applications is the signal-to-noise ratio of a device. *Noise* consists of random fluctuations in the output signal that are not correlated with the physical variable that is being measured (e.g., force or displacement). The *signal-to-noise ratio* is the ratio of the average value of a signal correlated to the variable being measured, to the noise in a system. Excellent sensors may exhibit a signal-to-noise ratio on the order of 10,000 : 1 or 1000 : 1, whereas poor sensors may exhibit a signal-to-noise ratio of 10 : 1 or even as low as 2 : 1.

One of the difficulties in direct comparison of sensors is that figures of merit such as resolution and signal-to-noise ratio are functions of several parameters that are not related to the properties of the material. For example, piezoelectric sensors for measuring force or displacement require electronics to convert a signal from the material to an electrical output. The signal-to-noise ratio of the sensor is dependent not only on the material properties but also on the properties of the electronics used to convert the output of the material to an electrical signal. Superior design of the electronics will result in a superior signal-to-noise ratio, even in the case of identical sensor materials.

1.6 ORGANIZATION OF THE BOOK

This book is written for advanced undergraduate students, graduate students, and practicing engineers who are interested in learning more about the field of smart materials. It is assumed that the reader has a basic knowledge of statics and dynamics as taught in a typical undergraduate engineering curriculum. Another main assumption is that the reader has a thorough understanding of differential equations, primarily for the solution of first and second-order ordinary differential equations. A knowledge of system dynamics, particularly the topic of Laplace transforms, is also assumed. A familiarity with vibrations, matrix analysis, and introductory mechanics and electrostatics is also helpful, although these topics are reviewed in the following chapters.

The book is organized into three main sections. The first section, consisting of Chapters 2 and 3, is a review of basic material that is required for understanding the material in the remaining chapters. One of the challenges in writing a book about smart material systems is that a diverse set of topics is required for understanding the material. As discussed earlier, this book focuses on materials that exhibit coupling between mechanical, electrical, and thermal domains. For this reason the material in Chapter 2 focuses on reviewing relevant topics in the mechanics of deformable bodies and basic electrostatics. These two topics form the basis of an understanding of the material models introduced in later chapters. After reviewing basic topics in mechanics and electrostatics, we introduce the concept of energy methods and their relationship to the development of equations of static and dynamic equilibrium of systems. Application of the energy approaches will yield a set of equations that define the static and dynamic behavior of a smart material system. Chapter 3 then focuses on general solutions for systems of second and first-order linear differential equations. These concepts lead to the topics of eigenvalue analysis for second-order systems of equations and state analysis for systems of first-order differential equations. These concepts are reviewed, along with fundamental results in the static and dynamic response of these systems.

The second section of the text, consisting of Chapters 4 through 7, presents models for the smart materials studied in this book. We begin each discussion by describing the basic physical mechanism that gives rise to the coupling exhibited by the material. This is followed by a discussion on how to model the material for the purpose of analysis and design of engineering systems. This analysis generally consists of discussing the means of representing the coupling inherent in the material and how this can be applied to systems-level analysis.

As discussed earlier, the text focuses on three types of materials: piezoelectric materials, shape memory alloys, and certain types of electroactive polymers. Chapters 4 and 5 present a detailed analysis of piezoelectric materials, starting with a discussion of their constitutive behavior in Chapter 4, which naturally leads to the development of what we call a transducer model of the material behavior. Transducer models are useful for the design of actuators and sensors with piezoelectric materials, and Chapter 4 concludes with a discussion of the basic properties of piezoelectric devices. Throughout the chapter there are a number of examples that highlight the basic characteristics of piezoelectric materials and devices in engineering applications. The chapter closes with a discussion of the nonlinear properties of these materials.

The transducer models of piezoelectric devices presented in Chapter 4 are followed by a thorough discussion of piezoelectric material systems in Chapter 5. The development of piezoelectric material systems is based on the use of energy methods for deriving the equations of static and dynamic equilibrium. Energy methods provide a convenient framework for analyzing the response of systems that incorporate piezoelectric materials. In Chapter 5 we focus on the basic properties of piezoelectric materials incorporated into beam and plate elements. Beams and plates are used to examine the properties of distributed spatial sensing and actuation using piezoelectric materials.

Shape memory alloys is the second set of materials studied in this book. In Chapter 6 we focus on the development of constitutive models for shape memory alloys and their use in understanding mechanical response as a function of temperature. The basic physics of shape memory alloys is reviewed and is followed by a series of models for shape memory behavior. Shape memory alloy materials are highly nonlinear, and much of the discussion in Chapter 6 revolves around modeling the nonlinear relationships between stress, strain, and temperature. Models of heat flux are introduced to couple the constitutive behavior of shape memory alloys to the time response. The chapter concludes with a discussion of actuator models for shape memory alloys.

In Chapter 7 we introduce the class of materials known as electroactive polymers and focus on the electromechanical properties of these materials. The chapter begins with an introduction to polymer materials, concentrating on how they differ from the ceramics and metals studied in earlier chapters. Three different types of electroactive polymers are studied. Dielectric elastomer actuators are discussed and models of their electromechanical response are presented. Conducting polymers are then introduced and transducer models of their actuation response are presented. Finally, transducer models of ionomeric polymer sensors and actuators are introduced. Throughout the chapter the examples are used to highlight the fundamental differences between electroactive polymers and the other types of smart materials studied in this book. Hopefully, this will give readers a sense of the capabilities of these materials compared to those of more traditional smart materials.

The third section of the book consists of 8 through 11 and concentrates on systems-level applications of the smart materials discussed in Chapters 4 through 7. The applications are chosen (with some bias by the author) primarily from the topics of structural analysis and control. Chapter 8 focuses on motion control applications of smart materials. Significant attempts are made to present the material such that the reader can distinguish the salient characteristics of different types of materials as they relate to motion control applications. Thus, the examples and charts are meant to highlight the differences in materials as they relate to the amount of motion and force obtained from these materials as well as their characteristic time response.

Chapter 9 follows with a discussion of using smart materials for active–passive vibration suppression. A significant amount of the chapter is devoted to the use of piezoelectric materials as active–passive dampers using electronic shunts and switched-state control. Applications of shape memory alloys as passive dampers are also explored.

Chapter 10 deals with the use of smart materials as elements of active vibration control systems. In this chapter we use dynamic models of piezoelectric materials to understand how the sensing and actuation properties of these materials are useful for active damping or active vibration suppression.

In the final chapter we discuss a topic that is central to the use of smart materials—power—and present a unified framework to understand the power requirements for smart material systems. The discussion is based on the fundamental properties of power flow in resistive and capacitive devices. An understanding of these topics leads to an understanding of how to analyze the power requirements for systems that incorporate piezoelectric materials, shape memory alloys, or electroactive polymers. The chapter concludes with a discussion of power electronics and recent advances in the development of efficient amplifiers and energy harvesters for smart material systems.

1.7 SUGGESTED COURSE OUTLINES

This book is intended to be used for a senior-level undergraduate or first-year graduate course in smart materials. It is assumed that students have had courses in deformable bodies, introductory mechanics, and engineering mathematics. Also useful, although not mandatory, are courses in introductory electrostatics, system dynamics, and introductory vibrations. The latter two courses would be especially useful to introduce Laplace transforms.

The book has been written to be suitable for either a one-or two-semester course in smart materials. The author has taught a one-semester course that focuses on introducing students to various types of smart materials and some applications of smart material systems. As an example, a one-semester 15-week course may utilize the book in the following way:

Topic	Text Material	Suggested Time (weeks)
Introduction/review	Chapters 1–3	1
Piezoelectric materials	Chapters 4	4
Shape memory alloys	Chapter 6	4
Electroactive polymers	Chapter 7	2–3
Smart material system applications	selected sections of Chapters 8–11	3–4

This outline would provide a basic understanding of the various types of smart materials, and their basic coupling behavior, and would allow the instructor to pick and choose several topic areas at the close of the course as application focus areas for students. The topic areas could also be chosen according to the interests of the instructor and students. This outline is also amenable to a set of laboratory experiments throughout the course. For example, a laboratory experiment could accompany each of the three material topic areas in Chapters 4, 6, and 7. A final experiment could be performed at the end of the course in one of the application focus areas (e.g., semiactive damping or piezoelectric actuation).

A two-semester course using the book could separate the material in a different manner. If the first-semester course is a prerequisite for the second semester, one suggested path would be to separate the material loosely into linear and nonlinear behavior. As an example, the first-semester course could focus on a detailed understanding of the analysis and design of systems that incorporate piezoelectric materials and linear electroactive polymers, whereas the second-semester course could focus on a detailed understanding of shape memory alloys and nonlinear materials that exhibit electromechanical coupling. In this format a suggested outline for the first semester course would be:

Topic	Text Material	Suggested Time (weeks)
Introduction/review	selected sections of Chapters 1–3	1
Piezoelectric transducers	selected sections of Chapter 4	4
Piezoelectric material systems	Chapter 5	4
Linear electroactive polymers	selected sections of Chapter 7	2–3
Applications	selection sections of Chapters 8–11	3–4

The suggested outline would give students a very detailed understanding of how to model piezoelectric materials, both as transducers and as actuation and sensing elements in structures. The material on linear electroactive polymers would allow them to compare the properties of piezoelectric materials to those of electroactive polymers that exhibit similar coupling properties. The application focus areas could be selections from topics such as motion control, shunt damping, or vibration suppression and energy harvesting.

The second-semester course would follow with a detailed development of shape memory materials and additional applications of smart material systems. A suggested outline for the second semester would be:

Topic	Text Material	Suggested Time (weeks)
Introduction/review	selected sections of Chapters 1–3	1
Shape memory alloys	Chapter 6	5
Nonlinear electromechanical materials	Sections 4.8, 7.1, and 7.2	3
Applications	selected sections of Chapters 8–11	6

The second semester would emphasize heavily the nonlinear constitutive properties of shape memory materials, electrostrictive materials, and nonlinear electroactive polymers. The application sections could be a culmination of the two semesters with certain focus areas emphasizing applications of materials with nonlinear constitutive properties and others emphasizing system-level design and control (Chapters 8 and 10). As with the one-semester course, laboratory experiments could be integrated throughout the two semesters to align with course topics.

1.8 UNITS, EXAMPLES, AND NOMENCLATURE

In this book we use SI units consistent with the meter–kilogram–second notation. For readability purposes and to decrease the possibility of errors in the examples or the homework problems, it was decided not to mix units between SI and English units or

Table 1.1 Common SI-to-English unit conversions

Type of Unit	To Convert:	To:	Divide by:
Mass	kilogram	pound-mass	
Force	newton	pound-force	4.448
Pressure	megapascal	psi	6.895×10^{-3}
Length	centimeter	inch	2.54
Length	micrometer (micron)	milli-inch (mil)	25.4

to include English units in certain portions of the book. A conversion chart between common SI units and English units is provided in Table 1.1.

There are numerous worked examples throughout the book. The primary purpose of the examples is to illustrate the analysis and computations discussed in the text. The secondary purpose of the examples is to provide the reader a feel for the numbers associated with common engineering analyses of smart materials and systems. For example, in Chapter 4 there are several examples that accompany the analyses of extensional and bending piezoelectric actuators. The values chosen for the physical parameters and size of the actuators is representative of typical applications, and the values obtained from the computations are representative of piezoelectric actuator performance.

The examples are presented to elucidate representative parameters for the materials discussed in the text. For this reason, computations are presented such that the values in the intermediate computations are listed in kilograms, meters, and seconds, while the final result is listed in the correct engineering unit for the calculation. As an example, consider the computation of the stress, T, produced by an applied force of 100 N over a circular area with radius 3 mm. The computation would be written

$$\mathrm{T} = \frac{100\ \mathrm{N}}{\pi (3 \times 10^{-3}\ \mathrm{m})^2} = 3.54\ \mathrm{MPa}.$$

Notice in this computation that the intermediate values are expressed in terms of newtons and meters, while the final value is expressed in the more traditional units of stress, megapascal. Using the conversion in Table 1.1, this value can be converted to pounds per square inch (psi) by dividing by 0.006895. The result would be 513 psi.

The decision to use SI units is fairly straightforward since most science and engineering textbooks today use this system of units. A more difficult decision in writing the book was to decide on a consistent nomenclature. The discipline of smart materials is very diverse, and papers on the subject use a range of nomenclature for parameters such as stress, strain, force, and displacement. In the author's opinion, the nomenclature used by a majority of the field can be separated into those based on the use of nomenclature that is consistent with standards for piezoelectric materials, and those that are consistent with the more traditional nomenclature used by the mechanics and materials community. For example, notation that is consistent with piezoelectric standards would label stress and strain as T and S, respectively, whereas the mechanics community would generally list these variables as σ and ϵ.

At the risk of alienating a wide community of mechanics and materials researchers, it was decided to utilize notation that is consistent with the piezoelectric standards as the basis for the book. This decision was based on the fact that a large set of topics deal with piezoelectric materials, and based on the organization of the book, piezoelectric materials are the first type of smart material that we will discuss in detail. It was decided to introduce this notation in the review material in Chapters 2 and 3 and then be consistent throughout the remainder of the book. This decision probably has the most impact on the discussion of shape memory alloy materials, since a majority of the nomenclature for this community is based on the traditional notation for mechanics and materials.

PROBLEMS

1.1. Identify 10 companies that sell smart materials such as piezoelectric ceramics or polymers, shape memory materials, or electroactive polymers.

1.2. Identify 10 companies that sell products that utilize smart materials.

1.3. Identify a company that sells either piezoelectric ceramics, shape memory alloys, or electroactive polymers. List five properties of the materials that they sell.

1.4. Identify a company that sells piezoelectric motors, and list five properties that they use to define their motor performance.

1.5. Find a recent newspaper article or online article that discusses smart materials or applications of smart materials.

NOTES

One of the most difficult aspects of writing a book about smart materials is finding a suitable definition for a smart material. Various definitions of smart materials can be found in journals dedicated to the topic. The reader is referred to two journals that publish papers in the discipline, the *Journal of Intelligent Material Systems and Structures* and *Smart Materials and Structures*. The definition offered in this book, "a material that converts energy between multiple physical domains," emphasizes the concept of energy conversion, which is a central theme in the book. Although this definition is by nature very broad, it is felt that it adequately represents the central topic of the book.

The background material for this chapter was drawn from several seminal works in the field of smart materials. One of the earliest papers on the topic of smart materials for vibration control is that of Bailey and Hubbard [1] whereas the authors described the use of piezoelectric materials to control the vibration of a beam, an experiment that has been repeated many times since. Another seminal work in the field of smart materials is a paper by Crawley and de Luis [2] which has come to be one of the most frequently cited articles in the field. Historical information on the development of piezoelectric materials was drawn from Fujishima [3], Ikeda [4], Jaffe et al. [5], and the *IEEE Standard on Piezoelectricity* [6].

There are additional textbooks on the subject of smart materials. Research monographs on the topic are those of Culshaw [7] and Thompson [8]. A textbook on the subject is that of Srinivasan and McFarland [9]. Clark et al.'s book [10] includes sections on the use of smart materials in adaptive structures. Finally, a recent book by Smith [11] is an excellent mathematical treatment of the subject that focuses on linear and nonlinear modeling of smart materials and smart material systems.

2

MODELING MECHANICAL AND ELECTRICAL SYSTEMS

The development of models for smart material systems requires a basic understanding of mechanics and electrostatics. In this chapter we review the governing equations for both static and dynamic mechanical systems as well as the governing equations of electrostatics. The concepts of constitutive behavior for elastic and dielectric materials are introduced and related to the solution for basic elements such as an axial bar, a bending beam, and a capacitor. The variational approach to modeling electromechanical systems is derived and used to develop the static and dynamic equations of state for electrical and mechanical systems. Understanding the governing expressions in mechanics and electrostatics will form the basis for developing the governing equations for smart material systems in later chapters.

2.1 FUNDAMENTAL RELATIONSHIPS IN MECHANICS AND ELECTROSTATICS

In this book we utilize *system models* for the analysis, design, and control of smart material systems. One of the first requirements for analysis, design, and control is the development of a set of equations that represent the response of the system. These equations can take the form of a set of algebraic expressions or a set of differential equations in space and time. For the purposes of this book we define a *static system* as one that is represented by a set of algebraic equations, whereas a *dynamic system* is defined as one that is represented as a set of differential equations in time.

In this book we represent the governing equations in a Cartesian coordinate system. Consider a representative volume of material (Figure 2.1) referenced to a Cartesian coordinate system defined by three orthogonal axes. These axes are denoted the 1, 2, or 3 directions. A vector of unit length that lies in the direction of one axis is denoted the *unit vector* for that direction and given the symbol $\hat{x}_i$, where $i = 1, 2, 3$. A point in the Cartesian coordinate system is defined by the vector $\mathbf{x}$. Expanding $\mathbf{x}$ into the three orthogonal directions yields the relationship

$$\mathbf{x} = x_1\hat{x}_1 + x_2\hat{x}_2 + x_3\hat{x}_3. \tag{2.1}$$

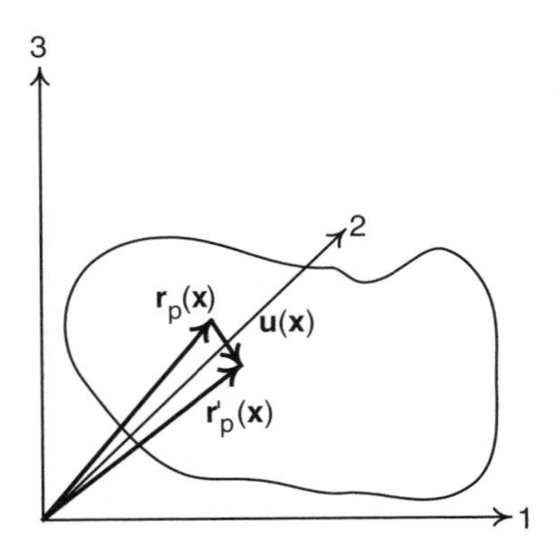

Figure 2.1 Arbitrary volume showing the definition of the position vectors and the displacement.

In certain instances we reduce the dimensionality of a problem, and the variables of interest in the analysis will be a function of only one or possibly two coordinate directions. In these cases it will be convenient to drop the subscript notation and simply refer to the coordinate directions as x, y, or z. In these cases the point vector $\mathbf{x}$ is denoted as

$$\mathbf{x} = x\hat{x} + y\hat{y} + z\hat{z}. \tag{2.2}$$

The notation used in the problem definition will be clear from the context of the analysis.

An important concept introduced in this chapter is the notion of the *state* of a material or a system, the set of all macroscopic properties that are pertinent to the analysis. One critical aspect of the state is that it is independent of the *path* that the system took to arrive at the set of parameters defined. In this chapter and later we often discuss the mechanical and electrical state of a material or system. As we shall see in the next two sections, the governing equations for the mechanics and electrostatics of a material are referenced to particular state variables that we define in the analysis. Defining the fundamental laws associated with the mechanics and electrostatics of a material is the topic of the following sections.

2.1.1 Mechanics of Materials

Consider an arbitrary volume of material whose position in space is referenced to a Cartesian coordinate system. The position vector to any point on a body can be written as $\mathbf{r}_p(\mathbf{x})$ (Figure 2.1). If we assume that this position vector changes to $\mathbf{r}'_p(\mathbf{x})$, the *displacement* vector $\mathbf{u}(\mathbf{x})$ is written as

$$\mathbf{u}(\mathbf{x}) = \mathbf{r}'_p(\mathbf{x}) - \mathbf{r}_p(\mathbf{x}). \tag{2.3}$$

In the case of static problems, the displacement at any point on the solid is written in component form as

$$\begin{aligned} u_1(x_1, x_2, x_3) &= u_1(x, y, z) = u_1 = u_x \\ u_2(x_1, x_2, x_3) &= u_2(x, y, z) = u_2 = u_y \\ u_3(x_1, x_2, x_3) &= u_3(x, y, z) = u_3 = u_z \end{aligned} \tag{2.4}$$

or in vector notation as

$$\mathbf{u} = \begin{bmatrix} u_1 \\ u_2 \\ u_3 \end{bmatrix}, \tag{2.5}$$

where the spatial dependence is implicit in the equation.

The first state variable that we will define is the *strain* of the solid. In this book we denote strain as S, and for arbitrary strains the strain–displacement relationship is written

$$\mathrm{S}_{ik} = \frac{1}{2}\left\{\frac{\partial u_i}{\partial x_k} + \frac{\partial u_k}{\partial x_i} + \sum_{l=1}^{3}\left[\left(\frac{\partial u_l}{\partial x_i}\right)\left(\frac{\partial u_l}{\partial x_k}\right)\right]\right\}. \tag{2.6}$$

The subscripts are indices that refer to the direction of the strain component. Consider a cube of material as shown in Figure 2.2. On each face of the cube we can define three strain components. One strain component is in the direction perpendicular to the plane of the face, and the two remaining strain components lie tangent to the plane of the face. The strain component perpendicular to the face is called the *normal strain* and the strain components tangent to the face are denoted the *shear* components. The common nomenclature is to define the 1 face as the face with a normal in the 1 coordinate direction. The remaining faces are defined in an identical manner. As shown in Figure 2.2, the first index of the strain subscript refers to the face on which the strain component acts, and the second index represents the component direction. Thus, S_{11}, S_{22}, and S_{33} are the normal strains and S_{ij}, $i \neq j$ are the shear strains.

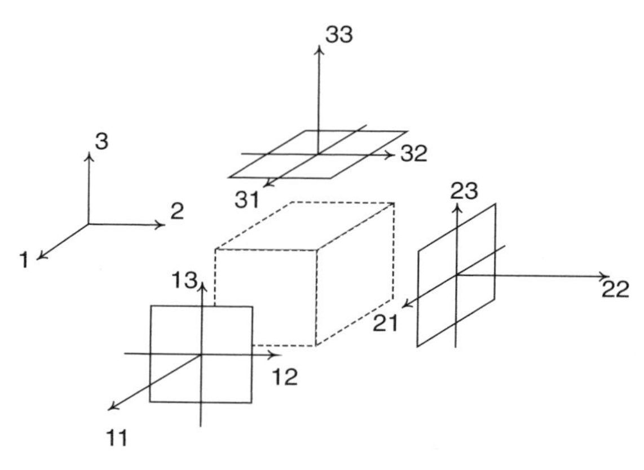

Figure 2.2 Cartesian coordinate system and the definition of the directions for stress and strain components.

For sufficiently small displacements the quadratic terms in equation (2.6) are neglected compared to the first-order terms and the strain–displacement relationships are

$$\mathrm{S}_{ik} = \frac{1}{2}\left(\frac{\partial u_i}{\partial x_k} + \frac{\partial u_k}{\partial x_i}\right). \tag{2.7}$$

Since strain is a ratio of the change in a length to its original length, the units of strain are m/m. In some instances the strain is written as a unitless parameter, but in this book we generally write strain in units of *percent strain*, which equals a strain of 0.01 m/m, or in units of *microstrain*, which is equivalent to 10^{-6} m/m.

Now consider the application of a surface force to the arbitrary volume shown in Figure 2.1. At the location of the force we define a small element that has a surface area ΔS. The force resultant acting on this surface is denoted $\Delta\mathbf{f}$, and the *stress* is defined as

$$\mathcal{T} = \lim_{\Delta S \to 0} \frac{\Delta \mathbf{f}}{\Delta S} = \frac{d\mathbf{f}}{dS}. \tag{2.8}$$

The units of stress are force per unit area, or, in SI notation, N/m^2.

Consider once again the cube shown in Figure 2.2. Denoting the stress component that acts on the face with a normal in the i direction as $\mathcal{T}_i,\ i = 1, 2, 3$, we can see that

$$\begin{aligned}
\mathcal{T}_1 &= \mathrm{T}_{11}\hat{x}_1 + \mathrm{T}_{12}\hat{x}_2 + \mathrm{T}_{13}\hat{x}_3 \\
\mathcal{T}_2 &= \mathrm{T}_{21}\hat{x}_1 + \mathrm{T}_{22}\hat{x}_2 + \mathrm{T}_{23}\hat{x}_3 \\
\mathcal{T}_3 &= \mathrm{T}_{31}\hat{x}_1 + \mathrm{T}_{32}\hat{x}_2 + \mathrm{T}_{33}\hat{x}_3,
\end{aligned} \tag{2.9}$$

where T_{ij} is the component of stress vector $\mathcal{T}_i$ with a component in the j direction.

This analysis demonstrates that the state of a solid element is defined by nine strain components, S_{ij}, and nine stress components, T_{ij}. The shear components of both the stress state and the strain state are shown to be symmetric: thus,

$$\begin{aligned}
\mathrm{T}_{ij} &= \mathrm{T}_{ji} \\
\mathrm{S}_{ij} &= \mathrm{S}_{ji},
\end{aligned} \tag{2.10}$$

and this symmetry reduces the number of independent stress and strain components from nine to six. The vector that defines the state of stress of the solid is denoted

$$\mathbf{T} = \begin{bmatrix} \mathrm{T}_{11} \\ \mathrm{T}_{22} \\ \mathrm{T}_{33} \\ \mathrm{T}_{23} \\ \mathrm{T}_{13} \\ \mathrm{T}_{12} \end{bmatrix}, \tag{2.11}$$

and the state of strain for the solid is expressed in a form that is analogous to equation (2.11):

$$\mathbf{S} = \begin{bmatrix} S_{11} \\ S_{22} \\ S_{33} \\ 2S_{23} \\ 2S_{13} \\ 2S_{12} \end{bmatrix}. \tag{2.12}$$

The stress and strain vectors are also written in *compact*, or *Voigt*, *notation* through the following definitions:

$$\begin{aligned} S_1 &= S_{11} & T_1 &= T_{11} \\ S_2 &= S_{22} & T_2 &= T_{22} \\ S_3 &= S_{33} & T_3 &= T_{33} \\ S_4 = S_{23} + S_{32} &= 2S_{23} & T_4 &= T_{23} = T_{32} \\ S_5 = S_{31} + S_{13} &= 2S_{13} & T_5 &= T_{31} = T_{13} \\ S_6 = S_{12} + S_{21} &= 2S_{12} & T_6 &= T_{12} = T_{21}. \end{aligned} \tag{2.13}$$

The strain–displacement relationships for small strains are also written in *operator notation* using the matrix expression

$$\mathbf{S} = \begin{bmatrix} S_1 \\ S_2 \\ S_3 \\ S_4 \\ S_5 \\ S_6 \end{bmatrix} = \begin{bmatrix} S_{11} \\ S_{22} \\ S_{33} \\ 2S_{23} \\ 2S_{13} \\ 2S_{12} \end{bmatrix} = \begin{bmatrix} \frac{\partial}{\partial x_1} & 0 & 0 \\ 0 & \frac{\partial}{\partial x_2} & 0 \\ 0 & 0 & \frac{\partial}{\partial x_3} \\ 0 & \frac{\partial}{\partial x_3} & \frac{\partial}{\partial x_2} \\ \frac{\partial}{\partial x_3} & 0 & \frac{\partial}{\partial x_1} \\ \frac{\partial}{\partial x_2} & \frac{\partial}{\partial x_1} & 0 \end{bmatrix} \begin{bmatrix} u_1 \\ u_2 \\ u_3 \end{bmatrix}, \tag{2.14}$$

which can be written as

$$\mathbf{S} = \mathrm{L_u}\mathbf{u}, \tag{2.15}$$

where

$$
\mathrm{L_u} = \begin{bmatrix} \dfrac{\partial}{\partial x_1} & 0 & 0 \\ 0 & \dfrac{\partial}{\partial x_2} & 0 \\ 0 & 0 & \dfrac{\partial}{\partial x_3} \\ 0 & \dfrac{\partial}{\partial x_3} & \dfrac{\partial}{\partial x_2} \\ \dfrac{\partial}{\partial x_3} & 0 & \dfrac{\partial}{\partial x_1} \\ \dfrac{\partial}{\partial x_2} & \dfrac{\partial}{\partial x_1} & 0 \end{bmatrix} \tag{2.16}
$$

is the differential operator that relates displacement to strain.

Example 2.1 A solid has the displacement field

$$
\begin{bmatrix} u_1 \\ u_2 \\ u_3 \end{bmatrix} = \begin{bmatrix} 0 \\ 0 \\ 2x_1^2 - 6x_1x_2 + 3x_2^2 + x_3 \end{bmatrix} \text{ mm.}
$$

Compute the strain vector (in compact form) from the displacement field.

Solution The strain vector is computed using the operator form of equation (2.14). Substituting the displacement field into the operator form of the displacement–strain relationships yields

$$
\begin{bmatrix} \mathrm{S}_1 \\ \mathrm{S}_2 \\ \mathrm{S}_3 \\ \mathrm{S}_4 \\ \mathrm{S}_5 \\ \mathrm{S}_6 \end{bmatrix} = \begin{bmatrix} \dfrac{\partial}{\partial x_1} & 0 & 0 \\ 0 & \dfrac{\partial}{\partial x_2} & 0 \\ 0 & 0 & \dfrac{\partial}{\partial x_3} \\ 0 & \dfrac{\partial}{\partial x_3} & \dfrac{\partial}{\partial x_2} \\ \dfrac{\partial}{\partial x_3} & 0 & \dfrac{\partial}{\partial x_1} \\ \dfrac{\partial}{\partial x_2} & \dfrac{\partial}{\partial x_1} & 0 \end{bmatrix} \begin{bmatrix} 0 \\ 0 \\ 2x_1^2 - 6x_1x_2 + 3x_2^2 + x_3 \end{bmatrix} \times 10^{-3} \text{ m}
$$

$$\begin{bmatrix} S_1 \\ S_2 \\ S_3 \\ S_4 \\ S_5 \\ S_6 \end{bmatrix} = \begin{bmatrix} 0 \\ 0 \\ 1 \\ -6x_1 + 6x_2 \\ 4x_1 - 6x_2 \\ 0 \end{bmatrix} \times 10^{-3} \text{ m/m}.$$

With this solution the strain can be computed at any point in the body.

Static equilibrium of the body shown in Figure 2.1 is enforced by noting that the summation of the volume integral of the body forces and the surface integral of the stress at the boundary must be equal to zero:

$$\int_{\text{Vol}} \mathbf{f}_V \, d\text{Vol} + \oint_{\text{Surf}} \mathcal{T} \cdot d\mathbf{S} = 0. \tag{2.17}$$

The surface integral in equation (2.17) is transformed into a volume integral through use of the divergence theorem,

$$\oint_{\text{Surf}} \boldsymbol{\mathcal{T}} \cdot d\mathbf{S} = \int_{\text{Vol}} \nabla \cdot \mathcal{T} \, d\text{Vol}, \tag{2.18}$$

which can be substituted into equation (2.17) to yield

$$\int_{\text{Vol}} (\mathbf{f}_V + \nabla \cdot \boldsymbol{\mathcal{T}}) \, d\text{Vol} = 0. \tag{2.19}$$

The symbol ∇ represents the operation

$$\nabla = \frac{\partial}{\partial x_1} x_1 + \frac{\partial}{\partial x_2} x_2 + \frac{\partial}{\partial x_3} x_3. \tag{2.20}$$

Applying the divergence operator to the stress yields

$$\nabla \cdot \mathcal{T} = \frac{\partial \mathcal{T}_1}{\partial x_1} + \frac{\partial \mathcal{T}_2}{\partial x_2} + \frac{\partial \mathcal{T}_3}{\partial x_3}. \tag{2.21}$$

Substituting the definition of $\mathcal{T}_i$ from equation (2.9) into equation (2.21) and inserting into equation (2.19) yields

$$\begin{aligned} \int_{\text{Vol}} (\mathbf{f}_V + \nabla \cdot \boldsymbol{\mathcal{T}}) \, d\text{Vol} = \int_{\text{Vol}} & \left(\frac{\partial T_{11}}{\partial x_1} + \frac{\partial T_{21}}{\partial x_2} + \frac{\partial T_{31}}{\partial x_3} + f_{V_1} \right) \hat{x}_1 \\ & + \left(\frac{\partial T_{12}}{\partial x_1} + \frac{\partial T_{22}}{\partial x_2} + \frac{\partial T_{32}}{\partial x_3} + f_{V_2} \right) \hat{x}_2 \\ & + \left(\frac{\partial T_{13}}{\partial x_1} + \frac{\partial T_{23}}{\partial x_2} + \frac{\partial T_{33}}{\partial x_3} + f_{V_3} \right) \hat{x}_3 \, d\text{Vol} = 0. \end{aligned} \tag{2.22}$$

Since the directions are independent, each term of the integrand must be equal to zero for the equation to be valid. Thus, the equations for static equilibrium can be written in indicial notation as

$$\mathrm{T}_{ji,j} + f_{V_i} = 0 \qquad i, j = 1, 2, 3. \tag{2.23}$$

The notation, j represents a partial derivative. Equation (2.23) is also written in matrix form:

$$\begin{bmatrix} \frac{\partial}{\partial x_1} & 0 & 0 & 0 & \frac{\partial}{\partial x_3} & \frac{\partial}{\partial x_2} \\ 0 & \frac{\partial}{\partial x_2} & 0 & \frac{\partial}{\partial x_3} & 0 & \frac{\partial}{\partial x_1} \\ 0 & 0 & \frac{\partial}{\partial x_3} & \frac{\partial}{\partial x_2} & \frac{\partial}{\partial x_1} & 0 \end{bmatrix} \begin{bmatrix} \mathrm{T}_{11} \\ \mathrm{T}_{22} \\ \mathrm{T}_{33} \\ \mathrm{T}_{23} \\ \mathrm{T}_{13} \\ \mathrm{T}_{12} \end{bmatrix} + \begin{bmatrix} f_{V_1} \\ f_{V_2} \\ f_{V_3} \end{bmatrix} = \begin{bmatrix} 0 \\ 0 \\ 0 \end{bmatrix}. \tag{2.24}$$

Equation (2.24) is written in compact form as

$$\mathrm{L}_{\mathrm{u}}'\mathbf{T} + \mathbf{f}_V = 0. \tag{2.25}$$

Equation (2.25) must be satisfied at each point on the solid. For a volume of material that has prescribed forces and displacements as shown in Figure 2.3, there are additional expressions that must be satisfied at the boundary. Define the region over which the surface stresses $\bar{\mathbf{t}}$ are prescribed as $\mathrm{S_t}$ and the region over which the displacements $\bar{\mathbf{u}}$ are prescribed as $\mathrm{S_u}$. Note that the value prescribed for the force or the displacement can be zero in these regions.

The conditions that must be satisfied at the boundary are

$$\mathbf{T} = \bar{\mathbf{t}} \qquad \text{on } \mathrm{S_t}, \tag{2.26}$$

which are the stress boundary conditions and

$$\mathbf{u} = \bar{\mathbf{u}} \qquad \text{on } \mathrm{S_u}, \tag{2.27}$$

which are the displacement, or kinematic, boundary conditions.

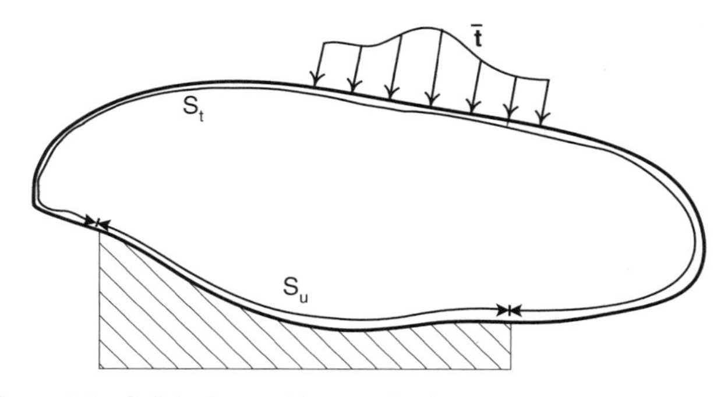

Figure 2.3 Solid volume with prescribed stresses and displacements.

Admissible stresses or *admissible displacements* can now be defined in relation to the equations of equilibrium and the boundary conditions. Admissible stresses are those that satisfy the equations of equilibrium, equation (2.25), and the stress boundary condition defined by equation (2.26). Similarly, kinematically admissible displacements are those that satisfy the equations of equilibrium and the kinematic boundary conditions specified in equation (2.27).

2.1.2 Linear Mechanical Constitutive Relationships

The next component of the mechanics model is the material law that relates the stress to the strain in the solid. For a linear elastic solid the strain is related to the stress through the expression

$$\mathrm{S}_{kl} = \mathcal{S}_{klmn}\mathrm{T}_{mn} \qquad k, l, m, n = 1, \ldots, 3, \tag{2.28}$$

or in compact notation,

$$\mathrm{S}_i = s_{ij}\mathrm{T}_j \qquad i, j = 1, \ldots, 6. \tag{2.29}$$

The material law expressed in compact notation is also written in matrix notation as

$$\mathbf{S} = \mathrm{s}\mathbf{T}. \tag{2.30}$$

In this book the matrix s is called the *compliance matrix* of the material. Equation (2.30) is also invertible, allowing the stress to be expressed as a linear function of the strain,

$$\mathbf{T} = \mathrm{c}\mathbf{S}, \tag{2.31}$$

where $\mathrm{c} = \mathrm{s}^{-1}$ is called the *modulus matrix*. The compliance and modulus matrices are both symmetric:

$$\mathrm{c} = \mathrm{c}' \qquad \mathrm{s} = \mathrm{s}'. \tag{2.32}$$

The form of the compliance (or modulus) matrix is a function of the material symmetry. In the most general case the compliance matrix takes the form

$$\mathrm{s} = \begin{bmatrix} s_{11} & s_{12} & s_{13} & s_{14} & s_{15} & s_{16} \\ s_{21} & s_{22} & s_{23} & s_{24} & s_{25} & s_{26} \\ s_{31} & s_{32} & s_{33} & s_{34} & s_{35} & s_{36} \\ s_{41} & s_{42} & s_{43} & s_{44} & s_{45} & s_{46} \\ s_{51} & s_{52} & s_{53} & s_{54} & s_{55} & s_{56} \\ s_{61} & s_{62} & s_{63} & s_{64} & s_{65} & s_{66} \end{bmatrix}, \tag{2.33}$$

which, by virture of symmetry, has 21 independent constants. An *orthotropic solid* has planes of symmetry such that

$$
\mathrm{s} = \begin{bmatrix} \frac{1}{Y_1} & -\frac{\nu_{12}}{Y_1} & -\frac{\nu_{13}}{Y_1} & 0 & 0 & 0 \\ -\frac{\nu_{21}}{Y_2} & \frac{1}{Y_2} & -\frac{\nu_{23}}{Y_2} & 0 & 0 & 0 \\ -\frac{\nu_{31}}{Y_3} & -\frac{\nu_{32}}{Y_3} & \frac{1}{Y_3} & 0 & 0 & 0 \\ 0 & 0 & 0 & \frac{1}{G_{23}} & 0 & 0 \\ 0 & 0 & 0 & 0 & \frac{1}{G_{13}} & 0 \\ 0 & 0 & 0 & 0 & 0 & \frac{1}{G_{12}} \end{bmatrix}, \tag{2.34}
$$

where $Y_i, i = 1, 2, 3$ are the *elastic modulus* values, ν_{ij} are the Poisson ratio values, and G_{ij} are the shear modulus parameters. Symmetry of the compliance matrix requires that

$$
\frac{\nu_{ij}}{Y_i} = \frac{\nu_{ji}}{Y_j} \qquad i, j = 1, 2, 3. \tag{2.35}
$$

A material whose constitutive properties do not depend on orientation is called *isotropic* and the compliance matrix takes the form

$$
\mathrm{s} = \frac{1}{Y} \begin{bmatrix} 1 & -\nu & -\nu & 0 & 0 & 0 \\ -\nu & 1 & -\nu & 0 & 0 & 0 \\ -\nu & -\nu & 1 & 0 & 0 & 0 \\ 0 & 0 & 0 & 2(1+\nu) & 0 & 0 \\ 0 & 0 & 0 & 0 & 2(1+\nu) & 0 \\ 0 & 0 & 0 & 0 & 0 & 2(1+\nu) \end{bmatrix}. \tag{2.36}
$$

The material properties of an isotropic material are defined by two parameters, the elastic modulus Y and the Poisson ratio ν. The compliance matrix for an isotropic material is inverted to yield the modulus matrix,

$$
\mathrm{c} = \frac{Y}{(1+\nu)(1-2\nu)} \begin{bmatrix} 1-\nu & \nu & \nu & 0 & 0 & 0 \\ \nu & 1-\nu & \nu & 0 & 0 & 0 \\ \nu & \nu & 1-\nu & 0 & 0 & 0 \\ 0 & 0 & 0 & (1-2\nu)/2 & 0 & 0 \\ 0 & 0 & 0 & 0 & (1-2\nu)/2 & 0 \\ 0 & 0 & 0 & 0 & 0 & (1-2\nu)/2 \end{bmatrix}. \tag{2.37}
$$

Example 2.2 (a) A material is said to be in a state of *plane stress* if

$$T_3 = T_4 = T_5 = 0.$$

Write the material law for an isotropic material that is in a state of plane stress. (b) Compute the state of strain for a material with an elastic modulus of 62 GPa and a Poisson ratio of 0.3 when the state of stress is

$$T_1 = 5\text{ MPa} \qquad T_2 = 4\text{ MPa} \qquad T_6 = 2\text{ MPa}.$$

Solution (a) The material law is obtained by substituting the assumptions regarding the stress state into equation (2.30) using equation (2.36) for the compliance matrix. The result is

$$\begin{aligned}
S_1 &= \frac{1}{Y}T_1 - \frac{\nu}{Y}T_2 \\
S_2 &= -\frac{\nu}{Y}T_1 + \frac{1}{Y}T_2 \\
S_3 &= -\frac{\nu}{Y}T_1 - \frac{\nu}{Y}T_2 \\
S_4 &= 0 \\
S_5 &= 0 \\
S_6 &= \frac{2(1+\nu)}{Y}T_6.
\end{aligned}$$

Note that the strain in the 3 direction is not equal to zero for an element in plane stress. Combining the components in the 1 and 2 directions, it is shown that

$$S_3 = -\frac{\nu}{1-\nu}(S_1 + S_2)$$

for an element in plane stress.

(b) Substituting the values stated in the problem into the material law derived in part (a) yields

$$\begin{aligned}
S_1 &= \frac{5\times 10^6\text{ Pa}}{62\times 10^9\text{ Pa}} - \frac{0.3}{62\times 10^9\text{ Pa}}4\times 10^6\text{ Pa} \\
S_2 &= -\frac{0.3}{62\times 10^9\text{ Pa}}5\times 10^6\text{ Pa} + \frac{4\times 10^6\text{ Pa}}{62\times 10^9\text{ Pa}} \\
S_3 &= -\frac{0.3}{62\times 10^9\text{ Pa}}5\times 10^6\text{ Pa} - \frac{0.3}{62\times 10^9\text{ Pa}}4\times 10^6\text{ Pa} \\
S_4 &= 0
\end{aligned}$$

$$S_5 = 0$$
$$S_6 = \frac{(2)(1+0.3)}{62 \times 10^9 \text{ Pa}} 2 \times 10^6 \text{ Pa},$$

which equals

$$\begin{aligned}
S_1 &= 61.2 \times 10^{-6} \text{ m/m} \\
S_2 &= 40.3 \times 10^{-6} \text{ m/m} \\
S_3 &= -43.5 \times 10^{-6} \text{ m/m} \\
S_4 &= 0 \\
S_5 &= 0 \\
S_6 &= 83.9 \times 10^{-6} \text{ m/m}.
\end{aligned}$$

As stated earlier, the unit 1×10^{-6} m/m is called a *microstrain* (μstrain).

Alternative forms of the governing equilibrium equations are obtained by combining equation (2.25) with the constitutive laws for a material. Assuming that the material is linear elastic, the relationship between stress and strain is expressed by equation (2.31). Substituting this expression into equation (2.25) yields

$$L_u' c\mathbf{S} + \bar{f}_V = 0. \tag{2.38}$$

The strain–displacement equation, equation (2.15), is now substituted in equation (2.38) to produce

$$L_u' c L_u \mathbf{u} + \bar{f}_V = 0. \tag{2.39}$$

The equilibrium expressions are now expressed in displacement form. Equation (2.39) represents a three-by-three set of equations that when solved yield a displacement vector that satisfies the equations of equilibrium. The boundary conditions expressed in equation (2.27) must also be satisfied for the solution to be admissible.

2.1.3 Electrostatics

In the preceding section we introduced the fundamental relationships for the mechanics of materials. Central to the discussion was the concept of stress and strain and the constitutive expressions that defined the relationships between these materials. The constitutive properties were then related to the boundary conditions through equilibrium expressions that must be satisfied to determine solutions to specified problems.

Many of the materials that we analyze herein have electronic properties in addition to their mechanical properties. Analysis of these materials will require a basic understanding of the concepts of electrostatics as they apply to the relationship between

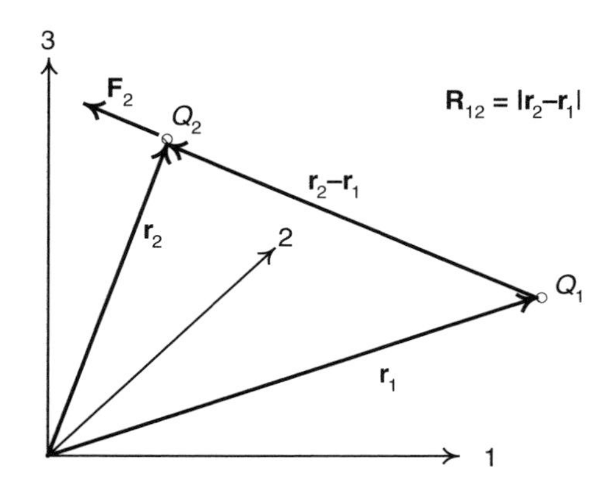

Figure 2.4 System of two point charges showing the electrostatic force vector.

electric fields. In our discussion of electrostatics we use the same procedure as in our discussion of mechanics. First we define the basic quantities of electrostatics and then define the governing equations that enable the solution of specific problems.

Let us begin by defining charge as the fundamental electrical quantity. In its simplest form, consider a *point charge* defined at a location in space in a Cartesian coordinate system. The size of the point charge is quantified by the amount of charge, Q_1, where charge is specified in the unit coulombs, which is given the symbol C. The vector from the origin of the coordinate system to the point charge is denoted **r**. Now consider the case where there are two point charges of size Q_1 and Q_2 located at points $\mathbf{r}_1$ and $\mathbf{r}_2$, respectively, as shown in Figure 2.4. Coulomb's law states that the magnitude of the force between any two objects in free space separated by a distance much larger than their dimension is

$$f = \frac{1}{4\pi \varepsilon_0} \frac{Q_1 Q_2}{R^2}, \tag{2.40}$$

where R^2 is the square of the distance between objects. The variable ε_0, the *permittivity of free space*, has the value 8.854×10^{-12} F/m. For the point charges shown in Figure 2.4, the force on point charge 2 due to the existence of point charge 1 is

$$\mathbf{f}_2 = \frac{Q_1 Q_2}{4\pi \varepsilon_0 R_{12}^2} \frac{\mathbf{r}_2 - \mathbf{r}_1}{|\mathbf{r}_2 - \mathbf{r}_1|}. \tag{2.41}$$

The force vector on point charge 2 is shown in Figure 2.4 for the case when the sign of the charges is the same. The force on point charge 1 is equal and opposite to the force on point charge 2; therefore, $\mathbf{f}_1 = -\mathbf{f}_2$.

For the moment let us define point charge 2 as a *test charge* and move the charge around the free space while keeping point charge 1 fixed. The force induced on point charge 2 is defined by equation (2.41). The magnitude of the force is proportional to the inverse of the distance between the charges squared and is always in the direction

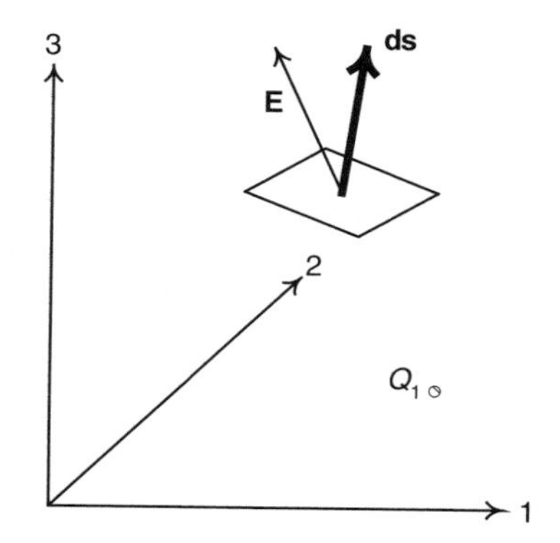

Figure 2.5 Electric field flux through a differential surface element.

of a vector that points from the location of Q_1 to Q_2. We define the *electric field intensity*, or simply the *electric field*, as the force $\mathbf{f}_2$ normalized with respect to the size of the test charge, which in this case is Q_2. Thus,

$$\mathbf{E} = \frac{\mathbf{f}_2}{Q_2} = \frac{Q_1}{4\pi\varepsilon_0 R_{12}^2} \frac{\mathbf{r}_2 - \mathbf{r}_1}{|\mathbf{r}_2 - \mathbf{r}_1|}. \tag{2.42}$$

The electric field has units of force per unit charge, N/C. Shortly we will define the unit *volts* (V), which is equivalent to the product of force and distance per unit charge, or N · m/C. Using these definitions, we can also define the units of electric field as volts per unit length, or V/m.

The electric field produced by a unit of charge is a vector quantity whose magnitude is proportional to the size of the point charge and inversely proportional to the distance from the point charge. The electric field vector points directly away from the point charge location along a radial line with the center at $\mathbf{r}_1$. The electric field generated by a point charge can be visualized as a set of electric field *lines* that emanate from the location of the charge. If we specify a surface at some distance away from the point charge and count the number of electric field lines that cross the surface (Figure 2.5) and normalize the result with respect to the surface area, we can define the *electric flux intensity*, or *electric displacement*, as the vector **D**. The direction of the vector **D** is the direction of the electric field lines that cross the surface. Faraday demonstrated that the electric displacement is related to the electric field in a free space through the expression

$$\mathbf{D} = \varepsilon_0 \mathbf{E} = \frac{Q_1}{4\pi R_{12}^2} \frac{\mathbf{r}_2 - \mathbf{r}_1}{|\mathbf{r}_2 - \mathbf{r}_1|}. \tag{2.43}$$

Consider a system that consists of multiple point charges Q_i located within a volume. The total charge is denoted Q and is expressed as

$$Q = \sum_{i=1}^{N_Q} Q_i. \tag{2.44}$$

Coulomb's law is linear and therefore the electric field and electric displacement at any point **x** within the space can be expressed as a summation of the electric fields and electric displacements due to each individual point charge:

$$\mathbf{E}(\mathbf{x}) = \sum_{i=1}^{N_Q} \frac{Q_i}{4\pi \varepsilon_0 R_{ix}^2} \frac{\mathbf{r}_x - \mathbf{r}_i}{|\mathbf{r}_x - \mathbf{r}_i|} \tag{2.45}$$

$$\mathbf{D}(\mathbf{x}) = \sum_{i=1}^{N_Q} \frac{Q_i}{4\pi R_{ix}^2} \frac{\mathbf{r}_x - \mathbf{r}_i}{|\mathbf{r}_x - \mathbf{r}_i|}. \tag{2.46}$$

Example 2.3 A fixed charge of $+Q$ coulombs has been placed in free space at $x_1 = a/2$ and a second fixed charge of $-Q$ coulombs has been placed at $x_1 = -a/2$. Determine the electric field at the location $(a/2, a/2)$.

Solution Since Coulomb's law is linear, we can compute the electric field to each charge separately and add them to obtain the combined electric field at the point. This is stated in equation (2.46). Defining the negative charge as Q_1 and the positive charge as Q_2, the position vectors for the problem are

$$\mathbf{r}_1 = -\frac{a}{2}\hat{x}_1$$

$$\mathbf{r}_2 = \frac{a}{2}\hat{x}_1$$

$$\mathbf{r}_x = \frac{a}{2}\hat{x}_1 + \frac{a}{2}\hat{x}_2.$$

Computing the electric field from the negative charge located at $(-a/2, 0)$ requires computation of

$$\mathbf{r}_x - \mathbf{r}_1 = \frac{a}{2}\hat{x}_1 + \frac{a}{2}\hat{x}_2 - \left(-\frac{a}{2}\hat{x}_1\right)$$

$$= a\hat{x}_1 + \frac{a}{2}\hat{x}_2.$$

The magnitude R_{1x} is

$$|\mathbf{r}_x - \mathbf{r}_1| = R_{1x} = \sqrt{a^2 + \frac{a^2}{4}} = \frac{a\sqrt{5}}{2}.$$

The direction of the electric field from charge 1 (the negative charge) to the test point is

$$\frac{\mathbf{r}_x - \mathbf{r}_1}{|\mathbf{r}_x - \mathbf{r}_1|} = \frac{a\hat{x}_1 + (a/2)\hat{x}_2}{a\sqrt{5}/2}$$

$$= \frac{2}{\sqrt{5}}\hat{x}_1 + \frac{1}{\sqrt{5}}\hat{x}_2.$$

The electric field at the test point due to charge 1 is

$$\mathbf{E}_{1x} = \frac{-Q}{4\pi\varepsilon_0(5a^2/4)}\left(\frac{2}{\sqrt{5}}\hat{x}_1 + \frac{1}{\sqrt{5}}\hat{x}_2\right)$$

$$= \frac{-Q}{a^2\pi\varepsilon_0}\left(\frac{2}{5\sqrt{5}}\hat{x}_1 + \frac{1}{5\sqrt{5}}\hat{x}_2\right).$$

The electric field due to charge 2 can be found in the same manner, or, by inspection,

$$\mathbf{E}_{2x} = \frac{Q}{a^2\pi\varepsilon_0}\hat{x}_2.$$

The electric field is the sum of the electric field due to the individual charges,

$$\mathbf{E} = \frac{Q}{a^2\pi\varepsilon_0}\left[-\frac{2}{5\sqrt{5}}\hat{x}_1 + \left(1 - \frac{1}{5\sqrt{5}}\right)\hat{x}_2\right]$$

$$= \frac{Q}{a^2\pi\varepsilon_0}\left(-\frac{2}{5\sqrt{5}}\hat{x}_1 + \frac{5\sqrt{5}-1}{5\sqrt{5}}\hat{x}_2\right).$$

If we compute the exact solution out to three decimal places, we have

$$\mathbf{E} = \frac{Q}{a^2\pi\varepsilon_0}(-0.179\hat{x}_1 + 0.911\hat{x}_2).$$

Comparing the total electric field vector with the field vector due only to charge 2, we see that the field due to charge 1 tends to bend the electric field line back toward the negative charge. The electric field still has the largest component in the $\hat{x}_2$ direction, due to the fact that the positive charge is closer, but there is still a noticeable effect, due to the negative charge at $(-a/2,0)$.

Now let us consider a system of point charges. If we draw a closed surface around this system of point charges, as shown in Figure 2.6, at any point on the surface we can define a differential unit of area whose unit normal is perpendicular to the surface at that point. We call this unit normal $d\mathbf{s}$. The electric displacement that is in the direction of the unit normal at point $\mathbf{x}$ is written $\mathbf{D}(\mathbf{x}) \cdot d\mathbf{s}$. One of the fundamental

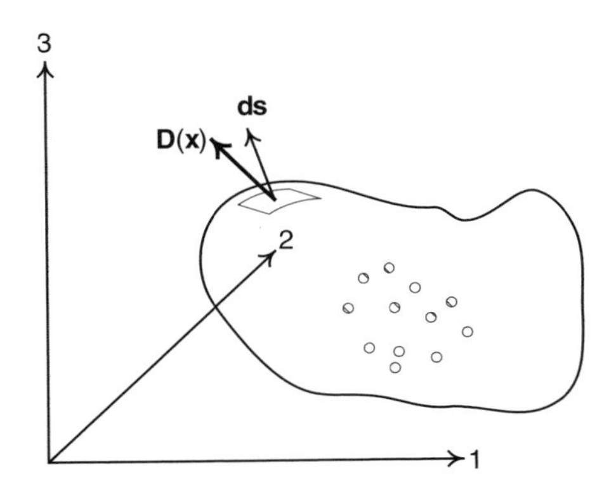

Figure 2.6 System of point charges illustrating the closed surface associated with applying Gauss's law.

laws of electrostatics, Gauss's law, states that the integral of $\mathbf{D}(\mathbf{x}) \cdot d\mathbf{s}$ over *any* closed surface is equal to the total charge enclosed within that surface. Mathematically, Gauss's law states that

$$\oint_{\text{Surf}} \mathbf{D}(\mathbf{x}) \cdot d\mathbf{s} = Q. \tag{2.47}$$

If we assume that the system of charges is not merely a set of discrete point charges but a continuous function of charge per unit volume, $\rho_v(\mathbf{x})$, the total charge within a volume is the volume integral

$$Q = \int_{\text{Vol}} \rho_v(\mathbf{x})\, d\text{Vol}, \tag{2.48}$$

and Gauss's law can be restated as

$$\oint_{\text{Surf}} \mathbf{D}(\mathbf{x}) \cdot d\mathbf{s} = \int_{\text{Vol}} \rho_v(\mathbf{x})\, d\text{Vol}. \tag{2.49}$$

Applying the divergence theorem to the surface integral of equation (2.49),

$$\oint_{\text{Surf}} \mathbf{D}(\mathbf{x}) \cdot d\mathbf{s} = \int_{\text{Vol}} \nabla \cdot \mathbf{D}(\mathbf{x})\, d\text{Vol}, \tag{2.50}$$

allows us to write Gauss's law as

$$\int_{\text{Vol}} \nabla \cdot \mathbf{D}(\mathbf{x})\, d\text{Vol} = \int_{\text{Vol}} \rho_v(\mathbf{x})\, d\text{Vol}. \tag{2.51}$$

Now that both sides of the equation are volume integrals, we can equate the integrands. This leads to an expression of the point form of the first of Maxwell's equations,

$$\nabla \cdot \mathbf{D}(\mathbf{x}) = \rho_v(\mathbf{x}). \tag{2.52}$$

Combining equations (2.52) and (2.20) yields a less compact version of Maxwell's first equation:

$$\frac{\partial D_1}{\partial x_1} + \frac{\partial D_2}{\partial x_2} + \frac{\partial D_3}{\partial x_3} = \rho_v(\mathbf{x}). \tag{2.53}$$

The charge density can also be related to the electric field by recalling that electric field and electric displacement are related through equation (2.43). Introducing equation (2.43) into equations (2.52) and (2.53) yields

$$\begin{aligned} \nabla \cdot \mathbf{E}(\mathbf{x}) &= \frac{\rho_v(\mathbf{x})}{\varepsilon_0} \\ \frac{\partial E_1}{\partial x_1} + \frac{\partial E_2}{\partial x_2} + \frac{\partial E_3}{\partial x_3} &= \frac{\rho_v(\mathbf{x})}{\varepsilon_0}. \end{aligned} \tag{2.54}$$

Example 2.4 The electric displacement field in free space over the range $-a < x_1 < a$ and $-b < x_2 < b$ is defined as

$$\mathbf{D}(\mathbf{x}) = \frac{\cosh(10x_1/a)}{\cosh(10)}\hat{x}_1 + \frac{\cosh(10x_2/b)}{\cosh(10)}\hat{x}_2.$$

Compute (a) the electric displacement vector at $(a,0)$ and (a,b), and (b) the expression for the volume charge using the point form of Gauss's law.

Solution (a) Evaluating the electric displacement vector at $(a,0)$ yields

$$\begin{aligned} \mathbf{D}(a,0) &= \frac{\cosh[10(a)/a]}{\cosh(10)}\hat{x}_1 + \frac{\cosh[10(0)/b]}{\cosh(10)}\hat{x}_2 \\ &= \hat{x}_1 + \frac{1}{\cosh(10)}\hat{x}_2 \end{aligned}$$

and evaluating at (a,b) yields

$$\begin{aligned} \mathbf{D}(a,b) &= \frac{\cosh[10(a)/a]}{\cosh(10)}\hat{x}_1 + \frac{\cosh[10(b)/b]}{\cosh(10)}\hat{x}_2 \\ &= \hat{x}_1 + \hat{x}_2. \end{aligned}$$

The term cosh(10) is on the order of 10,000; therefore, the electric displacement vector is approximately in the 1 unit direction at $(a,0)$. At the corner of the region the electric displacement vector makes a 45° angle to the horizontal axis.

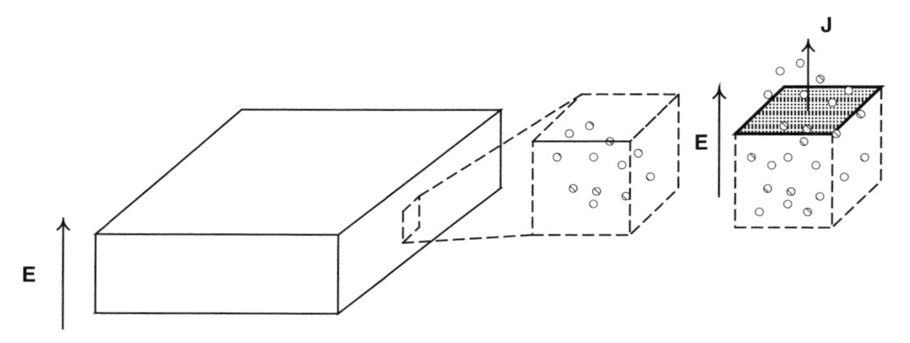

Figure 2.7 System of volume point charges conducting through a material.

(b) The point form of Gauss's law is shown in equation (2.52) and expanded in equation (2.53). Using the expanded form, we need to obtain the partial derivatives

$$\frac{\partial D_1}{\partial x_1} = \frac{10}{a}\frac{\sinh(10x_1/a)}{\cosh(10)}$$

$$\frac{\partial D_2}{\partial x_2} = \frac{10}{b}\frac{\sinh(10x_2/b)}{\cosh(10)}.$$

The volume charge density is the summation of the two terms:

$$\rho_v(\mathbf{x}) = \frac{10}{\cosh(10)}\left[\frac{\sinh(10x_1/a)}{a} + \frac{\sinh(10x_2/b)}{b}\right].$$

Thus far in our discussion we have focused on systems of fixed charges and related the charge distribution to electric field and electric displacement. Time-varying charge distributions produce electronic *current* through a material. Consider the material shown in Figure 2.7, which incorporates a volume charge that is depicted in the figure as a set of discrete point charges. If these charges are moving, the current is defined as the time rate of change of charge motion,

$$i(t) = \frac{dQ}{dt}. \tag{2.55}$$

The units of current are C/s or amperes (A). Drawing a differential surface with unit normal $d\mathbf{s}$ allows us to determine the amount of current passing through this surface per unit area. We call this quantity the *current density*, denote it as $\mathbf{J}$, and state it in units of current per area, A/m^2 in SI units. The differential amount of current that passes through the surface is defined by the dot product of the current density and the differential surface element:

$$di = \mathbf{J} \cdot d\mathbf{s}. \tag{2.56}$$

For a surface of finite area, the total current passing through the surface is equal to the integral over the surface, or

$$i = \int_{\text{Surf}} \mathbf{J} \cdot d\mathbf{s}, \tag{2.57}$$

where the time dependence of the current is implicit in the expression. If we integrate over a closed surface, the continuity of charge specifies that

$$\oint_{\text{Surf}} \mathbf{J} \cdot d\mathbf{s} = -\int_{\text{Vol}} \frac{\partial \rho_v}{\partial t}\, d\text{Vol}. \tag{2.58}$$

The negative sign can be understood by considering that the unit normal of the surface is defined as outward; therfore, a position current flow through the closed surface implies that the volume charge is *decreasing*. The negative sign in equation (2.58) reflects this definition.

The divergence theorem is applied to write both sides of equation (2.58) as a volume integral,

$$\int_{\text{Vol}} \nabla \cdot \mathbf{J}\, d\text{Vol} = -\int_{\text{Vol}} \frac{\partial \rho_v}{\partial t}\, d\text{Vol}. \tag{2.59}$$

which yields the definition of continuity of charge at a point,

$$\nabla \cdot \mathbf{J} = -\frac{\partial \rho_v}{\partial t}. \tag{2.60}$$

2.1.4 Electronic Constitutive Properties of Conducting and Insulating Materials

In Section 2.1.3 we stated the equations that govern electrostatics. These equations were derived in relation to an idealized medium, free space, which was characterized by a parameter called the permittivity of free space. In a manner similar to our discussion of mechanical systems, we now need to introduce expressions that relate the state variables of electrostatic systems to material properties of actual media.

Two types of electronic materials that we study in this book are conductors and insulators. *Conductors* are materials that consist of a large number of mobile charges that will move upon application of an electric field. The exact mechanism by which the charge moves is not central to our discussion, but this charge motion can be visualized as the flow of charged species within the material. *Insulators* are materials that consist of a large proportion of bound charge. Bound charge in insulators will reorient under the application of an electric field but will not exhibit the motion that is characteristic of conductors.

What is important to our discussion of smart materials is that the equations that govern the electrostatics of a material change depending on whether it is a conductor or an insulator. These expressions are derived by incorporating a constitutive law that defines the relationship between the electrical state variables of the system.

In a conductor the current density is related to the electric field through the *conductivity* of the material, σ:

$$\mathbf{J} = \sigma \mathbf{E}. \tag{2.61}$$

Equation (2.61) is a statement of Ohm's law expressed between current per area and electric field. The material parameter σ specifies the quality of the conducting material. As equation (2.61) demonstrates, higher conductivity indicates that a larger current flow will occur for a prescribed electric field. Conducting materials are used in all types of electronics applications. Examples of good conductors, those that have high conductivity, are copper and brass. These materials are often used as wiring and interconnects in electronics. In the field of smart materials, shape memory alloys are good conductors, and we will see that we can use a material's conductivity as a stimulus for thermomechanical actuation.

Insulators do not contain a large proportion of mobile charge but are very useful in electronic applications and, as we shall see, smart material applications. Insulators are characterized by an atomic structure that contains a large proportion of *bound charge*, charge that will not conduct through a material but will reorient in the presence of an applied electric field. Bound charge can be visualized as a pair of point charges of equal and opposite charge Q separated by a distance $\mathbf{d}$. At the center of the distance connecting the charge pair is a pivot, which is similar to a mechanical pivot but allows the charges to rotate about the point (Figure 2.8).

The pair of point charges separated by a distance is called an *electronic dipole*, or simply a *dipole*. Placing a dipole in an electric field will produce a motion of both charges but *will not result in charge conduction*, due to the fact that the charges are bound. The *dipole moment* associated with this pair of point charges is defined as

$$\mathbf{p} = Q\mathbf{d}, \tag{2.62}$$

where the vector $\mathbf{d}$ is defined as the vector from the negative charge to the positive charge. If there are n dipole moments per unit volume, the *polarization* of the material

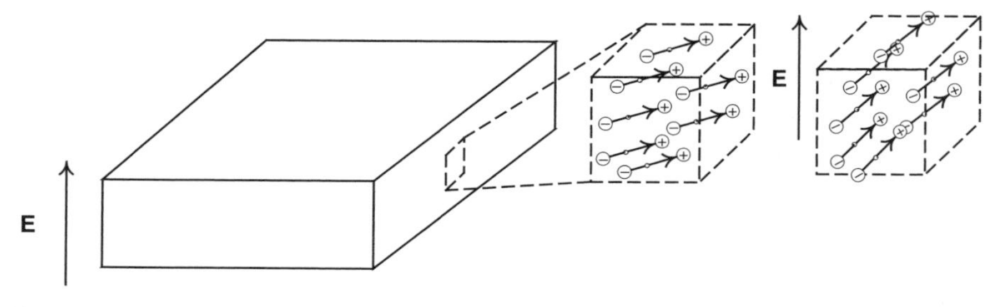

Figure 2.8 Representation of bound charge within a dielectric insulator as a group of dipoles.

is defined as

$$\mathbf{P} = \lim \frac{1}{\Delta \mathrm{Vol}} \sum_{i=1}^{n} \mathbf{p}_i. \tag{2.63}$$

The units of the dipole moment are C·m; therefore, the polarization (which is the number of dipole moments per unit volume) has units of charge per area, C/m^2.

Once again let us define a surface with unit normal $d\mathbf{s}$ that is pointing outward. Imagine that the application of an electric field causes n dipole moments to rotate such that charge crosses the surface. The differential amount of *bound charge*, Q_b, that crosses the surface is equal to

$$dQ_b = nQ\mathbf{d} \cdot d\mathbf{s} = \mathbf{P} \cdot d\mathbf{s}. \tag{2.64}$$

If we perform the integration over a closed surface, the total bound charge within the closed surface is equal to

$$Q_b = -\oint_{\mathrm{Surf}} \mathbf{P} \cdot d\mathbf{s}. \tag{2.65}$$

The question that we want to answer is: How does the existence of a polarization within the material change our expressions for the electrostatics of the material? To answer this question, consider a volume of material that contains both bound charge, Q_b, and free (or mobile) charge, Q. The total charge in the volume is

$$Q_t = Q_b + Q. \tag{2.66}$$

Applying Gauss's law to the volume to solve for the total charge, we have

$$Q_t = Q_b + Q = \oint_{\mathrm{Surf}} \varepsilon_0 \mathbf{E} \cdot d\mathbf{s}. \tag{2.67}$$

Substituting equation (2.65) into equation (2.67) and rearranging the terms yields

$$Q = \oint_{\mathrm{Surf}} (\varepsilon_0 \mathbf{E} + \mathbf{P}) \cdot d\mathbf{s}. \tag{2.68}$$

Equation (2.68) is an important expression because it tells us that polarization in the material can be thought of as an additional term of the electric displacement of the material. This can be expressed mathematically by rewriting the electric displacement in equation (2.47) as

$$\mathbf{D} = \varepsilon_0 \mathbf{E} + \mathbf{P}, \tag{2.69}$$

which we see makes it equivalent to equation (2.68).

In many materials the relationship between polarization and electric field is assumed to be a linear relationship. In this case we can write the polarization as the product of a constant and the electric field vector,

$$\mathbf{P} = (\varepsilon_R - 1)\,\varepsilon_0 \mathbf{E}, \tag{2.70}$$

where ε_R is defined as the *relative permittivity*, which is a unitless parameter that is always greater than 1. The somewhat strange way of writing the constant of proportionality in equation (2.70) is due to the fact that when we substitute the expression into equation (2.69), we obtain

$$\mathbf{D} = \varepsilon_0 \mathbf{E} + (\varepsilon_R - 1)\,\varepsilon_0 \mathbf{E} = \varepsilon_R \varepsilon_0 \mathbf{E}. \tag{2.71}$$

Defining the *permittivity of the material* as the product of the relative permittivity and the permittivity of free space,

$$\varepsilon = \varepsilon_R \varepsilon_0, \tag{2.72}$$

we can relationship between electric displacement and electric field as

$$\mathbf{D} = \varepsilon \mathbf{E}. \tag{2.73}$$

This analysis demonstrates that bound charge in a material can be treated mathematically as an increase in the permittivity, which, in turn, produces an increase in the electric displacement for an applied electric field. The relative permittivity is a material constant that can range from between 2 and 10 for some polymer materials to the range 1000 to 5000 for some ceramics. Smart materials that exhibit dielectric properties include piezoelectric materials such as the ones we study in upcoming chapters.

Consider once again a cube of material as shown in Figure 2.2. Defining a Cartesian coordinate system as we did for the analysis of stress and strain allows us to write a relationship between the electric displacement in three dimensions and the applied electric field. This expression is written in indicial notation,

$$\mathrm{D}_i = \varepsilon_{ij} \mathrm{E}_j, \tag{2.74}$$

where the indices range from 1 to 3. Equation (2.74) represents the constitutive relationship between electric field and electric displacement for a linear material. For an anisotropic dielectric material, equation (2.74) is written

$$\begin{aligned} \mathrm{D}_1 &= \varepsilon_{11}\mathrm{E}_1 + \varepsilon_{12}\mathrm{E}_2 + \varepsilon_{13}\mathrm{E}_3 \\ \mathrm{D}_2 &= \varepsilon_{21}\mathrm{E}_1 + \varepsilon_{22}\mathrm{E}_2 + \varepsilon_{23}\mathrm{E}_3 \\ \mathrm{D}_3 &= \varepsilon_{31}\mathrm{E}_1 + \varepsilon_{32}\mathrm{E}_2 + \varepsilon_{33}\mathrm{E}_3. \end{aligned} \tag{2.75}$$

In an anisotropic material an electric field applied in one coordinate direction can produce electric displacement in an orthogonal coordinate direction. In an isotropic dielectric material

$$\varepsilon_{ij} = 0 \qquad i \neq j, \tag{2.76}$$

and the constitutive relationship is written as

$$\begin{aligned} D_1 &= \varepsilon_{11} E_1 \\ D_2 &= \varepsilon_{22} E_2 \\ D_3 &= \varepsilon_{33} E_3. \end{aligned} \tag{2.77}$$

Sometimes the constitutive relationship for an isotropic linear dielectric material is written in a compact notation that uses only a single subscript:

$$D_k = \varepsilon_k E_k \qquad k = 1, 2, 3. \tag{2.78}$$

Example 2.5 The dielectric properties of an insulator material have been measured and shown to have slight anisotropy. The relative dielectric properties have been measured to be

$$\begin{aligned} \varepsilon_{11} = \varepsilon_{22} &= 1200 \\ \varepsilon_{23} = \varepsilon_{32} &= 150 \\ \varepsilon_{33} &= 1800. \end{aligned}$$

Compute the electric displacement to the applied electric fields (a) $\mathbf{E} = 1 \times 10^6 \hat{x}_1$ V/m and (b) $\mathbf{E} = 1 \times 10^6 \hat{x}_3$ V/m.

Solution (a) Applying equation (2.75) and substituting in the relative permittivity values and the electric field yields

$$\begin{aligned} D_1 &= (1200\varepsilon_0 \text{ F/m})(1 \times 10^6 \text{ V/m}) + (0)(0) + (0)(0) \\ D_2 &= (0)(1 \times 10^6 \text{ V/m}) + (1200\varepsilon_0 \text{ F/m})(0) + (150\varepsilon_0 \text{ F/m})(0) \\ D_3 &= (0)(1 \times 10^6 \text{ V/m}) + (150\varepsilon_0 \text{ F/m})(0) + (1800\varepsilon_0 \text{ F/m})(0). \end{aligned}$$

Performing the computations yields the electric displacement vector for the material:

$$\begin{aligned} D_1 &= 0.0106 \text{ C/m}^2 \\ D_2 &= 0 \\ D_3 &= 0. \end{aligned}$$

(b) Substituting in the electric field vector in the 3 direction produces

$$\begin{aligned} D_1 &= (1200\varepsilon_0 \text{ F/m})(0) + (0)(0) + (0)(1 \times 10^6 \text{ V/m}) \\ D_2 &= (0)(0) + (1200\varepsilon_0 \text{ F/m})(0) + (150\varepsilon_0 \text{ F/m})(1 \times 10^6 \text{ V/m}) \\ D_3 &= (0)(0) + (150\varepsilon_0 \text{ F/m})(0) + (1800\varepsilon_0 \text{ F/m})(1 \times 10^6 \text{ V/m}), \end{aligned}$$

which yields

$$\begin{aligned} D_1 &= 0 \\ D_2 &= 0.0013 \text{ C/m}^2 \\ D_3 &= 0.0159 \text{ C/m}^2. \end{aligned}$$

This example illustrates that the electric field applied in the direction that is isotropic (in this case, the 1 direction) produces electric displacement in the same direction. Applying the field in a direction that exhibits anisotropy (the 3 direction) produces an electric displacement vector that has components in orthogonal directions.

2.2 WORK AND ENERGY METHODS

The governing equations of mechanical and electrical systems were introduced in Section 2.1. The fundamental properties included a definition of the state variables associated with these two domains, the basic equilibrium laws, and a state of the constitutive relationships between the state variables. For a mechanical system the state variables are the stress and strain, while for an electronic system the state variables are electric field and electric displacement. For linear materials the state variables are related through constitutive relationships that are defined in terms of a fourth-order tensor in the case of mechanical systems and a second-order tensor in the case of electrical systems.

In subsequent chapters we will see that the constitutive relationships can form the basis for models of smart material devices. For example, in Chapter 4 we utilize the constitutive properties of electromechanical devices to derive basic properties of piezoelectric actuators and sensors. A similar approach is taken in Chapter 6 for the development of actuator models for shape memory alloys. In certain instances, however, alternative methods based on energy and work will be more efficient means of obtaining system models. For example, when deriving a model of a piezoelectric material integrated into an elastic structure, as we discuss in Chapter 5, it would be possible to use the governing laws described in Section 2.1 to obtain a system model, This is sometimes referred to as a *direct method*. In this book we develop models of systems using approaches based on energy and work principles, called *indirect methods*. Direct and indirect methods should yield the same result, but in this book indirect methods are the preferred approach because they are amenable to solution using computational techniques.

Before discussing modeling methods using the concepts of energy and work, we must define these concepts for mechanical and electrical systems. This is the focus of the following sections.

2.2.1 Mechanical Work

Consider a particle that is moving along a path as shown in Figure 2.9 and being acted upon by the prescribed force **f**. If the particle moves a differential amount along this

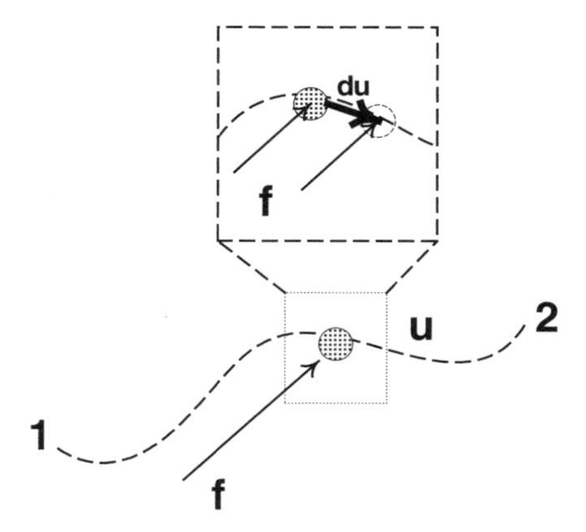

Figure 2.9 Concept of a differential unit of work applied to a moving particle.

path, $d\mathbf{u}_i$, the force produces a differential amount of work, $d\mathrm{W}$, defined by

$$d\mathrm{W} = \mathbf{f} \cdot d\mathbf{u}. \tag{2.79}$$

The total work on the particle from $\mathbf{u}_1$ to $\mathbf{u}_2$ is defined by the integral

$$\mathrm{W}_{12} = \int_{\mathbf{u}_1}^{\mathbf{u}_2} \mathbf{f} \cdot d\mathbf{u}. \tag{2.80}$$

The units of work are N·m, or joules, which is denoted by the symbol J.

An important feature of work, as compared to force and displacement, is that *work is a scalar quantity*, due to the fact that it is a dot product of the force vector with the displacement vector. Another important property of the dot product is that only the force that is in the direction of the motion performs work; force that is orthogonal to the direction of the motion does no work on the particle. Although work is a scalar quantity, the sign of the work is important. Positive work is work that is performed in the direction of motion, whereas negative work is work that is in a direction opposite the motion.

The force vector shown in Figure 2.9 is an example of a prescribed force that is independent of the particle location. In many instances, though, the force is a function of the particle displacement, $\mathbf{f} = \mathbf{f}(\mathbf{u})$, and the work performed on the particle is obtained from the expression

$$\mathrm{W}_{12} = \int_{\mathbf{u}_1}^{\mathbf{u}_2} \mathbf{f}(\mathbf{u}) \cdot d\mathbf{u}. \tag{2.81}$$

In this case the work expression can be integrated to yield a function of $\mathbf{u}$ that we define as an energy function U,

$$\mathrm{U} = \int_{\mathbf{u}_1}^{\mathbf{u}_2} \mathbf{f}(\mathbf{u}) \cdot d\mathbf{u}. \tag{2.82}$$

The integrand of equation (2.82) is defined as a differential amount of work,

$$dU = \mathbf{f} \cdot d\mathbf{u}. \tag{2.83}$$

Expanding equation (2.83) in a Cartesian coordinate system defined in the 1, 2, and 3 directions results in the expressions

$$\begin{aligned} dU &= (f_1\hat{x}_1 + f_2\hat{x}_2 + f_3\hat{x}_3) \cdot (du_1\hat{x}_1 + du_2\hat{x}_2 + du_3\hat{x}_3) \\ &= f_1\,du_1 + f_2\,du_2 + f_3\,du_3. \end{aligned} \tag{2.84}$$

The energy function U is a function of the displacement coordinates u_i; therefore, we can express the total differential of the function U as

$$dU = \frac{\partial U}{\partial u_1}du_1 + \frac{\partial U}{\partial u_2}du_2 + \frac{\partial U}{\partial u_3}du_3. \tag{2.85}$$

Equating equations (2.84) and (2.85), we see that

$$\begin{aligned} f_1 &= \frac{\partial U}{\partial u_1} \\ f_2 &= \frac{\partial U}{\partial u_2} \\ f_3 &= \frac{\partial U}{\partial u_3}. \end{aligned} \tag{2.86}$$

Equation (2.86) is an important relationship between the energy function and the forces that act on a system. The expressions state that the force in a particular coordinate direction is equal to the partial derivative of the energy function with respect to the displacement in that direction. This result provides another interpretation of a force that is dependent on the motion of the particle because it allows us to derive *all of the forces* from a single scalar energy function.

The relationship between energy and force can be visualized if we restrict ourselves to a single dimension. Let us assume for the moment that the motion of the particle consists of only a single dimension denoted by the displacement u. Figure 2.10*a* is a plot of a generic force-to-displacement relationship for a single dimension. The differential unit of energy is defined as the rectangle shown in the figure, which represents the quantity $f(u)$. Integrating this function from u_1 to u_2 yields the area under the curve as shown in Figure 2.10*b*.

The relationships described by equation (2.86) can be visualized by drawing a generic energy function U as shown in Figure 2.11. Since we are assuming that the system is a function of only a single dimension, equation (2.86) is reduced to

$$f = \frac{dU}{du}. \tag{2.87}$$

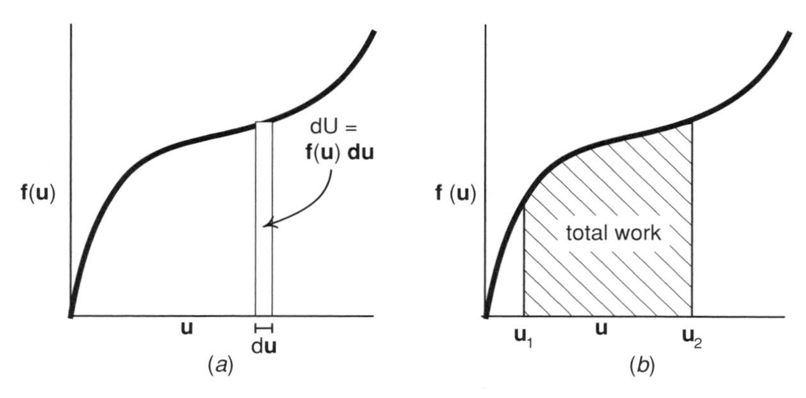

Figure 2.10 (*a*) Differential unit of energy in a generic force-to-displacement relationship; (*b*) total work performed as the area under the force–displacement curve.

The force is visualized as the slope of the curve at a particular point:

$$f(u_1) = \left.\frac{d\text{U}}{d\text{u}}\right|_{u_1}. \tag{2.88}$$

This relationship is also shown in Figure 2.11. It is interesting to note that in this example it is clear that the slope, and hence the force, varies as a function of u.

We will find that there are a number of energy functions that will be useful to us in our study of active materials. Another such function is the *potential energy* of a system, defined as the negative of the energy function U and denoted

$$-\text{V} = \text{U}. \tag{2.89}$$

The potential energy is defined in this manner due simply to the convention that the potential energy function of a system *decreases* when work is performed on the

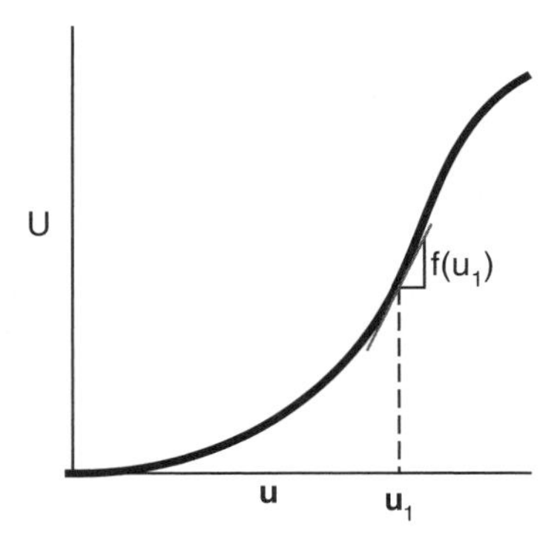

Figure 2.11 Relationship between energy and force.

system. The differential of the potential energy is defined as

$$dV = -\mathbf{f}(\mathbf{u}) \cdot d\mathbf{u}. \tag{2.90}$$

Combining the relationships in equation (2.86) with the definition of potential energy allows us to write a succinct relationship between force and potential energy:

$$\mathbf{f} = -\nabla V. \tag{2.91}$$

Equation (2.91) states that the force vector is the negative gradient of the potential function V.

Example 2.6 The potential energy function for a nonlinear spring is

$$V = \frac{1}{2}u^2 + \frac{1}{4}u^4. \tag{2.92}$$

Plot the potential energy function over the range -2 to 2 and determine the force at $u = 1$ and $u = -1.5$.

Solution Equation (2.87) defines the relationship between the internal force and the energy function U. Combining this expression with equation (2.89) yields

$$-\frac{dV}{dx} = f(u) = -u - u^3. \tag{2.93}$$

Substituting $u = 1$ and $u = -1.5$ into this expression, we have

$$f(1) = -1 - (1)^3 = -2 \tag{2.94}$$

$$f(-1.5) = 1.5 - (-1.5)^3 = 4.875. \tag{2.95}$$

Plots of the potential energy and force are shown in Figure 2.12. As shown, the force can be interpreted as the negative of the slope of the potential energy function at the elongation specified. As the potential energy function illustrates, the slope of V changes at $u = 0$; therefore, the sign of the force changes as the nonlinear spring changes from extension to compression.

The discussion so far has centered on the work done to a particle by an applied force. If we consider the situation where we have stress applied at the body of an elastic body (as shown in Figure 2.1), we define the work *per unit volume* performed by the applied stress as

$$\tilde{W}_{12} = \int_{S_1}^{S_2} \mathbf{T} \cdot d\mathbf{S}. \tag{2.96}$$

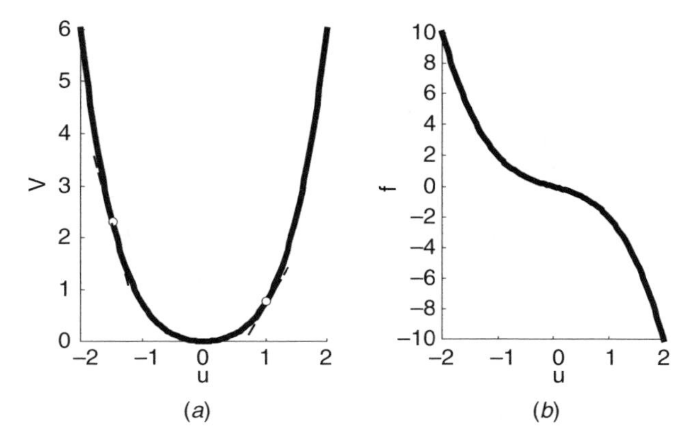

Figure 2.12 Plots of the (*a*) potential energy and (*b*) force associated with $V = \frac{1}{2}u^2 + \frac{1}{4}u^4$.

The integral is written in indicial notation as

$$\tilde{W}_{12} = \int_{S_1}^{S_2} T_i \, dS_i \tag{2.97}$$

and the units of this work term are joules per unit volume, J/m^3.

The energy expression for an elastic body is derived by incorporating the relationship between stress and strain into the work expression. For a general elastic body the stress is written as a function of the strain, **T**(**S**), and the integration in equation (2.96) can be expressed as

$$\tilde{U} = \int_{S_1}^{S_2} \mathbf{T}(\mathbf{S}) \cdot d\mathbf{S}. \tag{2.98}$$

For a linear elastic body the stress–strain relationship is written as equation (2.31), and the energy function is

$$\tilde{U} = \int_{S_1}^{S_2} c\mathbf{S} \cdot d\mathbf{S}. \tag{2.99}$$

The energy function for a linear elastic body can be integrated to yield

$$\tilde{U} = \frac{1}{2}\mathbf{S}'c\mathbf{S}\Big|_{S_1}^{S_2}, \tag{2.100}$$

which demonstrates that the energy function for a linear elastic body is a quadratic function of the strain.

2.2.2 Electrical Work

In Section 2.1.3 we introduced the fact that forces on electronic systems are produced due to the interaction of charged particles. This fact, stated by Coulomb's law, gives rise to the definition of electric field and electric displacement as the electrostatic state variables. Within this definition we defined the electric field as the force per unit charge produced by a charged particle. This definition was quantified by equation (2.42). Rewriting this definition, we can express the force produced on a test charge Q by an electric field as

$$\mathbf{f} = Q\mathbf{E}. \tag{2.101}$$

It stands to reason that this force will produce work as the test charge is moved through an electric field in the same way that a force on an uncharged particle will produce mechanical work as the particle traverses a path. Substituting equation (2.101) into the expression for work, equation (2.80), yields

$$W_{12} = \int_{\mathbf{u}_1}^{\mathbf{u}_2} Q\mathbf{E} \cdot d\mathbf{u}. \tag{2.102}$$

The charge Q is not a function of position, so it can be brought outside the integration. The scalar *electric potential*, V, is defined as the work performed per unit charge,

$$V_2 - V_1 = \frac{W_{12}}{Q} = \int_{\mathbf{u}_1}^{\mathbf{u}_2} \mathbf{E} \cdot d\mathbf{u}. \tag{2.103}$$

The units of electric potential are work per unit charge, J/C, or *volts*.

The term inside the integrand of equation (2.103) can be thought of as the differential unit of potential, dV, which can be expanded in a Cartesian frame of reference. By definition, we express

$$\begin{aligned} dV &= -\mathbf{E}(\mathbf{u}) \cdot d\mathbf{u} \\ &= -(E_1\hat{x}_1 + E_2\hat{x}_2 + E_3\hat{x}_3) \cdot (du_1\hat{x}_1 + du_2\hat{x}_2 + du_3\hat{x}_3) \\ &= -E_1\,du_1 - E_2\,du_2 - E_3\,du_3. \end{aligned} \tag{2.104}$$

The differential of the potential is also expanded in terms of the partial derivatives,

$$dV = \frac{\partial V}{\partial u_1}du_1 + \frac{\partial V}{\partial u_2}du_2 + \frac{\partial V}{\partial u_3}du_3. \tag{2.105}$$

Equating the terms in equations (2.104) and (2.105) yields the relationships

$$E_1 = -\frac{\partial V}{\partial u_1}$$

$$E_2 = -\frac{\partial V}{\partial u_2}$$
$$E_3 = -\frac{\partial V}{\partial u_3}. \tag{2.106}$$

The relationship between field and potential can be written in terms of the gradient operator,

$$\mathbf{E} = -\nabla V. \tag{2.107}$$

Comparing the equation (2.107) with equation (2.91), we see that there is a one-to-one correspondance between the concept of electrical potential and the potential function associated with a mechanical system. The force applied to a mechanical system can be expressed as the negative gradient of the mechanical potential function, and the force per unit charge, or electric field, can be expressed as the negative gradient of the electric potential. This relationship will become important in energy-based models of smart materials because we will find that we can express a single scalar potential function that describes the interactions between mechanical and electrical domains within the material. This will allow us to develop a unified framework for analyzing electromechanical interactions in certain materials.

Example 2.7 Compute the work required to move a charge of 0.1 C from the point $\mathbf{u}_1$ = (0,0.01,0) to $\mathbf{u}_2$ = (0.01,0.03,0) in a constant electric field $\mathbf{E} = -100\hat{x}_2$ V/m.

Solution Equation (2.102) is the expression for the work required to move a charge. The work per unit charge, or electric potential, required to move the charge is

$$\begin{aligned}
V_2 - V_1 &= \int_{(0,0.01,0)}^{(0.01,0.03,0)} (-100\hat{x}_2) \cdot (du_1\hat{x}_1 + du_2\hat{x}_2 + du_3\hat{x}_3) \\
&= -100 \int_{(0,0.01,0)}^{(0.01,0.03,0)} du_2 \\
&= -100 u_2 \Big|_{0.01}^{0.03} \\
&= -2 \text{ V}.
\end{aligned}$$

Note that the only component of the motion that contributes to the potential is the component in the $\hat{x}_2$ direction. The work is the product of the potential and the charge

$$W_{12} = (0.1 \text{ C})(-2 \text{ V}) = -0.2 \text{ J}.$$

Recall that a volt is defined as work per unit charge, or J/C; therefore, the units of charge multiplied by potential are units of work and energy, joules.

The electrical work associated with the application of an electric field at a point is expressed in terms of the applied field and the electric displacement. The electrical work per unit volume is

$$\tilde{W}_{12} = \int_{\mathbf{D}_1}^{\mathbf{D}_2} \mathbf{E} \cdot d\mathbf{D}. \tag{2.108}$$

This is written in indicial notation as

$$\tilde{W}_{12} = \int_{\mathbf{D}_1}^{\mathbf{D}_2} E_i \, d\mathrm{D}_i. \tag{2.109}$$

For the case in which the electric field is a function of the electric displacement, $\mathbf{E}(\mathbf{D})$, the stored electrical energy per unit volume is

$$\tilde{U} = \int_{\mathbf{D}_1}^{\mathbf{D}_2} \mathbf{E}(\mathbf{D}) \cdot d\mathbf{D}. \tag{2.110}$$

For a linear dielectric material the relationship between electric field and electric displacement is expressed by equation (2.73). The matrix of dielectric constants can be inverted to yield the relationship $\mathbf{E} = \varepsilon^{-1}\mathbf{D}$. This expression is substituted into equation (2.110) and the result is integrated:

$$\tilde{\mathrm{U}} = \frac{1}{2}\mathbf{D}'\varepsilon^{-1}\mathbf{D}\Big|_{\mathbf{D}_1}^{\mathbf{D}_2}. \tag{2.111}$$

As in the case of an elastic body, where the stored elastic energy is a quadratic function of the strain, the stored electric energy for a linear dielectric material is a quadratic function of the electric displacement.

2.3 BASIC MECHANICAL AND ELECTRICAL ELEMENTS

The theory discussed in Sections 2.1 and 2.2 allows us to develop equations that relate stress, strain, electric displacement, and electric field in different types of mechanical and electrical systems. We will see in upcoming chapters that there are a few basic elements that we will analyze many times in our analysis of smart material systems. In this section we use the governing equations discussed in Sections 2.1 and 2.2 to analyze these basic elements.

2.3.1 Axially Loaded Bars

Many of the smart material devices that we analyze in the succeeding chapters are modeled as an axially loaded mechanical member, or simply, *axial member*. Consider a bar in which the 1 direction is aligned along the length of the bar. We assume that the

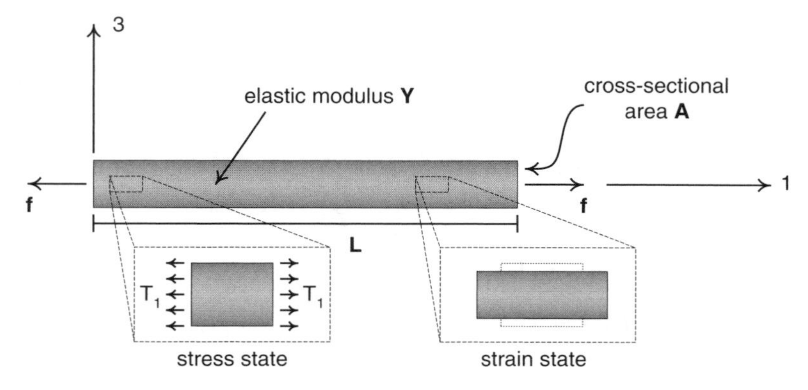

Figure 2.13 Axially loaded bar, illustrating the stress and strain states.

material properties of the bar are homogeneous: that they do not vary in any direction, the material is isotropic, and the cross-sectional area, A, of the bar does not change along the length. Furthermore, we assume that the only stress applied on the bar is T_1 and that all other stress components are zero, as shown in Figure 2.13. Neglecting body forces, the governing equations for equlibrium, equation (2.23), are

$$\frac{\partial T_1}{\partial x_1} = \frac{dT_1}{dx_1} = 0, \tag{2.112}$$

which, when integrated, indicate that the stress along the bar is constant. Equation (2.112) also indicates that the partial derivative can be written as a full derivative, due to the fact that the stress is only a function of x_1. Assuming that the force on the bar is evenly distributed, the stress is

$$T_1 = \frac{f}{A}. \tag{2.113}$$

Substituting the stress state into the constitutive relationships for a linear elastic material, equation (2.30), the strain state for the material is

$$\begin{aligned} S_1 &= \frac{1}{Y}T_1 = \frac{f}{YA} \\ S_2 &= -\frac{\nu}{Y}T_1 = -\nu\frac{f}{YA} \\ S_3 &= -\frac{\nu}{Y}T_1 = -\nu\frac{f}{YA}. \end{aligned} \tag{2.114}$$

For an axial bar, $S_2 = S_3 = -\nu S_1$. The displacement of the bar is computed from the strain–displacement relationship. Assuming small strain, equation (2.14) is reduced to

$$\mathrm{S}_1 = \frac{du_1}{dx_1} = \frac{f}{YA}. \tag{2.115}$$

The displacement is obtained from the integral

$$\int_0^L du_1 = \int_0^L \frac{f}{YA} dx_1, \tag{2.116}$$

which, when integrated, yields

$$u_1(L) - u_1(0) = \Delta u_1 = \frac{L}{YA} f. \tag{2.117}$$

Equation (2.117) is rewritten as a relationship between force and deflection as

$$f = \frac{YA}{L} \Delta u_1. \tag{2.118}$$

This expression illustrates that the force–deflection relationship for a linear elastic axial bar is also a linear relationship. The coefficient YA/L, called the *stiffness* of the bar, has the unit N/m. The stiffness can be increased by using a material with a higher modulus (e.g., switching from aluminum to steel) or increasing the cross-sectional area of the bar. The stiffness of the bar will decrease when the length increases.

The strain energy in the axial bar is expressed by equation (2.100) due to the assumption that the material is linear elastic. The strain energy in the axial bar is simply

$$\tilde{\mathrm{U}} = \frac{Y}{2} \mathrm{S}_1^2 = \frac{1}{2} \frac{f^2}{YA^2}. \tag{2.119}$$

The total energy stored in the bar is the product of the strain energy and the volume, AL,

$$\mathrm{U} = \frac{1}{2} \frac{L}{YA} f^2 = \frac{1}{2} (\Delta u_1)(f), \tag{2.120}$$

which demonstrates that the total energy stored in an axial bar by a uniform load is equal to one-half the product of the deflection and the force applied.

2.3.2 Bending Beams

A *beam* is a mechanical element that supports a load perpendicular to its length. The theory discussed in this section applies to beams whose length is much greater than either their width or their thickness. Under these circumstances we will be able to

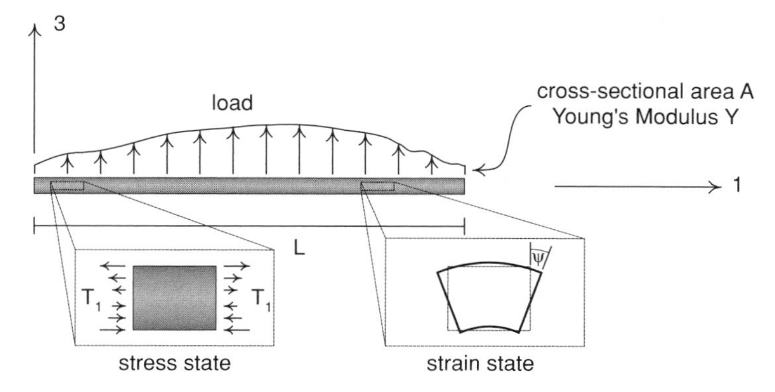

Figure 2.14 States of stress and strain for a bending beam.

derive a relationship between the applied load and the deflection of the beam and the stored mechanical energy.

The beam geometry that we consider is shown in Figure 2.14. We assume that the length of the beam is aligned with the 1 axis of the Cartesian coordinate system and that any loads are applied in the 3 direction. The material properties of the beam are assumed to be homogeneous. Making the kinematic assumptions that there are no dimensional changes in the 3 direction and that any section that is plane before deformation remains a plane section after deformation, displacement in the 1 direction is written as

$$u_1 = u_0 + \psi x_3, \tag{2.121}$$

where ψ is the slope of the plane as shown in Figure 2.14 and u_0 is the uniform axial displacement. Assuming small strains, the slope of the plane is approximately equal to the first derivative of the displacement in the 3 direction, u_3. Assuming that the uniform axial displacement is equal to zero yields

$$u_1 = x_3 \frac{du_3}{dx_1}. \tag{2.122}$$

The strain S_1 is obtained from equation (2.14):

$$S_1 = \frac{du_1}{dx_1} = x_3 \frac{d^2 u_3}{d^2 x_1}. \tag{2.123}$$

Assuming that each differential element of the material is under uniaxial stress, the strain in the 2 and 3 directions is

$$\begin{aligned} S_2 &= -\nu S_1 = -\nu x_3 \frac{d^2 u_3}{d^2 x_1} \\ S_3 &= -\nu S_1 = -\nu x_3 \frac{d^2 u_3}{d^2 x_1}, \end{aligned} \tag{2.124}$$

and the only nonzero stress is given by

$$T_1 = x_3 Y \frac{d^2 u_3}{d^2 x_1}. \tag{2.125}$$

The moment caused by the applied load is in equilibrium with the moment produced by the induced stress. The relationship is

$$M_2 = \int_A x_3 T_1 \, dA, \tag{2.126}$$

where dA is the differential area element $dx_2\, dx_3$. Substituting equation (2.125) into equation (2.126) and taking the terms that are independent of x_2 and x_3 out of the integral, we have

$$M_2 = \left(Y \frac{d^2 u_3}{d^2 x_1} \right) \int_A x_3^2 \, dA. \tag{2.127}$$

We denote the integral as the second area moment of inertia, in this case about the x_3 axis, and write the relationship as

$$M_2 = Y I_{33} \frac{d^2 u_3}{d^2 x_1}. \tag{2.128}$$

Equation (2.128) is a direct relationship between the moment due to applied loads and the resulting deflection.

Determining the deflection along the length of the beam requires an explicit definition of the moment and a set of prescribed boundary conditions. The first case to consider is the case in which the moment is constant along the length. Under this assumption, equation (2.128) can be integrated once to yield

$$Y I_{33} \frac{du_3}{dx_1} = M_2 x_1 + c_1, \tag{2.129}$$

where c_1 is a constant of integration. Integrating once more and dividing by YI_{33} yields an expression for the displacement:

$$u_3(x_1) = \frac{1}{Y I_{33}} \left(\frac{1}{2} M_2 x_1^2 + c_1 x_1 + c_2 \right). \tag{2.130}$$

The displacement of the beam in the 3 direction varies as a quadratic function along the length of the beam.

The exact expression for the displacement is a function of the boundary conditions. Two common boundary conditions are *clamped–free* and *pinned–pinned*. These two boundary conditions impose the following constraints on the displacement function:

$$\begin{aligned} &\text{clamped–free:} && u_3(0) = \left.\frac{du_3}{dx_1}\right|_{x_1=0} = 0 \\ &\text{pinned–pinned:} && u_3(0) = 0 u_3(L) = 0. \end{aligned} \tag{2.131}$$

Substituting the boundary conditions into equations (2.129) and (2.130) yields the displacement functions:

$$\begin{aligned} &\text{clamped–free:} && u_3(x_1) = \frac{M_2}{2YI_{33}} x_1^2 \\ &\text{pinned–pinned:} && u_3(x_1) = \frac{M_2}{2YI_{33}} \left(x_1^2 - Lx_1\right). \end{aligned} \tag{2.132}$$

Another loading condition that is common to the problems that we will study later is that of a point load applied along the length of the beam. First consider the case of a clamped–free beam with a point load applied at the free end. The moment induced by this load is

$$M_2(x_1) = f(L - x_1), \tag{2.133}$$

and the first integral of equation (2.128) is

$$YI_{33} \frac{du_3}{dx_1} = f\left(Lx_1 - \frac{1}{2}x_1^2\right) + c_1. \tag{2.134}$$

Integrating equation (2.134) once again yields

$$YI_{33} u_3(x_1) = f\left(\frac{1}{2}Lx_1^2 - \frac{1}{6}x_1^3\right) + c_1 x_1 + c_2. \tag{2.135}$$

Applying the boundary conditions for a clamped–free beam from equation (2.131) results in $c_1 = c_2 = 0$. Substituting this result into equation (2.135) and writing the expression for the deflection results in

$$u_3(x_1) = \frac{f}{YI_{33}} \left(\frac{1}{2}Lx_1^2 - \frac{1}{6}x_1^3\right). \tag{2.136}$$

The result of the analysis is that the deflection of a clamped–free beam with an end load is a cubic equation in x_1. This contrasts with the result for a constant moment, which was a quadratic expression. The stiffness of the beam at the point of loading is obtained by substituting $x_1 = L$ into equation (2.135) and solving for the force in terms of the displacement:

$$\frac{f}{u_3(L)} = 3\frac{YI_{33}}{L^3}. \tag{2.137}$$

Now consider the case in which the point load acts at the center of a pinned–pinned beam. In this case the moment induced by the point load is a piecewise continuous function of the form

$$M_2(x_1) = \begin{cases} \dfrac{f}{2}x_1 = YI_{33}\dfrac{d^2u_3}{d^2x_1} & 0 \le x_1 \le L/2 \\ \dfrac{f}{2}(L - x_1) = YI_{33}\dfrac{d^2u_3}{d^2x_1} & L/2 \le x_1 \le L. \end{cases} \tag{2.138}$$

Substituting the two expressions into the flexure equation and integrating once yields

$$2YI_{33}\frac{du_3}{dx_1} = \begin{cases} \dfrac{1}{2}fx_1^2 + c_1 & 0 \le x_1 \le L/2 \\ f\left(Lx_1 - \dfrac{1}{2}x_1^2\right) + c_3 & L/2 \le x_1 \le L. \end{cases} \tag{2.139}$$

Integrating a second time yields

$$2YI_{33}u_3 = \begin{cases} \dfrac{1}{6}fx_1^3 + c_1x_1 + c_2 & 0 \le x_1 \le L/2 \\ f\left(\dfrac{1}{2}Lx_1^2 - \dfrac{1}{6}x_1^3\right) + c_3x_1 + c_4 & L/2 \le x_1 \le L. \end{cases} \tag{2.140}$$

Notice that the solution requires solving for four constants of integration. Two of the integration constants are obtained from the boundary conditions for a pinned–pinned beam, $u_3(0) = u_3(L) = 0$. The remaining two constants of integration are obtained by ensuring continuity of the solutions at the location of the point load, $x_1 = L/2$. Applying these four conditions yields the following system of equations:

$$\begin{aligned} c_2 &= 0 \\ -c_3L - c_4 &= \frac{fL^3}{3} \\ c_1\frac{L}{2} + c_2 - c_3\frac{L}{2} - c_4 &= \frac{fL^3}{12} \\ c_1 - c_3 &= \frac{fL^2}{4}. \end{aligned} \tag{2.141}$$

The system of equations is solved to yield

$$\begin{aligned} c_1 &= -\frac{fL^2}{8} \\ c_2 &= 0 \\ c_3 &= -\frac{3fL^2}{8} \\ c_4 &= \frac{fL^3}{24}. \end{aligned} \tag{2.142}$$

The solutions for the constants of integration are substituted back into equation (2.140). After simplification, the displacement functions over the two regions of the beam are

$$u_3(x_1) = \begin{cases} \dfrac{f}{2YI_{33}}\left(\dfrac{1}{6}x_1^3 - \dfrac{L}{8}x_1^2\right) & 0 \le x_1 \le L/2 \\ \dfrac{f}{2YI_{33}}\left(-\dfrac{1}{6}x_1^3 + \dfrac{L}{2}x_1^2 - \dfrac{3L^2}{8}x_1 + \dfrac{L^3}{24}\right) & L/2 \le x_1 \le L. \end{cases} \tag{2.143}$$

An important feature of the solution for a pinned–pinned beam is that the deflection is expressed as a *piecewise* continuous function over the length of the beam. Enforcing the continuity conditions ensures that the solution and its first derivative are continuous at the location of the point load. At the application of the point load the second derivative of the solution is discontinuous and the solution in each of the two regions is described by equation (2.138). This contrasts with the results for the cases studied previously, that of a constant moment and a point load for a clamped–free beam, where the solution for the moment is a continuous function over the domain of the beam. This issue will be important in future chapters when we study techniques for approximating the solution of smart material systems using energy methods.

The stiffness at the centerpoint of the pinned–pinned beam is obtained by substituting $x_1 = L/2$ into equation (2.143) and solving for the ratio of the force to the deflection. The result is

$$\frac{f}{u_3(L/2)} = 48\frac{YI_{33}}{L^3}. \tag{2.144}$$

Comparing this result with equation (2.137), we see that a pinned–pinned beam has a stiffness that is 16 times larger than the stiffness of a cantilever beam with a point load at the free end.

To complete the analysis of bending beams, let us determine the strain energy function. Since we have assumed that each differential element is in a state of uniaxial stress and have neglected any shear stress in the beam, the strain energy per unit volume is equal to the strain energy for an axial bar. Substituting the expression for the bending strain, equation (2.123), into equation (2.119) yields

$$\tilde{U} = \frac{1}{2}x_3^2 Y \left(\frac{d^2u_3}{d^2x_1}\right)^2. \tag{2.145}$$

Note that the stored energy per unit volume changes through the thickness of the beam because the strain changes through the thickness. The total stored energy is obtained by integrating equation (2.145) throughout the volume. This integral is separated in the following fashion:

$$U = \int_0^L \frac{1}{2}Y \left(\frac{d^2u_3}{d^2x_1}\right)^2 \int_A x_3^2 \, dx_2 \, dx_3 \, dx_1. \tag{2.146}$$

The area integral is equal to our definition of the second area moment of inertia. Assuming that both the modulus and the area moment of inertial do not change along the length, we can write the total stored energy as

$$U = \frac{1}{2}YI_{33} \int_0^L \left(\frac{d^2u_3}{d^2x_1}\right)^2 dx_1. \tag{2.147}$$

Equation (2.147) is a general expression that applies to any beam in bending that is consistent with the assumptions of the analysis. The exact expression for the stored energy will be a function of the loading and the boundary conditions. The expression can be solved explicitly with the solutions for the displacement functions obtained in this section.

2.3.3 Capacitors

We now turn our attention to a basic electrical element, the *capacitor*. A capacitor is formed by placing a dielectric material between two conducting plates. Assuming that the interfacial material is a perfect dielectric, the application of a potential difference between the conductive plates produces the motion of bound charge within the dielectric and results in the storage of electrical energy. As we shall see in upcoming chapters, many of the smart materials we study in this book are dielectric materials that store energy under the application of an electric potential.

Consider a model system that consists of a dielectric material whose upper and lower surfaces are assumed to be perfect conductors (Figure 2.15). The dielectric is assumed to be homogeneous with a relative permittivity of ε_r; it is also assumed to be a perfect insulator with zero free charge. A potential is applied at both conductive faces, and the governing equations for electrostatics are used to determine the relationship between the applied potential, electric displacement, and electric field.

Assuming that the length and width of the capacitor are much larger than its thickness, the electric fields in the 1 and 2 directions can be ignored and we can assume that the only nonzero electric field is in the 3 direction. Under this assumption,

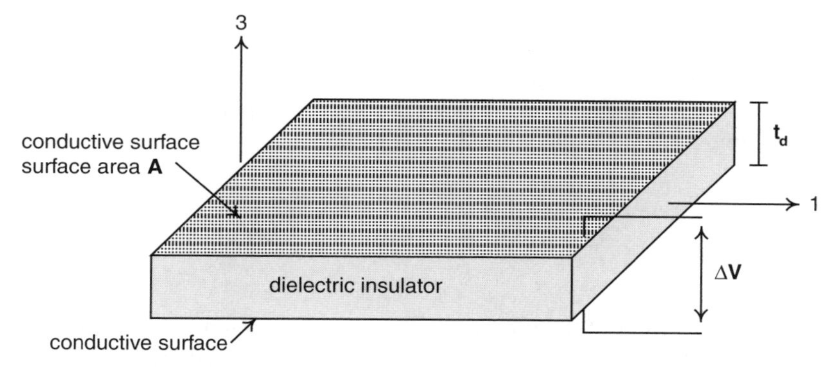

Figure 2.15 Ideal capacitor consisting of a dielectric insulator between two conductive surfaces.

equation (2.54) reduces to

$$\frac{d\mathrm{E}_3}{dx_3} = 0, \tag{2.148}$$

where the zero on the right-hand side is due to the fact that we assume a perfect insulator (i.e., the free charge is zero). Integrating equation (2.148) with respect to x_3 demonstrates that the electric field inside a capacitor is constant. For the moment, we denote this constant c_1:

$$\mathrm{E}_3 = c_1. \tag{2.149}$$

We have assumed that we are prescribing the electric potential at both conductor faces; therefore, we now apply equation (2.106),

$$\int_{-t_d/2}^{t_d/2} dV = -c_1 \int_{-t_d/2}^{t_d/2} du_3, \tag{2.150}$$

which after integration yields

$$V(t_d/2) - V(-t_d/2) = \Delta V = -c_1 t_d. \tag{2.151}$$

From this expression we can solve for the constant and substitute the result into equation (2.149):

$$\mathrm{E}_3 = -\frac{\Delta V}{t_d}. \tag{2.152}$$

This expression demonstrates that the electric field in a capacitor is constant and is equal to the potential difference divided by the thickness. The negative sign indicates that the field points from the location of positive potential to the negative potential, which is consistent with our definition of electric field. The amount of charge stored on the face of the capacitor is equal to the product of the surface area and the electric displacement (recall that electric displacement is charge per unit area). Generally, this quantity is written as the absolute value of the electric displacement to eliminate the negative sign:

$$Q = A|\mathrm{D}_3| = \varepsilon_3 A|\mathrm{E}_3| = \frac{\varepsilon_3 A}{t_d}\Delta V. \tag{2.153}$$

The coefficient in front of the potential difference in equation (2.153) is the *capacitance* of the material. The capacitance can be increased by using a material with a higher dielectric constant, a higher surface area, or by reducing the distance between electrodes.

The energy stored per unit volume in the dielectric material is computed from equation (2.111). Substituting $\mathrm{D}_3 = \varepsilon_3 \mathrm{E}_3$ into equation (2.111) yields

$$\tilde{\mathrm{U}} = \frac{1}{2}\frac{\varepsilon_3\, \Delta V^2}{t_d^2}. \tag{2.154}$$

The total energy stored is the product of equation (2.154) and the volume of the material, At_d,

$$\mathrm{U} = \frac{1}{2}\frac{\varepsilon_3 A}{t_d}\Delta V^2 = \frac{1}{2}Q\,\Delta V. \tag{2.155}$$

Thus, in an ideal insulator the stored energy is equal to one-half of the product of the stored charge and the applied potential.

2.3.4 Summary

In Section 2.3 we have described application of the governing laws to typical problems in the analysis of mechanical and electrical elements. As we shall see in upcoming chapters, these analyses will form the basis of many of the analyses that we perform for smart materials. For example, an analysis of axial bars is utilized when we model the motion of piezoelectric elements and develop equations that describe the transduction of piezoelectric devices. The assumption of uniaxial stress is utilized when we analyze the actuation properties of shape memory alloys. Many types of smart material transducers and smart material systems contain beamlike elements; therefore, we utilize the analysis of bending beams when we analyze electroactive polymer benders or piezoelectric bimorphs. These analyses will be based on the expressions for bending beams derived in this chapter.

2.4 ENERGY-BASED MODELING METHODS

The concepts of stress, strain, electric field, and electric displacement can be brought together under a single principle by considering the relationship between these state variables and the work and energy associated with a system. As discussed in Section 2.2, the product of mechanical force and mechanical displacement, or electric field and electric displacement, results in work being done to a body. The work performed on a body can be related to the stored energy, as we shall see, and this relationship will enable efficient methods of developing models of smart material systems.

Energy-based methods are based on the first law of thermodynamics, which states that a change in the *total internal energy* of a body is equal to the sum of the *work performed on the body* and the *heat transfer*. Denoting the total internal energy E (not to be confused with the electric field variable introduced earlier in the chapter), the work W, and the heat transfer Q, the first law is written as

$$dE = dW + dQ. \tag{2.156}$$

In its most basic form, the first law of thermodynamics is a statement of the balance of energy. In this book we utilize this concept to develop equations that govern the deformation and motion of smart material systems.

In Section 2.2 we were introduced to the relationship between work and energy for mechanical and electrical systems. In that discussion we saw that the concepts of work and energy are strongly interrelated. If we have prescribed forces, mechanical or electrical, applied to a body, they will perform work on that body. In the same manner, if the forces have a functional relationship to the mechanical or electrical response, the body will store energy internally, and this stored energy is quantified by a potential function V. The forces that produce this stored energy are related to the potential function through the gradient operator.

Another important result from Section 2.2 is that energy and work associated with electrical and mechanical systems are defined in terms of particular *state variables*. For example, work and energy associated with a mechanical system are defined in terms of force, stress, displacement, or strain. Mechanical work on a particle is quantified by a force acting through a distance, whereas the mechanical work (or stored energy) of an elastic body is defined in terms of strain and stress. Similarly, the work and energy of an electrical system are defined in terms of charge, electric field, and electric displacement.

For the moment, let us consider the first law, equation (2.156), applied to a system that does not exhibit heat transfer. In this case we can write that the change in internal energy is equivalent to the work done on the system,

$$dE = dW. \tag{2.157}$$

Assume that the internal energy and work performed on the system are expressed in terms of a set of *generalized state variables*. We denote the generalized state variables w_i to highlight the fact that the variable can represent a mechanical state variable or an electrical state variable. A critical feature of the generalized state variables is that they are *independent*. Assuming that there are N independent generalized states, then

$$
\begin{aligned}
d\mathrm{E}(w_1, \ldots, w_N) &= \frac{\partial E}{\partial w_1} dw_1 + \frac{\partial E}{\partial w_1} dw_2 + \cdots + \frac{\partial E}{\partial w_N} dw_N \\
d\mathrm{W}(w_1, \ldots, w_N) &= \frac{\partial W}{\partial w_1} dw_1 + \frac{\partial W}{\partial w_1} dw_2 + \cdots + \frac{\partial W}{\partial w_N} dw_N .
\end{aligned}
\tag{2.158}
$$

Substituting equation (2.158) into equation (2.157) produces the equality

$$
\frac{\partial E}{\partial w_1} dw_1 + \frac{\partial E}{\partial w_1} dw_2 + \cdots + \frac{\partial E}{\partial w_N} dw_N = \frac{\partial W}{\partial w_1} dw_1 + \frac{\partial W}{\partial w_1} dw_2 + \cdots + \frac{\partial W}{\partial w_N} dw_N .
\tag{2.159}
$$

Equation (2.159) must hold for *arbitary* changes in the state variables to maintain the energy balance stated by the first law. Assuming that the changes in the generalized states are arbitrary, and coupling this assumption to the fact that they are chosen to be independent of one another, means that all terms preceding dw_i on both sides of the equation must equal one another for the equation to be valid. Thus, use of the first law (ignoring heat transfer) results in the following set of equations:

$$
\begin{aligned}
\frac{\partial E}{\partial w_1} &= \frac{\partial W}{\partial w_1} \\
\frac{\partial E}{\partial w_2} &= \frac{\partial W}{\partial w_2} \\
&\vdots \\
\frac{\partial E}{\partial w_N} &= \frac{\partial W}{\partial w_N} .
\end{aligned}
\tag{2.160}
$$

The set of equations (2.160) must be satisfied for the energy balance specified by the first law to be maintained.

2.4.1 Variational Motion

The statement of energy balance in the first law leads to a set of equations that must be satisfied for differential changes in the generalized state variables. As introduced in Section 2.3, these changes in state variables are completely arbitrary. This definition of the change in the state variable, though, can be problematic if the change in the state variable violates any constraints associated with the system under examination. For example, displacement on the boundary of an elastic body may be constrained to be zero, and it is important that the differential change in the state variables be chosen such that this constraint is not violated. The need to define a set of small changes in the state variables that are consistent with the constraints of the problem

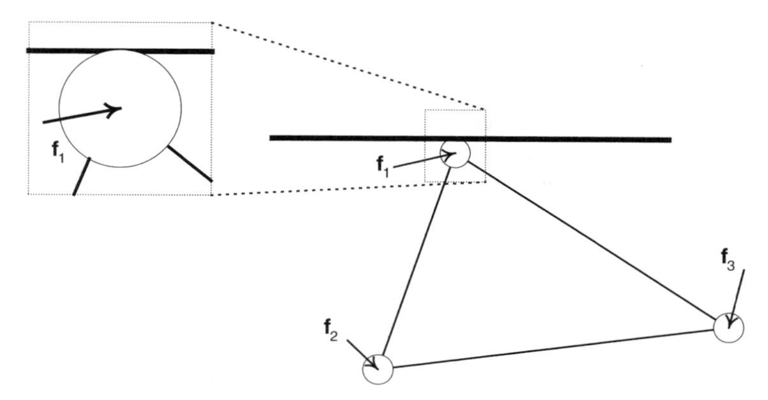

Figure 2.16 Three-link system with applied forces, with a geometric constraint on one of the nodes.

leads us to the concept of a *variation* of the state variable. A variational change in the state variable is a differential change that is consistent with the geometric constraints of the problem. In this book we denote a variational change by the variable δ; thus, a variational change in the state variable w_i is denoted δw_i. As is the case with a differential, we can take the variation of a vector. An example would be the variation of the vector of generalized state variables, $\mathbf{w}$, which is denoted $\delta\mathbf{w}$.

To illustrate the concept of a variational motion, consider the system shown in Figure 2.16, consisting of three rigid links connected at three nodes. There is an applied force at each node that is performing work on the system. One of the nodes is constrained to move in the x_2 direction while the other two nodes are unconstrained. Consider a differential motion of the node that lies against the frictionless constraint at the top of the figure. One potential differential motion is shown in Figure 2.17*a*. In this differential motion, we note that the node is moved through the constraint, and the connections between the node and the bars are not maintained.

Let's consider a differential motion that *does maintain the geometric constraints* of the problem. As shown in Figure 2.17*b*, we have a motion that is consistent with

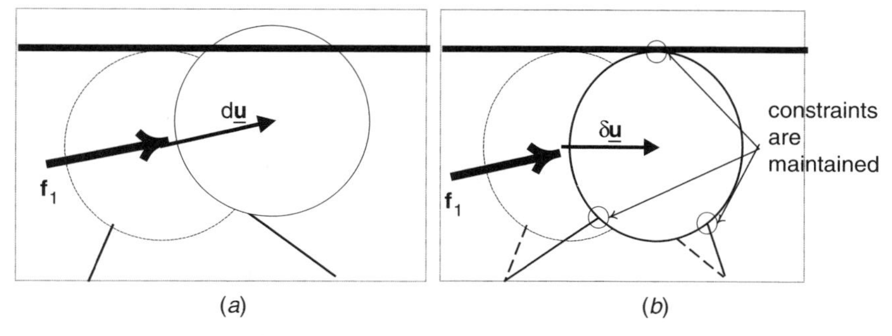

Figure 2.17 (*a*) Potential differential motion and (*b*) variational motion of a node. Note that the variational displacement is consistent with the geometric constraints of the problem.

the boundary constraint of the frictionless surface, and the connection between the node and the bars is maintained. The differential motions in Figure 2.17*a* and *b* are both valid, but we see that only the differential motion in Figure 2.17*b* is consistent with the constraints of the problem.

The *variational work* associated with a force $\mathbf{f}$ is the dot product of the force and the variational displacement, $\delta\mathbf{u}$,

$$\delta \mathrm{W} = \mathbf{f} \cdot \delta\mathbf{u}, \tag{2.161}$$

which represents the amount of work performed when the article undergoes a differential displacement that is consistent with the geometric constraints.

An important aspect of a variational motion is that constraint forces do not contribute to the variational work of a particle. Since all constraint forces are perpendicular to the motion, and the variational work is a dot product of the force and the variational displacement, constraint forces do not add to this function. This is beneficial to the development of equilibrium expressions for the system because the constraint forces need not be considered in the analysis of a problem.

The concept of *variational motion* also applies to the development of energy functions. Applying a variational displacement to a force that is dependent on $\mathbf{u}$ produces a variation in the energy, which is denoted

$$\delta \mathrm{U} = \mathbf{f}(\mathbf{u}) \cdot \delta\mathbf{u}. \tag{2.162}$$

2.5 VARIATIONAL PRINCIPLE OF SYSTEMS IN STATIC EQUILIBRIUM

The definitions of variational motion and its relationship to work enables the development of an efficient method for determining the equations that govern systems in static equilibrium. The fundamental law that governs the equilibrium of systems is

$$\mathbf{f} = 0. \tag{2.163}$$

Let us apply a variational displacement to the system in equilibrium. The resulting equation is

$$\mathbf{f} \cdot \delta\mathbf{u} = 0. \tag{2.164}$$

Let us separate the forces into four different types:

1. Externally applied mechanical forces $\mathbf{f}_M$ in which the force is prescribed
2. Externally applied electrical forces $\mathbf{f}_E$ in which the electric potential is prescribed
3. Mechanical forces that are internal to the system $\mathbf{f}_M(\mathbf{u})$ and have a functional dependence on the displacement

4. Electrical forces that are internal to the system $\mathbf{f}_E(\mathbf{u})$ and have a functional dependence on the displacement

Separating the forces into the four components and applying the variational displacement produces

$$\mathbf{f}_M \cdot \delta\mathbf{u} + \mathbf{f}_E \cdot \delta\mathbf{u} + \mathbf{f}_M(\mathbf{u}) \cdot \delta\mathbf{u} + \mathbf{f}_E(\mathbf{u}) \cdot \delta\mathbf{u} = 0. \tag{2.165}$$

Once again it is important to realize that this is now a scalar expression, whereas the original equilibrium expression, equation (2.163), is a vector-valued expression. We recognize the first and third terms as the mechanical work and the internal energy, respectively:

$$\begin{aligned} \delta \mathrm{W}_M &= \mathbf{f} \cdot \delta\mathbf{u} \\ \delta \mathrm{U}_M &= -\delta \mathrm{V}_M = \mathbf{f}(\mathbf{u}) \cdot \delta\mathbf{u}. \end{aligned} \tag{2.166}$$

The second and fourth terms in equation (2.165) can also be related to work and energy terms. Examining equation (2.103), we note that the work performed by the motion of charge in an electric field is equal to

$$\mathrm{W}_E = q(v_2 - v_1). \tag{2.167}$$

If the voltage is prescribed, the variational electrical work performed by the applied field is equal to

$$\delta \mathrm{W}_E = (v_2 - v_1)\,\delta q. \tag{2.168}$$

In the case in which the electric field can be derived from a potential function, the remaining term in equation (2.165) can be written as the variation in the electric potential function. Combining $\mathbf{f} = q\mathbf{E}$, we see that

$$\mathbf{f}_E(\mathbf{u}) \cdot \delta\mathbf{u} = q\mathbf{E}(\mathbf{u}) \cdot \delta\mathbf{u}. \tag{2.169}$$

The term

$$\mathbf{E}(\mathbf{u}) \cdot \delta\mathbf{u} = -\delta V, \tag{2.170}$$

according to equation (2.104). Let us define the electric potential function as

$$\mathbf{f}_E(\mathbf{u}) \cdot \delta\mathbf{u} = -\delta \mathrm{V}_E. \tag{2.171}$$

Substituting equations (2.166), (2.167), and (2.171) into equation (2.165) yields

$$\mathbf{f}_M \cdot \delta\mathbf{u} + (v_2 - v_1)\,\delta q - \delta \mathrm{V}_M - \delta \mathrm{V}_E = 0. \tag{2.172}$$

The first two terms on the left-hand side are recognized as the mechanical and electrical work; therefore, we can rewrite the expression as

$$\delta \mathrm{W}_M + \delta \mathrm{W}_E = \delta \mathrm{V}_M + \delta \mathrm{V}_E. \tag{2.173}$$

Now we see that the analysis produces an expression which states that the variation of the total work performed (mechanical and electrical) is equal to the variation of the total potential energy. This expression is analogous to our expression of the balance of energy expressed in the first law if we assume that the only energy term is the potential energy stored due to internal mechanical and electrical forces.

The variational principle stated in equation (2.173) is a powerful result, for several reasons. First, the principle is stated in terms of the *scalar* quantities of work and energy. In solving problems in mechanics or dynamics, mistakes are often introduced when manipulating vectors. When using the variational principle, these mistakes are often circumvented because all of the quantities of interest are expressed in terms of scalars. Second, it should be emphasized that satisfying the variational principle is equivalent to satisfying the governing equations of equilibrium. When both analyses are performed correctly, they yield identical results. One of the most important advantages of utilizing the variational approach rather than the vector form of the governing equations is that the work performed by constraint forces is identically equal to zero (recall the discussion of Figure 2.17). In using the vector form of the governing equations, it is often the case that the constraint forces must be computed to determine the solution to the problem. Using variational principles, these forces are ignored in the analysis because they perform no work. This often relieves some of the computational burden associated with finding a solution.

Finally, the derivation of the variational principle obscures somewhat the way in which the principle is used. The variational principle has been derived by first considering the equations that govern equilibrium and then deriving the variational relationship between work and internal energy. The variational principle is used in exactly the opposite manner. As we shall see in upcoming sections and later in the book, variational analysis *starts* with the expression of the variational principle, equation (2.173), and *the result is a derivation of the governing equations*. Thus, the analysis procedure begins by summing the various work terms and, after applying the variation to the state variables, results in the set of equations that must be satisfied for equilibrium. Before applying the principle in examples, we need to discuss the concept of generalized state variables in more detail. This is described in the following section.

2.5.1 Generalized State Variables

In our derivation of the variational principle, the variation has been applied to displacement of the mechanical system, $\mathbf{u}$, and the charge associated with the electrical system. For systems with multiple electrical elements, this is written as the vector $\mathbf{q}$. Thus, the variational principle has functional dependence on both the displacement vector and the vector of charge coordinates. In many instances it will be useful or required to rewrite the displacement vector in terms of a set of *generalized coordinates*, r_i, and to express the variational principle with respect to the generalized coordinates instead of with respect to the displacement vector. We define a set of generalized coordinates as the coordinates that locate a system with respect to a reference frame. Thus far in our discussion we have defined the displacement vector in a Cartesian frame of reference, but the generalized coordinates do not necessarily have to be expressed in one particular reference frame. They may have components in a Cartesian frame, a spherical frame of reference, or a mixture of multiple reference frames.

For the solution of problems in mechanics and in the mechanics of smart material systems, it is assumed that the variational principle is expressed as a set of generalized coordinates that are complete and independent. A set of coordinates is said to be *complete* if the coordinates chosen are sufficient to fix the locations of the parts of the system for an arbitrary configuration that is consistent with the geometric constraints. A set of coordinates is said to be *independent* if when all but any one of the coordinates is fixed, there remains a continuous range of values for the unfixed coordinate for all configurations, consistent with the geometric constraints.

Consider the case of specifying the generalized coordinates for a bar pivoting about a point. First let us assume that the bar is rigid and we define the displacement vector in terms of a Cartesian frame of reference that has the origin at the pivot point of the bar. Define u_1 as displacement in the 1 direction and u_2 as displacement in the 2 direction. The question is whether this choice of coordinates is complete and independent. To check completeness, we see determine u_1 and u_2 are sufficient to fix the displacement of the system for arbitrary configurations that are consistent with the geometric constraints. In this case the choice of coordinates is complete since specifying u_1 and u_2 will fix the location of the bar and the mass. To check whether the choice of coordinates is independent, we fix u_1 and determine if there is a continuous range of values for u_2 that are consistent with the geometric constraints. This check fails, because if we specify the location u_1, the location of u_2 is also fixed since the bar is assumed to be rigid. Thus, this choice of generalized coordinates is complete but not independent.

Now let us consider the case in which the bar is assumed to be elastic. The choice of the coordinates u_1 and u_2 is complete for the same rationale as for the case of a rigid bar. To check whether the coordinates are independent, fix u_1 and determine if there is a continuous range of values for u_2. In this case, u_2 is independent of u_1 since the elasticity of the bar allows us to vary u_2 even for the case in which u_1 is fixed. Thus, for the case of an elastic bar, the generalized coordinates u_1 and u_2 form a complete and independent set.

Two questions now arise. The first is how to choose generalized coordinates when the obvious choice does not form a complete and independent set. To answer this, let us return to the example of the rigid bar pivoting about a point. From the geometry of the problem we can write

$$\begin{aligned} u_1 &= l \sin \psi \\ u_2 &= -l \cos \psi. \end{aligned} \tag{2.174}$$

Since the length of the bar is fixed, we see that choosing ψ as the generalized coordinate forms a complete and independent set. Thus, for the case in which the bar is rigid, the generalized coordinate ψ can be used to specify the work and energy terms in the variational principle.

The second question that arises is how to choose the generalized coordinates when more than one set of variables is complete and independent. Returning to the example of the elastic bar with the end mass, we see that the choice of u_1 and u_2 is a complete and independent set; thus, these two coordinates could be used as the generalized coordinates of the problem. For the same reason, the choice of the coordinates l and θ could also form a complete and independent set. The decision to choose one set of generalized coordinates over another is somewhat problem dependent and is often dictated by the geometry of the problem. Unfortunately, there are no specific rules that define which is a better choice, but some of these issues are illustrated in upcoming examples, and experience in using the variational principle often helps in the decision process.

The definition of generalized coordinates allows us to rewrite the variational principle in terms of these variables. The displacement $\mathbf{u}$ is assumed to be a function of the N_r generalized coordinates, $\mathbf{u} = u(r_1, r_2, \ldots, r_N)$. The variation of the displacement is then written as

$$\delta \mathbf{u} = \frac{\partial \mathbf{u}}{\partial r_1}\delta r_1 + \frac{\partial \mathbf{u}}{\partial r_2}\delta r_2 + \cdots + \frac{\partial \mathbf{u}}{\partial r_N}\delta r_N. \tag{2.175}$$

Substituting this result into equation (2.172) yields

$$\sum_{i=1}^{N_r} \left(\mathbf{f}_M \cdot \frac{\partial \mathbf{u}}{\partial r_i} \right) \delta r_i + \sum_{j=1}^{N_q} v_j \, \delta q_j = \delta \mathrm{V}_E + \delta \mathrm{V}_M. \tag{2.176}$$

Let us denote the term in parentheses as the *generalized mechanical force*, $\mathcal{F}_i$,

$$\mathcal{F}_i = \mathbf{f}_M \cdot \frac{\partial \mathbf{u}}{\partial r_i}. \tag{2.177}$$

The term on the right-hand side is written as a variation of the total potential energy, $\delta \mathrm{V}_T$. After expressing displacement in terms of the generalized coordinates, the total potential energy is written as the function

$$\delta \mathrm{V}_M + \delta \mathrm{V}_E = \delta \mathrm{V}_T(r_1, \ldots, r_{N_r}, q_1, \ldots, q_{N_q}). \tag{2.178}$$

The variation of the total potential energy is written as

$$\begin{aligned}
\delta \mathrm{V}_T &= \frac{\partial \mathrm{V}_T}{\partial r_1}\delta r_1 + \cdots + \frac{\partial \mathrm{V}_T}{\partial r_{N_r}}\delta r_{N_r} + \frac{\partial \mathrm{V}_T}{\partial q_1}\delta q_1 + \cdots + \frac{\partial \mathrm{V}_T}{\partial q_{N_q}}\delta q_{N_q} \\
&= \sum_{i=1}^{N_r} \frac{\partial \mathrm{V}_T}{\partial r_i}\delta r_i + \sum_{j=1}^{N_q} \frac{\partial \mathrm{V}_T}{\partial q_j}\delta q_j.
\end{aligned} \tag{2.179}$$

Combining equations (2.177) and (2.179) with equation (2.176) produces an expression for the work and energy balance as a function of the variations in the generalized state variables r_i and q_j:

$$\sum_{i=1}^{N_r} \mathcal{F}_i \,\delta r_i + \sum_{j=1}^{N_q} v_j \,\delta q_j = \sum_{i=1}^{N_r} \frac{\partial \mathrm{V}_T}{\partial r_i} \delta r_i + \sum_{j=1}^{N_q} \frac{\partial \mathrm{V}_T}{\partial q_j} \delta q_j. \tag{2.180}$$

This expression can be rewritten as

$$\sum_{i=1}^{N_r} \left(\mathcal{F}_i - \frac{\partial \mathrm{V}_T}{\partial r_i} \right) \delta r_i + \sum_{j=1}^{N_q} \left(v_j - \frac{\partial \mathrm{V}_T}{\partial q_j} \right) \delta q_j = 0. \tag{2.181}$$

Recall that the generalized coordinates are assumed to be independent of one another. Under this condition, the only way for equation (2.181) to hold *for arbitrary variations in the generlized state variables* is for the following two sets of equations to be satisfied:

$$\begin{aligned} \mathcal{F}_i - \frac{\partial \mathrm{V}_T}{\partial r_i} &= 0 \qquad i = 1, \ldots, N_r \\ v_j - \frac{\partial \mathrm{V}_T}{\partial q_j} &= 0 \qquad j = 1, \ldots, N_q. \end{aligned} \tag{2.182}$$

Rewriting the expressions, we have

$$\begin{aligned} \mathcal{F}_i &= \frac{\partial \mathrm{V}_T}{\partial r_i} \qquad i = 1, \ldots, N_r \\ v_j &= \frac{\partial \mathrm{V}_T}{\partial q_j} \qquad j = 1, \ldots, N_q. \end{aligned} \tag{2.183}$$

The final result of this analysis is a set of $N_r + N_q$ governing equations in terms of the generalized state variables. These governing equations must be satisfied for equilibrium to be satisfied, which is equivalent to saying that satisfying the set of equations in equation (2.183) is identical to satisfying the energy balance expressed in the variational principle.

Example 2.8 Derive the equilibrium expressions for the system consisting of three springs as shown in Figure 2.18. Assume that the nodes are massless and that the coordinates are the displacement of the two nodes. The stiffness of the left and right springs is k and the stiffness of the middle spring is αk, where α is a positive constant.

Solution The displacements of the system are defined as u_1 and u_2. Writing the potential energy of the system requires that we add the potential energies for each of

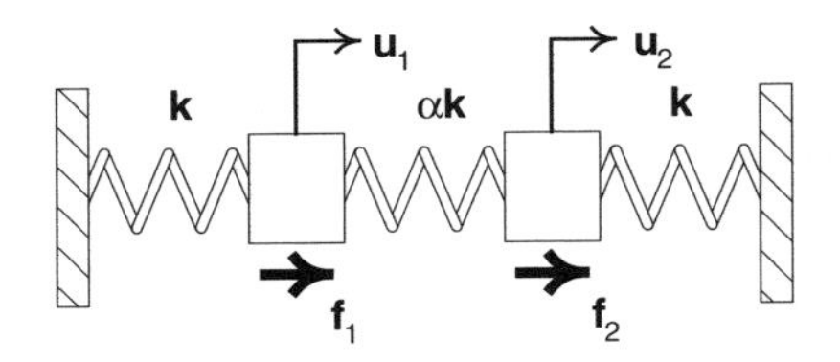

Figure 2.18 Three-spring system for static analysis.

the three springs:

$$\mathrm{V}_T = \frac{1}{2}ku_1^2 + \frac{1}{2}\alpha k(u_2 - u_1)^2 + \frac{1}{2}ku_2^2.$$

The variation of the mechanical work is defined as

$$\delta \mathrm{W}_M = f_1\,\delta u_1 + f_2\,\delta u_2.$$

Taking the variation of the potential energy function and combining it with the variation of the mechanical work yields

$$\delta \mathrm{V}_T + \delta \mathrm{W}_M = ku_1\,\delta u_1 + \alpha k(u_2 - u_1)\,\delta(u_2 - u_1) + ku_2\,\delta u_2 + f_1\,\delta u_1 + f_2\,\delta u_2.$$

Grouping the terms according to their variational displacement yields

$$[f_1 - ku_1 + \alpha k(u_2 - u_1)]\,\delta u_1 + [f_2 - ku_2 + \alpha k(u_1 - u_2)]\,\delta u_2 = 0.$$

Since the variational displacements are independent, the terms in brackets must each be equal to zero for the system to be in equilibrium for arbitrary choices of the variational displacements. This produces two equations for static equilibrium:

$$\begin{aligned} f_1 - ku_1 + \alpha k(u_2 - u_1) &= 0 \\ f_2 - ku_2 + \alpha k(u_1 - u_2) &= 0. \end{aligned}$$

These equations can be rewritten in matrix form as

$$\begin{pmatrix} f_1 \\ f_2 \end{pmatrix} = k \begin{bmatrix} 1+\alpha & -\alpha \\ -\alpha & 1+\alpha \end{bmatrix} \begin{pmatrix} \mathrm{u}_1 \\ \mathrm{u}_2 \end{pmatrix}.$$

Notice that the α term produces the off-diagonal coefficients in the matrix. Physically, this represents that fact that the middle spring *couples* the motion of the two nodes. If the middle spring was not there, $\alpha = 0$ and the two nodes would move independently. Increasing the stiffness of the middle spring compared to the other two springs ($\alpha \gg 1$) makes the system move as a rigid body.

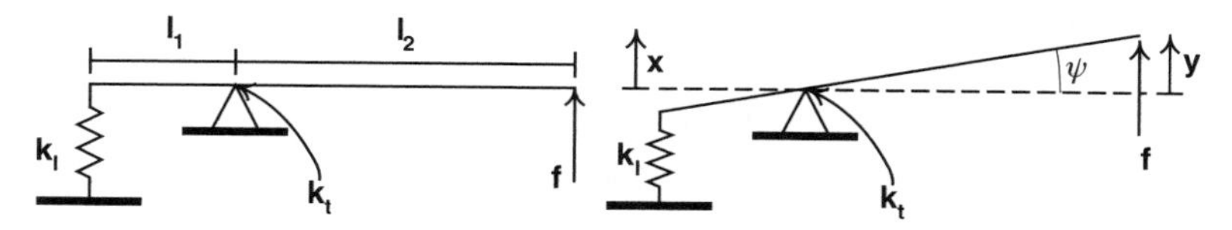

Figure 2.19 Mechanical lever with linear and torsional stiffness.

Example 2.9 A mechanical lever with linear and torsional stiffness is shown in Figure 2.19. The force is applied at the right end of the bar, which is at a distance l_2 from the pivot. A torsionsal spring with stiffness k_t is located at the pivot and a linear spring of stiffness k_l is located at the left end at a distance l_1 from the pivot. Derive the equlibrium equations for small angles ψ.

Solution The first step in applying the variational principle is to choose a set of independent coordinates. In this example we see that it is natural to choose three coordinates to represent the motion of the system: the motion at the left end of the bar, x, the motion at the right end of the bar, y, and the rotational angle ψ. Examining the geometry we see that there are kinematic constraints among these three coordinates,

$$x = -l_1 \sin\psi$$
$$y = l_2 \sin\psi$$
$$x = -\frac{l_1}{l_2} y.$$

Now we can write the potential energy terms as a function of our choice of coordinates,

$$V_T = \frac{1}{2} k_l x^2 + \frac{1}{2} k_t \psi^2,$$

where the first term is the potential energy due to the linear spring and the second term is the potential energy due to the rotational spring. The variation of the external work performed by the force is

$$W_M = f\,\delta y$$

To continue with the analysis, we need to choose a single coordinate to represent all of the work and energy terms. We have the freedom to choose the coordinate; therefore, it is best to choose the coordinate that will simplify the following analysis. Due to the fact that we have to take derivatives with respect to the coordinates, let us choose to represent the work and energy terms as a function of the rotational angle ψ. Substituting the expressions for x and y as a function of the rotational angle yields the potential energy term

$$V_T = \frac{1}{2} k_l (-l_1 \sin\psi)^2 + \frac{1}{2} k_t \psi^2.$$

Applying the variation to the energy and work terms produces

$$\begin{aligned}\delta \mathrm{V}_T &= k_l l_1^2 \sin\psi \cos\psi \delta\psi + k_t \psi \delta\psi \\ \delta \mathrm{W}_M &= f\, l_2 \cos\psi \delta\psi.\end{aligned}$$

Combining these two terms according to the variational principle, equation (2.173), yields

$$\left(fl_2 \cos\psi - k_l l_1^2 \sin\psi \cos\psi - k_t \psi\right) \delta\psi = 0.$$

For this expression to be valid, the term in parentheses must be equal to zero for arbitrary variational displacements. This leads to the equilibrium expression

$$k_l l_1^2 \sin\psi \cos\psi + k_t \psi = fl_2 \cos\psi.$$

For small angles, $\sin\psi \approx \psi$ and $\cos\psi \approx 1$, leading to the expressions

$$\left(k_l l_1^2 + k_t\right) \psi = fl_2.$$

This equation represents the equilibrium expression for small angles. Note that the stiffness of the system is a combination of the stiffness due to the torsional spring and the stiffness due to the liner spring. The stiffness due to the linear spring is modified by the square of the distance due to the lever.

2.6 VARIATIONAL PRINCIPLE OF DYNAMIC SYSTEMS

Until now we have dealt with systems that are assumed to be in static equilibrium. For a system to be in static equilibrium, the sum of the external forces must be equal to zero. We saw earlier in the chapter that the variational principle can be derived directly from this equilibrium statement, with the added benefit that we can pose the problem as a balance between variational work and the variation in potential energy.

We must extend this result if we are to work with systems whose coordinates are changing as a function of time. These systems, called *dynamic systems*, can also be analyzed using a variational approach, with one important addition. The important new information is contained within the path of the system from the initial time to the final time. We will find that the path of the coordinates as a function of time is the critical feature that allows us to determine the equations of motion for a dynamic system.

The analysis begins at the same point at which we began for a system in static equilibrium. The equilibrium expression is written as

$$\mathbf{f}(t) = \frac{d\mathbf{p}(t)}{dt}, \qquad (2.184)$$

where we must explicitly define the fact that the force and coordinates are time-dependent functions. Also, Newton's laws state that the sum of the forces in a dynamic system are equal to the time derivative of the linear momentum. The linear momentum is denoted $\mathbf{p}(t)$ in equation (2.184). To proceed with the analysis without too much confusion regarding notation, let us assume that all functions in equation (2.184) are functions of time and simply write the expression as

$$\mathbf{f} = \frac{d\mathbf{p}}{dt}, \tag{2.185}$$

for clarity. Equation (2.185) is rewritten in a manner similar to a system in static equilibrium, by subtracting the time derivative of the linear momentum from both sides of the expression:

$$\mathbf{f} - \frac{d\mathbf{p}}{dt} = 0. \tag{2.186}$$

Now we can interpret the momentum term as simply a force due to the motion of the particle.

Before proceeding with the derivation, we must reexamine our definition of the variational displacement. For systems in static equilibrium we defined the variational displacement as a differential motion that is consistent with the geometric boundary conditions. For dynamic systems we need to augment this definition with certain assumptions regarding the path that the variation displacement takes as a function of time. Consider a system that begins at the point $\mathbf{u}(t_1)$ at time t_1 and has the final position $\mathbf{u}(t_2)$ at time t_2. One interpretation of the solution of the dynamic equations of motion is that we are trying to find the path that the system takes as it moves from $\mathbf{u}(t_1)$ to $\mathbf{u}(t_2)$.

Consider a system that begins at $\mathbf{u}(t_1)$ and travels to $\mathbf{u}(t_2)$ through the solid path shown in Figure 2.20. Notice that there are an infinite number of paths that connect these two points in the coordinate space; the question arises: Why does the solid path represent the actual motion of the dynamic system? Let's consider applying a variational displacement to the actual path at any function of time. As stated, this

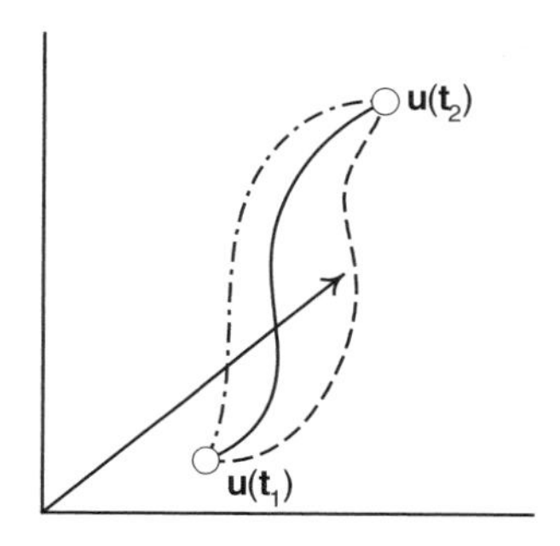

Figure 2.20 Paths associated with the variational displacements for a dynamic problem.

variational displacement is consistent with the geometric boundary conditions of the problem. Furthermore, we assume that *the variational displacement at time t_1 and t_2 is equal to zero*. Effectively, we are assuming that the initial and final positions of the system are fixed and cannot be varied. With this assumption we can state that

$$\delta \mathbf{u}(t_1) = \delta \mathbf{u}(t_2) = 0. \tag{2.187}$$

Continuing with the derivation, next we take the dot product of equation (2.186) with the variational displacement,

$$\mathbf{f} \cdot \delta \mathbf{u} - \frac{d\mathbf{p}}{dt} \cdot \delta \mathbf{u} = 0. \tag{2.188}$$

We recognize that the first term of equation (2.188) can be written as the variation of the total external work, $\delta \mathrm{W}_M + \delta \mathrm{W}_E$, and the variation of the total potential energy,$-\delta \mathrm{V}_T = -\delta \mathrm{V}_M - \delta \mathrm{V}_E$; thus, we can write

$$\delta \mathrm{W}_M + \delta \mathrm{W}_E - \delta \mathrm{V}_T - \frac{d\mathbf{p}}{dt} \cdot \delta \mathbf{u} = 0. \tag{2.189}$$

The question is how to eliminate the momentum term in equation (2.189). Under the assumption of Newtonian mechanics, the momentum can be written as

$$\mathbf{p}(t) = m\frac{d\mathbf{u}}{dt}, \tag{2.190}$$

where m is the mass. Substituting the momentum expression, equation (2.190), into equation (2.189), and integrating between t_1 and t_2 yields

$$\int_{t_1}^{t_2} (\delta \mathrm{W}_M + \delta \mathrm{W}_E - \delta \mathrm{V}_T)\, dt - \int_{t_1}^{t_2} m\dot{\mathbf{u}} \cdot \delta \mathbf{u}\, dt = 0, \tag{2.191}$$

where the overdot represents differentiation with respect to time. Applying integration by parts to the momentum term produces

$$\int_{t_1}^{t_2} m\dot{\mathbf{u}} \cdot \delta \mathbf{u}\, dt = m\dot{\mathbf{u}} \cdot \delta \mathbf{u}\Big|_{t_1}^{t_2} - \int_{t_1}^{t_2} m\dot{\mathbf{u}} \cdot \frac{d}{dt}\delta \mathbf{u}\, dt. \tag{2.192}$$

The terms evaluated at t_1 and t_2 are equal to zero by our definition of the variational displacement, and the term in the integral can be rewritten

$$-\int_{\mathrm{t}_1}^{\mathrm{t}_2} \mathrm{m}\dot{\mathbf{u}} \cdot \frac{\mathrm{d}}{\mathrm{dt}}\delta \mathbf{u}\, \mathrm{dt} = -\int_{\mathrm{t}_1}^{\mathrm{t}_2} \mathrm{m}\dot{\mathbf{u}} \cdot \delta \dot{\mathbf{u}}\, \mathrm{dt} \tag{2.193}$$

by virtue of the fact that variation and differentiation are interchangeable operations. Now we must realize that the term in the integral can be written as a variation of

$$\delta\left(\frac{m}{2}\dot{\mathbf{u}}\cdot\dot{\mathbf{u}}\right) = m\dot{\mathbf{u}}\cdot\delta\dot{\mathbf{u}}. \tag{2.194}$$

Denoting the term in parentheses on the left-hand side of the expression as a variation of the *kinetic energy*, T, we can write

$$\delta \mathrm{T} = m\dot{\mathbf{u}}\cdot\delta\dot{\mathbf{u}}. \tag{2.195}$$

Substituting equation (2.195) into equation (2.193) and then incorporating into the work and energy expression, equation (2.191), yields

$$\int_{t_1}^{t_2} (\delta \mathrm{W}_M + \delta \mathrm{W}_E - \delta \mathrm{V}_T + \delta \mathrm{T})\, dt = 0. \tag{2.196}$$

The term $\mathrm{T} - \mathrm{V_T}$ is called the *Lagrangian* and is given the symbol L. The variational operator is additive; therefore, we can write equation (2.196) as

$$\int_{t_1}^{t_2} \{\delta \mathrm{L} + \delta \mathrm{W}_M + \delta \mathrm{W}_E\}\, dt = 0. \tag{2.197}$$

Equation (2.196) is the variational principle for dynamic systems and is often called *Hamilton's principle*.

The variational principle for dynamic systems is equally powerful as the principle applied to systems in static equilibrium. Once again it transforms the problem of solving for the equations of motion from one involving vector terms to one that only involves scalar quantities. The additional terms that are required for dynamic analysis are incorporated in the kinetic energy of the Lagrangian. The kinetic energy can be visualized as the energy associated with the motion of the system, whereas the potential energy is the energy stored in the system.

Another benefit of the variational approach is that application of the method is almost identical to application of the method for systems in static equilibrium:

1. Choose a complete set of independent generalized coordinates and generalized velocities.
2. Write the potential and kinetic energy functions in terms of the generalized coordinates and find the variation $\delta \mathrm{V}_T$ and $\delta \mathrm{T}$.
3. Determine the work expression for each external force on the system and find the variation $\delta \mathrm{W}_M + \delta \mathrm{W}_E$.
4. Apply equation (2.196) and collect the terms associated with each independent variational displacement. Due to the fact that the variational displacements are independent, the coefficients that multiply the variational displacements must all be equal to zero. These coefficients are the equilibrium expressions for the system.

The concept of generalized coordinates applies to the variational principle for dynamic systems in the same manner as for the variational principle for systems in equilibrium. The displacement vector is written as a function of a set of generalized coordinate that form a complete set and are independent. Once this has been done, the Lagrangian can be written as

$$\begin{aligned} &\mathrm{L}(\dot{r}_1, \ldots, \dot{r}_{N_r}, r_1, \ldots, r_{N_r}, q_1, \ldots, q_{N_q}) \\ &\quad = \mathrm{T}(\dot{r}_1, \ldots, \dot{r}_{N_r}) - \mathrm{V}_T(r_1, \ldots, r_{N_r}, q_1, \ldots, q_{N_q}). \end{aligned} \tag{2.198}$$

The variation of the Lagrangian is

$$\delta \mathrm{L} = \sum_{i=1}^{N_r} \frac{\partial \mathrm{T}}{\partial \dot{r}_i} \delta \dot{r}_i + \sum_{i=1}^{N_r} \frac{\partial \mathrm{V}_T}{\partial r_i} \delta r_i + \sum_{j=1}^{N_q} \frac{\partial \mathrm{V}_T}{\partial q_j} \delta q_j. \tag{2.199}$$

Substituting the variation of the Lagrangian into equation (2.197) and combining it with the expressions for the generalized mechanical forces and the electrical work yields

$$\int_{t_1}^{t_2} \left[\sum_{i=1}^{N_r} \left(\mathcal{F}_i - \frac{\partial \mathrm{V}_T}{\partial r_i} \right) \delta r_i + \frac{\partial \mathrm{T}}{\partial \dot{r}_i} \delta \dot{r}_i + \sum_{j=1}^{N_q} \left(v_j - \frac{\partial \mathrm{V}_T}{\partial q_j} \right) \delta q_j \right] dt = 0. \tag{2.200}$$

The integrand of equation (2.200) contains variations of both the generalized coordinates and the time derivatives of the generalized coordinates, also called *generalized velocities*. Before the governing equations can be derived, all variations of the generalized velocities must be eliminated from the integrand. The generalized velocities are eliminated by applying integration by parts to the terms associated with $\delta \dot{r}_i$:

$$\int_{t_1}^{t_2} \frac{\partial \mathrm{T}}{\partial \dot{r}_i} \delta \dot{r}_i \, dt = \left. \frac{\partial \mathrm{T}}{\partial \dot{r}_i} \delta r_i \right|_{t_1}^{t_2} - \int_{t_1}^{t_2} \frac{d}{dt} \left(\frac{\partial \mathrm{T}}{\partial \dot{r}_i} \right) \delta r_i \, dt. \tag{2.201}$$

The first term on the right-hand side of equation (2.201) is equal to zero since the generalized coordinates are defined to be equal to zero at t_1 and t_2. Setting this term equal to zero, we can combine equations (2.200) and (2.201) and write the variational principle as

$$\int_{t_1}^{t_2} \left\{ \sum_{i=1}^{N_r} \left[\mathcal{F}_i - \frac{\partial \mathrm{V}_T}{\partial \dot{r}_i} - \frac{d}{dt} \left(\frac{\partial \mathrm{T}}{\partial \dot{r}_i} \right) \right] \delta r_i + \sum_{j=1}^{N_q} \left(v_j - \frac{\partial \mathrm{V}_T}{\partial q_j} \right) \delta q_j \right\} dt = 0. \tag{2.202}$$

Eliminating the generalized velocities allows us to obtain the governing equations from the variational principle. Since the variational coordinates δr_i and δq_j are independent, the integral is equal to zero if and only if the set of equations

$$\begin{aligned} \mathcal{F}_i - \frac{\partial \mathrm{V}_T}{\partial r_i} - \frac{d}{dt}\left(\frac{\partial \mathrm{T}}{\partial \dot{r}_i}\right) &= 0 \qquad i = 1, \ldots, N_r \\ v_j - \frac{\partial \mathrm{V}_T}{\partial q_j} &= 0 \qquad j = 1, \ldots, N_q \end{aligned} \tag{2.203}$$

are satisfied. These are the set of governing equations for the dynamic system. Rewrite the governing equations as

$$\begin{aligned} \mathcal{F}_i &= \frac{\partial \mathrm{V}_T}{\partial r_i} - \frac{d}{dt}\left(\frac{\partial \mathrm{T}}{\partial \dot{r}_i}\right) \qquad i = 1, \cdots, N_r \\ v_j &= \frac{\partial \mathrm{V}_T}{\partial q_j} \qquad j = 1, \cdots, N_q. \end{aligned} \tag{2.204}$$

The governing equations for a dynamic system are similar in form to the governing equations for a system in static equilibrium, equation (2.183), except for the additional time derivative of the kinetic energy term. This term represents the forces associated with the motion of the system. There are no time derivative terms in the governing equations for the electrical system since the kinetic energy does not have any functional dependence on the time derivative of the charge coordinates.

Example 2.10 Derive the equations of motion for the system shown in Figure 2.18 when a mass m is placed at each node.

Solution The coordinates of the system are defined as u_1 and u_2. The kinetic energy of the system is the summation of the kinetic energy terms for the two masses:

$$\mathrm{T} = \frac{1}{2}m\dot{u}_1^2 + \frac{1}{2}m\dot{u}_2^2.$$

The potential energy term is identical to that in Example 2.9:

$$\mathrm{V}_T = \frac{1}{2}ku_1^2 + \frac{1}{2}\alpha k\,(u_2 - u_1)^2 + \frac{1}{2}ku_2^2.$$

To apply equation (2.205), it is necessary to find the generalized mechanical forces and partial derivatives of the potential and kinetic energy functions. As in Example 2.9, the electrical work and energy terms are zero since there are no electrical elements. Finding the partial derivatives results in

$$\begin{aligned} \frac{\partial \mathrm{T}}{\partial \dot{u}_1} &= m\dot{u}_1 \\ \frac{\partial \mathrm{T}}{\partial \dot{u}_2} &= m\dot{u}_2 \end{aligned}$$

$$\frac{\partial \mathrm{V}_T}{\partial u_1} = ku_1 + \alpha k(u_2 - u_1)(-1)$$

$$\frac{\partial \mathrm{V}_T}{\partial u_2} = \alpha k(u_2 - u_1) + ku_2.$$

Taking the time derivative of the kinetic energy terms yields

$$\frac{d}{dt}\left(\frac{\partial \mathrm{T}}{\partial \dot{u}_1}\right) = m\ddot{u}_1$$

$$\frac{d}{dt}\left(\frac{\partial \mathrm{T}}{\partial \dot{u}_2}\right) = m\ddot{u}_2.$$

The generalized mechanical forces are

$$\mathcal{F}_1 = f_1$$
$$\mathcal{F}_2 = f_2.$$

Combining the terms according to equation (2.205) produces the governing equations

$$f_1 = ku_1 - \alpha k\,(u_2 - u_1) + m\ddot{u}_1$$

$$f_2 = \alpha k\,(u_2 - u_1) + ku_2 + m\ddot{u}_2.$$

The two equations can be rewritten with the forcing terms on the right-hand side as

$$m\ddot{u}_1 + k\,(1 + \alpha)\,u_1 - \alpha k u_2 = f_1$$

$$m\ddot{u}_2 - \alpha k u_1 + k\,(1 + \alpha)\,u_2 = f_2$$

and the two equations can also be rewritten in matrix form as

$$m\begin{bmatrix} 1 & 0 \\ 0 & 1 \end{bmatrix}\begin{pmatrix} \ddot{u}_1 \\ \ddot{u}_2 \end{pmatrix} + k\begin{bmatrix} 1+\alpha & -\alpha \\ -\alpha & 1+\alpha \end{bmatrix}\begin{pmatrix} u_1 \\ u_2 \end{pmatrix} = \begin{pmatrix} f_1 \\ f_2 \end{pmatrix},$$

which is a standard second-order form for vibrating systems.

2.7 CHAPTER SUMMARY

Analyzing smart material systems requires a basic understanding of mechanical, electrical, and thermal analysis. In this chapter certain fundamental topics in these three disciplines were presented as a review for upcoming chapters. In the field of mechanics, the definitions of stress and strain were presented and related to the constitutive properties of materials. Similarly, a review of electrostatics was based on the definitions of charge, electric potential, and electric field. These relationships were then used to define insulating and conducting materials. The review of work and energy methods presented for mechanical and electrical systems will serve as a precursor to discussions later when we develop equations of motion derived from variational principles of mechanics. One of the central features of work and energy methods is that the terms associated with the analysis are scalar quantities. This aspect of work and energy analysis often simplifies the procedures associated with finding equations of motion for smart material systems.

Defining the fundamental elements of mechanical and electrical analysis allowed us to study several representative problems in mechanical and electrostatics. The axial deformation of a bar was studied to highlight one-dimensional mechanics analysis. Beam analysis was also presented to demonstrate how the equations of mechanics could be used to derive expressions for the static displacement of beams for various boundary conditions. Common electrical elements such as a capacitor were studied using definitions from electrostatics. All of these basic elements are used later to analyze and design smart material systems.

In the final section of the chapter we reviewed variational methods for deriving equations of motion based on the work and energy concepts introduced earlier in the chapter. Variational approaches for static and dynamic systems were presented.

PROBLEMS

2.1. A solid has the displacement field

$$\begin{aligned} u_1 &= 6x_1 \\ u_2 &= 8x_2 \\ u_3 &= 3x_3^2. \end{aligned}$$

Determine the strain field in the material.

2.2. A solid has the displacement field

$$\begin{aligned} u_1 &= x_1^2 + x_2^2 \\ u_2 &= 2x_2x_1 \\ u_3 &= 0. \end{aligned}$$

Determine the strain field in the material.

2.3. Determine if the stress field

$$\mathrm{T}_{11} = 4x_1^2x_3 \qquad \mathrm{T}_{33} = \frac{4}{3}x_3^2 \qquad \mathrm{T}_{13} = -4x_1x_3^2$$

is in equilibrium when the body forces are assumed to be equal to zero.

2.4. A material is said to be in *plane strain* if

$$S_3 = S_4 = S_5 = 0.$$

(a) Write the stress–strain relationships for a linear elastic, isotropic material assumed to be in a state of plane strain.

(b) Compute the state of stress for an isotropic material with a modulus of 62 GPa and a Poisson's ratio of 0.3 if the strain state is

$$S_1 = 150\ \mu\text{strain} \qquad S_2 = 50\ \mu\text{strain} \qquad S_6 = -35\ \mu\text{strain}.$$

2.5. Compute the electrostatic force vector between a charge of 200 μC located at (0,2) and a second charge of -50 μC located at $(-1,-6)$ in a two-dimensional plane in free space. Draw a schematic of this problem, identifying the locations of the charged particles and the electrostatic force vector.

2.6. A linear spring of stiffness k has a charged particle of q coulombs fixed at each end. Determine the expression for the deflection of the spring at static equilibrium if the spring is constained to move in only the linear direction.

2.7. A set of fixed charges is located in free space as shown in Figure 2.21.

(a) Compute the electric field at the origin of the coordinate system.

(b) Compute the electric field at $(0, c/2)$ and $(0, -c/2)$.

2.8. The charge density profile at the interface between two materials is modeled as

$$\rho_v(x) = \alpha x e^{-\beta|x|}.$$

A representative plot of the charge density is shown in Figure 2.22.

(a) Compute the function for the electric displacement. Assume that the electric displacement is continuous at $x \to 0$.

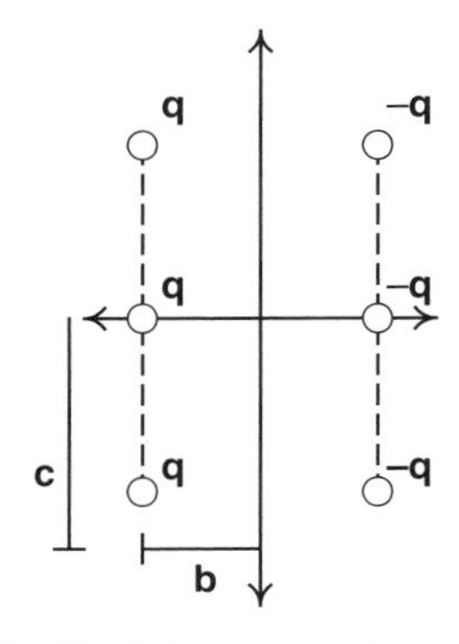

Figure 2.21 Fixed charges located in free space.

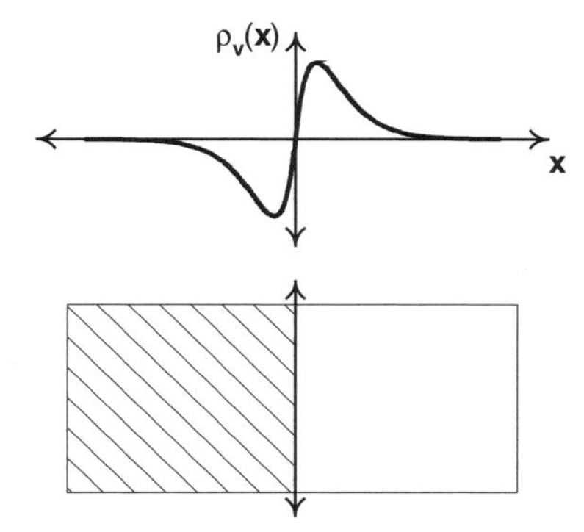

Figure 2.22 Charge density profile at the interface between two materials.

(b) Plot the charge density and electric displacement for $\alpha = 10$ and $\beta = 3$.

2.9. The charge density within the material shown in Figure 2.22 has the profile

$$\rho_v(x) = \alpha x e^{-\beta|x|}(1 - e^{-\lambda t}).$$

Compute the expression for flux in the x direction.

2.10. An electric field of 10 mV/m is applied to a conductive wire with a circular cross section. The wire has a diameter of 2 mm and a conductivity of 50 $(\Omega \cdot \mu\text{m})^{-1}$. Compute the current in the wire.

2.11. An isotropic dielectric material with the permittivity matrix

$$\varepsilon = \text{diag}(500, 500, 1500) \times 8.54 \times 10^{-12}\ \text{F/m}$$

has an applied electric field of

$$\mathbf{E} = 100\hat{x}_1 + 500\hat{x}_2\ \text{V/mm}.$$

Compute the electric displacement in the material.

2.12. Compute the work required to lift a 5-kg box from the ground to a height of 1.3 m.

2.13. A model for a nonlinear softening spring is

$$f(u) = -k \tan^{-1} \frac{u}{u_s},$$

where k/u_s represents the small displacement spring constant and u_s is the saturation displacement.

(a) Compute the energy function U and the potential energy function V for this spring.

(b) Plot the force versus displacement over the range -10 to 10 for the values $k = 100$ N/mm and $u_s = 3$ mm. Compute the work required to stretch the spring from 0 to 5 mm and illustrate this graphically on a plot of force versus displacement.

2.14. The potential energy function for a spring is found to be

$$\mathrm{V} = \frac{1}{2}k_1u_1^2 + k_2u_1u_2 + \frac{1}{2}k_3u_2^2 + \frac{1}{2}k_4u_3^2.$$

Determine the force vector for this spring.

2.15. Determine the displacement function for a cantilevered bending beam with a load applied at $x_1 = L_f$, where $L_f < L$. Note that the solution will be in the form of a piecewise continuous function.

2.16. Determine the displacement function for a cantilevered bending beam with moment M_1 applied at $x_1 = L_1$ and moment $-M_1$ applied at $x_1 = L_2$, where $L_2 > L_1$. Note that the solution will be in the form of a piecewise continuous function.

2.17. **(a)** Determine the expression for the stored energy of a cantilevered bending beam with a load applied at the free end.

(b) Repeat part (a) for a pinned–pinned beam with a load applied at the center.

2.18. **(a)** Compute the electric field in a capacitor of thickness 250 μm with an applied voltage difference of 100 V. The dielectric material in the capacitor has a relative dielectric constant of 850 and a surface area of 10 mm^2.

(b) Compute the charge stored in the capacitor with the properties given in part (a).

(c) Compute the stored energy in the capacitor with the properties given in part (a).

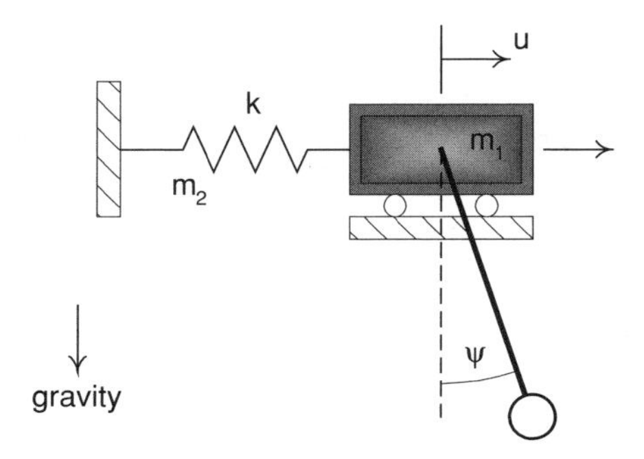

Figure 2.23 Two-mass mechanical system.

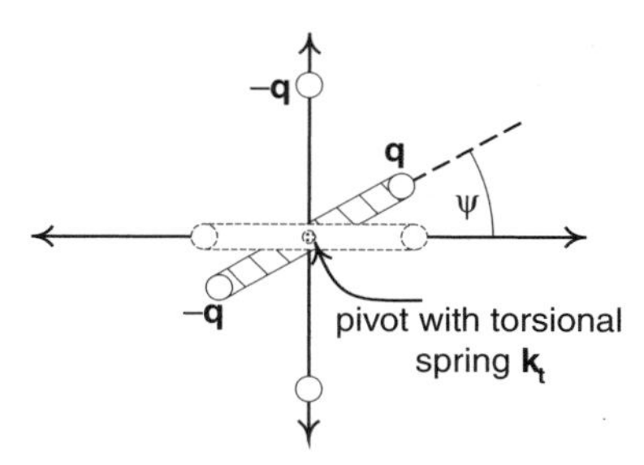

Figure 2.24 Rigid link.

2.19. Use the variational approach for static systems to derive the equations of motion for the mechanical system shown in Figure 2.23. Assume that the masses are both zero for this analysis. Use u and ψ as the generalized coordinates for the analysis.

2.20. Determine the governing equations for a system that has the potential energy and work expressions

$$\begin{aligned} \mathrm{V}_T &= \frac{1}{2}ku^2 + duq + \frac{1}{2C}q^2 \\ \mathrm{W}_M + \mathrm{W}_E &= fu + vq, \end{aligned}$$

where the generalized coordinates are u and q.

2.21. A rigid link of length $2a$ has two charges attached to its end (Figure 2.24). It is placed in a free space with two fixed charges. The fixed charge $-q$ is located at $(0,b)$ and the fixed charge $+q$ is located at $(0,-b)$, where $b > a$. At the center of the rigid link is a pivot that contains a linear torsional spring of spring constant k_t.

(a) Determine an expression for the potential energy of this system (ignoring gravity).

(b) Use the variational principle to determine the governing equations of static equilibrium.

2.22. Use the variational approach for dynamic systems to derive the equations of motion for the mechanical system shown in Figure 2.23. Use u and ψ as the generalized coordinates for the analysis.

2.23. Repeat Problem 2.20 including a kinetic energy term of the form $\mathrm{T} = \frac{1}{2}m\dot{u}^2$.

NOTES

The material in this chapter was drawn from several textbooks on the subjects of mechanics, electrostatics, and work and energy methods. The book by Gere and

Timoshenko [12] was used as a reference for the sections on mechanics of materials, as was to the text of Allen and Haisler [13]. References on work and energy methods included Pilkey and Wunderlich [14] and Reddy [15] for mechanical systems and the excellent text by Crandall et al. [16] for electromechanical systems. The latter text includes a thorough discussion of systems that incorporate both mechanical and electrical energy.

3

MATHEMATICAL REPRESENTATIONS OF SMART MATERIAL SYSTEMS

In Chapter 2 we saw that there are alternative methods for deriving governing equations for smart material systems. We considered the combined effects of mechanical and electrical forces and determined that the model for systems in static equilibrium or for dynamic systems could be derived from the governing equations or through variational methods based on the concepts of work and energy. In either case the model for a system in static equilibrium was found to be a set of algebraic equations in the generalized state variables. In the case of a dynamic system, the governing equations are a set of differential equations in time.

In this chapter we discuss methods of solving for the time and frequency response of the equations that govern the response of smart material systems. We first discuss solutions of equations that are described by algebraic equations that model the static response of a system. Next, we discuss the solution of the differential equations that describe a dynamic system. Both first- and second-order equations of motion are studied. Finally, we describe impedance methods used to solve for system response.

3.1 ALGEBRAIC EQUATIONS FOR SYSTEMS IN STATIC EQUILIBRIUM

The equations that govern the response of a system in static equilibrium are a set of algebraic expressions that relate the generalized state variables to the generalized forces that act on a system. In Chapter 2 we saw that the generalized state variables consist of generalized coordinates that describe the mechanical response and charge coordinates that describe the electrical response. To generalize the result for discussion in this chapter, we define the vector of generalized state variables as $\mathbf{r}$. The generalized forces are denoted $\mathbf{f}$ even though they could represent both mechanical and electrical forces that act on the system.

The most general relationship for the set of governing equations is a functional relationship between the generalized forces $\mathbf{f}$ and generalized state variables $\mathbf{r}$,

$$\mathbf{g}(\mathbf{r}) = \mathbf{f}. \tag{3.1}$$

The term $\mathbf{g}$ denotes a set of functions that represent the governing equations for the system. In most of the book we study systems whose governing equations are linear. In this case the system of equations is reduced to

$$\mathrm{K}\mathbf{r} = \mathrm{B}_f\mathbf{f}, \tag{3.2}$$

where K is called the *stiffness matrix* and B_f the *input matrix* or *influence matrix* of the system. If the stiffness matrix is nonsingular, the matrix inverse of the stiffness matrix exists and the generalized state variables can be computed from the expression

$$\mathbf{r} = \mathrm{K}^{-1}\mathrm{B}_f\mathbf{f}. \tag{3.3}$$

For low-order models (e.g., models with two or three generalized states), equation (3.3) can be solved by hand. For higher-order systems there are a number of efficient computer algorithms for solving matrix inverses for systems with hundreds and possibly thousands of generalized states.

In many problems it is useful to define an *observation matrix* or *output matrix* that relates the generalized states to an observed set of outputs. For example, it is often the case in smart material systems that certain sensors are used to measure, for example, the displacement of a device or at a particular location on a structure. The outputs of the system are some function of the generalized state variables, and the observation matrix defines this functional relationship. In the most general form, the relationship between the outputs observed, $\mathbf{y}$, and the generalized state variables is

$$\mathbf{y} = \mathbf{h}(\mathbf{r}). \tag{3.4}$$

In most cases in this book we assume that the outputs can be written as a linear combination of the generalized states; therefore, the outputs can be written in matrix form as

$$\mathbf{y} = \mathrm{H}_d\mathbf{r}. \tag{3.5}$$

Combining equations (3.3) and (3.5), we can write the *input–output* relationship between the applied forces and the outputs observed:

$$\mathbf{y} = \mathrm{H}_d\mathrm{K}^{-1}\mathrm{B}_f\mathbf{f}. \tag{3.6}$$

In a majority of cases, the number of outputs is smaller than the number of generalized states; therefore, the observation matrix H_d has more columns than rows.

3.2 SECOND-ORDER MODELS OF DYNAMIC SYSTEMS

Models of dynamic systems are represented as differential equations in time as discussed in detail in Chapter 2. In this book, models for dynamic systems are derived from application of the governing laws of mechanics or through the application of the variational principle for dynamic systems. As we saw in Chapter 2, these models result in the definition of a set of second-order equations that represent the balance of applied forces (or equivalently, work) with the forces due to the stored potential and kinetic energy of the system. A general relationship for a dynamic system is of the form

$$\mathbf{f} = \mathbf{g}\left(\ddot{\mathbf{r}}, \dot{\mathbf{r}}, \mathbf{r}\right). \tag{3.7}$$

If the equations are linear and there are no terms due to the first derivative of the generalized states, the equations of motion can be written

$$\mathrm{M}\ddot{\mathbf{r}}(t) + \mathrm{K}\mathbf{r}(t) = \mathrm{B}_f\mathbf{f}(t), \tag{3.8}$$

where M is called the *mass matrix* for the system. The mass matrix arises from the kinetic energy terms and represents forces due to the time derivative of the momentum. In many instances there are also forces due to viscous damping. These forces are represented as a force that is proportional to the first derivative of the generalized states and can be added into equation (3.8) as

$$\mathrm{M}\ddot{\mathbf{r}}(t) + \mathrm{D}_v\dot{\mathbf{r}}(t) + \mathrm{K}\mathbf{r}(t) = \mathrm{B}_f\mathbf{f}(t), \tag{3.9}$$

where D_v is the *viscous damping matrix* for the system.

Before solving the matrix set of equations, let's consider the case where we have only a single generalized state to illustrate the fundamental results associated with second-order systems. In the case in which there is no damping, the equations of motion are written as

$$m\ddot{r}(t) + kr(t) = f_o f(t), \tag{3.10}$$

where f_o represents the amplitude of the time-dependent force $f(t)$. The solution is generally found after normalizing the system to the mass,

$$\ddot{r}(t) + \frac{k}{m}r(t) = \frac{f_o}{m}f(t). \tag{3.11}$$

The ratio of the stiffness to the mass is denoted

$$\frac{k}{m} = \omega_n^2 \tag{3.12}$$

and is called the *undamped natural frequency* of the system. The importance of the undamped natural frequency is evident when we consider the solution of the homogeneous differential equation. Setting $f(t) = 0$, the solution is

$$r(t) = A\sin(\omega_n t + \phi), \tag{3.13}$$

where

$$A = \frac{1}{\omega_n}\sqrt{\omega_n^2 r(0)^2 + \dot{r}(0)^2}$$
$$\phi = \tan^{-1}\frac{\omega_n r(0)}{\dot{r}(0)}. \qquad (3.14)$$

Thus, the solution to an unforced second-order dynamic system is a harmonic function that oscillates at the undamped natural frequency. The amplitude and phase of the system is defined by the initial displacement and initial velocity.

The solution to a forced system depends on the type of forcing input. Typical forcing inputs are step functions and harmonic functions. The solutions to these two types of inputs are defined as

$$\text{Step:} \quad r(t) = \frac{f_0}{k}(1 - \cos\omega_n t) \qquad r(0) = \dot{r}(0) = 0$$
$$\text{Harmonic:} \quad r(t) = \frac{\dot{r}(0)}{\omega_n}\sin\omega_n t + \left(r_0(0) - \frac{f_0 m}{\omega_n^2 - \omega^2}\right)\cos\omega_n t + \frac{f_0/m}{\omega_n^2 - \omega^2}\cos\omega t \qquad (3.15)$$

The general solution to a forcing input is defined in terms of the *convolution integral*,

$$r(t) = \frac{f_0}{m\omega_n}\int_0^t f(t-\tau)\sin\omega_n\tau\, d\tau. \qquad (3.16)$$

When viscous damping is present in a system, the mass normalized equations of motion are

$$\ddot{r}(t) + 2\zeta\omega_n\dot{r}(t) + \omega_n^2 r(t) = \frac{f_o}{m}f(t). \qquad (3.17)$$

The variable ζ is the *damping ratio* of the system and is related to the amount of viscous damping. For most systems we study in this book, the damping ratio is a positive value. The form of the solution for a damped system depends on the value of ζ. In this book we study systems that have a limited amount of viscous damping, and generally the damping ratio will be much less than 1. For any system in which $\zeta < 1$, the homogeneous solution is

$$r(t) = Ae^{-\zeta\omega_n t}\sin(\omega_d t + \phi) \qquad (3.18)$$

where

$$\omega_d = \omega_n\sqrt{1-\zeta^2}$$

$$A = \sqrt{\frac{(\dot{r}(0) + \zeta\omega_n r(0))^2 + (r(0)\omega_d)^2}{\omega_d^2}}$$

$$\phi = \tan^{-1} \frac{r(0)\omega_d}{f(0) + \zeta\omega_n r(0)}$$

in terms of the initial conditions. As we see from the solution, a damped second-order system will also oscillate at the damped natural frequency. The primary difference is that the amplitude of the system will decay with time due to the term $e^{-\zeta\omega_n t}$. The rate of decay will increase as ζ becomes larger. As $\zeta \to 0$, the solution will approach the solution of the undamped system. The solutions for common types of forcing functions are (for zero initial conditions)

$$\text{Step:} \quad r(t) = \frac{f_0}{k} - \frac{f_0}{k\sqrt{1-\zeta^2}} e^{-\zeta\omega_n t} \cos(\omega_d t - \phi) \quad \phi = \tan^{-1} \frac{\zeta}{\sqrt{1-\zeta^2}}$$

$$\text{Harmonic:} \quad r(t) = x\cos(\omega t - \theta)$$

$$x = \frac{f_0 m}{\sqrt{(\omega_n^2 - \omega^2)^2 + (2\zeta\omega_n\omega)^2}}$$

$$\theta = \tan^{-1} \frac{2\zeta\omega_n\omega}{\omega_n^2 - \omega^2} \tag{3.19}$$

and the general solution is written in terms of the convolution integral:

$$r(t) = \frac{f_0}{m\omega_d} \int_0^t f(t-\tau) e^{-\zeta\omega_n\tau} \sin\omega_d\tau \, d\tau. \tag{3.20}$$

Return now to second-order systems with multiple degrees of freedom as modeled by equation (3.8). The solution for the homogeneous undamped multiple-degree-of-freedom (MDOF) case is obtained by assuming a solution of the form

$$\mathbf{r}(t) = \mathbf{V}e^{j\omega t}, \tag{3.21}$$

where $\mathbf{V}$ is a vector of unknown coefficients. Substituting equation (3.21) into equation (3.8) yields

$$\left(\mathrm{K} - \omega^2\mathrm{M}\right)\mathbf{V}e^{j\omega t} = 0. \tag{3.22}$$

The only nontrivial solution to equation (3.22) is the case in which

$$\left|\mathrm{K} - \omega^2\mathrm{M}\right| = 0. \tag{3.23}$$

Solving for the determinant is equivalent to the solution of a *symmetric eigenvalue problem*, which yields N_v eigenvalues ω_{ni}^2 and corresponding eigenvectors $\mathbf{V}_i$. Generally, the eigenvalues are ordered such that $\omega_{n1} < \omega_{n2} < \cdots$. The solution to the

MDOF problem can be cast as the solution to a set of SDOF problems by forming the matrix

$$\mathrm{P} = \begin{bmatrix} \mathbf{V}_1 & \mathbf{V}_2 & \cdots & \mathbf{V}_{N_v} \end{bmatrix}, \tag{3.24}$$

and substituting the coordinate transformation $\mathbf{r}(t) = \mathrm{P}\eta(t)$ into equation (3.8). The result is

$$\mathrm{MP}\ddot{\eta}(t) + \mathrm{KP}\eta(t) = \mathrm{B}_f\mathbf{f}(t). \tag{3.25}$$

If the eigenvectors are normalized such that $\mathbf{V}_i\mathbf{M}\mathbf{V}_j = \delta_{ij}$, we can premultiply equation (3.25) by P':

$$\mathrm{P}'\mathrm{MP}\ddot{\eta}(t) + \mathrm{P}'\mathrm{KP}\eta(t) = \mathrm{P}'\mathrm{B}_f\mathbf{f}(t). \tag{3.26}$$

Due to the normalization of the eigenvectors,

$$\begin{aligned} \mathrm{P}'\mathrm{MP} &= I \\ \mathrm{P}'\mathrm{KP} &= \Lambda = \operatorname{diag}\left(\omega_{n1}^2, \omega_{n2}^2, \ldots\right) \\ \mathrm{P}'\mathrm{B}_f &= \Phi \end{aligned} \tag{3.27}$$

and the equations of motion can be written as a set of *uncoupled* second-order equations:

$$\ddot{\eta}_i(t) + \omega_{ni}^2\eta(t) = \sum_{j=1}^{N_f} \Phi_{ij} f_j(t). \tag{3.28}$$

The term Φ_{ij} is the (i, j)th element of the matrix Φ.

Decoupling the equations of motion is a significant result because it allows the solution of the multiple-degree-of-freedom system to be obtained by applying the results for single-degree-of-freedom systems. Once the MDOF system has been written as a set of decoupled equations, as in equation (3.28), the results discussed previously in this section can be applied to solve each of the equations separately for $r_i(t)$. Once this is completed, the coordinate transformation $\mathbf{r}(t) = \mathrm{P}\eta(t)$ is applied to obtain the complete solution in the coordinates of the original system.

Models that incorporate viscous damping as shown in equation (3.9) can also be decoupled if the viscous damping matrix is decoupled by the eigenvectors of the undamped system. For systems with light damping this assumption is often made because it greatly simplifies the analysis. Additionally, the model of viscous damping is often added into the decoupled equations because an exact model of damping is not available or there are experimental data that allow one to estimate the damping coefficient. Under the assumption that

$$\mathrm{P}'\mathrm{DP} = \operatorname{diag}(2\zeta_i\omega_{ni}), \tag{3.29}$$

the decoupled equations of motion are

$$\ddot{\eta}_i(t) + 2\zeta_i \omega_{ni} \dot{\eta}_i + \omega_{ni}^2 \eta(t) = \sum_{j=1}^{N_f} \Phi_{ij} f_j(t). \tag{3.30}$$

As in the case of the undamped system, the equations for each of the transformed coordinates are solved separately, and the total result can be obtained by applying the coordinate transformation $\mathbf{r}(t) = \mathrm{P}\eta(t)$.

3.3 FIRST-ORDER MODELS OF DYNAMIC SYSTEMS

Many of the systems that we discuss are amenable to being modeled as second-order differential equations, as discussed in Section 3.2. Modeling a dynamic system as a set of second-order equations is often desirable because it provides insight into the vibrational characteristics of the system. Another common way to represent models of dynamic systems is in *first-order* or *state variable form*. First-order form is often desirable when analyzing the control of smart material systems.

A general representation of a first-order model is

$$\frac{d\mathbf{z}(t)}{dt} = \mathbf{g}(\mathbf{z}(t), \mathbf{w}(t), t), \tag{3.31}$$

where $\mathbf{z}(t)$ is a vector of *states* and $\mathbf{w}(t)$ is a vector of *inputs*. If the right-hand side of equation (3.31) can be written as a linear combination of the states and the inputs, we can write

$$\frac{d\mathbf{z}(t)}{dt} = \mathrm{A}(t)\mathbf{z}(t) + \mathrm{B}(t)\mathbf{w}(t), \tag{3.32}$$

where the notation in equation (3.32) implies that the coefficients on the right-hand side are explicit functions of time. The matrix $\mathrm{A}(t)$ is the *state matrix* and $\mathrm{B}(t)$ is the *input matrix*.

If the coefficients on the right-hand side of equation (3.32) are independent of time, the state variable equations are written

$$\frac{d\mathbf{z}(t)}{dt} = \mathrm{A}\mathbf{z}(t) + \mathrm{B}\mathbf{w}(t). \tag{3.33}$$

Equation (3.33) is the linear time-invariant (LTI) form of the state equations.

The LTI state equations for a dynamic system can be solved to obtain an expression for the states as a function of time:

$$\mathbf{z}(t) = e^{\mathrm{A}t}\mathbf{z}(0) + \int_0^t e^{\mathrm{A}(t-\lambda)}\mathrm{B}\mathbf{u}(\lambda)\, d\lambda, \tag{3.34}$$

where

$$e^{At} = \mathcal{L}^{-1}(sI - A)^{-1} \tag{3.35}$$

is the *state transition matrix*. As shown in equation (3.35), this is computed from the inverse Laplace transform of a matrix constructed from $(sI - A)^{-1}$.

3.3.1 Transformation of Second-Order Models to First-Order Form

Second-order models of dynamic systems can be transformed to first-order form by defining a relationship between the states, $\mathbf{z}$, of the first-order system and the generalized states of the second-order model, $\mathbf{r}$. Defining two sets of states as

$$\begin{aligned} \mathbf{z}_1 &= \mathbf{r} \\ \mathbf{z}_2 &= \dot{\mathbf{r}} \end{aligned} \tag{3.36}$$

and substituting into equation (3.9) yields

$$M\dot{\mathbf{z}}_2 + D_v\mathbf{z}_2 + K\mathbf{z}_1 = B_f\mathbf{f}. \tag{3.37}$$

Premultiplying equation (3.37) by the inverse of the mass matrix and solving for $\dot{\mathbf{z}}_2$ produces the expression

$$\dot{\mathbf{z}}_2 = -M^{-1}K\mathbf{z}_1 - M^{-1}D_v\mathbf{z}_2 + M^{-1}B_f\mathbf{f}. \tag{3.38}$$

Equation (3.38) is one of the two first-order equations that are required to model the second-order system. The remaining equation is derived from the definition of the state variables:

$$\dot{\mathbf{z}}_1 = \mathbf{z}_2. \tag{3.39}$$

The first-order form of the equations is written by combining equations (3.38) and (3.39). The expressions are written in matrix form as

$$\begin{pmatrix} \dot{\mathbf{z}}_1 \\ \dot{\mathbf{z}}_2 \end{pmatrix} = \begin{bmatrix} 0 & I \\ -M^{-1}K & -M^{-1}D_v \end{bmatrix} \begin{pmatrix} \mathbf{z}_1 \\ \mathbf{z}_2 \end{pmatrix} + \begin{bmatrix} 0 \\ M^{-1}B_f \end{bmatrix} \mathbf{f}. \tag{3.40}$$

Comparing equation (3.40) with equation (3.33), we see that

$$\begin{aligned} A &= \begin{bmatrix} 0 & I \\ -M^{-1}K & -M^{-1}D_v \end{bmatrix} \\ B &= \begin{bmatrix} 0 \\ M^{-1}B_f \end{bmatrix} \end{aligned} \tag{3.41}$$

is the relationship between the matrices that represent the second-order system and state and input matrices of first-order form.

3.3.2 Output Equations for State Variable Models

Equations (3.31), (3.32), and (3.33) represent expressions for the states of a dynamic system. It is rare that all of the internal system states can be observed directly at the output of the system; therefore, we must define a second set of equations that expresses the internal states that can be measured at the output. For a general state variable model, the output equations can again be represented in general function form:

$$\mathbf{y}(t) = \mathbf{h}\left(\mathbf{z}(t), \mathbf{w}(t), t\right). \tag{3.42}$$

If the observed outputs can be written as a linear combination of the states and the inputs, we can write the output expression as

$$\mathbf{y}(t) = \mathrm{C}(t)\mathbf{z}(t) + \mathrm{D}(t)\mathbf{w}(t), \tag{3.43}$$

where $\mathrm{C}(t)$ and $\mathrm{D}(t)$ are called the *observation* or *output matrix* and the *direct transmission matrix*, respectively. If the coefficient matrices are time invariant, the output expressions are written as

$$\mathbf{y}(t) = \mathrm{C}\mathbf{z}(t) + \mathrm{D}\mathbf{w}(t). \tag{3.44}$$

The dimension and definitions of matrices C and D are equivalent to the case of time-varying systems.

Expressions for the outputs as a function of the states for an LTI system are obtained by combining equations (3.34) and (3.44):

$$\mathbf{y}(t) = \mathrm{C}e^{\mathrm{A}t}\mathbf{z}(0) + \int_0^t \mathrm{C}e^{\mathrm{A}(t-\lambda)}\mathrm{B}\mathbf{w}(\lambda)\,d\lambda + \mathrm{D}\mathbf{w}(t). \tag{3.45}$$

Example 3.1 Derive the state equations for a mass–spring–damper oscillator with a force applied to the mass as shown in Figure 3.1. Write the state equations and the output equations assuming that the position and input force are the observed variables.

Solution Drawing a free-body diagram and applying Newton's second law yields

$$m\ddot{u}(t) = -ku(t) - c\dot{u}(t) + f(t). \tag{3.46}$$

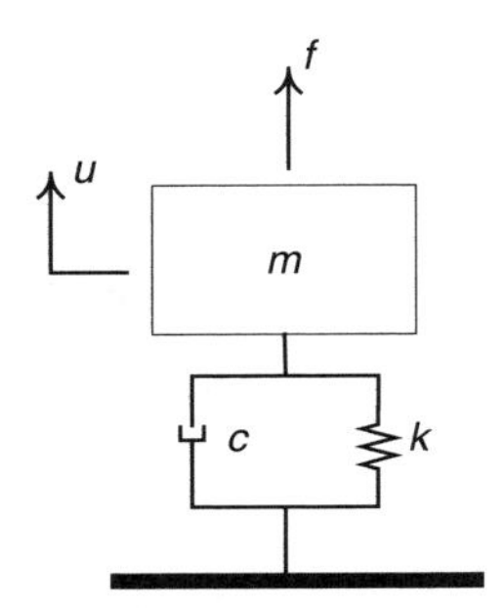

Figure 3.1 Mass–spring–damper oscillator with a force applied.

Dividing by the mass yields

$$\ddot{u}(t) = -\frac{k}{m}u(t) - \frac{c}{m}\dot{u}(t) + \frac{1}{m}f(t). \tag{3.47}$$

Defining the state variables as

$$z_1(t) = u(t) \tag{3.48}$$
$$z_2(t) = \dot{u}(t) \tag{3.49}$$

and substituting into equation (3.47) produces an expression for the first derivative of $z_2(t)$:

$$\dot{z}_2(t) = -\frac{k}{m}z_1(t) - \frac{c}{m}z_2(t) + \frac{1}{m}f(t). \tag{3.50}$$

The equation for the derivative of the first state is obtained from the state definitions:

$$\dot{z}_1(t) = z_2(t). \tag{3.51}$$

Combining equations (3.50) and (3.51) into a matrix expression yields

$$\begin{pmatrix} \dot{z}_1(t) \\ \dot{z}_2(t) \end{pmatrix} = \begin{bmatrix} 0 & 1 \\ -k/m & -c/m \end{bmatrix} \begin{pmatrix} z_1(t) \\ z_2(t) \end{pmatrix} + \begin{bmatrix} 0 \\ 1/m \end{bmatrix} f(t). \tag{3.52}$$

The output equations are written as

$$y_1(t) = z_1(t) \tag{3.53}$$
$$y_2(t) = f(t). \tag{3.54}$$

Combining these expressions into one matrix expression produces

$$\mathbf{y}(t) = \begin{bmatrix} 1 & 0 \\ 0 & 0 \end{bmatrix} \begin{pmatrix} \mathbf{z}_1(t) \\ \mathbf{z}_2(t) \end{pmatrix} + \begin{bmatrix} 0 \\ 1 \end{bmatrix} \mathbf{u}(t). \tag{3.55}$$

The state matrices are determined from equations (3.52) and (3.55) as

$$A = \begin{bmatrix} 0 & 1 \\ -k/m & -c/m \end{bmatrix} \tag{3.56}$$

$$B = \begin{bmatrix} 0 \\ 1/m \end{bmatrix} \tag{3.57}$$

$$C = \begin{bmatrix} 1 & 0 \\ 0 & 0 \end{bmatrix} \tag{3.58}$$

$$D = \begin{bmatrix} 0 \\ 1 \end{bmatrix}. \tag{3.59}$$

3.4 INPUT–OUTPUT MODELS AND FREQUENCY RESPONSE

First and second-order models of a dynamic system contain information about the internal states of the system. Earlier we have shown how these models are derived from the governing equations or through application of a variational principle. In some cases a model that includes information about the internal states of the system is not necessary or not possible. In these instances we can derive an *input–output model* that contains only information about how the outputs of a system will change as a function of the inputs. Such input–output models are generally expressed in the Laplace or *frequency domain*. In this section we derive a general formulation for an input–output *transfer function* of an LTI system, and in Section 3.5 we transform the result into the frequency domain.

Consider again the state variable equations for an LTI system and the associated output equations (3.33) and (3.44). Transforming equation (3.33) into the Laplace domain assuming zero initial conditions yields

$$s\mathbf{z}(s) = A\mathbf{z}(s) + B\mathbf{w}(s). \tag{3.60}$$

Combining like terms on the left-hand side and premultiplying by $(sI - A)^{-1}$ produces

$$\mathbf{z}(s) = (sI - A)^{-1}B\mathbf{w}(s). \tag{3.61}$$

Transforming equation (3.44) into the Laplace domain yields

$$\mathbf{y}(s) = C\mathbf{z}(s) + D\mathbf{w}(s). \tag{3.62}$$

Substituting equation (3.61) into equation (3.62) and combining terms yields

$$\mathbf{y}(\mathsf{s}) = \left[\mathrm{C}(\mathsf{s}\mathrm{I} - \mathrm{A})^{-1}\mathrm{B} + \mathrm{D}\right]\mathbf{w}(\mathsf{s}). \tag{3.63}$$

Equation (3.63) represents the general expression for a matrix of transfer functions between the inputs $\mathbf{w}(t)$ and the outputs $\mathbf{y}(t)$. Note that variables related to the internal states are not contained explicitly within the equation for the transfer functions. The number of internal states does influence this equation, though, through the size of the matrix $(\mathsf{s}\mathrm{I} - \mathrm{A})^{-1}$.

Any relationship between the ith output and the jth input can be determined from the expression

$$y_i(\mathsf{s}) = \left[\mathrm{C}_i(\mathsf{s}\mathrm{I} - \mathrm{A})^{-1}\mathrm{B}_j + \mathrm{D}_{ij}\right] w_j(\mathsf{s}), \tag{3.64}$$

where C_i is the ith row of C, B_j is the jth column of B, and D_{ij} is the (i, j)th element of D. Since the input–output relationship in equation (3.64) is a single function, we can write $y_i(\mathsf{s})/w_j(\mathsf{s})$ as a ratio of Laplace polynomials:

$$\frac{y_i(\mathsf{s})}{w_j(\mathsf{s})} = \mathrm{C}_i(\mathsf{s}\mathrm{I} - \mathrm{A})^{-1}\mathrm{B}_j + \mathrm{D}_{ij} = \frac{b_o\mathsf{s}^m + b_1\mathsf{s}^{m-1} + \cdots + b_{m-1}\mathsf{s} + b_m}{\mathsf{s}^n + a_1\mathsf{s}^{n-1} + \cdots + a_{n-1}\mathsf{s} + a_n}. \tag{3.65}$$

The roots of the denominator polynomial are called the *poles* of the system, and the roots of the numerator polynomial are the *zeros*. There are two important properties to note about the single input–output transfer function expressed as a ratio of Laplace polynomials. First, the number of terms in the denominator will be equal to or less than the number of internal states of the dynamic system. When there are no pole–zero cancellations between the numerator and denominator, $n = k$, where k is the number of internal states. If pole–zero cancellations occur, $n = k - pz$, where pz is the number of pole–zero cancellations. Second, the relative order of the numerator and denominator is related directly to the existence of the direct transmission term in the state variable model. A nonzero direct transmission term, D_{ij}, will produce a transfer function in which the order of the numerator is equal to the order of the denominator (i.e., $m = n$). In this case the transfer function is called *proper*. In the case in which $D_{ij} = 0$, the order m will be less than the order n and the transfer function is called *strictly proper*.

Example 3.2 Determine the transfer function matrix for the mass–spring–damper system introduced in Example 3.1.

Solution The state matrices for the mass–spring–damper system discussed in Example 3.1 are shown in equations (3.56) to (3.59). The matrix of transfer functions is determined from the expression $C\,(\mathsf{s}I - A)\,B + D$. The matrix expression $\mathsf{s}I - A$

is

$$\mathsf{s}I - A = \begin{bmatrix} \mathsf{s} & -1 \\ k/m & \mathsf{s} + c/m \end{bmatrix}. \tag{3.66}$$

The matrix inverse is

$$(\mathsf{s}I - A)^{-1} = \frac{1}{\mathsf{s}^2 + (c/m)\mathsf{s} + k/m} \begin{bmatrix} \mathsf{s} + c/m & 1 \\ -k/m & \mathsf{s} \end{bmatrix}. \tag{3.67}$$

The transfer function is obtained from

$$C\,(\mathsf{s}I - A)\,B + D = \frac{1}{\mathsf{s}^2 + (c/m)\mathsf{s} + k/m} \begin{bmatrix} 1 & 0 \\ 0 & 0 \end{bmatrix} \begin{bmatrix} \mathsf{s} + c/m & 1 \\ -k/m & \mathsf{s} \end{bmatrix} \begin{bmatrix} 0 \\ 1/m \end{bmatrix} + \begin{bmatrix} 0 \\ 1 \end{bmatrix}. \tag{3.68}$$

Multiplying out the expression on the right-hand side yields

$$C\,(\mathsf{s}I - A)\,B + D = \begin{bmatrix} \dfrac{1/m}{\mathsf{s}^2 + (c/m)\mathsf{s} + k/m} \\ 1 \end{bmatrix}. \tag{3.69}$$

The dimensions of the transfer function matrix are 2×1, which matches the fact that there are two outputs and one input to this system. The transfer function matrix can be placed into a form that is familiar to vibration analysis through the substitutions

$$\frac{c}{m} = 2\zeta\omega_n \tag{3.70}$$

$$\sqrt{\frac{k}{m}} = \omega_n, \tag{3.71}$$

where ζ is the nondimensional damping ratio and ω_n is the natural frequency in rad/s. With these definitions, the solution can be written as

$$C\,(\mathsf{s}I - A)\,B + D = \begin{bmatrix} \dfrac{1/m}{\mathsf{s}^2 + 2\zeta\omega_n\mathsf{s} + \omega_n^2} \\ 1 \end{bmatrix}. \tag{3.72}$$

3.4.1 Frequency Response

An input–output representation of an LTI dynamic system leads to the concept of the frequency response. Consider an LTI dynamic system modeled as a matrix of transfer functions,

$$\mathbf{y}(\mathsf{s}) = \mathrm{H}(\mathsf{s})\mathbf{w}(\mathsf{s}), \tag{3.73}$$

where $\mathrm{H}(\mathsf{s})$ is obtained from equation (3.63):

$$\mathrm{H}(\mathsf{s}) = \mathrm{C}(\mathsf{s}\mathrm{I} - \mathrm{A})^{-1}\mathrm{B} + \mathrm{D}. \tag{3.74}$$

The matrix $\mathrm{H}(\mathsf{s})$ is the matrix of input–output transfer functions as defined by the state variable representation of the system. Consider a harmonic input of the form

$$w_j(t) = W_j \sin \omega t, \tag{3.75}$$

where W_j is an amplitude of the jth input and ω is the frequency of the harmonic excitation. All other inputs are assumed to be equal to zero. The expression for the ith output can be written as

$$y_i(\mathsf{s}) = \mathrm{H}_{ij}(\mathsf{s})w_j(\mathsf{s}). \tag{3.76}$$

Assuming that the system is asymptotically stable, the steady-state output to a harmonic excitation can be written as

$$y_i(t \to \infty) = |\mathrm{H}_{ij}(j\omega)|W_j \sin(\omega t + \angle \mathrm{H}_{ij}(j\omega)). \tag{3.77}$$

The term $\mathrm{H}_{ij}(j\omega)$ is a complex-valued expression that can be written in real and imaginary terms as

$$\mathrm{H}_{ij}(j\omega) = \Re\left\{\mathrm{H}_{ij}(j\omega)\right\} + \Im\left\{\mathrm{H}_{ij}(j\omega)\right\}. \tag{3.78}$$

The magnitude and phase can then be determined from

$$|\mathrm{H}_{ij}(j\omega)| = \sqrt{\Re\{\mathrm{H}_{ij}(j\omega)\}^2 + \Im\{\mathrm{H}_{ij}(j\omega)\}^2} \tag{3.79}$$

$$\angle \mathrm{H}_{ij}(j\omega) = \tan^{-1} \frac{\Im\{\mathrm{H}_{ij}(j\omega)\}}{\Re\left\{\mathrm{H}_{ij}(j\omega)\right\}}. \tag{3.80}$$

Equation (3.77) illustrates three important results in linear system theory:

1. The steady-state response of an asymptotically stable system oscillates at the same frequency as the frequency of the input.
2. The amplitude of the output is scaled by the magnitude of the transfer function evaluated at $\mathsf{s} = j\omega$.
3. The phase of the output is shifted by the phase of the transfer function evaluated at $\mathsf{s} = j\omega$.

These results emphasize the importance of the magnitude and phase of the input–output frequency response. Evaluating the input–output transfer function at $\mathsf{s} = j\omega$ allows us to determine the amplitude and phase of the output relative to the input. For an asymptotically stable system, this explicitly determines the steady-state response of the system.

Example 3.3 Show that the frequency response $ku(j\omega)/f(j\omega)$ in Example 3.2 can be written as a nondimensional function of the damping ratio and the frequency ratio $\Omega = \omega/\omega_n$.

Solution The solution to Example 3.2 can be written in the Laplace domain as

$$\begin{pmatrix} y(\mathsf{s}) \\ f(\mathsf{s}) \end{pmatrix} = \begin{bmatrix} \dfrac{1/m}{\mathsf{s}^2 + 2\zeta\omega_n\mathsf{s} + \omega_n^2} \\ 1 \end{bmatrix} f(\mathsf{s}). \tag{3.81}$$

The input–output transfer function $u(j\omega)/f(j\omega)$ is equal to the first row of the transfer function matrix:

$$\frac{u(\mathsf{s})}{f(\mathsf{s})} = \frac{1/m}{\mathsf{s}^2 + 2\zeta\omega_n\mathsf{s} + \omega_n^2}. \tag{3.82}$$

The frequency response is obtained by substituting $\mathsf{s} = j\omega$ into expression (3.82) and combining terms:

$$\frac{u(j\omega)}{f(j\omega)} = \frac{1/m}{\omega_n^2 - \omega^2 + j2\zeta\omega_n\omega}. \tag{3.83}$$

Substituting the parameter $\Omega = \omega/\omega_n$ into the expression and dividing through by ω_n^2 yields

$$\frac{u(j\omega)}{f(j\omega)} = \frac{1}{m\omega_n^2}\frac{1}{1 - \Omega^2 + j2\zeta\Omega}. \tag{3.84}$$

Recalling that $\omega_n^2 = k/m$, we can multiply both sides by the stiffness to produce the nondimensional expression

$$\frac{ku(j\omega)}{f(j\omega)} = \frac{1}{1 - \Omega^2 + j2\zeta\Omega}. \tag{3.85}$$

This result demonstrates that the right-hand side of the expression is a nondimensional function of the damping ratio and the frequency ratio.

Example 3.4 Plot the magnitude and phase of the frequency response $ku(j\omega)/f(j\omega)$ for $\zeta = 0.01, 0.05, 0.10, 0.30$, and 0.707. Discuss the nature of the response for $\Omega << 1$, $\Omega \approx 1$, and $\Omega >> 1$.

Solution Plotting the frequency response requires that we obtain an expression for the magnitude and phase of the transfer function. This is obtained by first writing the transfer function in real and imaginary components as shown in equation (3.78). Multiplying the transfer function by the complex conjugate of the denominator yields

$$\frac{ku(j\omega)}{f(j\omega)} = \frac{1-\Omega^2}{\left(1-\Omega^2\right)^2 + 4\zeta^2\Omega^2} - j\frac{2\zeta\Omega}{\left(1-\Omega^2\right)^2 + 4\zeta^2\Omega^2}. \tag{3.86}$$

The magnitude of the frequency response is obtained from equation (3.79) and the phase is obtained from equation (3.80):

$$\left|\frac{ku(j\omega)}{f(j\omega)}\right| = \frac{1}{\sqrt{\left(1-\Omega^2\right)^2 + 4\zeta^2\Omega^2}} \tag{3.87}$$

$$\angle\frac{ku(j\omega)}{f(j\omega)} = \frac{-2\zeta\Omega}{1-\Omega^2}. \tag{3.88}$$

A plot of the frequency response magnitude and phase is shown in Figure 3.2.

The frequency response can be separated into three distinct regions. At excitation frequencies well below the resonance frequency of the system ($\Omega \ll 1$), the magnitude of the frequency response function is flat and the phase is approximately equal to 0°.

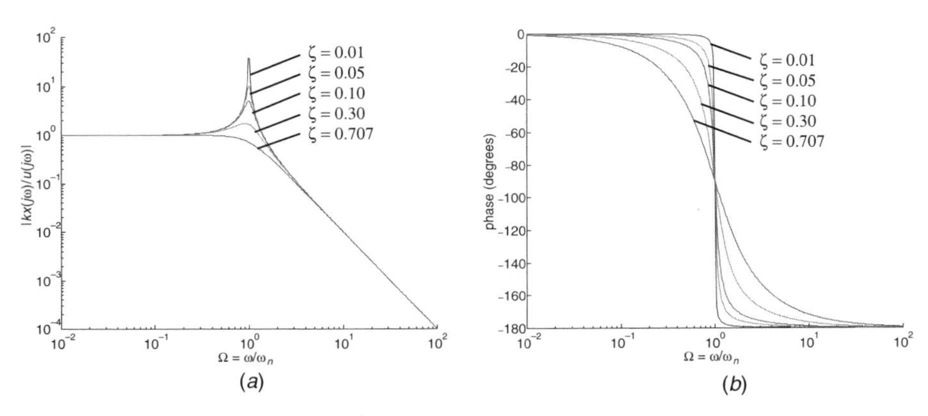

Figure 3.2 (*a*) Magnitude and (*b*) phase of a mass–spring–damper system as a function of nondimensional frequency and damping ratio.

In this frequency region,

$$\left|\frac{ku(j\omega)}{f(j\omega)}\right| \approx 1 \qquad \Omega \ll 1; \tag{3.89}$$

therefore, the amplitude $u(j\omega)/f(j\omega) \approx 1/k$ when the excitation frequency is much less than the natural frequency. Furthermore, the input and output waveform are approximately in phase at these frequencies. Notice that the damping ratio, and hence the damping coefficient and mass, do not influence the amplitude of the response at frequencies well below resonance. The magnitude of the response becomes amplified when the excitation frequency approaches the natural frequency of the system. The magnitude and phase at this frequency are

$$\left|\frac{ku(j\omega_n)}{f(j\omega_n)}\right| = \frac{1}{2\zeta} \tag{3.90}$$

$$\angle \frac{ku(j\omega_n)}{f(j\omega_n)} \to \pm\infty. \tag{3.91}$$

These results demonstrate that the damping ratio strongly influences the amplitude at resonance. Smaller values of ζ will produce a larger resonant amplitude.

The relationship between damping and the response at resonance is due to the competing physical processes within the system. At resonance, the force due to the spring stiffness and the force due to the inertial acceleration of the mass cancel one another out. Thus, the only force that resists motion at the resonance frequency is the force associated with the damping in the system. This result is general and highlights an important relationship between system response and damping. Energy dissipation strongly influences the response of the system near resonance frequencies. The smaller the energy dissipation, the larger the resonant amplification.

As the excitation frequency becomes much larger than the resonant frequency ($\Omega \gg 1$), the magnitude and phase approach

$$\left|\frac{ku(j\omega)}{f(j\omega)}\right| \to \frac{1}{\Omega^2} \tag{3.92}$$

$$\angle \frac{ku(j\omega)}{f(j\omega)} \to \frac{-2\zeta}{-\Omega}. \tag{3.93}$$

The expression for the magnitude approaches a small number as the frequency ratio becomes large, indicating that the displacement amplitude will decrease as the excitation frequency becomes large with respect to the natural frequency. As shown in Figure 3.2*b*, the phase approaches -180° as the frequency ratio becomes much larger than 1. (Note that the signs in the phase expression have been retained to emphasize quadrant associated with the inverse tangent.) The phase response illustrates that the input and output will be of opposite sign at frequencies much higher than the resonant frequency.

A second interpretation of the frequency response of an LTI system is obtained by considering an impulse input applied to the system at time zero. In this case, the input to the system is modeled as a delta function

$$w_j(t) = \delta(t). \tag{3.94}$$

The Laplace transform of an impulse input at time zero is

$$\mathcal{L}\{\delta(t)\} = 1. \tag{3.95}$$

Substituting this result into equation (3.76) yields the result

$$y_i(\mathsf{s}) = \mathrm{H}_{ij}(\mathsf{s}). \tag{3.96}$$

The time response $y_i(t)$ can be determined from the inverse Laplace transform of equation (3.96),

$$y_i(t) = \mathcal{L}^{-1}\{y_i(\mathsf{s})\} = \mathcal{L}^{-1}\{\mathrm{H}_{ij}(\mathsf{s})\}. \tag{3.97}$$

This result demonstrates that the impulse response of an LTI system is equivalent to the inverse Laplace transform of the input–output transfer function. This result is used continuously; therefore, we generally assign the symbol

$$h_{ij}(t) = \mathcal{L}^{-1}\{\mathrm{H}_{ij}(\mathsf{s})\} \tag{3.98}$$

to designate the *impulse response* between the ith output and jth input.

The importance of this result can be understood by examining the inverse Laplace transform of equation (3.76). If we apply the inverse Laplace transform to equation (3.76), we obtain the following expression for $y_i(t)$ through the convolution theorem:

$$y_i(t) = \mathcal{L}^{-1}\{y_i(\mathsf{s})\} = \mathcal{L}^{-1}\{\mathrm{H}_{ij}(\mathsf{s})w_j(\mathsf{s})\} = \int_0^t h_{ij}(t-\lambda)w_j(\lambda)d\lambda. \tag{3.99}$$

Note that equation (3.99) applies to *any* deterministic function $w_j(t)$. Thus, the response to any deterministic input can be determined by convolving the impulse response and the input function. As discussed above, the impulse response is related to the frequency response of the dynamic system; therefore, having an expression for the frequency response is equivalent to saying that the response to any deterministic input can be obtained from convolution. This is a very powerful result for LTI systems.

Comparing this result to equation (3.45), we see that the impulse response can be determined directly from the state variable representation as

$$h_{ij}(t) = \mathrm{C}_i e^{\mathrm{A}t}\mathrm{B}_j. \tag{3.100}$$

This result provides a link between the state variable representation and the impulse response of an LTI system.

3.5 IMPEDANCE AND ADMITTANCE MODELS

One type of input–output model that bears additional attention is an *impedance model* of a dynamic system. This type of model will be useful for the purpose of gaining insight into the concepts of power and energy transfer associated with smart material systems. As highlighted in Chapter 2, certain analogies can be made between the generalized state variables that model electrical and mechanical systems. For example, the concept of mechanical force and stress in a mechanical system is analogous to the applied voltage or applied electric field in an electrical system. Similarly, displacement and strain in a mechanical system have analogous relationships to charge or electric displacement in an electrical system.

The concept of analogous state variables come together when discussing the work or energy of a system. As discussed in Chapter 2, the work or stored energy associated with a system is related to an integral of the force and displacement for a mechanical system or equivalently, charge and voltage for an electrical system. The time rate of change of work is the *power* associated with a system. It has units of joules per second or watts (W). The instantaneous mechanical power is defined as

$$\Pi_M(t) = f(t)\dot{u}(t), \tag{3.101}$$

and the instantaneous electrical power is defined as

$$\Pi_E(t) = v(t)\dot{q}(t) = v(t)i(t), \tag{3.102}$$

where $i(t) = \dot{q}(t)$ is the *current*. The *average power* over a defined time interval T is defined as

$$\begin{aligned} < \Pi_M > &= \frac{1}{T}\int_t^{t+T} f(\tau)\dot{u}(\tau)\,d\tau \\ < \Pi_E > &= \frac{1}{T}\int_t^{t+T} v(\tau)i(\tau)\,d\tau. \end{aligned} \tag{3.103}$$

The average power at the location of an input to a system can be related to the input–output response of an LTI model. The analogy between force and voltage and velocity and current can be extended by considering an LTI model which has as the input a generalized force, which we denote ϕ, and generalized *flux*, which we denote ψ. The force could be either a mechanical force or an applied voltage, and the flux can be either a velocity or current. The input–output relationship of an LTI model is expressed in equation (3.73) as a matrix of Laplace domain transfer functions. If we assume that the input to this system is the generalized force, ϕ, and the output is the generalized flux term, ψ, the input–output response is written in the frequency domain as

$$\psi = \mathrm{Z}^{-1}(j\omega)\phi, \tag{3.104}$$

where $\mathrm{Z}^{-1}(j\omega)$ is the inverse of the *impedance matrix* $\mathrm{Z}(j\omega)$. The matrix of transfer functions $\mathrm{Z}^{-1}(j\omega)$ is also called the *admittance matrix*. Multiplying both sides of equation (3.104) by the impedance matrix produces an expression between the applied forces and the flux response,

$$\phi = \mathrm{Z}(j\omega)\psi. \tag{3.105}$$

The physical significance of the impedance matrix is analyzed by considering the case of a harmonic input to the ith flux input:

$$\psi_i(t) = \Psi_i \sin \omega t. \tag{3.106}$$

The steady-state output of the force at the ith location to this input (assuming that the system is asymptotically stable) is

$$\phi_i(t) = |\mathrm{Z}_{ii}|\, \Psi_i \sin(\omega t + \angle \mathrm{Z}_{ii}). \tag{3.107}$$

The instantaneous power is the product of the flux and the force:

$$\begin{aligned} \Pi(t) = \psi_i(t)\phi_i(t) &= |\mathrm{Z}_{ii}|\, \Psi_i^2 \left[\frac{1-\cos 2\omega t}{2} \cos \angle \mathrm{Z}_{ii} + \frac{1}{2} \sin 2\omega t \sin \angle \mathrm{Z}_{ii} \right] \\ &= \Psi_i^2 \left[\frac{1-\cos 2\omega t}{2} \Re(\mathrm{Z}_{ij}) + \frac{1}{2} \sin 2\omega t \Im(\mathrm{Z}_{ij}) \right]. \end{aligned} \tag{3.108}$$

This form of the instantaneous power provides insight into the meaning of the impedance. The term that multiplies the real part of the impedance is bounded between 0 and 1, whereas the term that multiplies the imaginary part of the impedance has a mean of zero and is bounded by ± 1. The first term in brackets in equation (3.108) is called the *real impedance*; the second term is called the *reactive impedance*. The real impedance is proportional to the real component of Z_{ij}, and the reactive impedance is proportional to the imaginary component. Computing the average power over a single period, we see that

$$< \Pi >= \frac{1}{T} \int_0^T \psi_i(t)\phi_i(t)dt = \frac{1}{2} \Psi_i^2 \Re(\mathrm{Z}_{ii}). \tag{3.109}$$

Thus, the average power is proportional to the real part of the impedance function Z_{ii}. As shown in equation (3.108), the reactive component of the impedance is proportional to the imaginary component of Z_{ii}, and the contribution of the reactive component to the average power is zero.

The notion of real and reactive impedance is often related to the *power flow* at the input–output location. The impedance function allows us to analyze the power flow directly. A real-valued impedance indicates that the system is dissipative at that frequency, whereas an impedance that is purely imaginary indicates that the system exhibits energy storage and the power will oscillate between the source and the system at the terminal location. For an impedance that consists of real and imaginary components, the relative amplitude of the terms is directly related to the amount of energy dissipation and energy storage at the terminals.

Example 3.5 Derive the admittance and impedance functions for the mass–spring–damper oscillator shown in Figure 3.1. Nondimensionalize the expressions in terms of the frequency ratio ω/ω_n. Plot the magnitude and phase of the impedance function and discuss the physical signficance of regions in which the impedance is low.

Solution As shown in equation (3.104), the admittance is the ratio of the flux to the force. In a mechanical system, this reduces to the ratio of the velocity of the mass, $\dot{x}(t)$, to the input force, $u(t)$. The ratio of velocity to force can be derived using the state variable representation shown in equation (3.52) and the output equations

$$\underline{y}(t) = \begin{bmatrix} 0 & 1 \end{bmatrix} \begin{pmatrix} \underline{z}_1(t) \\ \underline{z}_2(t) \end{pmatrix}. \tag{3.110}$$

Applying equation (3.63) yields

$$\mathcal{L}\{\dot{u}(t)\} = \mathsf{s}u(\mathsf{s}) = \frac{\mathsf{s}/m}{\mathsf{s}^2 + (c/m)\mathsf{s} + k/m} f(s). \tag{3.111}$$

The admittance function is then written as

$$A(\mathsf{s}) = \frac{\mathsf{s}u(\mathsf{s})}{f(\mathsf{s})} = \frac{\mathsf{s}/m}{\mathsf{s}^2 + (c/m)\mathsf{s} + k/m}. \tag{3.112}$$

Note that the subscripts have been dropped for convenience since there is only a single input and a single output [i.e., $A(\mathsf{s}) = A_{11}(\mathsf{s})$]. Substituting the definitions $c/m = 2\zeta\omega_n$ and $k/m = \omega_n^2$, we can write the frequency response of the admittance function as

$$A(j\omega) = \frac{j\omega/m}{\omega_n^2 - \omega^2 + j2\zeta\omega\omega_n}. \tag{3.113}$$

The expression can be nondimensionalized through the substitution $\Omega = \omega/\omega_n$ as

$$A(j\omega) = \frac{1}{m\omega_n} \frac{j\Omega}{1 - \Omega^2 + j2\zeta\Omega}. \tag{3.114}$$

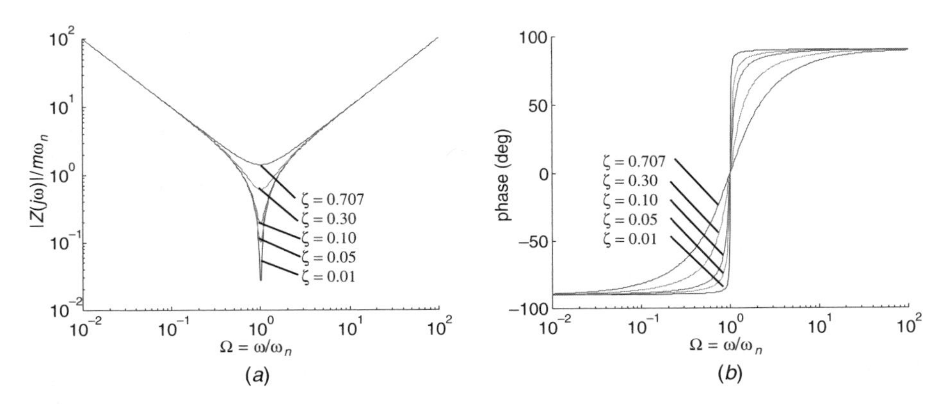

Figure 3.3 (*a*) Magnitude and (*b*) phase of the impedance function for a mass–spring–damper system as a function of damping ratio and nondimensional frequency.

Checking units, we see that the units of $m\omega_n$ are s/kg, which is correct for a ratio of velocity to force.

The impedance function is simply the inverse of the admittance; therefore,

$$Z(j\omega) = A(j\omega)^{-1} = m\omega_n \frac{2\zeta\Omega + j\left(\Omega^2 - 1\right)}{\Omega}. \tag{3.115}$$

This result is obtained by multiplying the numerator and denominator by j and rearranging terms. The magnitude and phase can be determined by writing the impedance as a sum of real and imaginary terms:

$$\frac{1}{m\omega_n} Z(j\omega) = 2\zeta + j\frac{\Omega^2 - 1}{\Omega} \tag{3.116}$$

and computing the magnitude and phase using equations (3.79) and (3.80). The plots are shown in Figure 3.3, where we see that the impedance is low in the frequency range in which $\omega \approx \omega_n$. This frequency range corresponds to the frequency range of the mechanical resonance of the system. The impedance at this frequency is only a function of the damping ratio. This can be quantified by computing

$$\frac{1}{m\omega_n} Z(j\omega_n) = 2\zeta. \tag{3.117}$$

This expression illustrates that when the excitation frequency is equivalent to the natural frequency, the impedance is only a function of the energy dissipation in the system. Once again this is due to the fact that the forces due to the spring stiffness and the inertial motion cancel one another at this frequency. A low impedance implies that it would be easy to move this system at this frequency (i.e., the system is soft when the excitation frequency is equal to the natural frequency).

3.5.1 System Impedance Models and Terminal Constraints

The concept of impedance can be applied to the analysis of systems with multiple inputs and outputs. The expression for a system impedance, equation (3.105), can be written in matrix notation as

$$\begin{Bmatrix} \phi_1 \\ \vdots \\ \phi_n \end{Bmatrix} = \begin{bmatrix} Z_{11} & \cdots & Z_{1n} \\ \vdots & \ddots & \vdots \\ Z_{n1} & \cdots & Z_{nn} \end{bmatrix} \begin{Bmatrix} \psi_1 \\ \vdots \\ \psi_n \end{Bmatrix}. \tag{3.118}$$

The system-level impedance model is a convenient framework for analyzing the relationships between forces and fluxes in a dynamic system. When a terminal constraint exists between a force and a flux, the impedance model expressed in equation (3.118) can be modified to determine the effect of this terminal constraint on the remaining force and flux terms. We define a terminal constraint as an explicit relationship between a single force–flux term,

$$\phi_k = -Z_c \psi_k, \tag{3.119}$$

where the constraint is expressed as an impedance $-Z_c$. The negative sign is chosen for convenience. Expanding equation (3.118) to include the terminal constraint, we have

$$\begin{aligned} \phi_1 &= Z_{11}\psi_1 + \cdots + Z_{1k}\psi_k + \cdots + Z_{1n}\psi_n \\ &\vdots \\ \phi_k &= Z_{k1}\psi_1 + \cdots + Z_{kk}\psi_k + \cdots + Z_{kn}\psi_n \\ &\vdots \\ \phi_n &= Z_{n1}\psi_1 + \cdots + Z_{nk}\psi_k + \cdots + Z_{nn}\psi_n. \end{aligned} \tag{3.120}$$

Substituting equation (3.119) into equation (3.121) and solving the kth equation for ψ_k yields

$$\psi_k = -\frac{Z_{k1}}{Z_c + Z_{kk}}\psi_1 - \cdots - \frac{Z_{kn}}{Z_c + Z_{kk}}\psi_n. \tag{3.121}$$

Substituting this expression into the remaining $n - 1$ equations produces

$$\phi_1 = \left(Z_{11} - \frac{Z_{1k}Z_{k1}}{Z_c + Z_{kk}}\right)\psi_1 + \cdots + \left(Z_{1n} - \frac{Z_{1k}Z_{kn}}{Z_c + Z_{kk}}\right)\psi_n \tag{3.122}$$

$$\vdots$$

$$\phi_n = \left(Z_{n1} - \frac{Z_{nk}Z_{k1}}{Z_c + Z_{kk}}\right)\psi_1 + \cdots + \left(Z_{nn} - \frac{Z_{nk}Z_{kn}}{Z_c + Z_{kk}}\right)\psi_n. \tag{3.123}$$

Examining equation (3.122), we see that a general expression for the ith force when a constraint exists at the kth location is

$$\phi_i = \sum_{m=1}^{n} \left(Z_{im} - \frac{Z_{ik} Z_{km}}{Z_c + Z_{kk}} \right) \psi_m \qquad m \neq k. \tag{3.124}$$

This result indicates that, in general, every force–flux relationship is affected by a terminal constraint at a single location in the system.

Two particular types of terminal constraints are *zero-force constraints* and *zero-flux constraints*. A zero-force constraint can be determined by setting $Z_c = 0$ in equation (3.119) and rewriting the remaining $n - 1$ transduction equations. The result is

$$\phi_i^{\phi_k} = \sum_{j=1}^{n} \left(Z_{ij} - \frac{Z_{ik} Z_{kj}}{Z_{kk}} \right) \psi_j \qquad j \neq k. \tag{3.125}$$

The superscript on the force term indicates that a zero-force constraint exists at location k. We can rewrite equation (3.125) as

$$\phi_i^{\phi_k} = \sum_{m=1}^{n} Z_{ij} \left(1 - \frac{Z_{ik} Z_{kj}}{Z_{ij} Z_{kk}} \right) \psi_j \qquad j \neq k \tag{3.126}$$

and denote

$$\frac{Z_{ik} Z_{kj}}{Z_{ij} Z_{kk}} = \mathrm{K} \tag{3.127}$$

as the *generalized coupling coefficient*. With this definition we can rewrite equation (3.126) as

$$\phi_i^{\phi_k} = \sum_{j=1}^{n} Z_{ij} \left(1 - \mathrm{K} \right) \psi_j \qquad j \neq k. \tag{3.128}$$

This definition makes it clear that the coupling coefficient describes how much the impedance function changes upon the introduction of a zero-force constraint at location k.

Zero-flux constraints are imposed by letting $Z_c \to \infty$ and substituting the result into equation (3.124). Letting the constraint impedance approach infinity produces the result

$$\phi_i^{\psi_k} = \sum_{j=1}^{n} Z_{ij} \psi_j \qquad j \neq k. \tag{3.129}$$

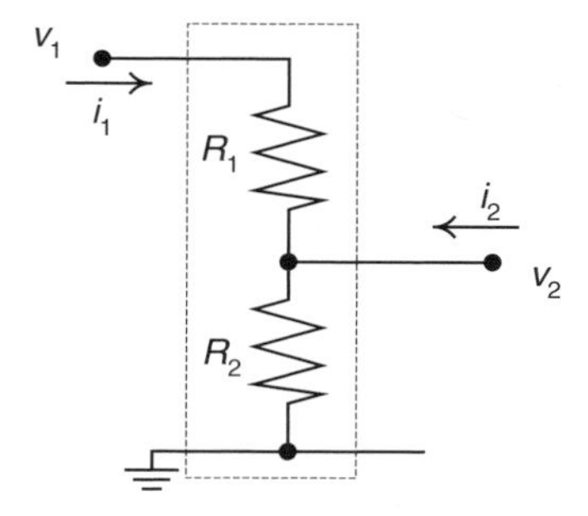

Figure 3.4 Voltage divider with input voltage and current and output voltage and current.

This result demonstrates that the impedance terms at the remaining $n-1$ locations are not changed by the introduction of a zero-flux constraint at location k.

Another interpretation of the coupling coefficient is related to the change in the impedance from a zero-force to a zero-flux constraint at location k. The coupling coefficient is also equal to the difference in the zero-flux and zero-force impedance divided by the original impedance,

$$\frac{\phi_i^{\psi_k} - \phi_i^{\phi_k}}{Z_{im}} = \frac{Z_{im} - Z_{im}(1-\text{K})}{Z_{im}} = \text{K}. \tag{3.130}$$

We will find in future chapters that the concept of coupling coefficients is related directly both to material properties and to the interaction of an active material with the external system.

Example 3.6 Consider the voltage divider shown in Figure 3.4 with input voltage, v_1, and input current, i_1, and output voltage and current, v_2 and i_2, respectively. Determine the impedance model of this circuit.

Solution The voltage divider has two forces and two flux terms; therefore, the impedance model will be a 2×2 system. The individual impedance terms can be obtained by setting one of the flux terms equal to zero and determining the corresponding force terms.

The impedance terms Z_{11} and Z_{21} can be obtained by setting the current i_2 equal to zero and determining the voltage as a function of the current i_1. When $i_2 = 0$, the voltage terms are

$$\begin{aligned} v_1 &= (R_1 + R_2)\, i_1 \\ v_2 &= R_2 i_1 \end{aligned}$$

and the impedance terms can be written as

$$\begin{aligned} Z_{11} &= R_1 + R_2 \\ Z_{12} &= R_2. \end{aligned}$$

Setting the current i_1 equal to zero produces a zero voltage drop over R_1 and the two voltages are equivalent. Thus,

$$\begin{aligned} v_1 &= R_2 i_2 \\ v_2 &= R_2 i_2 \end{aligned}$$

and

$$\begin{aligned} Z_{12} &= R_2 \\ Z_{22} &= R_2 \end{aligned}$$

Combining the impedance terms yields the matrix

$$Z = \begin{bmatrix} R_1 + R_2 & R_2 \\ R_2 & R_2 \end{bmatrix}. \tag{3.131}$$

Example 3.7 Determine (a) the coupling coefficients of the system introduced in Example 3.5 and (b) the impedance Z_{11} when there is a zero-force constraint at terminal 2.

Solution (a) The voltage divider discussed in Example 3.5, has two force–flux terms; therefore, $n = 2$. For $i = 1, k = 2$, and $m = 1$, equation (3.127) is reduced to

$$K = \frac{Z_{12}Z_{21}}{Z_{11}Z_{22}} = \frac{R_2^2}{R_2(R_1 + R_2)} = \frac{R_2}{R_1 + R_2}. \tag{3.132}$$

The coupling coefficient when $i = 2, k = 1$, and $m = 2$ is

$$K = \frac{Z_{21}Z_{12}}{Z_{22}Z_{11}} = \frac{R_2^2}{R_2(R_1 + R_2)} = \frac{R_2}{R_1 + R_2}. \tag{3.133}$$

This is consistent with the result that there are $n - 1$ coupling coefficients for a system that has n terminals.

(b) The impedance Z_{11} when a zero-force constraint exists at terminal 2 can be computed from equation (3.128):

$$\begin{aligned} Z_{11}^{\phi_2} &= Z_{11}(1 - K) \\ &= (R_1 + R_2)\left(1 - \frac{R_2}{R_1 + R_2}\right) \\ &= (R_1 + R_2)\frac{R_1}{R_1 + R_2} \\ &= R_1. \end{aligned} \tag{3.134}$$

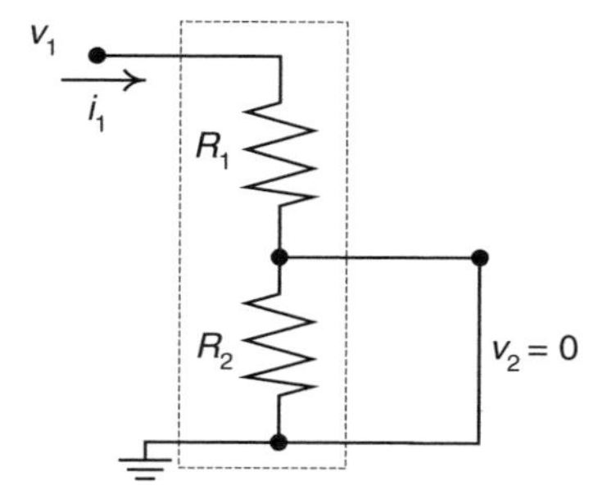

Figure 3.5 Voltage divider with a zero-force (voltage) constraint at terminal 2.

This result is reasonable because a zero-force constraint at terminal 2 is equivalent to connecting the terminal to ground. Therefore, the impedance of the resulting circuit is simply the resistance R_1 (see Figure 3.5).

Example 3.8 Terminal 2 of the voltage divider is connected to a digital voltmeter that draws no current, as shown in Figure 3.6. Determine the ratio of the output voltage to the input voltage with this constraint at terminal 2.

Solution The digital voltmeter is assumed to draw no current; therefore, it imposes a zero-flux constraint at terminal 2. With a zero-flux constraint at the terminal, the transduction equations reduce to

$$\begin{aligned} v_1 &= (R_1 + R_2)i_1 \\ v_2 &= R_2 i_1 \end{aligned} \tag{3.135}$$

Solving for ψ_1 from the first expression and substituting it into the second expression yields

$$v_2 = \frac{R_2}{R_1 + R_2} v_1, \tag{3.136}$$

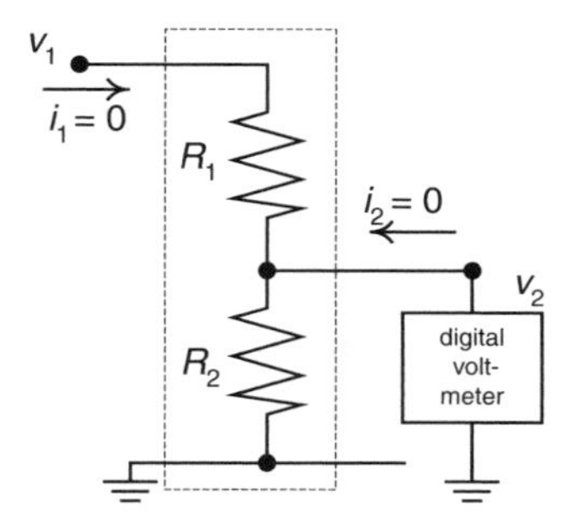

Figure 3.6 Voltage divider with a measurement device that draws zero flux (current).

which can be solved to yield the ratio

$$\frac{v_2}{v_1} = \frac{R_2}{R_1 + R_2}. \tag{3.137}$$

This is the common result for a voltage divider, except that this analysis highlights the fact that the typical equation for a voltage divider implicitly assumes that zero current is being drawn at the output terminal.

3.6 CHAPTER SUMMARY

Several approaches to modeling systems were presented in this chapter. The static response of systems is obtained through the solution of a set of algebraic equations. Dynamic systems are modeled as a set of differential equations that must be solved to determine the response of the system as a function of time. The basic properties of a second-order mass–spring–damper system were analyzed. Second-order equations of motion were presented as a means of representing structural material systems. A discussion of eigenvalue analysis was presented and related to the solution of systems modeled with second-order equations of motion. Systems modeled in second-order form can also be transformed to first-order form, also called state variable form. State variable representations of the differential equations enable alternative methods of solving for the time response of the system. Although the two methods will yield identical results when solved properly, it is generally agreed that second-order representations yield physical insight into solution of the differential equations, while state variable methods are generally superior when analyzing feedback control. In the final section of the chapter we discussed impedance methods for system analysis. Although impedance methods are limited by the fact that they are used only for frequency-domain analysis, they do yield physical insight into the solution of a number of problems relevant for analysis of smart material systems.

PROBLEMS

3.1. The equations of equilibrium for a system in static equilibrium are

$$\begin{bmatrix} 10 & -1 & -2 \\ -1 & 15 & -4 \\ -2 & -4 & 12 \end{bmatrix} \begin{pmatrix} v_1 \\ v_2 \\ v_3 \end{pmatrix} = \begin{bmatrix} 1 & 0 \\ 0 & 1 \\ 1 & -1 \end{bmatrix} \begin{pmatrix} f_1 \\ f_2 \end{pmatrix}.$$

(a) Compute the static response $\mathbf{v}$ for this system of equations.

(b) Compute the input–output relationship for the output $y = [1 \quad -1 \quad 2]\mathbf{v}$.

3.2. The state equations of equilibrium for the system studied in Example 2.8 is

$$\begin{pmatrix} f_1 \\ f_2 \end{pmatrix} = k \begin{bmatrix} 1+\alpha & -\alpha \\ -\alpha & 1+\alpha \end{bmatrix} \begin{pmatrix} \mathrm{u}_1 \\ \mathrm{u}_2 \end{pmatrix}.$$

Determine the expressions for the static response of the system and analyze the response as $\alpha \to 0$ and $\alpha \to \infty$. Relate this analysis to the system drawing in Figure 2.18.

3.3. The equations of motion for a mass–spring system are

$$\begin{bmatrix} 2 & 0 \\ 0 & 1 \end{bmatrix} \begin{pmatrix} \ddot{u}_1(t) \\ \ddot{u}_2(t) \end{pmatrix} + \begin{bmatrix} 5 & -2 \\ -2 & 3 \end{bmatrix} \begin{pmatrix} u_1(t) \\ u_2(t) \end{pmatrix} = \begin{bmatrix} 1 \\ 2 \end{bmatrix} f.$$

(a) Compute the eigenvalues and eigenvectors of the system.

(b) Compute the response $u_1(t)$ and $u_2(t)$ to a unit step input at time zero.

(c) Plot the results from part (b).

3.4. The equations of motion for a mass-spring system are

$$\begin{bmatrix} 2 & 0 & 0 \\ 0 & 3 & 0 \\ 0 & 0 & 2 \end{bmatrix} \begin{pmatrix} \ddot{u}_1(t) \\ \ddot{u}_2(t) \\ \ddot{u}_3(t) \end{pmatrix} + \begin{bmatrix} 10 & -1 & -2 \\ -1 & 5 & -1 \\ -2 & -1 & 3 \end{bmatrix} \begin{pmatrix} u_1(t) \\ u_2(t) \\ u_3(t) \end{pmatrix} = \begin{bmatrix} 0 \\ 0 \\ 0 \end{bmatrix} f(t).$$

(a) Compute the eigenvalues and eigenvectors of the system.

(b) Compute the response of each coordinate to an initial condition $\mathbf{u}(0) = [1 \quad 2 \quad -1]'$.

3.5. Repeat Problem 3 including a damping matrix in which each modal coordinate has a damping ratio of 0.01, 0.05, and 0.30. Plot the results.

3.6. Repeat Problem 4 including a damping matrix in which each modal coordinate has a damping ratio of 0.01, 0.05, and 0.30. Plot the results.

3.7. Transform the second-order equations from Problem 3 into first-order form.

3.8. The first-order equations that govern a system are

$$\begin{pmatrix} \dot{z}_1(t) \\ \dot{z}_2(t) \end{pmatrix} = \begin{bmatrix} -3 & 1 \\ 0 & -1 \end{bmatrix} \begin{pmatrix} z_1(t) \\ z_2(t) \end{pmatrix} + \begin{bmatrix} -2 \\ 1 \end{bmatrix} f(t).$$

(a) Solve for the response to the initial condition $\mathbf{z}(0) = [1 \quad -1]'$ when $f(t) = 0$.

(b) Solve for the state response when the input is a unit step at time zero and the initial conditions are zero.

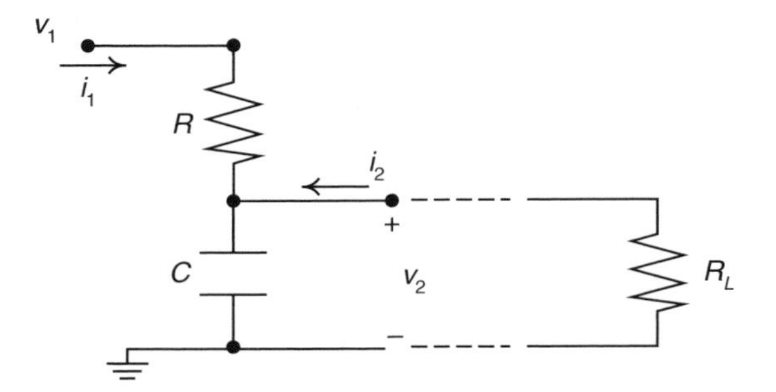

Figure 3.7 Passive resistive–capacitive circuit.

(c) Solve for the frequency response when $y(t) = [1 \quad 1]$.

(d) Solve for the impulse response function of the output.

3.9. The first-order equations that govern a system are

$$\begin{pmatrix} \dot{z}_1(t) \\ \dot{z}_2(t) \end{pmatrix} = \begin{bmatrix} 0 & 1 \\ -10 & -1 \end{bmatrix} \begin{pmatrix} z_1(t) \\ z_2(t) \end{pmatrix} + \begin{bmatrix} 0 \\ 1 \end{bmatrix} f(t).$$

(a) Solve for the response due to the initial conditions $\mathbf{z}(0) = [1 \quad -1]'$ when $f(t) = 0$.

(b) Solve the the state response when the input is a unit step at time zero and the initial conditions.

(c) Solve for the frequency response when $y(t) = [1 \quad 1]$.

(d) Solve for the impulse response function of the output.

3.10. (a) Determine the impedance matrix for the system shown in Figure 3.7.

(b) Determine the impedance v_1/i_1 when the resistor R_L is connected across the terminal at v_2.

(c) Determine the impedance v_1/i_1 for the limits $R_L \to 0$ and $R_L \to \infty$. Relate these conditions to the zero-force and zero-flux conditions.

3.11. The impedance function for a mechanical system is

$$Z_m(j\Omega) = \frac{1 - \Omega^2 + j2\zeta\Omega}{j\Omega},$$

where $\Omega = \omega/\omega_n$ is the ratio of the driving frequency to the excitation frequency.

(a) Determine the average power at $\Omega = 0.1$, 1 and 10.

(b) Determine the magnitude of the reactive power at $\Omega = 0.1$, 1 and 10.

NOTES

References on second-order systems are described in any number of textbooks on vibrations and vibration analysis. A general text on the engineering analysis of vibrating systems is that of Inman [17]. This book includes derivations of the response functions for single oscillators with various damping mechanisms and a discussion of multiple-degree-of-freedom systems. Details on the mathematical foundations of vibration analysis can also be found in a number of texts, such as that of Meirovitch [18] and his more recent text [19]. These books derive the relationships between eigenvalue problems and the solution of systems of equations that represent vibrating systems.

The foundations of first-order analysis are also discussed in a number of textbooks on systems theory and control theory. A good overview of systems theory is that of Chen [20], which contains the mathematical foundations of eigenvalue analysis for first-order systems. Others who discuss linear system theory as it relates to control are Frieldland [21] and Franklin et al. [22]. Their books include examples of transforming equations of motion into first-order form for dynamic analysis. Research on impedance analysis is fading fast since the inception of modern computer-aided design techniques, but Beranek [23] presented very clear examples of how to use systems theory and impedance analyses to understand the response of dynamic systems.

4

PIEZOELECTRIC MATERIALS

The first class of smart materials that we study in detail are *piezoelectric materials*. As discussed in Chapter 1, piezoelectric materials are used widely in transducers such as ultrasonic transmitters and receivers, sonar for underwater applications, and as actuators for precision positioning devices. We focus on development of the constitutive equations and the application of these equations to the basic operating modes of piezoelectric devices.

4.1 ELECTROMECHANICAL COUPLING IN PIEZOELECTRIC DEVICES: ONE-DIMENSIONAL MODEL

Piezoelectric materials exhibit *electromechanical coupling*, which is useful for the design of devices for sensing and actuation. The coupling is exhibited in the fact that piezoelectric materials produce an electrical displacement when a mechanical stress is applied and can produce mechanical strain under the application of an electric field. Due to the fact that the mechanical-to-electrical coupling was discovered first, this property is termed the *direct effect*, while the electrical-to-mechanical coupling is termed the *converse piezoelectric effect*. It is also known that piezoelectric materials exhibit a thermomechanical coupling called the *pyroelectric effect*, although in this chapter we concentrate on development of the constitutive equations and basic mechanisms of electromechanical coupling.

4.1.1 Direct Piezoelectric Effect

Consider a specimen of elastic material that has mechanical stress applied to the two opposing faces and is constrained to move only in the direction of the applied stress, T. This state of loading can be approximated using a tensile specimen that is common in mechanical testing (Figure 4.1). Applying stress to the material will produce elongation in the direction of the applied load, and under the assumption that the material is in a state of uniaxial strain, the *strain*, S, is defined as the total

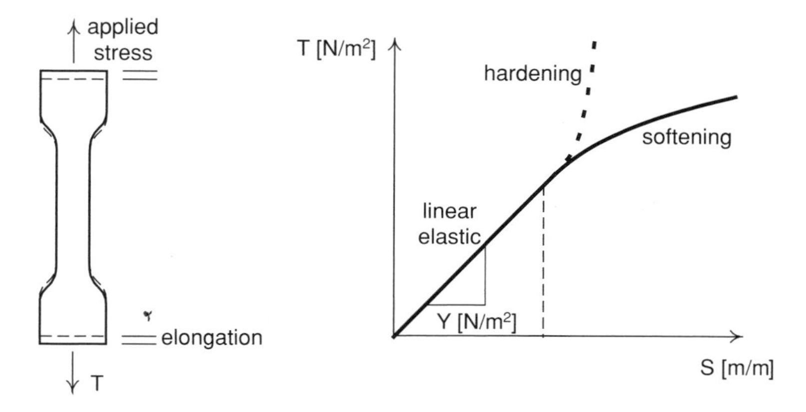

Figure 4.1 Representative stress–strain behavior for an elastic material.

elongation divided by the original length of the specimen. At low values of applied stress the strain response will be linear until a critical stress, at which the material will begin to yield. In the linear elastic region the slope of the stress–strain curve is constant. The slope of the line is called the *modulus*, or *Young's modulus*, and in this book the modulus is denoted Y and has units of units N/m^2. The stress–strain relationship in this region is

$$S = \frac{1}{Y}T = sT, \tag{4.1}$$

where s, the reciprocal of the modulus, is called the *mechanical compliance* (m^2/N). Above the critical stress the slope of the stress–strain curve changes as a function of applied load. A softening material will exhibit a decreasing slope as the stress is increased, whereas a hardening material will exhibit an increasing slope for stress values above the critical stress.

Now consider the case when a piezoelectric material is being subjected to an applied stress. In addition to elongating like an elastic material, a piezoelectric material will produce a charge flow at electrodes placed at the two ends of the specimen. This charge flow is caused by the motion of *electric dipoles* within the material. The application of external stress causes the charged particles to move, creating an apparent charge flow that can be measured at the two electrodes. The charge produced divided by the area of the electrodes is the *electric displacement*, which has units of C/m^2. Applying an increasing stress level will produce an increase in the rotation of the electric dipoles and an increase in the electric displacement. Over a certain range of applied mechanical stress, there is a linear relationship between applied stress and measured electric displacement. The slope of the curve, called the *piezoelectric strain coefficient* (Figure 4.2), is denoted by the variable d. Expressing this relationship in

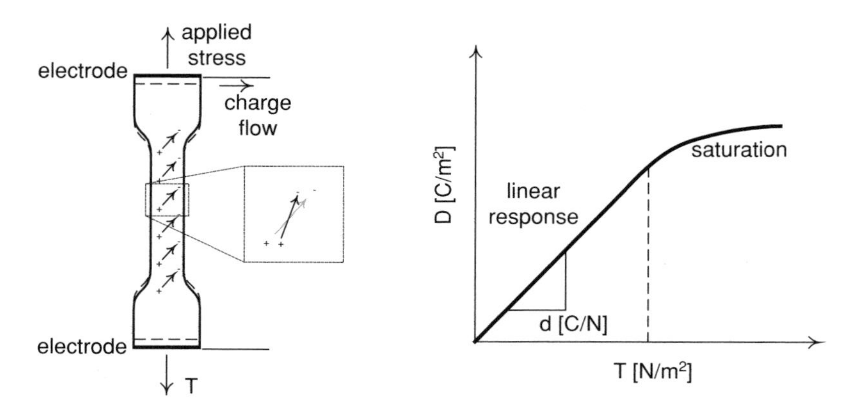

Figure 4.2 (*a*) Direct piezoelectric effect; (*b*) relationship between stress and electric displacement in a piezoelectric material.

a proportionality, we have

$$D = dT, \tag{4.2}$$

where D is the electric displacement (C/m^2) and d is the piezoelectric strain coefficient (C/N). At sufficient levels of applied stress, the relationship between stress and electric displacement will become nonlinear due to saturation of electric dipole motion (Figure 4.2). For the majority of this chapter we concern ourselves only with the linear response of the material; the nonlinear response is analyzed in the final section.

4.1.2 Converse Effect

The direct piezoelectric effect described in Section 4.1.1 is the relationship between an applied mechanical load and the electrical response of the material. Piezoelectric materials also exhibit a reciprocal effect in which an applied electric field will produce a mechanical response. Consider the application of a constant potential across the electrodes of the piezoelectric material as shown in Figure 4.3. Under the assumption that the piezoelectric material is a perfect insulator, the applied potential produces an electric field in the material, E, which is equal to the applied field divided by the distance between the electrodes (see Chapter 2 for a more complete discussion of ideal capacitors). The units of electric field are V/m. The application of an electric field to the material will produce attractions between the applied charge and the electric dipoles. Dipole rotation will occur and an electric displacement will be measured at the electrodes of the material. At sufficiently low values of the applied field, the relationship between E and D will be linear and the constant of proportionality, called the *dielectric permittivity*, has the unit F/m. The relationship between field and electric

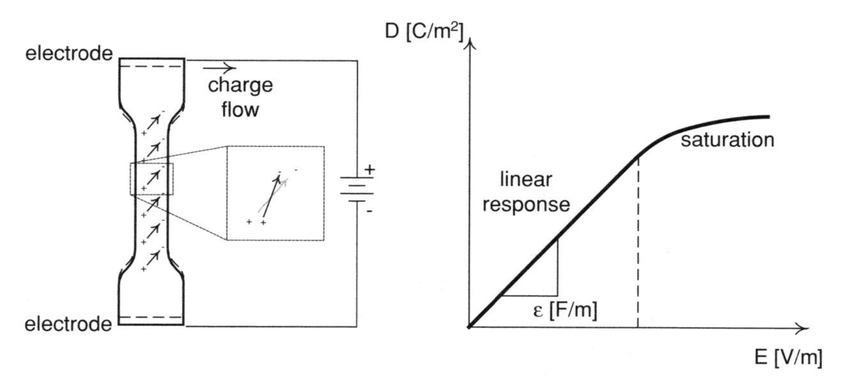

Figure 4.3 Relationship between applied electric field and the electric displacement in piezoelectric material.

displacement in the linear regime is

$$D = \varepsilon E. \tag{4.3}$$

As is the case with an applied stress, the application of an increasingly high electric field will eventually result in saturation of the dipole motion and produce a nonlinear relationship between the applied field and electric displacement. The converse piezoelectric effect is quantified by the relationship between the applied field and mechanical strain. For a direct piezoelectric effect, application of a stress produced dipole rotation and apparent charge flow. Upon application of an electric field, dipole rotation will occur and produce a strain in the material (Figure 4.4). Applying sufficiently low values of electric field we would see a linear relationship between the applied field and mechanical strain. Remarkably enough, the slope of the field-to-strain relationship would be equal to the piezoelectric strain coefficient, as shown in

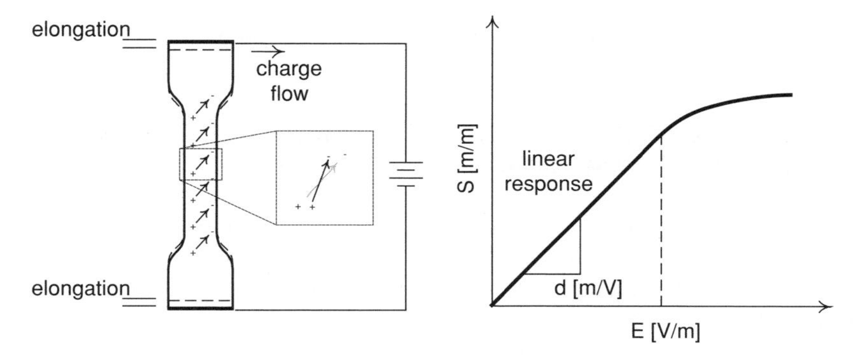

Figure 4.4 Relationship between electric field and strain in a piezoelectric material.

Figure 4.4. Expressing this as an equation, we have

$$S = dE. \tag{4.4}$$

In this expression, the piezoelectric strain coefficient has the unit m/V. Equation (4.4) is an expression of the *converse effect* for a linear piezoelectric material.

Example 4.1 Consider a piezoelectric material with a piezoelectric strain coefficient of 550×10^{-12} m/V and a mechanical compliance of 20×10^{-12} m^2/N. The material has a square geometry with a side length of 7 mm. Compute (a) the strain produced by a force of 100 N applied to the face of the material when the applied electric field is zero, and (b) the electric field required to produce an equivalent amount of strain when the applied stress is equal to zero.

Solution (a) Compute the stress applied to the face of the material,

$$T = \frac{100 \text{ N}}{(7 \times 10^{-3} \text{ m})(7 \times 10^{-3} \text{ m})} = 2.04 \text{ MPa}.$$

The strain is computed using equation (4.1),

$$S = (20 \times 10^{-12} \text{ m}^2/\text{N})(2.04 \times 10^6 \text{ Pa}) = 40.8 \times 10^{-6} \text{ m/m}.$$

The units of $\times 10^{-6}$ m/m are often called *microstrain*; a stress of 2.04 MPa produces 40.8 microstrain in the piezoelectric material.

(b) The electric field required to produce the same strain in the material is computed using the equations for the converse effect. Solving equation (4.4) for E yields

$$E = \frac{S}{d} = \frac{40.8 \times 10^{-6} \text{ m/m}}{550 \times 10^{-12} \text{ m/V}} = 74.2 \text{ kV/m}.$$

This value is well within the electric field limits for a typical piezoelectric material.

4.2 PHYSICAL BASIS FOR ELECTROMECHANICAL COUPLING IN PIEZOELECTRIC MATERIALS

The physical basis for piezoelectricity in solids is widely studied by physicists and materials scientists. Although a detailed discussion of these properties is not the focus of this book, it is important to understand the basic principles of piezoelectricity for a deeper understanding of the concepts introduced.

Most piezoelectric materials belong to a class of *crystalline* solids. Crystals are solids in which the atoms are arranged in a single pattern repeated throughout the body. Crystalline materials are highly ordered, and an understanding of the bulk properties of the material can begin by understanding the properties of the crystals repeated throughout the solid. The individual crystals in a solid can be thought of as building blocks for the material. Joining crystals together produces a three-dimensional arrangement of the crystals called a *unit cell*.

One of the most important properties of a unit cell in relation to piezoelectricity is the *polarity* of the unit cell structure. Crystallographers have studied the structure of unit cells and classified them into a set of 32 crystal classes or *point groups*. Each point group is characterized by a particular arrangement of the consituent atoms. Of these 32 point groups, 10 have been shown to exhibit a *polar axis* in which there is a net separation between positive charges in the crystal and their associated negative charges. This separation of charge produces an *electric dipole*, which can give rise to piezoelectricity.

As discussed in Chapter 2, an electric dipole can be visualized by imagining a positive charge and a negative charge separated by a distance with a pin in the center. Placing this electric dipole in an electric field will produce attraction between opposite charges and will result in rotation of the dipole. If we think of the dipole as being "attached" to the surrounding material (which is an acceptable visual but not the reality), we can easily imagine that this dipole rotation will produce strain in the surrounding structure. This is the physical basis for the converse piezoelectric effect discussed in Section 4.1.

Similarly, if a mechanical strain is applied to the material, one can envision that the dipole in the crystal will rotate. The motion of charge in the unit cell structure will produce an apparent charge flow which can be measured at the face of the material. Electrodes placed at the material faces will measure a charge flow, or current, due to the rotation of the electric dipoles. This is the physical basis for the direct piezoelectric effect (Figure 4.5).

4.2.1 Manufacturing of Piezoelectric Materials

Piezoelectricity is a phenomenon that is present in a number of natural materials. As discussed in Chapter 1, the phenomenon of piezoelectricity was first discovered

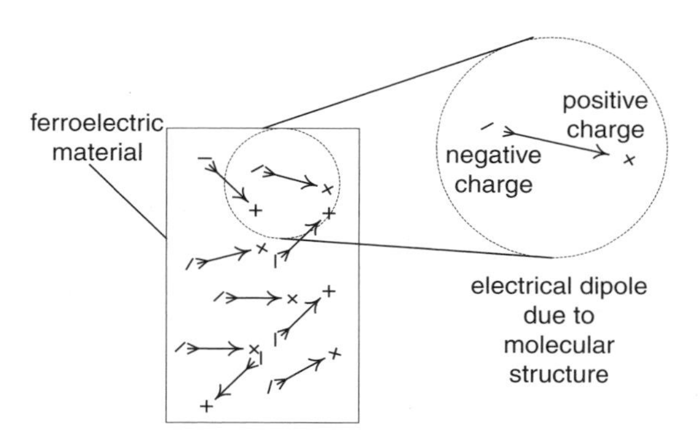

Figure 4.5 Electric dipoles that lead to electromechanical coupling in piezoelectric materials.

in a natural crystal called Rochelle salt in the late nineteenth century. For a number of years the only piezoelectric materials that were studied were natural crystals that exhibited only weak piezoelectricity. It was not until the mid-twentieth century that synthetic piezoelectric materials with increased coupling properties enabled practical applications.

The manufacture of synthetic piezoelectric materials typically begins with the constituent materials in powder form. A typical mixture of materials that exhibit piezoelectric properties are lead (with the chemical symbol Pb), zirconium (Zr), and titanium (Ti). These materials produce the common piezoelectric material lead–zirconium–titinate, typically referred to as PZT. Other types of piezoelectric materials are barium titinate and sodium–potassium niobates.

The processing of a piezoelectric ceramic typically begins by heating the powders to temperatures in the range 1200 to 1500°C. The heated materials are then formed and dimensioned with conventional methods such as grinding or abrasive media. The result of this process is generally a wafer of dimensions on the order of a few centimeters on two sides and thicknesses in the range 100 to 300 μm. Electrodes are placed on the wafers by painting a thin silver paint onto the surface. The resulting wafer can be cut with a diamond saw or joined with other layers to produce a multilayer device.

As discussed in Section 4.1, the piezoelectric effect is strongly coupled to the existence of electric dipoles in the crystal structure of the ceramic. Generally, after processing the raw material does not exhibit strong piezoelectric properties, due to the fact that the electric dipoles in the material are pointing in random directions. Thus, the net dipole properties of the material are very small at the conclusion of the fabrication process. The orientation of the individual electric dipoles in a piezoelectric material must be aligned for the material to exhibit strong electromechanical coupling.

The dipoles are oriented with respect to one another through a process called *poling*. Poling requires that the piezoelectric material be heated up above its *Curie temperature* and then placed in a strong electric field (typically, 2000 V/mm). The combination of heating and electric field produces motion of the electronic dipoles. Heating the material allows the dipoles to rotate freely, since the material is softer at higher temperatures. The electric field produces an alignment of the dipoles along the direction of the electric field as shown in Figure 4.6. Quickly reducing the temperature and removing the electric field produces a material whose electric dipoles are oriented in the same direction. This direction is referred to as the *poling direction* of the material.

Orienting the dipoles has the effect of enhancing the piezoelectric effect in the material. Now an applied electric field will produce similar rotations throughout the material. This results in a summation of strain due to the applied field. Conversely, we see that strain induced in a particular direction will produce a summation of apparent charge flow in the material, resulting in an increase in the charge output of the material.

The basic properties of a piezoelectric material are expressed mathematically as a relationship between two mechanical variables, stress and strain, and two electrical variables, electric field and electric displacement. The direct and converse piezoelectric effects are written as the set of linear equations in equations (4.1) to (4.4). The

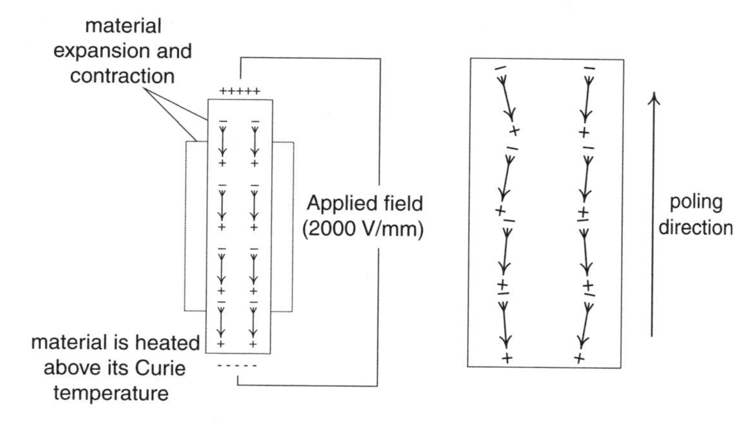

Figure 4.6 Poling process associated with piezoelectric materials.

expressions for the direct and converse piezoelectric effect can be combined into one matrix expression by writing the relationship between strain and electric displacement as a function of applied stress and applied field:

$$\begin{Bmatrix} \mathrm{S} \\ \mathrm{D} \end{Bmatrix} = \begin{bmatrix} s & d \\ d & \varepsilon \end{bmatrix} \begin{Bmatrix} \mathrm{T} \\ \mathrm{E} \end{Bmatrix}. \tag{4.5}$$

The top partition of equation (4.5) represents an equation for the converse piezoelectric effect, whereas the bottom partition represents an expression of the direct effect.

Writing the expressions as a matrix highlights some fundamental concepts of piezoelectric materials. Examining the matrix expression, we see that the on-diagonal terms represent the constitutive relationships of a mechanical and an electrical material, respectively. For example, the (1,1) term in the matrix, s, represents the mechanical constitutive relationship between stress and strain, whereas the (2,2) term ε represents the electrical constitutive equation. These constitutive relationships would exist in a material that was either purely elastic or purely dielectric.

The electromechanical coupling in the material is represented by the off-diagonal terms of equation (4.5). A larger off-diagonal term will result in a material that produces more strain for an applied electric field and more electric displacment for an applied mechanical stress. For these reasons, the piezoelectric strain coefficient is an important parameter for comparing the relative strength of different types of piezoelectric materials. In the limit as d approaches zero, we are left with a material that exhibits very little electromechanical coupling. Examining equation (4.5) we see that the coefficient matrix is symmetric. The symmetry is not simply a coincidence, we will see that symmetry in the coefficient matrix represents reciprocity between the electromechanical transductions mechanisms in the material. This will naturally arise when we discuss the energy formulation of the piezoelectric consitutive equations in Chapter 5.

There is no reason why equation (4.5) has to be expressed with stress and electric field as the independent variables and strain and electric displacement as the dependent variables. Equation (4.5) can be inverted to write the expressions with stress and field as the dependent variables and strain and electric displacement as the independent variables. Taking the inverse of the 2×2 matrix produces the expression

$$\begin{Bmatrix} \mathrm{T} \\ \mathrm{E} \end{Bmatrix} = \frac{1}{s\varepsilon - d^2} \begin{bmatrix} \varepsilon & -d \\ -d & s \end{bmatrix} \begin{Bmatrix} \mathrm{S} \\ \mathrm{D} \end{Bmatrix}. \tag{4.6}$$

The determinant can be incorporated into the matrix to produce

$$\begin{Bmatrix} \mathrm{T} \\ \mathrm{E} \end{Bmatrix} = \begin{bmatrix} \dfrac{1}{s}\left(\dfrac{1}{1 - d^2/s\varepsilon}\right) & -\dfrac{d/s\varepsilon}{1 - d^2/s\varepsilon} \\ -\dfrac{d/s\varepsilon}{1 - d^2/s\varepsilon} & \dfrac{1}{\varepsilon}\left(\dfrac{1}{1 - d^2/s\varepsilon}\right) \end{bmatrix} \begin{Bmatrix} \mathrm{S} \\ \mathrm{D} \end{Bmatrix}. \tag{4.7}$$

The term $d^2/s\varepsilon$ appears quite often in an analysis of piezoelectric materials. The square root of this term is called the *piezoelectric coupling coefficient* and is denoted

$$k = \frac{d}{\sqrt{s\varepsilon}}. \tag{4.8}$$

An important property of the piezoelectric coupling coefficient is that it is always positive and bounded between 0 and 1. The bounds on the coupling coefficient are related to the energy conversion properties in the piezoelectric material, and the bounds of 0 and 1 represent the fact that only a fraction of the energy is converted between mechanical and electrical domains. The piezoelectric coupling coefficient quantifies the electromechanical energy conversion. The rationale for these bounds will become clearer when we derive the constitutive equations from energy principles in Chapter 5.

Substituting the definition of the piezoelectric coupling coefficient into equation (4.7) yields

$$\begin{Bmatrix} \mathrm{T} \\ \mathrm{E} \end{Bmatrix} = \begin{bmatrix} \dfrac{1}{s}\dfrac{1}{1 - k^2} & -\dfrac{1}{d}\dfrac{k^2}{1 - k^2} \\ -\dfrac{1}{d}\dfrac{k^2}{1 - k^2} & \dfrac{1}{\varepsilon}\dfrac{1}{1 - k^2} \end{bmatrix} \begin{Bmatrix} \mathrm{S} \\ \mathrm{D} \end{Bmatrix}. \tag{4.9}$$

Simplifying the expression yields

$$\begin{Bmatrix} \mathrm{T} \\ \mathrm{E} \end{Bmatrix} = \frac{1}{1 - k^2} \begin{bmatrix} s^{-1} & -d^{-1}k^2 \\ -d^{-1}k^2 & \varepsilon^{-1} \end{bmatrix} \begin{Bmatrix} \mathrm{S} \\ \mathrm{D} \end{Bmatrix}. \tag{4.10}$$

The fact that $0 < k^2 < 1$ implies that the term $1/(1 - k^2)$ must be greater than 1.

Example 4.2 A coupling coefficient of $k = 0.6$ has been measured for the piezoelectric material considered in Example 4.1. Compute the dielectric permittivity of the sample.

Solution Solving equation (4.8) for dielectric permittivity yields

$$\varepsilon = \frac{d^2}{sk^2}.$$

Substituting the values for the piezoelectric strain coefficient and mechanical compliance into the expression yields

$$\varepsilon = \frac{(550 \times 10^{-12}\ \text{m/V})^2}{(20 \times 10^{-12}\ \text{m}^2/\text{N})(0.6^2)} = 42.0 \times 10^{-9}\ \text{F/m}.$$

The dielectric permittivity is often quoted in reference to the permittivity of a vacuum, $\varepsilon_o = 8.85 \times 10^{-12}$ F/m. The relative permittivity is defined as

$$\varepsilon_r = \frac{\varepsilon}{\varepsilon_o} = \frac{42.0 \times 10^{-9}\ \text{F/m}}{8.85 \times 10^{-12}\ \text{F/m}} = 4747$$

and is a nondimensional quantity.

4.2.2 Effect of Mechanical and Electrical Boundary Conditions

Electromechanical coupling in piezoelectric devices gives rise to the fact that the properties of the material are also a function of the mechanical and electrical boundary conditions. Consider again our piezoelectric cube in which we are measuring the mechanical compliance s by applying a known stress and measuring the induced strain. An important parameter in the test setup is the *electrical* boundary condition that exists between the opposing faces. Assume for a moment that we have a short-circuit condition in which the faces of the piezoelectric cube are connected directly, as shown in Figure 4.7. This electrical boundary condition results in a zero field across the faces of the material but does allow charge to flow from the positive terminal to the negative terminal. Substituting the condition $\text{E} = 0$ into equation (4.5) results in the expressions

$$\text{S} = s\text{T} \tag{4.11}$$

$$\text{D} = d\text{T}. \tag{4.12}$$

Now consider performing the same experiment when the electrical terminals are open such that no charge can flow between the faces of the material. In this experiment the electrical displacement $\text{D} = 0$, and the constitutive relationship in equation (4.9)

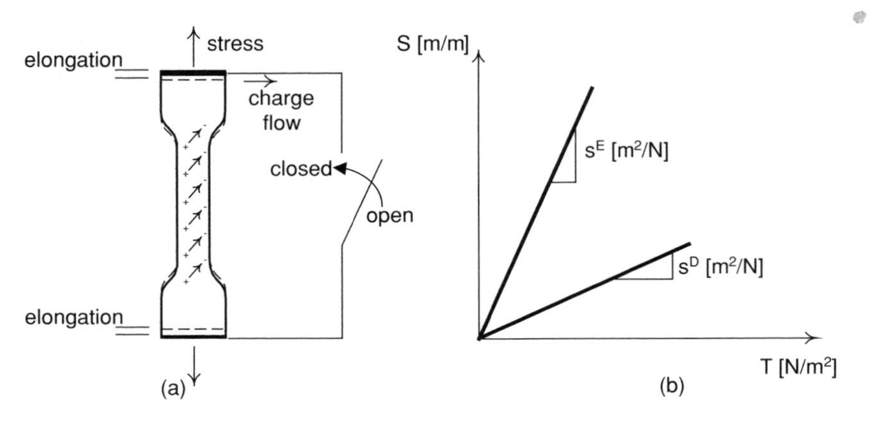

Figure 4.7 (*a*) Stress–strain tests on a piezoelectric material; (*b*) stress–strain relationships as a function of boundary conditions.

reduces to

$$\mathrm{T} = \frac{1}{s(1-k^2)}\mathrm{S} \tag{4.13}$$

$$\mathrm{E} = \frac{k^2}{d(1-k^2)}\mathrm{S}. \tag{4.14}$$

Inverting equation (4.13) we see that

$$\mathrm{S} = \begin{cases} s\mathrm{T} & \text{short circuit} \quad (4.15) \\ s(1-k^2)\mathrm{T} & \text{open circuit.} \quad (4.16) \end{cases}$$

The result demonstrates that *the mechanical compliance changes when the electrical boundary condition is changed*. The fact that $k^2 > 0$ indicates that the mechanical compliance decreases when the electrical boundary condition is changed from a short-circuit to an open-circuit condition. For this reason it is improper to refer to the mechanical compliance without specifying the electrical boundary condition.

It is convention to adopt a superscript to denote the boundary condition associated with the measurement of a particular mechanical or electrical property. The superscript E or D denotes a constant electric field and constant electric displacement, respectively, for a mechanical property. Rewriting equations (4.15) and (4.16) using this notation produces

$$\mathrm{S} = \begin{cases} s^{\mathrm{E}}\mathrm{T} & \text{short circuit} \quad (4.17) \\ s^{\mathrm{E}}(1-k^2)\mathrm{T} & \text{open circuit.} \quad (4.18) \end{cases}$$

The fact that equation (4.18) was derived assuming an open circuit (D = 0), we can write a relationship between the short-circuit mechanical compliance and open-circuit mechanical compliance as

$$s^{\mathrm{D}} = s^{\mathrm{E}}(1 - k^2). \tag{4.19}$$

An analogous relationship exists for specifying electrical quantities such as the dielectric permittivity ε. The relationship between electrical displacement and applied field changes depending on mechanical boundary conditions. A stress-free (T = 0) condition is achieved by applying a field without mechanical constraints placed at the boundary of the piezoelectric material, whereas a strain-free (S = 0) condition is achieved by clamping both ends of the material such that there is zero motion. Performing an analysis similar to the one presented for the electrical boundary conditions, we arrive at the conclusion that

$$\varepsilon^{\mathrm{S}} = \varepsilon^{\mathrm{T}}(1 - k^2). \tag{4.20}$$

Note that the piezoelectric strain coefficient is independent of the mechanical or electrical boundary conditions.

Example 4.3 Determine the percentage change in the mechanical compliance of a piezoelectric material when the electrical boundary conditions are changed from short circuit to open circuit.

Solution The percentage change in the compliance is

$$\%\ \text{change} = 100 \times \frac{s^{\mathrm{D}} - s^{\mathrm{E}}}{s^{\mathrm{E}}}.$$

Substituting the result from equation (4.19) yields

$$\%\ \text{change} = 100 \times \frac{s^{\mathrm{E}}(1 - k^2) - s^{\mathrm{E}}}{s^{\mathrm{E}}} = -100k^2.$$

Thus, a material with a coupling coefficiency of $k = 0.5$ would be able to change the compliance by 25%. The negative sign indicates that short-circuit compliance is smaller than open-circuit compliance.

4.2.3 Interpretation of the Piezoelectric Coupling Coefficient

The piezoelectric coupling coefficient k plays an important role in the analysis of piezoelectric materials. Mathematically, it is related to the inverse of the matrix that relates the strain, electric displacement, electric field, and stress. The definition of the coupling coefficient, equation (4.8), demonstrates that it is related to all three piezoelectric material properties: the compliance, permittivity, and strain coefficient.

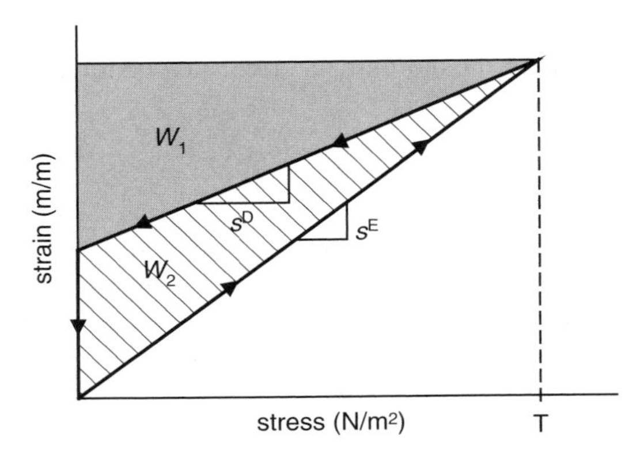

Figure 4.8 Work and energy interpretation of the piezoelectric coupling coefficient.

Physically, we showed in Section 4.2.2 that the coupling coefficient quantifies the change in mechanical (electrical) compliance when the electrical (mechanical) boundary conditions are changed.

There is another interpretation of the piezoelectric coupling coefficient that highlights its relationship to the energy stored in a piezoelectric material due to electromechanical coupling. Consider a piezoelectric material that is in a short-circuit condition such that the mechanical compliance is represented by s^{E}. Application of a stress T to the material produces the stress–strain response illustrated in Figure 4.8. The work performed during this deformation is represented by the shaded and crosshatched areas in the figure,

$$W_1 + W_2 = \frac{1}{2} s^{\mathrm{E}} \mathrm{T}^2. \tag{4.21}$$

At this point in the cycle the electrical boundary conditions are changed from short circuit to open circuit and the applied stress is reduced to zero. The resulting work is represented by the shaded region in Figure 4.8 and is equal to

$$W_1 = \frac{1}{2} s^{\mathrm{D}} \mathrm{T}^2. \tag{4.22}$$

It is clear from Figure 4.8 that the amount of work performed during the application of stress is different from the amount of work performed during the removal of stress, due to the change in the compliance of the material from short-circuit to open-circuit boundary conditions. To complete the cycle, we assume that an ideal removal of the strain is performed at a zero-stress state to return to the initial state of the material.

The difference between the energy stored during stress application and the energy return during the removal of stress is equal to W_2. Forming the ratio of this energy term to the total energy stored $W_1 + W_2$, we have

$$\frac{W_2}{W_1 + W_2} = \frac{\frac{1}{2}s^{\mathrm{E}}\mathrm{T}^2 - \frac{1}{2}s^{\mathrm{D}}\mathrm{T}^2}{\frac{1}{2}s^{\mathrm{E}}\mathrm{T}^2} = \frac{s^{\mathrm{E}} - s^{\mathrm{D}}}{s^{\mathrm{E}}}. \tag{4.23}$$

Substituting the relationship between open- and short-circuit mechanical compliance, equation (4.19), into equation (4.23), produces

$$\frac{W_2}{W_1 + W_2} = \frac{s^{\mathrm{E}} - s^{\mathrm{E}}(1 - k^2)}{s^{\mathrm{E}}} = k^2. \tag{4.24}$$

This analysis provides a visual interpretation of the piezoelectric coupling coefficient and introduces a relationship between k and the energy storage properties of the material. The square of the coupling coefficient was shown to be equal to the ratio of the energy remaining in the piezoelectric material after a complete cycle to the total work performed or energy stored during the application of stress. The smaller the remaining energy, the smaller the coupling coefficient of the material.

4.3 CONSTITUTIVE EQUATIONS FOR LINEAR PIEZOELECTRIC MATERIAL

In Section 4.2 we introduced the fundamental concept of a piezoelectric material. We saw that electromechanical coupling was parameterized by three variables: the mechanical compliance, the dielectric permittivity, and the piezoelectric strain coefficient. The direct piezoelectric effect, as well as the converse piezoelectric effect, could be expressed as a relationship between stress, strain, electric field, and electric displacement. The expressions were in terms of the three material parameters, s, ε, and d. The mechanical compliance and electrical permittivity were shown to be functions of the electrical and mechanical boundary condition, respectively, and the boundary condition needed to be specified when writing these parameters.

In this section we generalize this result to the case of an arbitrary volume of piezoelectric material. The result will be a general expression that relates the stress, strain, electric field, and electric displacement within the material in all three directions. As we will see, the relationships will be expressed in terms of matrices that represent the mechanical compliance matrix, dielectric permittivity matrix, and matrix of piezoelectric strain coefficients.

Consider once again a cube of piezoelectric material, although in this discussion we make no assumptions regarding the direction in which the electric field is applied or the directions in which the material is producing stress or strain. We define a coordinate system in which three directions are specified numerically, and we use the common convention that the 3 direction is aligned along the poling axis of the material (Figure 4.9).

We see from the figure that there are three directions in which we can apply an electric field. We label these directions E_i, where $i = 1, 2, 3$, and express these fields

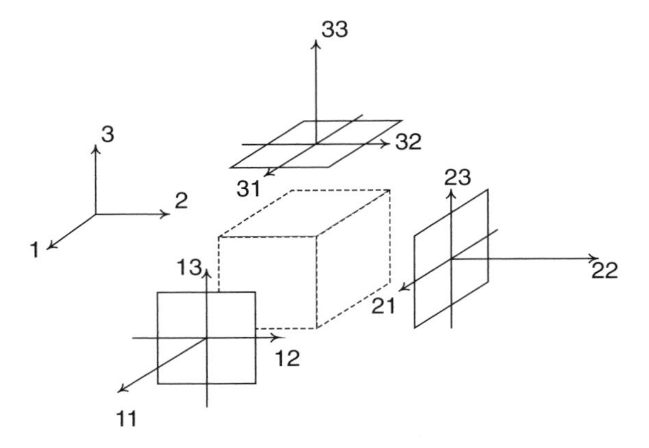

Figure 4.9 Piezoelectric cube indicating the coordinate axes of the three-dimensional analysis.

in terms of the electric field vector:

$$\underline{\mathrm{E}} = \begin{Bmatrix} \mathrm{E}_1 \\ \mathrm{E}_2 \\ \mathrm{E}_3 \end{Bmatrix}. \tag{4.25}$$

Similarly, we note that there are three directions in which we can produce electric displacement within the material. These directions are expressed in terms of the vector

$$\underline{\mathrm{D}} = \begin{Bmatrix} \mathrm{D}_1 \\ \mathrm{D}_2 \\ \mathrm{D}_3 \end{Bmatrix}. \tag{4.26}$$

The fact that there are three directions associated with the electric field and three associated with the electric displacement means that the general relationship between the variables takes the form

$$\mathrm{D}_1 = \varepsilon_{11}^{\mathrm{T}}\mathrm{E}_1 + \varepsilon_{12}^{\mathrm{T}}\mathrm{E}_2 + \varepsilon_{13}^{\mathrm{T}}\mathrm{E}_3 \tag{4.27}$$
$$\mathrm{D}_2 = \varepsilon_{21}^{\mathrm{T}}\mathrm{E}_1 + \varepsilon_{22}^{\mathrm{T}}\mathrm{E}_2 + \varepsilon_{23}^{\mathrm{T}}\mathrm{E}_3 \tag{4.28}$$
$$\mathrm{D}_3 = \varepsilon_{31}^{\mathrm{T}}\mathrm{E}_1 + \varepsilon_{32}^{\mathrm{T}}\mathrm{E}_2 + \varepsilon_{33}^{\mathrm{T}}\mathrm{E}_3. \tag{4.29}$$

These expressions can be stated concisely in indicial notation:

$$\mathrm{D}_m = \varepsilon_{mn}^{\mathrm{T}}\mathrm{E}_n. \tag{4.30}$$

The equations that relate strain to stress in the three-dimensional case can be derived in a similar fashion. In the case of a general state of stress and strain for the cube of material, we see that nine terms are required for complete specification. The components of stress and strain that are normal to the surfaces of the cube are denoted T_{11}, T_{22}, T_{33} and S_{11}, S_{22}, S_{33}, respectively. There are six shear components, T_{12}, T_{13}, T_{23}, T_{21}, T_{32}, T_{31} and S_{12}, S_{13}, S_{23}, S_{21}, S_{32}, S_{31}. For a linear elastic material, we can relate the strain to the stress with the tensor expression,

$$S_{ij} = \mathcal{S}^{E}_{ijkl} T_{kl}, \tag{4.31}$$

where the tensor $\mathcal{S}_{ijkl}$ represents 81 mechanical compliance terms.

The last step in writing the three-dimensional constitutive relationships is to specify the coupling between the electrical and mechanical variables. In the most general case, we see that the nine states of strain are related to the three applied electric fields through the expression

$$S_{ij} = \mathcal{D}_{ijn} E_n \tag{4.32}$$

and the three electric displacement terms are related to the mechanical stress through the expression

$$D_m = \mathcal{D}_{mkl} T_{kl}. \tag{4.33}$$

Combining the previous four expressions, we can write the complete set of constitutive equations for a linear piezoelectric material:

$$S_{ij} = \mathcal{S}^{E}_{ijkl} T_{kl} + \mathcal{D}_{ijn} E_n \tag{4.34}$$

$$D_m = \mathcal{D}_{mkl} T_{kl} + \varepsilon^{T}_{mn} E_n. \tag{4.35}$$

The complete set of equations are defined by 81 mechanical compliance constants, 27 piezoelectric strain coefficient values, and 9 dielectric permittivities.

4.3.1 Compact Notation for Piezoelectric Constitutive Equations

Equations (4.34) and (4.35) represent the full set of constitutive relationships for a piezoelectric material. They are written in tensor, or indicial, notation, due to the fact they represent a relationship between the nine mechanical field variables (either stress or strain) and the three variables associated with the electric properties.

For analyses of piezoelectric materials based on first principles, it is always wise to use the indicial form of the constitutive equations. Recall from Chapter 2 that we can use a more compact form of the constitutive equations that allows us to write the constitutive equations in matrix form without the need for indicial notation. The compact form of the constitutive relationships is based on the fact that the stress and strain tensors are symmetric; therefore,

$$T_{ij} = T_{ji} \tag{4.36}$$

$$S_{ij} = S_{ji}. \tag{4.37}$$

With the symmetry of the stress and strain tensor in mind, we realize that instead of nine independent elements of stress and strain, we really only have six independent elements. Thus, we can define the following set of new stress and strain components:

$$S_1 = S_{11} \qquad T_1 = T_{11} \tag{4.38}$$
$$S_2 = S_{22} \qquad T_2 = T_{22} \tag{4.39}$$
$$S_3 = S_{33} \qquad T_3 = T_{33} \tag{4.40}$$
$$S_4 = S_{23} + S_{32} \qquad T_4 = T_{23} = T_{32} \tag{4.41}$$
$$S_5 = S_{31} + S_{13} \qquad T_5 = T_{31} = T_{13} \tag{4.42}$$
$$S_6 = S_{12} + S_{21} \qquad T_6 = T_{12} = T_{21}. \tag{4.43}$$

With these definitions, we can write the constitutive equations in a much more compact notation:

$$S_i = s_{ij}^{E} T_j + d_{ik} E_k \tag{4.44}$$
$$D_m = d_{mj} T_j + \varepsilon_{mk}^{T} E_n, \tag{4.45}$$

where i and j take on values between 1 and 6, and m and n take on values between 1 and 3. This notation highlights that there are only 36 independent elastic constants, 18 piezoelectric strain coefficients, and 9 dielectric permittivity values that characterize a piezoelectric material. We can expand these equations into the form

$$\begin{Bmatrix} S_1 \\ S_2 \\ S_3 \\ S_4 \\ S_5 \\ S_6 \end{Bmatrix} = \begin{bmatrix} s_{11} & s_{12} & s_{13} & s_{14} & s_{15} & s_{16} \\ s_{21} & s_{22} & s_{23} & s_{24} & s_{25} & s_{26} \\ s_{31} & s_{32} & s_{33} & s_{34} & s_{35} & s_{36} \\ s_{41} & s_{42} & s_{43} & s_{44} & s_{45} & s_{46} \\ s_{51} & s_{52} & s_{53} & s_{54} & s_{55} & s_{56} \\ s_{61} & s_{62} & s_{63} & s_{64} & s_{65} & s_{66} \end{bmatrix} \begin{Bmatrix} T_1 \\ T_2 \\ T_3 \\ T_4 \\ T_5 \\ T_6 \end{Bmatrix} + \begin{bmatrix} d_{11} & d_{12} & d_{13} \\ d_{21} & d_{22} & d_{23} \\ d_{31} & d_{32} & d_{33} \\ d_{41} & d_{42} & d_{43} \\ d_{51} & d_{52} & d_{53} \\ d_{61} & d_{62} & d_{63} \end{bmatrix} \begin{Bmatrix} E_1 \\ E_2 \\ E_3 \end{Bmatrix}$$

$$\begin{Bmatrix} D_1 \\ D_2 \\ D_3 \end{Bmatrix} = \begin{bmatrix} d_{11} & d_{12} & d_{13} & d_{14} & d_{15} & d_{16} \\ d_{21} & d_{22} & d_{23} & d_{24} & d_{25} & d_{26} \\ d_{31} & d_{32} & d_{33} & d_{34} & d_{35} & d_{36} \end{bmatrix} \begin{Bmatrix} T_1 \\ T_2 \\ T_3 \\ T_4 \\ T_5 \\ T_6 \end{Bmatrix} + \begin{bmatrix} \varepsilon_{11} & \varepsilon_{12} & \varepsilon_{13} \\ \varepsilon_{21} & \varepsilon_{22} & \varepsilon_{23} \\ \varepsilon_{31} & \varepsilon_{32} & \varepsilon_{33} \end{bmatrix} \begin{Bmatrix} E_1 \\ E_2 \\ E_3 \end{Bmatrix}. \tag{4.46}$$

Visualizing the expression in this manner, we see that we can write the compact form of the constitutive equations as a matrix expression,

$$\begin{aligned} \underline{S} &= \mathbf{s}^{E}\, \underline{T} + \mathbf{d}'\, \underline{E} \\ \underline{D} &= \mathbf{d}\, \underline{T} + \varepsilon^{T}\, \underline{E}, \end{aligned} \tag{4.47}$$

where $\mathbf{s}^{E}$ is a 6×6 matrix of compliance coefficients, $\mathbf{d}$ is a 3×6 matrix of piezoelectric strain coefficients, and ε^{T} is a 3×3 matrix of dielectric permittivity values. The prime notation denotes a matrix transpose. Equations (4.46) and (4.47) represent the full constitutive relationships for a linear piezoelectric material. In compact notation we see that there are 63 coefficients that must be specified to relate stress, strain, electric field, and electric displacement.

The number of variables required to specify the constitutive properties of piezoelectric materials are reduced significantly by considering the symmetry associated with the elastic, electrical, and electromechanical properties. Many common piezoelectrics are *orthotropic materials*, for which the compliance elements

$$\begin{aligned} s_{ij} = s_{ji} = 0 \qquad i = 1, 2, 3 \;\; j = 4, 5, 6 \\ s_{45} = s_{46} = s_{56} = s_{65} = 0 \end{aligned} \tag{4.48}$$

For an orthotropic material the compliance matrix reduces to the form

$$\mathrm{s}^{\mathrm{E}} = \begin{bmatrix} \frac{1}{Y_1^{\mathrm{E}}} & -\frac{\nu_{12}}{Y_1^{\mathrm{E}}} & -\frac{\nu_{13}}{Y_1^{\mathrm{E}}} & 0 & 0 & 0 \\ -\frac{\nu_{21}}{Y_2^{\mathrm{E}}} & \frac{1}{Y_2^{\mathrm{E}}} & -\frac{\nu_{23}}{Y_2^{\mathrm{E}}} & 0 & 0 & 0 \\ -\frac{\nu_{31}}{Y_3^{\mathrm{E}}} & -\frac{\nu_{32}}{Y_3^{\mathrm{E}}} & \frac{1}{Y_3^{\mathrm{E}}} & 0 & 0 & 0 \\ 0 & 0 & 0 & \frac{1}{G_{23}^{\mathrm{E}}} & 0 & 0 \\ 0 & 0 & 0 & 0 & \frac{1}{G_{13}^{\mathrm{E}}} & 0 \\ 0 & 0 & 0 & 0 & 0 & \frac{1}{G_{12}^{\mathrm{E}}} \end{bmatrix}, \tag{4.49}$$

where Y_i^{E}, $i = 1, 2, 3$ are the short-circuit elastic moduli in the 1, 2, and 3 directions, respectively; the ν_{ij} are Poisson's ratio of transverse strain in the j direction to the axial strain in the i direction when stressed in the i direction; and G_{23}^{E}, G_{13}^{E}, and G_{12}^{E} are the short-circuit shear moduli. The symmetry of the compliance matrix requires that

$$\frac{\nu_{ij}}{Y_i^{\mathrm{E}}} = \frac{\nu_{ji}}{Y_j^{\mathrm{E}}} \qquad i, j = 1, 2, 3. \tag{4.50}$$

Piezoelectric materials exhibit a plane of symmetry such that the elastic moduli in the 1 and 2 directions are equal,

$$Y_1^{\mathrm{E}} = Y_2^{\mathrm{E}}, \tag{4.51}$$

and therefore the compliance matrix s^E is reduced to

$$\mathbf{s}^{\mathrm{E}} = \begin{bmatrix} \frac{1}{Y_1^{\mathrm{E}}} & -\frac{\nu_{12}}{Y_1^{\mathrm{E}}} & -\frac{\nu_{13}}{Y_1^{\mathrm{E}}} & 0 & 0 & 0 \\ -\frac{\nu_{12}}{Y_1^{\mathrm{E}}} & \frac{1}{Y_1^{\mathrm{E}}} & -\frac{\nu_{23}}{Y_1^{\mathrm{E}}} & 0 & 0 & 0 \\ -\frac{\nu_{31}}{Y_3^{\mathrm{E}}} & -\frac{\nu_{32}}{Y_3^{\mathrm{E}}} & \frac{1}{Y_3^{\mathrm{E}}} & 0 & 0 & 0 \\ 0 & 0 & 0 & \frac{1}{G_{23}^{\mathrm{E}}} & 0 & 0 \\ 0 & 0 & 0 & 0 & \frac{1}{G_{13}^{\mathrm{E}}} & 0 \\ 0 & 0 & 0 & 0 & 0 & \frac{1}{G_{12}^{\mathrm{E}}} \end{bmatrix}. \tag{4.52}$$

Symmetry within the crystal structure of the piezoelectric produces similar reductions in the number of electromechanical and electrical parameters. Since electric fields applied in a particular direction will not produce electric displacements in orthogonal directions, the permittivity matrix reduces to a diagonal matrix of the form

$$\varepsilon = \begin{bmatrix} \varepsilon_{11} & 0 & 0 \\ 0 & \varepsilon_{22} & 0 \\ 0 & 0 & \varepsilon_{33} \end{bmatrix}. \tag{4.53}$$

Similarly, the strain coefficient matrix for typical piezoelectric materials reduces to

$$\mathbf{d} = \begin{bmatrix} 0 & 0 & 0 & 0 & d_{15} & 0 \\ 0 & 0 & 0 & d_{24} & 0 & 0 \\ d_{13} & d_{23} & d_{33} & 0 & 0 & 0 \end{bmatrix}. \tag{4.54}$$

Further reductions in the number of independent coefficients can occur when the material exhibits symmetry such that $d_{13} = d_{23}$ and $d_{15} = d_{24}$.

Combining equations (4.52), (4.53), and (4.54) allows us to rewrite the constitutive properties of the piezoelectric material as

$$
\begin{Bmatrix} S_1 \\ S_2 \\ S_3 \\ S_4 \\ S_5 \\ S_6 \end{Bmatrix} = \begin{bmatrix} \frac{1}{Y_1^E} & -\frac{\nu_{12}}{Y_1^E} & -\frac{\nu_{13}}{Y_1^E} & 0 & 0 & 0 \\ -\frac{\nu_{12}}{Y_1^E} & \frac{1}{Y_1^E} & -\frac{\nu_{23}}{Y_1^E} & 0 & 0 & 0 \\ -\frac{\nu_{31}}{Y_3^E} & -\frac{\nu_{32}}{Y_3^E} & \frac{1}{Y_3^E} & 0 & 0 & 0 \\ 0 & 0 & 0 & \frac{1}{G_{23}^E} & 0 & 0 \\ 0 & 0 & 0 & 0 & \frac{1}{G_{13}^E} & 0 \\ 0 & 0 & 0 & 0 & 0 & \frac{1}{G_{12}^E} \end{bmatrix} \begin{Bmatrix} T_1 \\ T_2 \\ T_3 \\ T_4 \\ T_5 \\ T_6 \end{Bmatrix} + \begin{bmatrix} 0 & 0 & d_{13} \\ 0 & 0 & d_{23} \\ 0 & 0 & d_{33} \\ 0 & d_{24} & 0 \\ d_{15} & 0 & 0 \\ 0 & 0 & 0 \end{bmatrix} \begin{Bmatrix} E_1 \\ E_2 \\ E_3 \end{Bmatrix}
$$

$$
\begin{Bmatrix} D_1 \\ D_2 \\ D_3 \end{Bmatrix} = \begin{bmatrix} 0 & 0 & 0 & 0 & d_{15} & 0 \\ 0 & 0 & 0 & d_{24} & 0 & 0 \\ d_{13} & d_{23} & d_{33} & 0 & 0 & 0 \end{bmatrix} \begin{Bmatrix} T_1 \\ T_2 \\ T_3 \\ T_4 \\ T_5 \\ T_6 \end{Bmatrix} + \begin{bmatrix} \varepsilon_{11} & 0 & 0 \\ 0 & \varepsilon_{22} & 0 \\ 0 & 0 & \varepsilon_{33} \end{bmatrix} \begin{Bmatrix} E_1 \\ E_2 \\ E_3 \end{Bmatrix}. \tag{4.55}
$$

Earlier in the chapter the piezoelectric coupling coefficient was introduced as a means of comparing the quality of a piezoelectric material and as a means of quantifying the electromechanical energy conversion properties. In actuality, there is not a single coupling coefficient for a piezoelectric material, but a group of coupling coefficients that are a function of the elastic boundary conditions imposed on the material. The definition of the coupling coefficient is

$$
k_{ij} = \frac{d_{ij}}{\sqrt{\varepsilon_{ii}^T s_{jj}^E}}. \tag{4.56}
$$

The material properties of a piezoelectric material can be obtained from information provided by vendors. Table 4.1 is a list of representative material properties for two different types of piezoelectric ceramics and a piezoelectric polymer film.

4.4 COMMON OPERATING MODES OF A PIEZOELECTRIC TRANSDUCER

In analyzing of systems with piezoelectric materials, it is wise to begin with the full constitutive relationships in either indicial notation or in compact matrix notation (i.e., Voigt notation). In many applications, though, we can reduce the full constitutive

Table 4.1 Representative piezoelectric material properties

Property	Unit	Symbol	APC 856	PZT-5H	PVDF
Relative dielectric constant	unitless	ε_r	4100	3800	12–13
Curie temperature	°C	T_c	150	250	
Coupling coefficient	unitless	k_{33}	0.73	0.75	
		k_{31}	0.36		0.12
		k_{15}	0.65		
Strain coefficient	10^{-12} C/N or m/V	d_{33}	620	650	−33
		$-d_{31}$	260	320	−23
		d_{15}	710		
Elastic compliance	10^{-12} m^2/N	s_{11}^E	15	16.1	250–500
		s_{33}^E	17	20	
Density	g/cm^3	ρ	7.5	7.8	1.78

relationships to a smaller subset of relationships, due to the assumptions associated with the problem at hand. In this section we examine some common problems in the development of piezoelectric transducers for use as sensors or actuators. For each problem we first simplify the full constitutive relationships into a smaller subset of expressions that will allow us to derive a set of relationships among force, displacement, charge, and voltage. For the purpose of this discussion, we denote these relationships the *transducer equations* for the piezoelectric device. The primary difference between constitutive relationships and transducer relationships is that the latter are a function of both the device geometry and the material parameters of the transducer.

4.4.1 33 Operating Mode

One of the most common operating modes of a piezoelectric device is the direction along the axis of polarization. As discussed earlier, the convention with piezoelectric materials is to align the 3 axis of the material in the direction of the polarization vector for the material. Let us assume that we have a small plate of piezoelectric material in which the area and thickness are defined as in Figure 4.10. Thus, the 1 and 2 directions are in the plane of the transducer. If we make the following assumptions

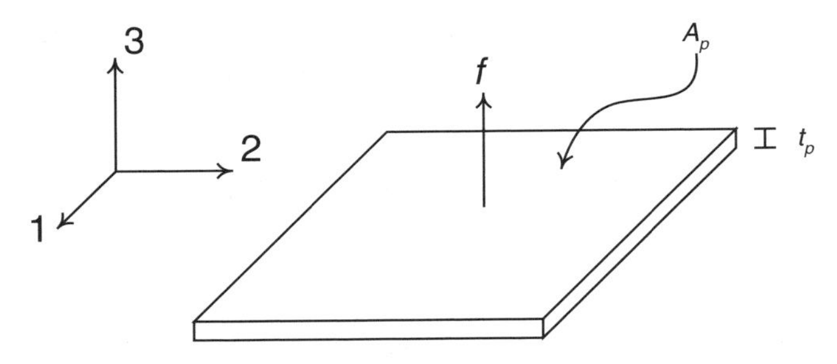

Figure 4.10 Piezoelectric plate.

regarding the state of stress and electric field within the material:

$$\begin{aligned} \mathrm{T_1} = \mathrm{E_1} &= 0 \\ \mathrm{T_2} = \mathrm{E_2} &= 0 \\ \mathrm{T_4} &= 0 \\ \mathrm{T_5} &= 0 \\ \mathrm{T_6} &= 0, \end{aligned} \tag{4.57}$$

the only nonzero stress and electric field are in the 3 direction. Under these assumptions, equation (4.55) is reduced to

$$\begin{aligned} \mathrm{S_1} &= -\frac{\nu_{13}}{Y_1^{\mathrm{E}}}\mathrm{T_3} + d_{13}\mathrm{E_3} \\ \mathrm{S_2} &= -\frac{\nu_{23}}{Y_1^{\mathrm{E}}}\mathrm{T_3} + d_{23}\mathrm{E_3} \\ \mathrm{S_3} &= \frac{1}{Y_3^{\mathrm{E}}}\mathrm{T_3} + d_{33}\mathrm{E_3} \\ \mathrm{D_3} &= d_{33}\mathrm{T_3} + \varepsilon_{33}^{\mathrm{T}}\mathrm{E_3}. \end{aligned} \tag{4.58}$$

These four equations define the state of strain and electric displacement in the piezoelectric material as a function of the stress and electric field prescribed. In this form of the constitutive equations, the stress and electric field are the *independent variables*, and strain and electric displacement are the *dependent variables*.

The expressions in equation (4.58) can be used in multiple ways to solve problems relating to piezoelectric materials and devices. The strain and electric displacement can be solved for by defining the state of stress, $\mathrm{T_3}$, and the electric field, $\mathrm{E_3}$. If one dependent variable (strain or electric displacement) in any direction is prescribed, a specific relationship between stress and electric can be specified and substituted into the remaining expressions. In this case we eliminate one of the independent variables in the equations. Similarly, if two dependent variables are prescribed, the stress and electric field are specified and the remaining dependent variables can be solved for in terms of the variables prescribed.

In most cases, for a 33 piezoelectric device we are interested in analyzing the state of the piezoelectric material in the 3 direction. In this case we prescribe electrical and mechanical boundary conditions by specifying $\mathrm{S_3}$ and $\mathrm{D_3}$ and solve for the stress and electric field, or we simply prescribe the stress and electric field and solve for the strain and electric displacement in all directions. For these analyses we can ignore the first two expressions in equation (4.58) and write the constitutive equations as

$$\begin{aligned} \mathrm{S_3} &= \frac{1}{Y_3^{\mathrm{E}}}\mathrm{T_3} + d_{33}\mathrm{E_3} \\ \mathrm{D_3} &= d_{33}\mathrm{T_3} + \varepsilon_{33}^{\mathrm{T}}\mathrm{E_3}. \end{aligned} \tag{4.59}$$

The expressions in equation (4.59) can be related directly back to the discussion of piezoelectricity in the first sections of this chapter. Earlier in the chapter the electromechanical coupling properties of piezoelectric materials were described in a single dimension, and it was shown that they could be written as a coupled relationship between stress, strain, electric field, and electric displacement. It is now clear that the full constitutive equations for a piezoelectric material are reduced to a set of equations that are identical to the one-dimensional analysis discussed earlier in the chapter by properly choosing the mechanical and electrical boundary conditions. In a manner similar to the discussion at the outset of the chapter, the constitutive relationships for the 33 operating mode can be written as a matrix expression:

$$\begin{pmatrix} \mathrm{S}_3 \\ \mathrm{D}_3 \end{pmatrix} = \begin{bmatrix} \dfrac{1}{Y_3^{\mathrm{E}}} & d_{33} \\ d_{33} & \varepsilon_{33}^{\mathrm{T}} \end{bmatrix} \begin{pmatrix} \mathrm{T}_3 \\ \mathrm{E}_3 \end{pmatrix}. \tag{4.60}$$

The piezoelectric coupling coefficient for this set of boundary conditions is

$$k_{33} = \frac{d_{33}\sqrt{Y_3^{\mathrm{E}}}}{\sqrt{\varepsilon_{33}^{\mathrm{T}}}} = \frac{d_{33}}{\sqrt{s_{33}^{\mathrm{E}}\varepsilon_{33}^{\mathrm{T}}}}. \tag{4.61}$$

The reduced constitutive equations can be used to define several important design parameters for a device operating in the 33 mode. The *free strain* is defined as the strain produced when there is no resistance stress on the material. Under this mechanical boundary condition, we assume that $\mathrm{T}_3 = 0$ and the strain produced is

$$\begin{aligned} \mathrm{S}_1|_{\mathrm{T}_3=0} &= d_{13}\mathrm{E}_3 \\ \mathrm{S}_2|_{\mathrm{T}_3=0} &= d_{23}\mathrm{E}_3 \\ \mathrm{S}_3|_{\mathrm{T}_3=0} &= d_{33}\mathrm{E}_3. \end{aligned} \tag{4.62}$$

The *blocked stress* is the stress produced by the material when the strain S_3 is constrained to be zero. With this assumption, $\mathrm{S}_3 = 0$, and the stress induced is

$$\mathrm{T}_3|_{\mathrm{S}_3=0} = -Y_3^{\mathrm{E}} d_{33}\mathrm{E}_3. \tag{4.63}$$

Blocking the strain of the material in the 3 direction produces strain in the orthogonal directions. The strain can be computed by substituting equation (4.63) into the first two expressions in equation (4.58),

$$\begin{aligned} \mathrm{S}_1|_{\mathrm{S}_3=0} &= \left(\nu_{13}\frac{Y_3^{\mathrm{E}}}{Y_1^{\mathrm{E}}}d_{33} + d_{13}\right)\mathrm{E}_3 \\ \mathrm{S}_2|_{\mathrm{S}_3=0} &= \left(\nu_{23}\frac{Y_3^{\mathrm{E}}}{Y_1^{\mathrm{E}}}d_{33} + d_{23}\right)\mathrm{E}_3. \end{aligned} \tag{4.64}$$

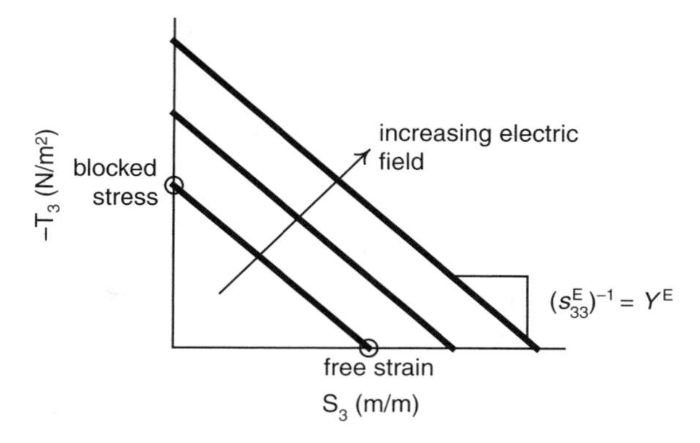

Figure 4.11 Stress–strain relationship as a function of the electric field.

The free strain and blocked stress are two common ways of characterizing the performance of a material. A general expression for the relationship between stress and strain in the material as a function of the electric field is

$$T_3 = Y_3^E S_3 - Y_3^E d_{33} E_3. \tag{4.65}$$

This relationship can be plotted to yield a design curve that indicates the amount of stress that can be produced by the material as a function of the strain and electric field. Typically, the plot is shown as a relationship between the stress *produced by the material*, which in our notation is actually $-T_3$. Thus, if we plot $-T_3$ as a function of S_3 and E_3, the curve is as shown in Figure 4.11. We see that the function is a straight line with a slope equal to the short-circuit elastic modulus. The intercept along the $S_3 = 0$ line is the blocked stress, and the intercept along the $T_3 = 0$ line is the free strain. For any point in between these two extremes, we see that the stress and strain are always less than these values. Thus, the blocked stress represents the maximum achievable stress, and the free strain represents the maximum achievable strain. Figure 4.11 also illustrates that the operating curves for the piezoelectric device are increased as a function of electric field. Increasing the applied electric field increases the achievable stress and strain linearly. Thus, the maximum stress and strain are produced at a maximum electric field.

Plotting the relationship between induced stress and strain in a piezoelectric device as shown in Figure 4.11 allows us to visualize an important property of piezoelectric materials. For a linear elastic material with linear piezoelectric properties, plotting the induced stress as a function of the induced strain produces a plot as shown in Figure 4.11. The area under any of the curves in Figure 4.11 is equal to the maximum energy per unit volume that can be produced by the device. This area is computed by multiplying one half of the blocked stress with the free strain, and is denoted the *volumetric energy density*, or sometimes simply the *energy density* of the material.

Denoting the volumetric energy density as E_v, we have

$$E_v = \frac{1}{2}\left(-\, \mathrm{T}_3|_{\mathrm{S}_3=0}\right)\left(\mathrm{S}_3|_{\mathrm{T}_3=0}\right) = \frac{1}{2}Y_3^{\mathrm{E}} d_{33}^2 \mathrm{E}_3^2 \tag{4.66}$$

for an orthotropic piezoelectric material. The energy density is an important figure of merit when comparing different types of piezoelectric materials and when comparing different materials with one another. Equation (4.66) illustrates that the energy density is an intrinsic property of the material since it does not depend on geometry. At equivalent electric fields we can form a figure of merit, $\frac{1}{2}Y^{\mathrm{E}}d_{33}^2$ and assess the relative ability of different materials to do mechanical work. A higher value of $\frac{1}{2}Y^{\mathrm{E}}d_{33}^2$ indicates that a material can perform more mechanical work at the same electric fields. This does not mean that it is necessarily a better material in every aspect: It may require much larger voltages, it may not work over a large temperature range, and so on, but it does indicate that the material has better intrinsic properties as an electromechanical actuator.

Example 4.4 Two piezoelectric materials are being considered for a mechanical actuation device. The first material being considered has a short-circuit modulus of 55 GPa and a piezoelectric strain coefficience of 425 pm/V. The second material being considered is a softer material that has a short-circuit modulus of 43 GPa but a strain coefficient of 450 pm/V. The maximum electric field that can be applied to the first material is 1.5 MV/m, while the second material is stable up to electric field values of 3 MV/m. (a) Plot the relationship between stress and strain induced for both of these materials at maximum applied electric field. (b) Compute the figure of merit $\frac{1}{2}Y^{\mathrm{E}}d_{33}^2$ for both materials. (c) Compute the energy density at maximum field.

Solution (a) Plotting the relationship between induced stress and strain requires computation of the blocked stress and free strain in both sets of materials. The free strain is computed from equation (4.62):

$$\text{material 1:} \quad \mathrm{S}_3|_{\mathrm{T}_3=0} = (425 \times 10^{-12}\ \mathrm{m/V})(1.5 \times 10^6\ \mathrm{V/m}) = 637.5\ \mu\text{strain}$$

$$\text{material 2:} \quad \mathrm{S}_3|_{\mathrm{T}_3=0} = (450 \times 10^{-12}\ \mathrm{m/V})(3 \times 10^6\ \mathrm{V/m}) = 1350\ \mu\text{strain.}$$

The blocked stress is computed from equation (4.63) or by noting that the blocked stress is equal to the product of the short-circuit elastic modulus and the free strain. The results are:

$$\text{material 1:} \quad -\,\mathrm{T}_3|_{\mathrm{S}_3=0} = (55 \times 10^9\ \mathrm{N/m^2})(637.5 \times 10^{-6}\ \mathrm{m/m}) = 35.1\ \mathrm{MPa}$$

$$\text{material 2:} \quad -\,\mathrm{T}_3|_{\mathrm{S}_3=0} = (43 \times 10^9\ \mathrm{N/m^2})(1350 \times 10^{-6}\ \mathrm{m/m}) = 58.1\ \mathrm{MPa.}$$

These results are plotted in Figure 4.12 using the computed values.

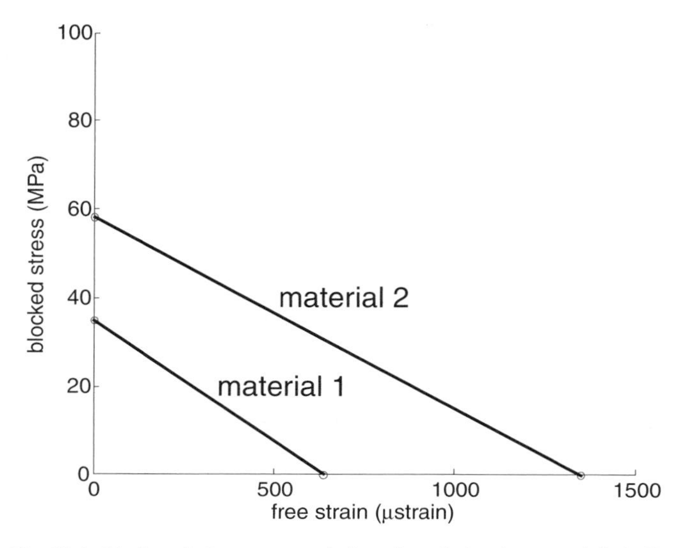

Figure 4.12 Plot of induced stress versus induced strain for the materials in Example 4.4.

(b) The figure of merit $\frac{1}{2}Y^{\mathrm{E}}d_{33}^2$ is equal to:

$$\begin{aligned}
\text{material 1:} \quad \frac{1}{2}Y^{\mathrm{E}}d_{33}^2 &= \frac{1}{2}(55 \times 10^9 \text{ N/m}^2)(425 \times 10^{-12} \text{ m/V})^2 \\
&= 4.97 \times 10^{-9} \text{ J/m}^3/(\text{V/m})^2 \\
\text{material 2:} \quad \frac{1}{2}Y^{\mathrm{E}}d_{33}^2 &= \frac{1}{2}(43 \times 10^9 \text{ N/m}^2)(450 \times 10^{-12} \text{ m/V})^2 \\
&= 4.35 \times 10^{-9} \text{ J/m}^3/(\text{V/m})^2.
\end{aligned}$$

(c) The maximum energy density is the result of the calculation in part (b) multiplied by the maximum electric field applied to the material:

$$\begin{aligned}
E_v &= [4.97 \times 10^{-9} \text{ J/m}^3/(\text{V/m})^2](1.5 \times 10^6 \text{ V/m})^2 = 11.18 \text{ kJ/m}^3 \\
E_v &= [4.35 \times 10^{-9} \text{ J/m}^3/(\text{V/m})^2](3 \times 10^6 \text{ V/m})^2 = 39.2 \text{ kJ/m}^3.
\end{aligned}$$

The volumetric energy density computations match the graphical results shown in Figure 4.12. The curve that has the largest area (material 2) has the largest energy density.

4.4.2 Transducer Equations for a 33 Piezoelectric Device

Often when designing a device that uses a 33-mode piezoelectric transducer, we do not want to work from the constitutive relationships but from a set of relationships that relate directly the force, displacement, voltage, and charge. To derive a relationship among force, displacement, voltage, and charge, first relate these variables to the field variables of the material. Assuming that the strain in the 3 direction is uniform,

$$S_3 = \frac{u_3}{t_p}. \tag{4.67}$$

If the stress is uniform over the surface, then

$$T_3 = \frac{f}{A_p}. \tag{4.68}$$

A uniform electric field is related to the applied potential, v, through the expression

$$E_3 = \frac{v}{t_p}. \tag{4.69}$$

Assuming a uniform electrode surface, the electric displacement can be related to the charge through

$$D_3 = \frac{q}{A_p}. \tag{4.70}$$

Substituting expressions (4.67) to (4.70) into equation (4.59) yields

$$\frac{u_3}{t_p} = \frac{1}{Y_3^E}\frac{f}{A_p} + d_{33}\frac{v}{t_p} \tag{4.71}$$

$$\frac{q}{A_p} = d_{33}\frac{f}{A_p} + \varepsilon_{33}^T\frac{v}{t_p}. \tag{4.72}$$

Rearranging the expressions produces the transducer relationships for a 33-mode piezoelectric device:

$$\begin{aligned} u_3 &= \frac{t_p}{Y_3^E A_p} f + d_{33}\, v \\ q &= d_{33}\, f + \frac{\varepsilon_{33}^T A_p}{t_p}\, v. \end{aligned} \tag{4.73}$$

Important design information can be derived from equation (4.73). The *free displacement*, δ_o, of the transducer is equal to

$$\delta_o = u_3|_{f=0} = d_{33} v \tag{4.74}$$

and is the displacement of the material in the 3 direction when there is no restraining force. Similarly, the blocked force is

$$f_{bl} = -\ f|_{u_3=0} = d_{33}Y_3^E \frac{A_p}{t_p} v. \tag{4.75}$$

Relationships can also be determined for the design of piezoelectric sensors. The charge produced by an applied force is equal to

$$q = d_{33} f \tag{4.76}$$

when a circuit has been designed to produce zero potential across the material. Similarly, the strain-sensing properties of the material are defined by the expression

$$q = d_{33}Y_3^E \frac{A_p}{t_p} u_3 \tag{4.77}$$

when the potential is also zero.

Example 4.5 Compute the blocked force and free displacement produced by a piezoelectric device with a length and width of 2 mm and a thickness of 0.25 mm. The applied voltage is 50 V. For the calculation, use the piezoelectric properties for APC 856.

Solution The expression for the blocked force is given in equation (4.75). To compute the blocked force, we need to know the mechanical compliance with zero field and the piezoelectric strain coefficient in the 33 direction. Both of these parameters are given in Table 4.1:

$$\begin{aligned} s_{33}^E &= \frac{1}{Y_3^E} = 17 \times 10^{-12}\ \mathrm{m^2/N} \\ d_{33} &= 620 \times 10^{-12}\ \mathrm{m/V}. \end{aligned}$$

Substituting into equation (4.75), we have

$$f_{bl} = \frac{(620 \times 10^{-12}\ \mathrm{m/V})(2 \times 10^{-3}\ \mathrm{m})^2}{(17 \times 10^{-12}\ \mathrm{m^2/N})(0.25 \times 10^{-3}\ \mathrm{m})} (50\ \mathrm{V}) = 29\ \mathrm{N}.$$

The free displacement can be computed from equation (4.74):

$$\delta_o = (620 \times 10^{-12}\ \mathrm{m/V})\ (50\ \mathrm{V}) = 31 \times 10^{-9}\ \mathrm{m} = 31\ \mathrm{nm}.$$

4.4.3 Piezoelectric Stack Actuator

From our derivation of the transducer equations for a piezoelectric material operating in the 33 mode, we realize that the amount of strain in the piezoelectric material at a specified electric field is limited by the piezoelectric strain coefficient, d_{33}. For a piezoelectric material with a strain coefficient on the order of 300 to 700 pm/V and a maximum field of 1 to 2 MV/m, a straightforward computation will demonstrate that the strain of the material in the 3 direction is on the order of 0.1 to 0.2%. For transducers of thickness on the order of 250 μm, displacement in the 3 direction is only on the order of tens to hundreds of nanometers. Although displacements in this range can be very useful for precise positioning applications, it is often of interest to develop a piezoelectric device that achieves one to two orders of magnitude more displacement than that of a single piezoelectric plate.

The need to increase the displacement of a piezoelectric device has lead to the development of *piezoelectric stack* actuators. As the name implies, a piezoelectric stack consists of multiple layers of piezoelectric plates placed on top of one another. The electrical connections of the device are wired such that the same voltage (and electric field) is placed across each layer. The stack geometry produces an amplification of the displacement since each layer (ideally) will displace the same amount. Also important is the fact that the force associated with the stack will be equivalent to the force of a single layer (again in the ideal case). Thus, a multilayer piezoelectric stack is a means of producing larger displacements than a single layer without a reduction in the achievable force.

The properties of a piezoelectric stack are analyzed by extending the results for a 33-mode piezoelectric device. The equations that relate force, displacement, charge, and voltage in a 33-mode transducer are shown in equation (4.73). Consider a device that consists of n layers of piezoelectric material connected electrically in parallel but mechanically in series, as shown in Figure 4.13. The displacement of the stack, u_s, is equal to the sum of the displacements of the individual layers; therefore, we have

$$u_s = nu_3 = \frac{nt_p}{Y_3^E A_p} f + nd_{33}v. \tag{4.78}$$

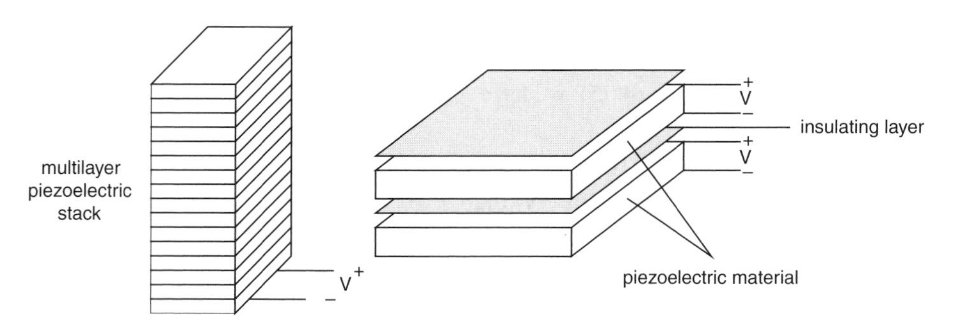

Figure 4.13 Piezoelectric stack actuator.

If we define the total length of the stack as $L_s = nt_p$, then

$$u_s = \frac{L_s}{Y_3^{\mathrm{E}} A_p} f + \frac{d_{33} L_s}{t_p} v. \tag{4.79}$$

The coefficient in front of the force term is recognized to be the inverse of the short-circuit mechanical stiffness. The coefficient in front of the voltage term indicates that we can achieve greater displacements by increase the ratio of L_s to t_p, which is equivalent to saying that the displacement is increased by increasing the number of layers in the stack.

Examining the electrical connections we see that the layers are connected in parallel. From basic electrical theory we know that parallel connection of capacitors produces a summation of the charge output of each layer, therefore the total charge output of the piezoelectric stack is

$$q_s = nq = nd_{33}\, f + n\frac{\varepsilon_{33}^{\mathrm{T}} A_p}{t_p}\, v. \tag{4.80}$$

Substituting the definition of the stack length into the expression yields

$$q_s = \frac{d_{33} L_s}{t_p}\, f + n\frac{\varepsilon_{33}^{\mathrm{T}} A_p}{t_p}\, v. \tag{4.81}$$

The coefficient in front of the voltage is recognized as the total stress-free capacitance of the device, which is simply a sum of the stress-free capacitance of the individual layers.

Equations (4.79) and (4.81) represent the transducer equations for a piezoelectric stack. These equations were derived with several assumptions, namely, that the mechanical and electric properties of the interfacial material are negligible, and that the material properties of each layer are identical. Setting the force and displacement equal to zero, respectively, allows us to derive the expressions for the free displacement of the stack and the blocked force:

$$\delta_o = d_{33}\frac{L_s}{t_p}\, v \tag{4.82}$$

$$f_{\mathrm{bl}} = d_{33} Y_3^{\mathrm{E}} \frac{A_p}{t_p} v. \tag{4.83}$$

Equations (4.82) and (4.83) are useful expressions for determining the geometry required for a stack to produce a specific amount of blocked force and free displacement.

4.4.4 Piezoelectric Stack Actuating a Linear Elastic Load

A common application of a piezoelectric stack is motion control in which a piezoelectric stack is applying a force against an object that can be modeled as an elastic load. The simplest model of an elastic object is a linear spring, as shown in Figure 4.14*a*.

To facilitate the analysis, rewrite equation (4.79) in the form

$$u_s = \frac{1}{k_s^E} f + u_o v, \tag{4.84}$$

where $k_s^E = Y_3^E A/L_s$ is the short-circuit stiffness of the piezoelectric stack and u_o is the free displacement of the stack per unit voltage input. For the case shown in Figure 4.14, $u_s = u$ and $f = -k_l u$, where k_l is the stiffness of the load. Substituting these expressions into equation (4.84), we have

$$u = -\frac{k_l}{k_s^E} u + u_o v. \tag{4.85}$$

Solving for the displacement as a function of the applied voltage, we have

$$u = \frac{u_o}{1 + k_l/k_s^E} v = \frac{\delta_o}{1 + k_l/k_s^E}. \tag{4.86}$$

From equation (4.86) we see that if the stiffness of the load is much less than the stiffness of the piezoelectric, $k_l/k_s^E \ll 1$ and the displacement is approximately equal to the free displacement of the stack. If the stiffness of the stack is much lower than the stiffness of the load, $k_l/k_s^E \gg 1$ and the displacement of the stack is much smaller than the free displacement. A plot of the normalized displacement as a function of

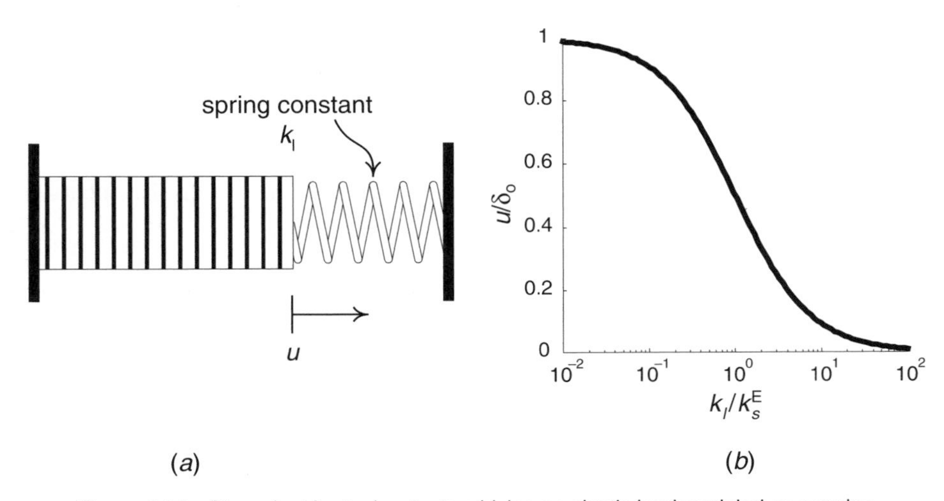

Figure 4.14 Piezoelectric stack actuator driving an elastic load modeled as a spring.

k_l/k_s^{E} is shown in Figure 4.14*b*. From the plot we see that the displacement becomes $\frac{1}{2}\delta_o$ when the stiffness of the load is equal to the stiffness of the actuator. This is sometimes referred to as the *stiffness match point* due to the fact that stiffness of the load and actuator are equal.

The stiffness match point is also important in the analysis of the force and work transferred from the piezoelectric to the load. If we solve for the blocked force from equation (4.84), we see that

$$f_{\mathrm{bl}} = \delta_o k_s^{\mathrm{E}}. \tag{4.87}$$

The force applied by the actuator to the load is $f = -k_l u$; therefore, we can solve for the ratio of the force to the blocked force as

$$\frac{f}{f_{\mathrm{bl}}} = \frac{k_l \delta_o / \left(1 + k_l/k_s^{\mathrm{E}}\right)}{\delta_o k_s^{\mathrm{E}}} = \frac{k_l/k_s^{\mathrm{E}}}{1 + k_l/k_s^{\mathrm{E}}}. \tag{4.88}$$

This function is plotted in Figure 4.15*a*. From the plot we see that the force output follows a trend that is opposite to the displacement. When the load is much softer than the actuator, $k_l/k_s^{\mathrm{E}} \ll 1$ and the output force is much less than the blocked force. In the opposite extreme we see that the force is nearly equivalent to the blocked force of the stack.

A final metric that is of importance is the output work of the stack. Recall that the work is defined as the produce of the force and displacement. For the piezoelectric stack we can define it as $W = fu$. The maximum work output of the device is equal to the product of the blocked force and the free displacement. If we normalize the output work with respect to the product of the blocked force and free displacement,

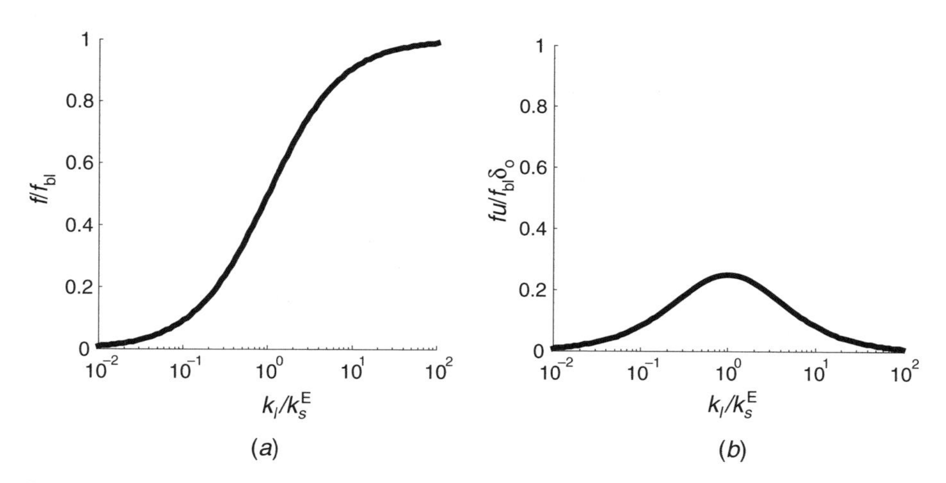

Figure 4.15 (*a*) Output force of a piezoelectric stack normalized with respect to the blocked force; (*b*) output work of a piezoelectric stack normalized with respect to the maximum possible work.

we can show that

$$\frac{fu}{f_{\text{bl}}\delta_o} = \frac{k_l/k_s^{\text{E}}}{\left(1 + k_l/k_s^{\text{E}}\right)^2}. \tag{4.89}$$

This function is plotted in Figure 4.15b and we see that the maximum output work is one-fourth of the product of the blocked force and the free displacement. Furthermore, the result demonstrates that this maximum occurs at the stiffness match between the actuator and the load. This is an important number to remember because it provides a straightforward method of computing the maximum energy transfer between the actuator and the elastic load if the blocked force and the free displacement are known. Often, the specifications from a manufacturer will list the blocked force and free displacement of the transducer. The maximum achievable energy transfer is then estimated as one-fourth of the product of f_{bl} and δ_o.

These results emphasize that there are three operating regimes when using the actuator to produce force and displacement on an elastic load. In the regime in which the actuator stiffness is much lower than the load stiffness, $k_l/k_s^{\text{E}} \ll 1$, and we see that very little load transfer occurs between the transducer and the load. This is often desirable when using the material as a *sensor* because it indicates that very little force is transferred to the load from the transducer. Thus, the load cannot "feel" the presence of the smart material, and its motion is not going to be affected by the presence of the transducer. At the opposite extreme we see that the motion of the transducer is not affected by the presence of the load; therefore, it is often desirable for $k_s^{\text{E}} \gg k_l$ for applications in motion control where the objective is to achieve maximum displacement in the piezoelectric actuator and the load. The regime in which $k_l \approx k_s^{\text{E}}$ is typically desirable when the application requires maximizing energy transfer between the transducer and the load. This is often the case when one is trying to design a system that dissipates energy in the load using a piezoelectric transducer.

Example 4.6 An application in motion control requires that a piezoelectric actuator produce 90 μm of displacement in a structural element that has a stiffness of 3 N/μm. The applications engineer has chosen a piezoelectric stack that produces a free displacement of 100 μm. Determine (a) the stiffness required to achieve 90 μm of displacement in the load, and (b) the amount of force produced in the load for this stiffness value.

Solution (a) From equation (4.86) we can solve for the stiffness ratio as a function of the free displacement,

$$\frac{k_l}{k_s^{\text{E}}} = \frac{\delta_o}{u} - 1,$$

and substitute in the values given in the problem:

$$\frac{k_l}{k_s^{\mathrm{E}}} = \frac{100\ \mu\mathrm{m}}{90\ \mu\mathrm{m}} - 1 = 0.11.$$

Solving for k_s^{E} yields

$$k_s^{\mathrm{E}} = \frac{3\ \mathrm{N}/\mu\mathrm{m}}{0.11} = 27.3\ \mathrm{N}/\mu\mathrm{m}.$$

(b) To compute the blocked force, we see from equation (4.87) that the blocked force is simply equal to the product of the free displacement and the stiffness; therefore,

$$f_{\mathrm{bl}} = (100\ \mu\mathrm{m})(27.3\ \mathrm{N}/\mu\mathrm{m}) = 2730\ \mathrm{N}.$$

The amount of force applied to the load can be computed from equation (4.88) as

$$f = (2730\ \mathrm{N})\,\frac{0.11}{1 + 0.11} = 270.5\ \mathrm{N}.$$

This design can be visualized by drawing the relationship between force and displacement for the stack and placing the *load line* that represents the stiffness of the load. This is illustrated in Figure 4.16, where the solid line represents the relationship between force and displacement for the actuator. It is a linear relationship with intercepts at 2730 N and 100 μm. The stiffness of the load is represented by the dashed line, which intercepts the force–displacement curve at a point defined by a displacement of 90 μm and 270 N. This is represented by a square in the figure.

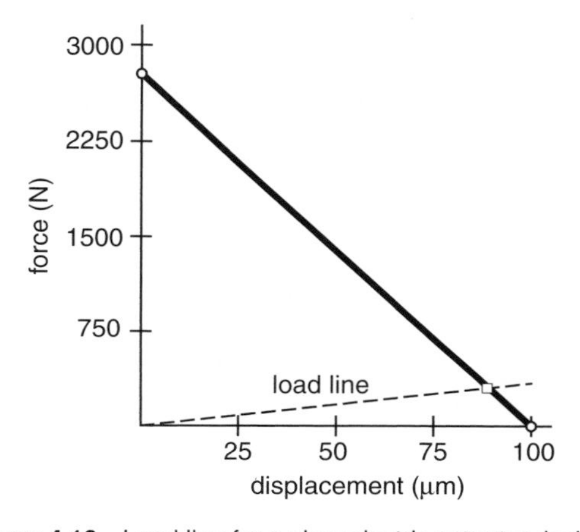

Figure 4.16 Load line for a piezoelectric actuator design.

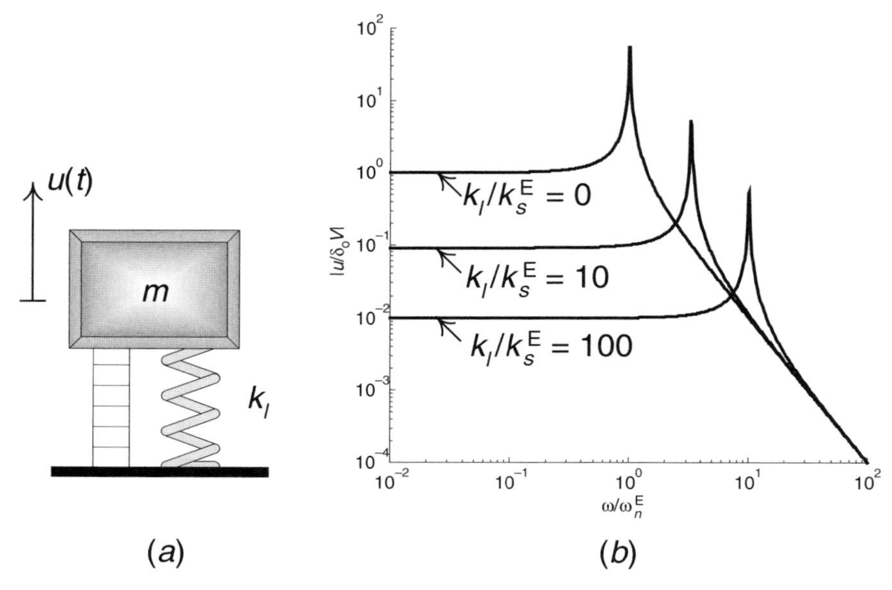

Figure 4.17 Piezoelectric stack actuating a mass–spring system.

The previous analysis concentrated on the static response of a piezoelectric stack actuator exciting a system modeled as an elastic spring. If we generalize the analysis such that the load is the mass–spring system shown in Figure 4.17*a*, the force applied to the piezoelectric stack is equal to

$$f(t) = -m\ddot{u}(t) - k_l u(t), \tag{4.90}$$

and equation (4.84) can be rewritten

$$u(t) = \frac{1}{k_s^{\mathrm{E}}}[-m\ddot{u}(t) - k_l u(t)] + u_o v(t). \tag{4.91}$$

Rewriting equation (4.91) as a mass–spring system, we have

$$m\ddot{u}(t) + \left(k_l + k_s^{\mathrm{E}}\right) u(t) = u_o k_s^{\mathrm{E}} v(t). \tag{4.92}$$

Taking the Laplace transform of equation (4.92) (assuming zero initial conditions) and rewriting as the transfer function $u(\mathsf{s})/v(\mathsf{s})$ yields

$$\frac{u(\mathsf{s})}{u_o v(\mathsf{s})} = \frac{k_s^{\mathrm{E}}/m}{\mathsf{s}^2 + \left(k_l + k_s^{\mathrm{E}}\right)/m}. \tag{4.93}$$

Denoting the ratio k_s^{E}/m as the square of the short-circuit natural frequency $\omega_n^{\mathrm{E}^2}$, we can write the transfer function in the frequency domain as

$$\frac{u(\omega)}{u_o v(\omega)} = \frac{1}{1 + k_l/k_s^{\mathrm{E}} - \left(\omega/\omega_n^{\mathrm{E}}\right)^2}. \tag{4.94}$$

The low-frequency asymptote of the frequency response as $\omega/\omega_n^{\mathrm{E}} \rightarrow 0$ matches the analysis previously performed for an elastic spring. This is expected because the analysis for an elastic spring is equivalent to the present analysis with $m = 0$.

A plot of the magnitude of the frequency response for three values of stiffness ratio is shown in Figure 4.17*b*. The low-frequency response is the quasistatic response which matches that of the analysis for an elastic spring. We note that the quasistatic response is flat, which indicates that the harmonic response at frequencies well below resonance has a peak amplitude equal to the static response. The sharp rise in the magnitude is associated with the resonant response of the system. The resonant frequency increases with increasing stiffness ratio, due to the additional stiffness of the load spring.

4.5 DYNAMIC FORCE AND MOTION SENSING

Piezoelectric materials are widely used as transducers for sensing motion and force, and the equations derived in Section 4.4 are readily adopted to analyze dynamic sensing properties. Consider the diagram shown in Figure 4.18*a*, in which a single layer of piezoelectric material is utilized in the 33 direction as a force sensor. The transducer is modeled as a mass with mass m and a time-dependent applied force $f(t)$. The transducer equations for a single layer of material are shown in equation (4.73). If the term $k_p^{\mathrm{E}} = Y_3^{\mathrm{E}} A_p/t_p$ is recognized as the short-circuit piezoelectric stiffness and we assume that the piezoelectric element has a short-circuit boundary condition

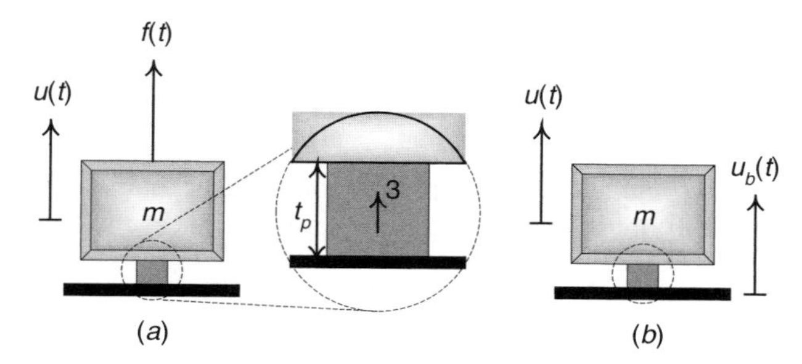

Figure 4.18 (*a*) Piezoelectric force sensor; (*b*) piezoelectric accelerometer.

$[v(t) = 0]$, the equations are written

$$u(t) = \frac{1}{k_p^{\mathrm{E}}} [f(t) - m\ddot{u}(t)]$$
$$q(t) = d_{33} \; [f(t) - m\ddot{u}(t)] . \tag{4.95}$$

Rewriting the top equation as a second-order differential equation and transforming into the Laplace domain yields

$$\frac{u(\mathrm{s})}{f(\mathrm{s})} = \frac{1}{m\mathrm{s}^2 + k_p^{\mathrm{E}}}. \tag{4.96}$$

The second equation can be rewritten as

$$q(\mathrm{s}) = d_{33} k_p^{\mathrm{E}} u(\mathrm{s}) = d_{33} \frac{k_p^{\mathrm{E}}}{m\mathrm{s}^2 + k_p^{\mathrm{E}}} f(\mathrm{s}). \tag{4.97}$$

Denoting the ratio of the short-circuit stiffness to the mass as the square of the short-circuit natural frequency, $\omega_n^{\mathrm{E}^2} = k_p^{\mathrm{E}}/m$, the charge output to an applied force can be written

$$\frac{q(\mathrm{s})}{f(\mathrm{s})} = d_{33} \frac{\omega_n^{\mathrm{E}^2}}{\mathrm{s}^2 + \omega_n^{\mathrm{E}^2}}. \tag{4.98}$$

Equation (4.98) illustrates that the quasistatic sensitivity of the force sensor is equal to the d_{33} coefficent of the material, and that the ratio of the stiffness to the mass limits the sensitivity at frequencies above the short-circuit natural frequency.

Piezoelectric materials are also used widely as motion sensors. A model system for a piezoelectric motion sensor is shown in Figure 4.18*b*. The base is assumed to have a prescribed motion $u_b(t)$, and the motion of the mass is denoted $u(t)$. In this configuration the strain in the polarization direction is equal to

$$\mathrm{S}_3 = \frac{u(t) - u_b(t)}{t_p}, \tag{4.99}$$

and the force applied to the piezoelectric is

$$f(t) = -m\ddot{u}(t). \tag{4.100}$$

Using equation (4.99) in the development of the transducer relationships and substituting equation (4.100) into the resulting expressions produces the equations

$$u(t) - u_b(t) = \frac{1}{k_p^{\mathrm{E}}}(-m\ddot{u}(t))$$
$$q(t) = d_{33}(-m\ddot{u}(t)). \tag{4.101}$$

Equation (4.101) assumes that the signal conditioning circuit produces a short-circuit (E = 0) condition across the piezoelectric.

Transforming the expressions in equation (4.101) into the Laplace domain and solving for the relationship between $u(\mathsf{s})$ and $u_b(\mathsf{s})$ yields

$$u(\mathsf{s}) = \frac{\omega_n^{\mathrm{E}^2}}{\mathsf{s}^2 + \omega_n^{\mathrm{E}^2}} u_b(\mathsf{s}). \tag{4.102}$$

Substituting equation (4.102) into the charge expression yields

$$q(\mathsf{s}) = -md_{33}\frac{\omega_n^{\mathrm{E}^2}}{\mathsf{s}^2 + \omega_n^{\mathrm{E}^2}}[\mathsf{s}^2 u_b(\mathsf{s})]. \tag{4.103}$$

The term $\mathsf{s}^2 u_b(\mathsf{s})$ is the Laplace transform of the base acceleration. The frequency response of $q(\mathsf{s})/[\mathsf{s}^2 u_b(\mathsf{s})]$ is a constant at low frequencies, exhibits a resonance at ω_n^{E}, and is then reduced above the resonance frequency. At low frequencies the charge output of the piezoelectric is approximated by

$$\frac{q(t)}{\ddot{u}_b(t)} = -md_{33}, \tag{4.104}$$

which is the sensitivity of the motion sensor. This analysis explains why a piezoelectric material is used to sense acceleration. The charge output of the piezoelectric material is proportional to the acceleration of the base through the *moving mass* m and the piezoelectric strain coefficient d_{33}.

Example 4.7 A piezoelectric plate with side dimensions of 2 mm × 2 mm and a thickness of 0.50 mm is being considered for the design of an accelerometer that measures into the low ultrasonic range (>20 kHz). The material being considered is APC 856, and the poling direction is in the direction of the motion. The design incorporates a 6-g moving mass. Compute the low-frequency sensitivity of the accelerometer and the short-circuit natural frequency.

Solution The low-frequency sensitivity of the accelerometer is obtained from equation (4.104). Assuming that the moving mass is 6 g, the low-frequency sensitivity is

$$\frac{q(t)}{\ddot{u}_b(t)} = (6 \times 10^{-3}\ \mathrm{kg})(620 \times 10^{-12}\ \mathrm{C/N}) = 3.72\ \mathrm{pC/m \cdot s^2}.$$

Computing the natural frequency requires that we compute the stiffness of the piezoelectric element. The short-circuit stiffness is computed from

$$k_p^{\mathrm{E}} = \frac{A}{s_{33}^{\mathrm{E}} t} = \frac{(2 \times 10^{-3}\ \mathrm{m})(2 \times 10^{-3}\ \mathrm{m})}{(17 \times 10^{-12}\ \mathrm{m^2/N})(0.5 \times 10^{-3}\ \mathrm{m})} = 470.6\ \mathrm{N}/\mu\mathrm{m}$$

and the short-circuit natural frequency is

$$\omega_n^{\mathrm{E}} = \sqrt{\frac{470.6 \times 10^6 \ \mathrm{N/m^2}}{6 \times 10^{-3} \ \mathrm{kg}}} = 2.8 \times 10^5 \ \mathrm{rad/s}.$$

The natural frequency in hertz is

$$f_n^{\mathrm{E}} = \frac{2.8 \times 10^5 \ \mathrm{rad/s}}{2\pi} = 44.6 \ \mathrm{kHz}.$$

4.6 31 OPERATING MODE OF A PIEZOELECTRIC DEVICE

Another common use of a piezoelectric device is to apply an electric field in the 3 direction but utilize the stress and strain produced in the 1 direction to create extension or bending in the material. In the use of the 31 operating mode of a piezoelectric, we assume that

$$\begin{aligned} \mathrm{E}_1 &= 0 \\ \mathrm{T}_3 &= 0 \\ \mathrm{T}_2 = \mathrm{E}_2 &= 0 \\ \mathrm{T}_4 &= 0 \\ \mathrm{T}_5 &= 0 \\ \mathrm{T}_6 &= 0. \end{aligned} \tag{4.105}$$

Under these assumptions, the constitutive equations are reduced to

$$\begin{aligned} \mathrm{S}_1 &= \frac{1}{Y_1^{\mathrm{E}}}\mathrm{T}_1 + d_{13}\mathrm{E}_3 \\ \mathrm{S}_2 &= -\frac{\nu_{21}}{Y_1^{\mathrm{E}}}\mathrm{T}_1 + d_{23}\mathrm{E}_3 \\ \mathrm{S}_3 &= -\frac{\nu_{31}}{Y_1^{\mathrm{E}}}\mathrm{T}_1 + d_{33}\mathrm{E}_3 \\ \mathrm{D}_3 &= d_{31}\mathrm{T}_1 + \varepsilon_{33}^{\mathrm{T}}\mathrm{E}_3, \end{aligned} \tag{4.106}$$

where the piezoelectric strain coefficients are symmetric (i.e., $d_{31} = d_{13}$). As is the case with a 33 transducer, the reduced constitutive equations for a 31 transducer have two independent variables and four dependent variables. In general, we focus our analysis on the two equations

$$\begin{aligned} \mathrm{S}_1 &= \frac{1}{Y_1^{\mathrm{E}}}\mathrm{T}_1 + d_{13}\mathrm{E}_3 \\ \mathrm{D}_3 &= d_{13}\mathrm{T}_1 + \varepsilon_{33}^{\mathrm{T}}\mathrm{E}_3. \end{aligned} \tag{4.107}$$

The coupling coefficient of a 31 piezoelectric transducer is

$$k_{31} = \frac{d_{13}\sqrt{Y_1^{\mathrm{E}}}}{\sqrt{\varepsilon_{33}^{\mathrm{T}}}} = \frac{d_{13}}{\sqrt{s_{11}^{\mathrm{E}}\varepsilon_{33}^{\mathrm{T}}}}. \tag{4.108}$$

The blocked stress and free strain of the transducer in the 1 direction can be derived in the same manner as equations (4.63) and (4.62):

$$\begin{aligned} \mathrm{T}_1|_{\mathrm{S}_1=0} &= -d_{13}Y_1^{\mathrm{E}}\mathrm{E}_3 \\ \mathrm{S}_1|_{\mathrm{T}_1=0} &= d_{13}\mathrm{E}_3. \end{aligned} \tag{4.109}$$

The transducer equations for a single piezoelectric element operating in the 31 mode can be derived by substituting the relationships

$$\mathrm{S}_1 = \frac{u_1}{L_p} \tag{4.110}$$

$$\mathrm{T}_1 = \frac{f}{w_p t_p} \tag{4.111}$$

$$\mathrm{E}_3 = \frac{v}{t_p} \tag{4.112}$$

$$\mathrm{D}_3 = \frac{q}{w_p L_p} \tag{4.113}$$

into equation (4.107):

$$\begin{aligned} \frac{u_1}{L_p} &= \frac{f}{Y_1^{\mathrm{E}} w_p t_p} + d_{13}\,\frac{v}{t_p} \\ \frac{q}{wL_p} &= d_{31}\,\frac{f}{w_p t_p} + \varepsilon_{33}^{\mathrm{T}}\,\frac{v}{t_p}. \end{aligned} \tag{4.114}$$

Rearranging the terms yields

$$\begin{aligned} u_1 &= \frac{L_p}{Y_1^{\mathrm{E}} w_p t_p}\,f + \frac{d_{13}L_p}{t_p}\,v \\ q &= \frac{d_{31}L_p}{t_p}\,f + \frac{\varepsilon_{33}^{\mathrm{T}} w_p L_p}{t_p}\,v. \end{aligned} \tag{4.115}$$

From these expressions we can solve for the free displacement and blocked force:

$$\delta_o = \frac{d_{13}L_p}{t_p}v \tag{4.116}$$

$$f_{\mathrm{bl}} = d_{13}Y_1^{\mathrm{E}} w_p v. \tag{4.117}$$

Comparing equations (4.74) and (4.116), we see that the free displacement of a 31-mode transducer can be scaled by increasing the length-to-thickness ratio, Lp/t_p, whereas the free displacement in the 33 mode is only a function of the piezoelectric strain coefficient.

Example 4.8 Compute the free displacement in the 3 and 1 directions for a piezoelectric transducer of length 10 mm, width 3 mm, and thickness 0.25 mm. The applied voltage of 100 V is in the polarization direction. Use the values for APC 856 for the computation.

Solution Using equation (4.74), we can write the free displacement as

$$\delta_{o3} = (620 \times 10^{-12} \text{ m/V})(100 \text{ V}) = 62 \times 10^{-9} \text{ m} = 62 \text{ nm}.$$

Using Table 4.1, we see that the piezoelectric strain coefficient $d_{13} = -260 \times 10^{-12}$ m/V. The free displacement in the 1 direction can be computed from equation (4.116):

$$\begin{aligned} \delta_{o1} &= \frac{(-260 \times 10^{-12} \text{ m/V})(10 \times 10^{-3} \text{ m})}{0.25 \times 10^{-3} \text{ m}} (100 \text{ V}) \\ &= -1.04 \times 10^{-6} \text{ m} \\ &= 1040 \text{ nm}. \end{aligned}$$

Comparing the results, we see that the free displacement in the 1 direction is substantially higher than the free displacement in the 3 direction for the same applied voltage even though the piezoelectric strain coefficient d_{13} is less than half that of d_{33}. The reason is that the free displacement in the 1 direction is *amplified* by the geometry of the transducer. Specifically, the length-to-thickness ratio of the transducer is 40; therefore, the strain in the 1 direction produces larger displacements.

4.6.1 Extensional 31 Piezoelectric Devices

Piezoelectric materials are often used in a multilayer composite as extensional or bending actuators. The composite consists of one or two layers of piezoelectric material and an inactive substrate made from a material such as brass, aluminum, or steel. The poling direction of the piezoelectric material is parallel with the thickness direction of the piezoelectric layer and the desired extension is perpendicular to the poling direction. Therefore, the 31 mode of the piezoelectric material is utilized in these applications. In a composite extensional actuator, the amount of strain and stress produced is a function not only of the piezoelectric material properties, but also the properties of the inactive layer.

In this section, expressions for the strain and stress produced by these composite actuators are derived as a function of the piezoelectric material properties and the material properties of the inactive layer. The derivation will focus on a typical composite

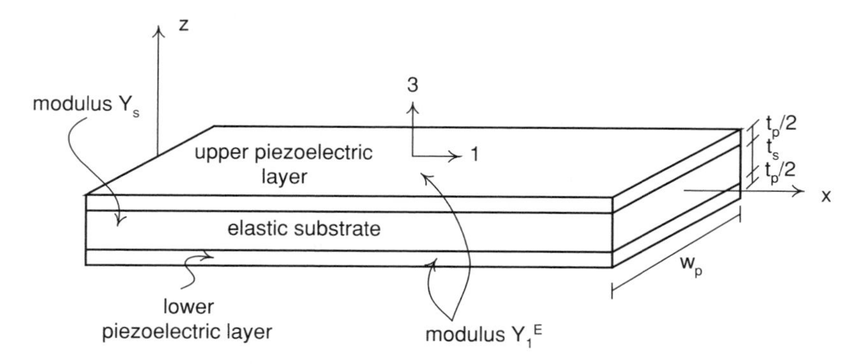

Figure 4.19 Composite actuator consisting of an elastic substrate and two piezoelectric layers.

layup that consists of piezoelectric layers attached to the surfaces of an inactive substrate as shown in Figure 4.19. The elastic substrate has a thickness t_s and a modulus of Y_s. Each piezoelectric layer has a thickness of $t_p/2$ and a short-circuit modulus of Y_1^E. For simplicity we assume that the layers are symmetric about the neutral axis of the composite and that the active and inactive layers are equal in width. The width of the piezoelectric materials and the substrate is denoted w_p.

Consider the case in which the voltage applied to the piezoelectric layers is aligned with the poling direction of both piezoelectric layers (Figure 4.20). Without a substrate and with no restraining force, the strain in the piezoelectric layers would be equal $d_{13}\mathrm{E}_3$. To determine the strain produced in the piezoelectric composite, first write the constitutive relationships for each of the three layers within the composite. These are

$$
\mathrm{S}_1 = \begin{cases} \dfrac{1}{Y_1^{\mathrm{E}}}\mathrm{T}_1 + d_{13}\mathrm{E}_3 & \dfrac{t_s}{2} \leq z \leq \dfrac{1}{2}(t_s + t_p) \\ \dfrac{1}{Y_s}\mathrm{T}_1 & -\dfrac{t_s}{2} \leq z \leq \dfrac{t_s}{2} \\ \dfrac{1}{Y_1^{\mathrm{E}}}\mathrm{T}_1 + d_{13}\mathrm{E}_3 & -\dfrac{1}{2}(t_s + t_p) \leq z \leq -\dfrac{t_s}{2} \end{cases} \tag{4.118}
$$

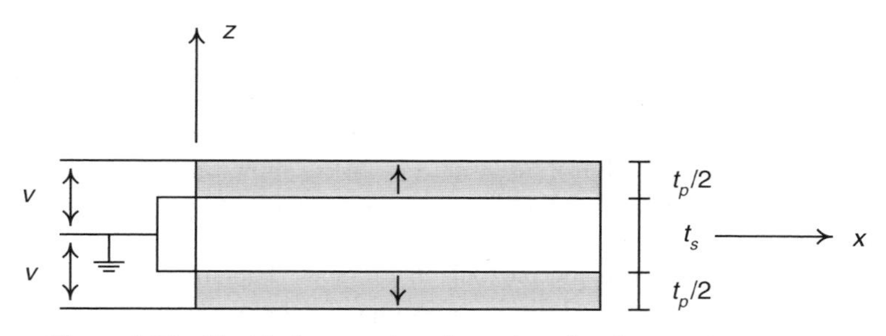

Figure 4.20 Electrical connections for a piezoelectric extender actuator.

Multiplying by the modulus values and integrating over the y and z directions for their respective domains produces the expressions

$$\frac{w_p t_p}{2} Y_1^E S_1 = \int_{y,z} T_1 \, dy \, dz + \frac{w_p t_p}{2} Y_1^E d_{13} E_3 \tag{4.119}$$

$$w_p t_s Y_s S_1 = \int_{y,z} T_1 \, dy \, dz \tag{4.120}$$

$$\frac{w_p t_p}{2} Y_1^E S_1 = \int_{y,z} T_1 \, dy \, dz + \frac{w_p t_p}{2} Y_1^E d_{13} E_3. \tag{4.121}$$

Assuming that the strain in all three regions of the beam is the same, which is equivalent to assuming that there is a perfect bond and no slipping at the boundaries, these three expressions can be added to obtain the equation

$$\left(Y_s w_p t_s + Y_1^E w_p t_p\right) S_1 = \int_{y,z} T_1 \, dy \, dz + Y_1^E w_p t_p d_{13} E_3. \tag{4.122}$$

The externally applied force f is in equilibrium with the stress resultant. If the force applied externally is equal to zero, the expression for the strain can be solved from equation (4.122):

$$S_1 = \frac{Y_1^E t_p}{Y_s t_s + Y_1^E t_p} d_{13} E_3. \tag{4.123}$$

The term $d_{13}E_3$ is recognized as the free strain associated with the piezoelectric material if there were no substrate layer. The coefficient modifying the free strain can be rewritten as a nondimensional expression by dividing the numerator and denominator by $Y_1^E t_p$:

$$\frac{S_1}{d_{13} E_3} = \frac{1}{1 + \Psi_e}, \tag{4.124}$$

where

$$\Psi_e = \frac{Y_s t_s}{Y_1^E t_p}. \tag{4.125}$$

Equation (4.125) illustrates that the variation in the free strain of the piezoelectric extender is a function of the relative stiffness between the piezoelectric layer and the substrate layer, Ψ_e. A plot of equation (4.124) is shown in Figure 4.21. As the relative stiffness approaches zero, indicating that the stiffness of the piezoelectric is large compared to the stiffness of the substrate layer, the free strain of the composite approaches the free strain of the piezoelectric layers. The stiffness match point at which $\Psi_e = 1$ produces a free strain in the composite that is one-half that of the free strain in the piezoelectric layers. Increasing the stiffness of the substrate layer such

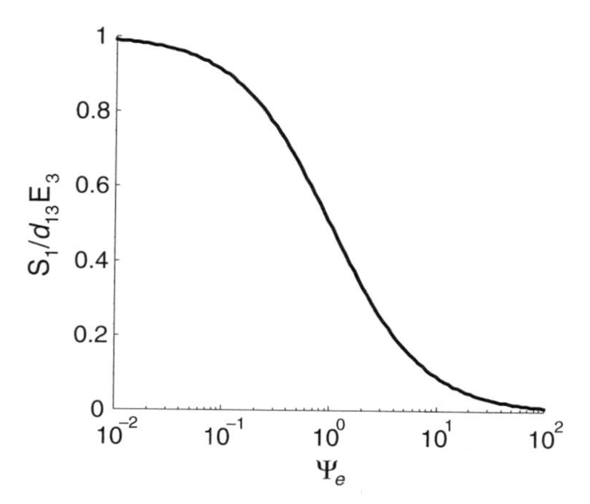

Figure 4.21 Variation in free strain for a piezoelectric extender as a function of the relative stiffness parameter.

that $\Psi_e \gg 1$ produces an extensional actuator whose free strain is much less than the free strain of the piezoelectric layers.

The deflection u_1 of a piezoelectric extender of total length L can be expressed as a function of the voltage by noting that $S_1 = u_1/L_p$ and that the electric field is equal to the applied voltage v divided by the piezoelectric layer thickness $t_p/2$. Substituting these relationships into equation (4.125) produces the expression for the deflection:

$$u_1 = \frac{2}{1+\Psi_e}\left(\frac{d_{13}L_p}{t_p}v\right). \tag{4.126}$$

Replacing the stress resultant in equation (4.122) with an applied force f divided by the area of the composite and solving for the blocked force yields

$$f_{\mathrm{bl}} = 2Y_1^{\mathrm{E}} w_p d_{13} v. \tag{4.127}$$

The trade-off between force and deflection for an extensional actuator is shown in Figure 4.22, which illustrates that the relative stiffness parameter will change the deflection of the extensional actuator but not the blocked force. The blocked force is equivalent to that of a piezoelectric layers combined, but the deflection is reduced as the relative stiffness parameter increases from much smaller than 1 to much larger than 1.

Example 4.9 A piezoelectric extensional actuator is fabricated from two 0.25-mm layers of PZT-5H and a single layer of 0.25-mm brass shim. Compute the free strain in the device when the applied electric field is 0.5 MV/m.

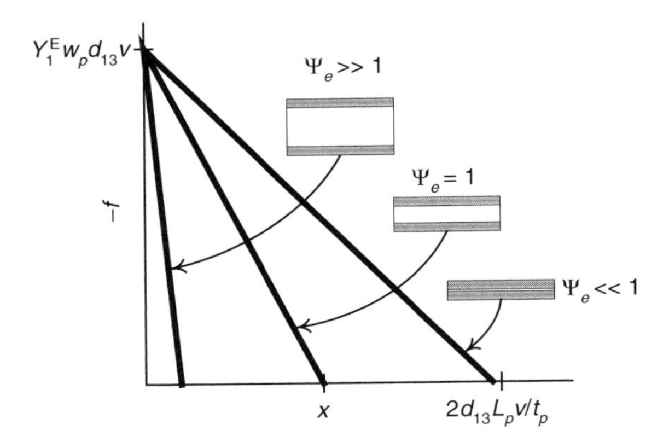

Figure 4.22 Force–deflection trade-off for an extensional actuator as a function of the relative stiffness parameter.

Solution The expression for the free strain is shown in equation (4.124) normalized with respect to the free strain in the unconstrained piezoelectric layers. The free strain in the unconstrained piezoelectric layers is

$$
\begin{aligned}
d_{13}\mathrm{E}_3 &= (320 \times 10^{-12}\ \mathrm{m/V})(0.5 \times 10^6\ \mathrm{V/m}) \\
&= 160\ \mu\mathrm{strain}.
\end{aligned}
$$

Brass shim is assumed to have a modulus of 117 GPa. Recognizing that t_p in equation (4.125) is the *total* thickness of the piezoelectric layers, we can compute the relative stiffness parameter:

$$
\begin{aligned}
\Psi_e &= \frac{(117 \times 10^9\ \mathrm{N/m^2})(0.25 \times 10^{-3}\ \mathrm{m})}{(50 \times 10^9\ \mathrm{N/m^2})(0.5 \times 10^{-3}\ \mathrm{m})} \\
&= 1.17.
\end{aligned}
$$

The free strain in the composite extensional actuator is computed from equation (4.124):

$$
\mathrm{S}_1 = \frac{160\ \mu\mathrm{strain}}{1 + 1.17} = 73.7\ \ \mu\mathrm{strain}.
$$

4.6.2 Bending 31 Piezoelectric Devices

Although the preceding development illustrates how a composite piezoelectric device is useful for extensional actuation, the primary use of 31-multilayer piezoelectric actuators is as a *bending* device. A three-layer device in which the piezoelectric layers are fixed to the outer surfaces of an inactive substrate is typically called a *bimorph actuator*. The electrical connections of a bimorph actuator are chosen such

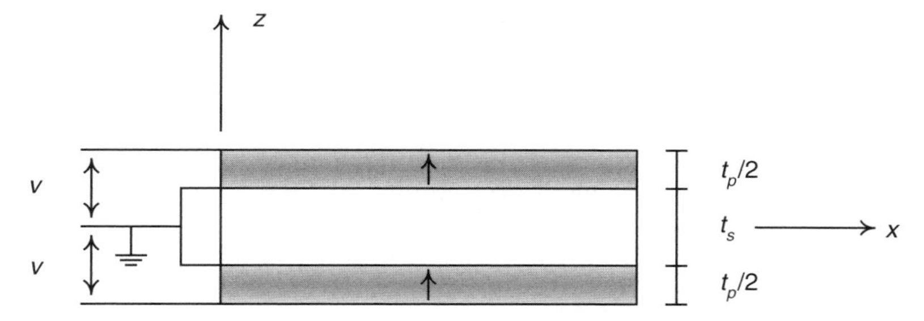

Figure 4.23 Electrical connections for a piezoelectric bimorph.

that the electric field is in the same direction as the poling direction in one of the layers, whereas in the second layer the electric field is in the direction opposite the poling direction. This is illustrated in Figure 4.23.

Application of an electric field to both layers produces extension in one of the layers and contraction in the other. The net result is a bending of the material. Assuming a perfect bond between the inactive layer and the piezoelectric layers, and assuming that the piezoelectric layers are symmetric about the neutral axis of the composite, the bending will result in the deformed shape shown in Figure 4.24.

Under the assumption that the field is in the poling direction in the top layer and opposite to the poling direction in the bottom layer, we can write the constitutive equations of the composite as

$$\mathrm{S}_1(z) = \begin{cases} \dfrac{1}{Y_1^{\mathrm{E}}}\mathrm{T}_1(z) + d_{13}\mathrm{E}_3 & \dfrac{t_s}{2} \leq z \leq \dfrac{1}{2}(t_s + t_p) \\ \dfrac{1}{Y_s}\mathrm{T}_1(z) & -\dfrac{t_s}{2} \leq z \leq \dfrac{t_s}{2} \\ \dfrac{1}{Y_1^{\mathrm{E}}}\mathrm{T}_1(z) - d_{13}\mathrm{E}_3 & -\dfrac{1}{2}(t_s + t_p) \leq z \leq -\dfrac{t_s}{2}. \end{cases} \tag{4.128}$$

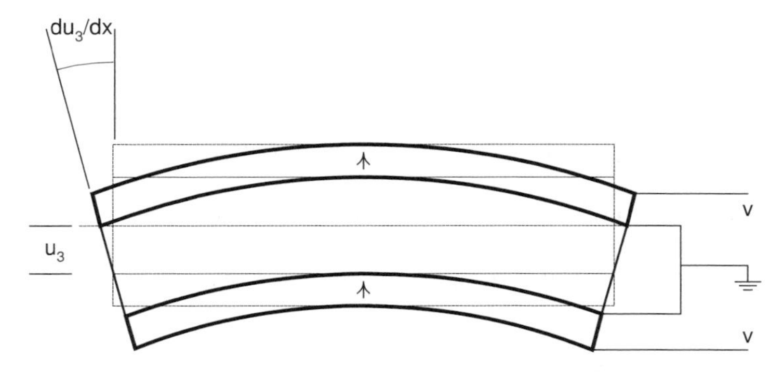

Figure 4.24 Bending induced in a symmetric piezoelectric bimorph.

Assuming that the Euler–Bernoulli beam assumptions are valid, the relationship between the strain and the curvature κ is

$$S_1(z) = \kappa z. \tag{4.129}$$

Substituting equation (4.129) into the constitutive relations and rewriting, we obtain

$$\begin{aligned} Y_1^E(\kappa z) &= T_1(z) + Y_1^E d_{13} E_3 \\ Y_s(\kappa z) &= T_1(z) \\ Y_1^E(\kappa z) &= T_1(z) - Y_1^E d_{13} E_3. \end{aligned} \tag{4.130}$$

The equilibrium expressions for the moment are obtained by multiplying equation (4.130) by z and integrating over the domain in y and z. The result is

$$\begin{aligned} Y_1^E w_p \kappa \left(\frac{t_p^3}{24} + \frac{t_p^2 t_s}{8} + \frac{t_p t_s^2}{8} \right) &= \int_{y,z} z T_1 \, dy \, dz + Y_1^E w_p d_{13} \left(\frac{t_p^2}{8} + \frac{t_p t_s}{4} \right) E_3 \\ Y_s \kappa \frac{w_p t_s^3}{12} &= \int_{y,z} z T_1 \, dy \, dz \\ Y_1^E w_p \kappa \left(\frac{t_p^3}{24} + \frac{t_p^2 t_s}{8} + \frac{t_p t_s^2}{8} \right) &= \int_{y,z} z T_1 \, dy \, dz + Y_1^E w_p d_{13} \left(\frac{t_p^2}{8} + \frac{t_p t_s}{4} \right) E_3. \end{aligned} \tag{4.131}$$

Adding the results from the three domains together yields

$$\begin{aligned} &Y_1^E w_p \kappa \left(\frac{t_p^3}{12} + \frac{t_p^2 t_s}{4} + \frac{t_p t_s^2}{4} \right) + Y_s \kappa \frac{w_p t_s^3}{12} \\ &= \int_{y,z} z T_1(z) \, dy \, dz + Y_1^E w_p d_{13} \left(\frac{t_p^2}{4} + \frac{t_p t_s}{2} \right) E_3 \end{aligned} \tag{4.132}$$

The integration of the stress component on the right-hand side of the expression is the moment resultant from externally applied loads. If this moment resultant is zero, we can solve for the curvature as a function of

$$\kappa = \frac{Y_1^E \left(t_p^2/4 + t_p t_s/2 \right)}{Y_1^E \left(t_p^3/12 + t_p^2 t_s/4 + t_p t_s^2/4 \right) + Y_s \left(t_s^3/12 \right)} d_{13} E_3. \tag{4.133}$$

A nondimensional expression for the curvature of the composite beam due to piezoelectric actuation is obtained by dividing the numerator and denominator by the inertia per unit width, $Y_1^E t_p^3/12$, and making the substitution $\tau = t_s/t_p$. The result is

$$\kappa \frac{t_s}{2 d_{13} E_3} = \frac{3\tau/2 + 3\tau^2}{1 + 3\tau + 3\tau^2 + \left(Y_s/Y_1^E \right) \tau^3}. \tag{4.134}$$

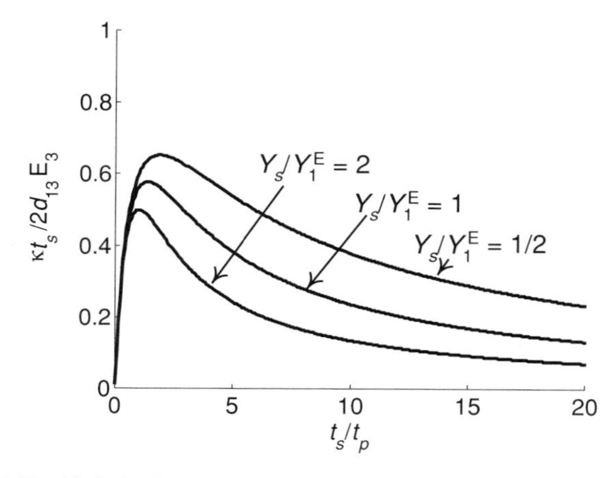

Figure 4.25 Variation in nondimensional curvature for a composite bimorph.

A plot of equation (4.134) is shown in Figure 4.25 for three different values of Y_s/Y_1^E. For a constant value of t_s, we see that the curvature will reach a maximum at a specific value of t_s/t_p. Increasing the substrate thickness relative to the piezoelectric layer thickness will produce a decrease in the induced curvature.

The nondimensional expression in equation (4.134) has physical significance if we examine the strain induced through the thickness of the bimorph. The strain induced at the interface between the substrate and the piezoelectric layers is equal to $\kappa t_s/2$; therefore, we can write the strain at the interface as a normalized expression:

$$\frac{S_1|_{z=t_s/2}}{d_{13}E_3} = \frac{3\tau/2 + 3\tau^2}{1 + 3\tau + 3\tau^2 + \left(Y_s/Y_1^E\right)\tau^3}. \tag{4.135}$$

The plot in Figure 4.25 can now be examined as the ratio of the induced bending strain to the extensional strain induced in the piezoelectric by the application of the electric field. As expected, this value is always less than 1. At large values of τ, we note that the induced strain is small due to the fact that the substrate layer is much thicker than the piezoelectric layers. As small values of τ, the induced strain at the interface becomes very small because the thickness of the substrate layer is small and the interface is becoming very close to the neutral axis of the composite bimorph.

The strain at the outer surface of the composite bimorph can also be obtained by evaluating S_1 at $z = \frac{1}{2}(t_s + t_p)$. The result in nondimensional form is

$$\frac{S_1|_{z=t_s/2+t_p/2}}{d_{13}E_3} = \frac{(3\tau/2 + 3\tau^2)(\tau + 1)}{\tau\left[1 + 3\tau + 3\tau^2 + \left(Y_s/Y_1^E\right)\tau^3\right]}. \tag{4.136}$$

A plot of equation (4.136) for a value of $Y_s/Y_1^E = 1$ is shown in Figure 4.26. The solid curve illustrates the variation in strain at the outer fibers of the composite

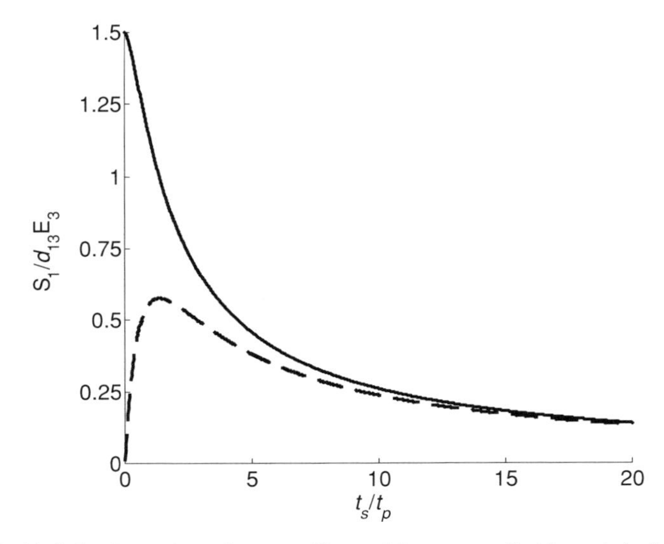

Figure 4.26 Variation in strain at the outer fibers of the composite bimorph (solid) and at the substrate–piezoelectric interface (dashed) for $Y_S/Y_1^E = 1$.

bimorph normalized with respect to the free strain produced in extension, $d_{13}E_3$. Also plotted is the normalized strain at the interface between the substrate and the piezoelectric layer (the dashed curve). The figure illustrates that these two values converge for large values of τ, due to the fact that the thickness of the piezoelectric layer becomes small. There is a large difference in the induced strain for small values of τ because the thickness of the piezoelectric layer is large compared to the substrate thickness.

Example 4.10 A symmetric piezoelectric bimorph is constructed from 2-mm-thick brass shim with piezoelectric thicknesses of 0.25 mm for each layer. Plot the variation in the strain as a function of thickness through the bimorph and label the strain at the substrate–piezoelectric interface and at the outer fibers of the bimorph. The piezoelectric material is APC 856 and the applied field is assumed to be 0.5 MV/m.

Solution The variation in the strain through the thickness is given by equation (4.129), and the expression for the nondimensional curvature is obtained from equation (4.134). The free strain in extension is computed from

$$d_{13}E_3 = (260 \times 10^{-12}\ \mathrm{m/V})(0.5 \times 10^6\ \mathrm{V/m}) = 130\ \mu\mathrm{strain}.$$

The thickness ratio in the bimorph is computed to be

$$\tau = \frac{2\ \mathrm{mm}}{0.5\ \mathrm{mm}} = 4.$$

The nondimensional curvature is computed from equation (4.134):

$$\kappa \frac{t_s}{2d_{13}E_3} = \frac{(3)(4/2) + (3)(4^2)}{1 + (3)(4) + (3)(4^2) + (117\text{ GPa}/66.7\text{ GPa})(4^3)}$$
$$= 0.3117.$$

The curvature is computed from

$$\kappa = (0.3117)\left(\frac{2}{2 \times 10^{-3}\text{ m}}\right) 130 \times 10^{-6}\text{ m/m}$$
$$= 0.0405\text{ m}^{-1}.$$

The strain through the thickness is computed from equation (4.129). The two strain values of interest are

$$S_1|_{z=t_s/2} = (0.0405\text{ m}^{-1})\frac{2 \times 10^{-3}\text{ m}}{2}$$
$$= 40.5\ \mu\text{strain}.$$
$$S_1|_{z=t_s/2+t_p/2} = (0.0405\text{ m}^{-1})\left(\frac{2 \times 10^{-3}\text{ m} + 0.35 \times 10^{-3}\text{ m}}{2}\right)$$
$$= 45.6\ \mu\text{strain}.$$

The results are illustrated in Figure 4.27. The diagonal line represents the strain through the thickness of the bimorph with the values labeled at the outer fibers and at the substrate–piezoelectric interface. Also shown to scale is the free extensional strain produced by the piezoelectric layers at the electric field value specified. Note that the free strain is approximately three times that of the maximum strain in the bimorph.

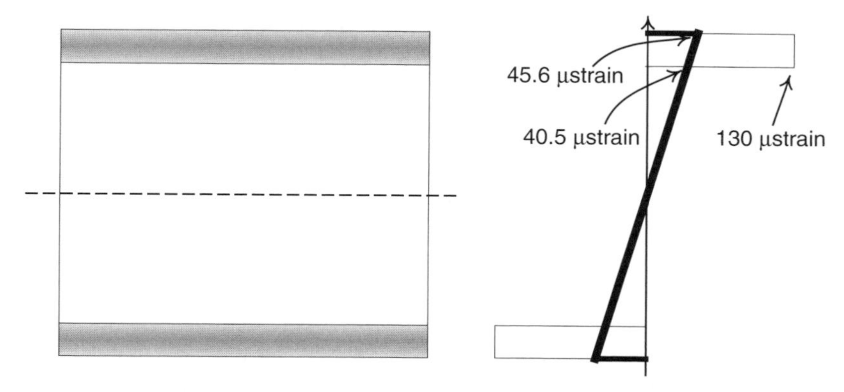

Figure 4.27 Variation in strain through the thickness of a symmetric bimorph for example 4.10.

This is due to the fact that the piezoelectric layers must overcome the bending stiffness of the inactive substrate to produce curvature in the bimorph.

4.6.3 Transducer Equations for a Piezoelectric Bimorph

The equations that relate force, deflection, and voltage for a symmetric piezoelectric bimorph can be derived from the analysis presented in Section 4.6.2 and the application of basic principles of mechanics. Let us rewrite equation (4.132) as

$$(EI)_c \kappa = M_e(x) + M_p \mathrm{E}_3, \tag{4.137}$$

where $(EI)_c$ is the bending stiffness of the composite,

$$(EI)_c = Y_1^{\mathrm{E}} w_p \left(\frac{t_p^3}{12} + \frac{t_p^2 t_s}{4} + \frac{t_p t_s^2}{4} \right) + Y_s \frac{w_p t_s^3}{12}, \tag{4.138}$$

$M_e(x)$ is the moment applied by external loads, and M_p is the moment applied by the piezoelectric layers per unit electric field,

$$M_p = Y_1^{\mathrm{E}} w_p d_{13} \left(\frac{t_p^2}{4} + \frac{t_p t_s}{2} \right). \tag{4.139}$$

From basic mechanics we know that the curvature is related to the displacement through the relationship

$$\kappa = \frac{d^2 u_3(x)}{dx^2}. \tag{4.140}$$

Combining equations (4.137) and (4.140), we can write the differential equation for the deflection of the beam as

$$\frac{d^2 u_3(x)}{dx^2} = \frac{M_e(x)}{(EI)_c} + \frac{M_p}{(EI)_c} \mathrm{E}_3. \tag{4.141}$$

Integrating equation (4.141) once yields the expression for the slope:

$$\frac{du_3(x)}{dx} = \frac{1}{(EI)_c} \int_0^x M_e(\xi)\, d\xi + \frac{M_p}{(EI)_c} x \mathrm{E}_3 + C_1, \tag{4.142}$$

where C_1 is the constant of integration. Integrating once more produces an expression for the deflection:

$$u_3(x) = \frac{1}{(EI)_c} \int_0^x \int_0^\zeta M_e(\xi)\, d\xi\, d\zeta + \frac{1}{2} \frac{M_p}{(EI)_c} x^2 \mathrm{E}_3 + C_1 x + C_2. \tag{4.143}$$

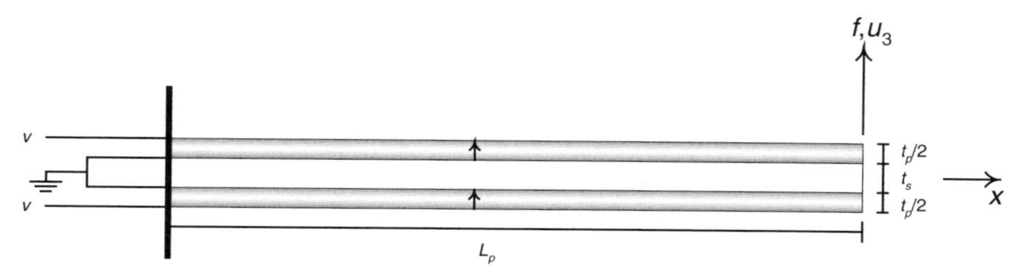

Figure 4.28 Piezoelectric cantilever beam.

Equation (4.143) is a general expression for the displacement. To use the result to compute the relationship between force, deflection, and applied voltage, we need to make an assumption about the geometry of the problem. One of the most common geometries is a cantilevered bimorph in which one end of the piezoelectric composite is fixed and the other end is free, as shown in Figure 4.28. Application of the geometric boundary conditions yields the relationships

$$\begin{aligned} \frac{du_3(x)}{dx}(0) &= C_1 = 0 \\ u_3(0) &= C_2 = 0. \end{aligned} \tag{4.144}$$

Thus, for a cantilevered beam we can write the expression for the deflection as

$$u_3(x) = \frac{1}{(EI)_c} \int_0^x \int_0^\zeta M_e(\xi)\, d\xi\, d\zeta + \frac{1}{2} \frac{M_p}{(EI)_c} x^2 \mathrm{E}_3. \tag{4.145}$$

The moment associated with a force at the tip is

$$M_e(x) = f(L_p - x), \tag{4.146}$$

and the double integration of the applied moment produces the expression

$$u_3(x) = \frac{1}{6(EI)_c} \left(3L_p x^2 - x^3\right) f + \frac{1}{2} \frac{M_p}{(EI)_c} x^2 \mathrm{E}_3. \tag{4.147}$$

The expressions for a bimorph transducer are obtained by substituting the relationship $\mathrm{E}_3 = 2v/t_p$ into the expression and evaluating the result at $x = L_p$,

$$u_3(L) = \frac{L_p^3}{3(EI)_c} f + \frac{M_p L_p^2}{t_p (EI)_c} v. \tag{4.148}$$

The coefficient in front of the force term is noted to be the inverse of the mechanical stiffness of the beam. The term multiplying the voltage is the deflection per unit volt for the composite bimorph.

A typical assumption when designing transducers using piezoelectric bimorphs is that the substrate layer is negligibly thick compared to the thickness of the piezoelectric layers. The transducer equations for this type of "ideal" bimorph is analyzed by letting $t_s \to 0$ in equation (4.148). Letting $t_s \to 0$ in equation (4.148) results in the transducer equations

$$u_3(L) = \frac{4L_p^3}{Y_1^E w_p t_p^3} f + 3d_{13} \frac{L_p^2}{t_p^2} v. \tag{4.149}$$

The blocked force and free deflection of an ideal bimorph are obtained from equation (4.149):

$$\begin{aligned} \delta_o &= 3d_{13} \frac{L_p^2}{t_p^2} v \\ f_{\text{bl}} &= \frac{3}{4} Y_1^E d_{13} w_p \frac{t_p}{L_p} v. \end{aligned} \tag{4.150}$$

This result demonstrates the fundamental trade-off in using an ideal bimorph actuator for actuation. The free deflection is proportional to L_p^2/t_p^2, while the blocked force is inversely proportional to L_p/t_p. Increasing the ratio of the length to thickness will produce a trade-off in the blocked force and the free deflection. The blocked force can be increased independent of the free deflection by increasing the width of the actuator. This concept is illustrated in Figure 4.29.

Example 4.11 A cantilievered piezoelectric bimorph with negligible substrate thickness is being designed for a new medical device. The device requires 0.2 mm of free displacement at the tip for a maximum applied voltage of 50 V to each layer. Each layer of the piezoelectric material is 0.25 mm thick and is made from APC 856.

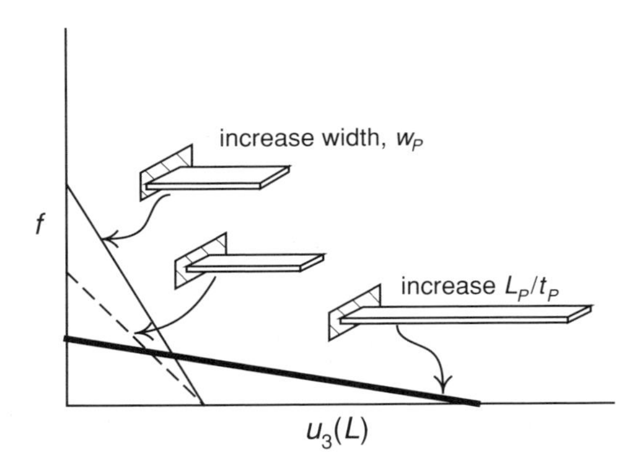

Figure 4.29 Trade-offs in the force–deflection curve of an ideal piezoelectric bimorph.

Determine (a) the length required to achieve 0.2 mm of tip deflection, and (b) the blocked force of the actuator for a width of 5 mm.

Solution (a) The length required to obtain 0.2 mm of free displacement at the tip is computed from equation (4.150):

$$L_p = t_p \sqrt{\frac{\delta_o}{3 d_{13} v}}.$$

substituting in the values from the problem gives us

$$\begin{aligned} L_p &= (0.5 \times 10^{-3}\ \text{m}) \sqrt{\frac{0.2 \times 10^{-3}\ \text{m}}{(3)(260 \times 10^{-12}\ \text{m/V})(50)\ \text{V}}} \\ &= 35.8\ \text{mm}. \end{aligned}$$

(b) The blocked force for a width of 5 mm is obtained using the second expression in equation (4.150):

$$\begin{aligned} f_{\text{bl}} &= \frac{3}{4}(66.7 \times 10^{9}\ \text{N/m}^2)(260 \times 10^{-12}\ \text{m/V})(5 \times 10^{-3}\ \text{m}) \\ &\quad \times \left(\frac{0.5 \times 10^{-3}\ \text{m}}{35.8 \times 10^{-3}\ \text{m}}\right)(50\ \text{V}) \\ &= 45.4\ \text{mN}. \end{aligned}$$

The geometric and material properties illustrated in Example 4.11 are very typical of piezoelectric bimorphs. From the example we can see that piezoelectric bimorphs produce much lower forces than a stack piezoelectric device, due to the mechanical amplification associated with beam bending. The drawback of a piezoelectric bimorph is that the force is much lower, due to the fact that the stiffness of a actuator is much smaller than that of a piezoelectric stack. For these reasons, piezoelectric bimorphs are typically used in applications that require larger deflections but smaller forces than those obtained with a piezoelectric stack.

4.6.4 Piezoelectric Bimorphs Including Substrate Effects

The transducer equations for a piezoelectric bimorph are modified when the substrate thickness becomes appreciable compared to the thickness of the piezoelectric layers. In the preceding discussion it was demonstrated that the curvature of the beam changed as a function of the ratio of the substrate thickness to the active layer thickness, t_s/t_p. The expressions for blocked force and free deflection are derived by considering the effects of the substrate layer on the transducer equations.

If the general expressions for M_a and $(EI)_c$ are retained in the analysis, the expressions for free deflection and blocked force are

$$\delta_0 = 3d_{13}\frac{L_p^2}{t_p^2}f_1(\tau)v$$
$$f_{bl} = \frac{3}{4}c_{11}^E d_{13} w_p \frac{t_p}{L_p} f_2(\tau)v, \tag{4.151}$$

where

$$f_1(\tau) = \frac{1+2\tau}{1+3\tau+3\tau^2+\left(Y_s/Y_1^E\right)\tau^3}$$
$$f_2(\tau) = 1+2\tau. \tag{4.152}$$

Note that the expressions for blocked force and free deflection for the case of a nonnegligible substrate thickness are written as the product of the coefficients for the ideal case and a nondimensional expression that is a function of the thickness ratio t_s/t_p. Both of the nondimensional expressions are equal to 1 in the limiting case of $\tau \to 0$; therefore, the expressions in equation (4.151) are reduced to the expressions for an ideal bimorph when the substrate thickness becomes negligible compared to the active layer thickness. The expression in equation (4.152) can be thought of as a deviation from the ideal case of a negligible substrate.

Increasing the substrate layer thickness compared to the thickness of the piezoelectric layers increases the blocked force and decreases the free deflection. The blocked force is increased by a factor of 2τ, where $\tau = t_s/t_p$. The reduction in the free deflection is quantified in Figure 4.30 for three values of the modulus ratio. For thickness ratios of less than 1/100, the free deflection is approximately equal to the free deflection of an ideal bimorph. Increasing the thickness ratio to 1/10 reduces the free deflection to approximately 90% of the value for an ideal bimorph. The effects of modulus ratio become more evident as the thickness ratio is increased. At thickness ratios of 1 or greater, the free deflection is reduced to between 20 and 40% of the value for an ideal bimorph, depending on the modulus ratio between the piezoelectric material and the substrate.

Example 4.12 Using the values from Example 4.11, compute the change in blocked force and free deflection if the two piezoelectric layers are placed on a 0.5-mm-thick substrate made of aluminum, brass, and steel. Assume that the modulus values for these materials are 69, 117, and 210 GPa, respectively.

Solution In the solution to Example 4.11, the blocked force and free deflection were computed to be 45.4 mN and 0.2 mm, respectively. These results are the solution for an "ideal" bimorph and can be used as a basis for computing the results with a nonnegligible substrate. To determine the effect of the substrate, first compute the

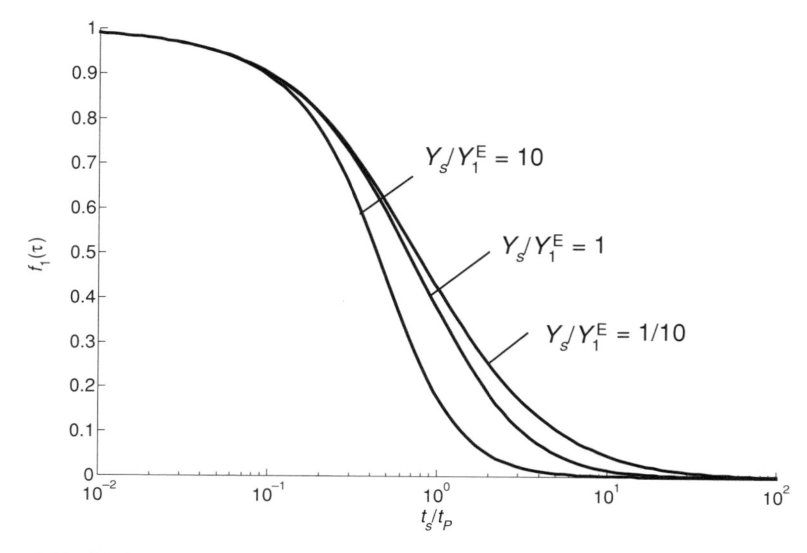

Figure 4.30 Variation in the nondimensional modifier to the free deflection as a function of the thickness ratio.

thickness ratio:

$$\frac{t_s}{t_p} = \frac{(2)(0.25 \text{ mm})}{0.5 \text{ mm}} = 1.$$

The change in blocked force is equal for all three substrate layers since the nondimensional modifier in equation (4.151) is not a function of the substrate modulus. The change is obtained by evaluating $f_2(\tau)$ in equation (4.152):

$$f_2(\tau) = 1 + (2)(1) = 3.$$

Therefore, the blocked force for all three substrate layers is equal to (3) (45.4 mN) = 136.2 mN. The reduction in free deflection is computed by evaluating the function $f_1(\tau)$ in equation (4.152). The modulus ratios for all three substrate layers are

$$\begin{aligned}
&\text{aluminum:} \quad \frac{Y_s}{Y_1^E} = \frac{69 \text{ GPa}}{66.7 \text{ GPa}} = 1.0345 \\
&\text{brass:} \quad \frac{Y_s}{Y_1^E} = \frac{117 \text{ GPa}}{66.7 \text{ GPa}} = 1.7541 \\
&\text{steel:} \quad \frac{Y_s}{Y_1^E} = \frac{210 \text{ GPa}}{66.7 \text{ GPa}} = 3.1484.
\end{aligned}$$

Substituting these values into $f_1(\tau)$ produces

$$\begin{array}{ll} \text{aluminum:} & f_1(1) = 0.3734 \\ \text{brass:} & f_1(1) = 0.3427 \\ \text{steel:} & f_1(1) = 0.2956. \end{array}$$

The amount of deflection was 0.2 mm with a negligible substrate; therefore, the free deflection for the various substrates are

$$\begin{array}{ll} \text{aluminum:} & \delta_o = (0.3734)(0.2\ \text{mm}) = 74.5\ \mu\text{m} \\ \text{brass:} & \delta_o = (0.3427)(0.2\ \text{mm}) = 68.5\ \mu\text{m} \\ \text{steel:} & \delta_o = (0.2956)(0.2\ \text{mm}) = 59.1\ \mu\text{m}. \end{array}$$

4.7 TRANSDUCER COMPARISON

In the preceding sections we have transformed the constitutive equations for a linear piezoelectric material into equations that relate the force, deflection, and voltage of the transducer. We have seen that this transformation introduces the geometry of the transducer into the expressions and produces expressions that can be used to compute relevant engineering parameters such as the maximum force or maximum deflection that are produced by the transducer. In both cases the *transducer* expressions can be used to analyze the trade-off in force and deflection.

The expressions also allow us to compare the performance of piezoelectric stacks operating in the 33 mode and piezoelectric bimorphs operating in the 31 mode. In general, the primary difference between the operating modes is the force–deflection trade-off associated with the transducer. Piezoelectric stacks are able to produce higher forces than piezoelectric bimorphs but generally produce smaller deflections for a transducer of similar size. This trade-off is a result of the fact that arranging piezoelectric materials in a bimorph configuration produces a displacement amplification similar to that of a mechanical lever.

The force–deflection trade-offs inherent in piezoelectric stacks and bimorphs can be analyzed by realizing that the tranducer equations for both actuators have the same form:

$$u = \frac{1}{k_p^{\mathrm{E}}} f + u_o v, \tag{4.153}$$

where k_p^{E} is the short-circuit piezoelectric stiffness and u_o is the free deflection per unit voltage. The expressions for these parameters for both a piezoelectric stack and a piezoelectric bimorph are shown in Table 4.2. The free deflection and blocked force can be expressed as

$$\begin{aligned} \delta_o &= u_o v \\ f_{\mathrm{bl}} &= k_p^{\mathrm{E}} u_o v = k_p^{\mathrm{E}} \delta_o. \end{aligned} \tag{4.154}$$

Table 4.2 Comparison of transducer properties for a piezoelectric stack and an ideal piezoelectric bimorph

	Stack	Cantilevered Bimorph
Short-circuit stiffness	$\dfrac{Y_3^{\mathrm{E}} A_p}{L_s}$	$\dfrac{Y_1^{\mathrm{E}} w_p t_p^3}{4L_p^3}$
Displacement/voltage	$d_{33}\dfrac{L_s}{t_p}$	$3d_{13}\dfrac{L_p^2}{t_p^2}$
Blocked force/voltage	$Y_3^{\mathrm{E}} d_{33}\dfrac{A_p}{t_p}$	$\dfrac{3}{4} Y_1^{\mathrm{E}} d_{13}\dfrac{w_p t_p}{L_p}$

The expressions for the blocked force and free deflection of stacks and bimorphs are also listed in Table 4.2.

Generalizing the transducer equations also allows us to compare other aspects of transducer performance. A parameter that is often of interest in transducer design is the time response of the actuator. Piezoelectric materials are often utilized because of their fast response to changes in voltage. This allows them to be used in applications that require fast positioning. Equation (4.153) allows us to quantify the time response and compare the response speed between piezoelectric stacks and bimorphs. Assuming that the resistance force on the piezoelectric element is due to an inertial load with mass m, the equation of motion for the system can be written

$$m\ddot{u}(t) + k_p^{\mathrm{E}} u(t) = k_p^{\mathrm{E}} u_o v(t). \tag{4.155}$$

Dividing by the mass allows us to write the equation in the familiar form of a single-degree-of-freedom oscillator:

$$\ddot{u}(t) + \omega_n^{\mathrm{E}^2} u(t) = \omega_n^{\mathrm{E}^2} u_o v(t), \tag{4.156}$$

where $\omega_n^{\mathrm{E}} = k_p^{\mathrm{E}}/m$ is the short-circuit natural frequency of the oscillator for a short-circuit electrical boundary condition; the superscript notation is dropped for convenience. Equation (4.156) represents the equation of motion for an undamped oscillator. The simplest method of adding energy dissipation to the equations is to add a linear, velocity-dependent damping term,

$$\ddot{u}(t) + 2\zeta\omega_n^{\mathrm{E}}\dot{u}(t) + \omega_n^{\mathrm{E}^2} u(t) = \omega_n^{\mathrm{E}^2} u_o v(t), \tag{4.157}$$

where ζ is the *damping ratio* that represents the energy dissipation in the transducer.

A parameter of interest in design is the speed at which the transducer will respond to a step change in the applied voltage. The inertial forces and damping forces will impede the mechanical response and produce a delay in the step response of the transducer. Writing the transducer equation as a single-degree-of-freedom damped oscillator allows us to utilize well-known results in controls and linear systems theory to quantify the delay in transducer response.

The response to a step change in potential can be solved with a variety of methods; including Laplace transforms and the convolution integral (discussed in Chapter 2). Using Laplace transforms, we write (assuming the initial conditions are zero)

$$\left(\mathrm{s}^2 + 2\zeta\omega_n^{\mathrm{E}}\mathrm{s} + \omega_n^{\mathrm{E}^2}\right) u(\mathrm{s}) = \omega_n^{\mathrm{E}^2} u_o v(\mathrm{s}). \tag{4.158}$$

Solving for the ratio $u(\mathrm{s})/v(\mathrm{s})$ yields

$$\frac{u(\mathrm{s})}{v(\mathrm{s})} = u_o \frac{\omega_n^{\mathrm{E}^2}}{\mathrm{s}^2 + 2\zeta\omega_n^{\mathrm{E}}\mathrm{s} + \omega_n^{\mathrm{E}^2}}. \tag{4.159}$$

The Laplace transform of a step voltage input is $v(\mathrm{s}) = V/s$. Substituting this result into equation (4.159) and finding the inverse Laplace transform produces

$$\frac{u(t)}{\delta_o} = 1 - \frac{\omega_n^{\mathrm{E}}}{\omega_d^{\mathrm{E}}} e^{-\zeta\omega_n^{\mathrm{E}} t} \sin\left(\omega_d^{\mathrm{E}} t + \phi\right), \tag{4.160}$$

where $\omega_d^{\mathrm{E}} = \omega_n^{\mathrm{E}}\sqrt{1-\zeta^2}$ and $\phi = \cos^{-1}\zeta$. Equation (4.160) assumes that the damping ratio of the system is less than 1, which is typical for most applications in which damping is not specifically designed into the device.

The transducer response to a step change in potential is affected strongly by the variation in energy dissipation. Figure 4.31*a* is a plot of the step response for three values of the damping ratio. We see that an undamped system will exhibit a peak response that is equal to $2\delta_o$, and increasing the damping ratio will produce a decrease in the peak response. The number of oscillations that occur until the response decays to the steady-state value also decreases as the damping in the system increases.

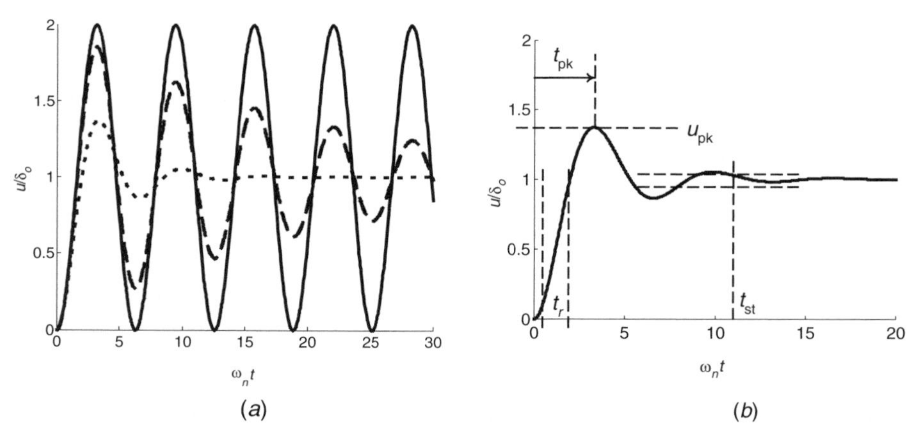

Figure 4.31 (*a*) Transducer step response for three damping ratios: $\zeta = 0$ (solid), $\zeta = 0.05$ (dashed), and $\zeta = 0.3$ (dotted). (*b*) Representation step response.

The response of a damped oscillator to a step input is often characterized by four parameters:

1. The peak response, u_{pk}, the maximum output over all time
2. The time required to reach the peak response, t_{pk}
3. The rise time, t_r, the time required for the output to go from 10% to 90% of its final value.
4. The setting time, t_{st}, the time required for the response to decay to within a prescribed boundary (typically, ±2%) of its steady-state value

These parameters are illustrated in Figure 4.31*b* for a representative step response. Expressions for these parameters have been derived and are written as

$$\begin{aligned} u_{pk} &= 1 + e^{-\zeta\pi/\sqrt{1-\zeta^2}} \\ t_{pk} &= \frac{\pi}{\omega_n^E\sqrt{1-\zeta^2}} \\ t_r &\approx \frac{1.8}{\omega_n^E} \\ t_{st} &= \frac{4}{\zeta\omega_n^E}. \end{aligned} \tag{4.161}$$

These expressions are useful for estimating the time response characteristics of a piezoelectric actuator.

Example 4.13 A piezoelectric stack actuator with a square cross section is being designed using PZT-5H piezoceramic. The positioning application requires the actuator to move a 300-g load with a free displacement of 30 μm. The rise time for the actuator must be less than 0.2 ms. Assuming that the maximum electric field is 1 MV/m, compute the geometry required to obtain these design specifications.

Solution The free deflection of a piezoelectric stack is obtained from equation (4.82). Replacing v/t_p with the maximum electric field of 1 MV/m and solving for the stack length, we have

$$\begin{aligned} L_s &= \frac{30 \times 10^{-6}\ \text{m}}{(650 \times 10^{-12}\ \text{m/V})(1 \times 10^6\ \text{V/m})} \\ &= 46.2\ \text{mm}. \end{aligned}$$

Using the approximations for a second-order oscillator, equation (4.161), the natural frequency required to obtain a 0.2-ms response time is

$$\omega_n^E = \frac{1.8}{0.0002} = 9000\ \text{rad/s}. \tag{4.162}$$

The short-circuit actuator stiffness that is required to obtain this natural frequency is

$$k_p^{\mathrm{E}} = (9000\ \mathrm{rad/s})^2\,(0.3\ \mathrm{kg}) \\ = 24{,}300{,}000\ \mathrm{N/m}.$$

The cross-sectional area that produces this stiffness is obtained from the expression in Table 4.2:

$$A_p = \frac{L_s k_p^{\mathrm{E}}}{Y_3^{\mathrm{E}}} = \frac{(46.2 \times 10^{-3}\ \mathrm{m})(24.3 \times 10^{6}\ \mathrm{N/m})}{62.1 \times 10^{9}\ \mathrm{N/m^2}} \\ = 1.808 \times 10^{-5}\ \mathrm{m}^2.$$

Since the cross-sectional geometry is square, the side length of the actuator is

$$w_p = \sqrt{1.808 \times 10^{-5}\ \mathrm{m}^2} = 4.3\ \mathrm{mm}.$$

The actuator geometry that meets the specifications has a side length of 4.3 mm and a length of 46.2 mm.

4.7.1 Energy Comparisons

In addition to having differences in the time response to step changes in voltage, piezoelectric stacks and bimorphs exhibit important differences in the energy output of the transducers. Recall that the energy, or work, of a device is defined as the product of force and displacement. One of the primary results of this chapter is that actuator geometry can be used to vary the force–deflection trade-offs in a piezoelectric device. A useful comparison of the transducers is to compare the amount of energy or work that can be performed as a function of actuator configuration and actuator geometry.

A useful metric for actuator comparison is the peak energy or work that can be performed as a function of voltage applied:

$$E_{\mathrm{pk}} = \frac{1}{2} f_{\mathrm{bl}} \delta_o. \tag{4.163}$$

The *volumetric energy density* is the peak energy normalized with respect to the actuator volume:

$$E_v = \frac{f_{\mathrm{bl}} \delta_o / 2}{\mathrm{volume}}. \tag{4.164}$$

The units of volumetric energy density are $\mathrm{J/m^3}$.

Using the values for free displacement and blocked force listed in Table 4.2, we can write the volumetric energy density of a piezoelectric stack as

$$E_v = \frac{d_{33}^2 Y_3^{\mathrm{E}} A_p L_s/2}{A_p L_s} \frac{v^2}{t_p^2}. \tag{4.165}$$

Noting that $v/t_p = E_3$, we can write

$$E_v = \frac{1}{2} d_{33}^2 Y_3^{\mathrm{E}} E_3^2 = \frac{1}{2} \left(d_{33} Y_3^{\mathrm{E}} \mathrm{E}_3\right) \left(d_{33} \mathrm{E}_3\right). \tag{4.166}$$

Equation (4.166) is identical to the expression for the energy density of the material in the 33 operating mode. An important attribute of an ideal piezoelectric stack is that there is no reduction in energy density by amplifying the strain through parallel arrangement of the individual piezoelectric layers.

Performing the same analysis for a cantilevered piezoelectric bimorph, we obtain

$$E_v = \frac{9}{8} \frac{d_{13}^2 Y_1^{\mathrm{E}} w_p L_p}{w_p L_p} \frac{v^2}{t_p^2}. \tag{4.167}$$

Recalling that for our definition of the bimorph geometry, $E_3 = 2v/t_p$, we can rewrite equation (4.167) in the form

$$E_v = \frac{9}{16} \left(\frac{1}{2} d_{13}^2 Y_1^{\mathrm{E}} E_3^2 \right) = \frac{9}{16} \left[\frac{1}{2} \left(d_{13} Y_1^{\mathrm{E}} E_3\right) \left(d_{13} E_3\right) \right]. \tag{4.168}$$

Equation (4.168) demonstrates that in the case of a piezoelectric bimorph, the energy density is equal to only 9/16, or approximately 56%, of the energy density of the piezoelectric material operated in the 31 mode. The reduction in volumetric energy density is due to the fact that amplifying the displacement through bending actuation is equivalent to a compliant mechanical amplifier. The compliance in the amplifier reduces the achievable energy density of the device. Comparing the results for a stack actuator to those for a cantilevered bimorph, we note that the energy density of a bimorph is reduced further by the fact that d_{13} is usually a factor of 2 or 3 lower than d_{33}. Accounting for the reduction in the strain coefficient, we see that the energy density of a piezoelectric bimorph might only be 10 to 20% of the energy density of a stack actuator. The reduction in strain coefficient in the 13 direction is offset somewhat by the increase in elastic modulus in the 1 direction.

The energy density of stacks and bimorphs fabricated from various types of piezoelectric material can be computed using equations (4.166) and (4.168). Table 4.3 lists the extensional and bending energy density values for various types of common piezoelectric material. In all cases the energy density of the stack is approximately five to eight times greater than the energy density of a bimorph fabricated from identical material at the same electric field. All results listed in Table 4.3 are for an electric field

Table 4.3 Energy density of different types of piezoelectric materials in extensional and bending mode at an electric field of 1 MV/m

Company		d_{33} (pm/V)	d_{13} (pm/V)	Y_3^E (GPa)	Y_1^E (GPa)	E_v: Extensional (kJ/m^3)	E_v: Bending (kJ/m^3)
Piezo Systems	PSI-5A4E	390	190	52	66	4.0	0.7
	PSI-5H4E	650	320	50	62	10.6	1.8
American Piezo	APC 840	290	125	68	80	2.9	0.4
	APC 850	400	175	54	63	4.3	0.5
	APC 856	620	260	45	58	8.6	1.1
Kinetic Ceramics	PZWT100	370	170	48	62	3.3	0.5
TRS Ceramics	PMN-PT	2250	1050	12	17	30.4	5.3
	TRSHK1 HD	750	360	57	65	16.0	2.4

of 1 MV/m. The energy density at other electric fields can be obtained by multiplying the value listed in the table by the square of the applied electric field in MV/m.

One material type stands out in Table 4.3, due to its high piezoelectric strain coefficients. The material, PMN-PT, is a *single-crystal piezoelectric* that exhibits large piezoelectric strain coefficients and large coupling coefficients (>90%). The large strain coefficient is offset somewhat by the fact that single-crystal ceramics are softer than their polycrystalline counterparts. The energy density of single-crystal materials is generally three to five times larger than that of a conventional ceramic. As of the writing of this book, single-crystal materials were also more expensive than conventional materials and were generally thought of as a good solution for high-end applications of piezoelectric materials where large strain (>0.5%) and good coupling properties were required.

The values listed in Table 4.3 are ideal values that do not account for certain limitations in the fabrication or operation of the material. For example, the values listed for piezoelectric stacks do not incorporate nonideal behavior introduced by inactive electrodes or insulating material. More important, these values do not reflect the inactive mass associated with important components such as the housing or preload springs. Adding the mass of inactive components can reduce the actual energy density by a factor of 3 to 5 compared to the energy density of the material itself. These issues are less important for bimorph actuators, which in many types of operation do not require preloading or casing.

4.8 ELECTROSTRICTIVE MATERIALS

Thus far in our discussion of piezoelectric materials we have assumed that the constitutive relationships in the field variables are linear functions. In the constitutive equations for the mechanical properties, the stress and strain were related by a matrix of elastic constants, and for the electrical properties it is assumed that the electric displacement and electric field are also related through a matrix of dielectric constants.

As discussed in Chapter 3, the linear constitutive relationships are derived from an energy formulation which assumes a quadratic relationship in the energy function. Another fundamental assumption is that the electromechanical coupling properties are also linear. The result of this assumption is that the coupling between electrical and mechanical domains is also modeled as a linear matrix of constants.

Electrostrictive materials are those in which the electromechanical coupling is represented by the quadratic relationship between strain and electric field. In indicial notation the strain–field relationships are written

$$\mathrm{S}_{ij} = M_{ijmn}\mathrm{E}_m\mathrm{E}_n. \tag{4.169}$$

The variable M_{ijmn} is a fourth-rank tensor of electrostriction coefficients. In the case in which the applied electric field is only in a single direction, the constitutive relationships are

$$\mathrm{S}_{ij} = M_{ijn}\mathrm{E}_n^2. \tag{4.170}$$

The quadratic relationship between applied field and strain produces a response that is fundamentally different from that of a piezoelectric material. Linear coupling between strain and field produces a mechanical response that will change polarity when the polarity of the electric field is changed. For example, a piezoelectric material with a positive strain coefficient will produce positive strain when the electric field is positive and negative strain when the electric field is negative. A quadratic strain–electric field relationship will produce strain in only a single direction. A positive electrostrictive coefficient will produce positive strain when the field is positive but will also produce positive strain when the polarity of the electric field is changed. This physical response is due to the quadratic field relationship in equation (4.170). The difference between the strain response of piezoelectric and electrostrictive materials is shown in Figure 4.32. The strain response of the piezoelectric material is, of course,

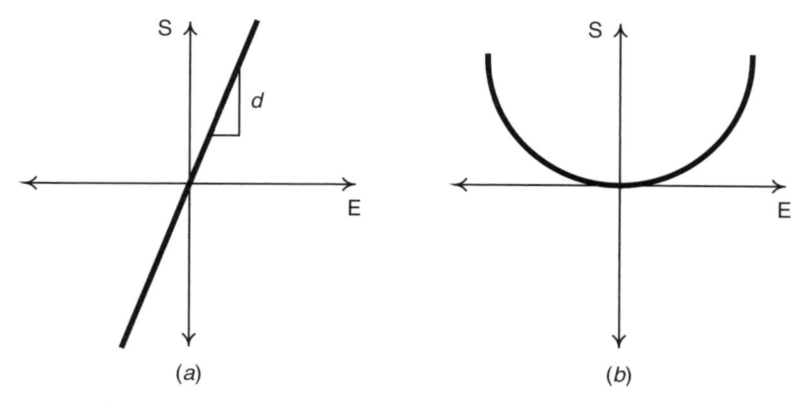

Figure 4.32 Representative strain responses for (*a*) piezoelectric and (*b*) electrostrictive materials.

linear where the slope is equal to the strain coefficient of the material. The parabola represents the quadratic strain response of the electrostrictive material. Although the curves are only representative, the crossing of the two curves is intential since a quadratic function will always produce a higher value at some value of the applied electric field. The exact value in which the two responses are identical is a function of piezoelectric and electrostrictive coefficients.

4.8.1 One-Dimensional Analysis

To understand the basic properties of electrostrictive materials and compare them to piezoelectric materials, let us consider an analysis in which the applied electric field is in only one direction and that we are only interested in the strains in a single direction. In this case we can drop the subscript notation in equation (4.170) and simply write the strain–electric field relationship as

$$\mathrm{S} = M\mathrm{E}^2, \tag{4.171}$$

where M is the electrostrictive coefficient in the direction of interest.

Example 4.14 An electrostrictive ceramic has an electrostrictive coefficient of $8 \times 10^{-17}\ \mathrm{m^2/V^2}$. A material sample has a thickness of 0.05 mm. Compute the strain induced by the application of 100 V.

Solution The electric field induced by the applied voltage is

$$\mathrm{E} = \frac{100\ \mathrm{V}}{0.05 \times 10^{-3}\ \mathrm{m}} = 2\ \mathrm{MV/m}.$$

The electrostrictive strain is computed from equation (4.171),

$$\mathrm{S} = (8 \times 10^{-17}\ \mathrm{m^2/V^2})(2 \times 10^6\ \mathrm{V/m})^2 = 320\ \mu\mathrm{strain}.$$

The quadratic relationship between applied field and mechanical strain is sometimes problematic for the development of devices using electrostrictive materials. For example, in applications for motion control, it is often desirable to have a linear relationship between applied electric field and strain since it simplifies the design of actuators and motors. One method of transforming the quadratic relationship of an electrostrictive material into an equivalent linear relationship is to apply a *biased* electric field that consists of the sum of a direct-current (dc) value and an alternating-current (ac) component:

$$\mathrm{E} = \mathrm{E_{dc}} + \mathrm{E_{ac}}. \tag{4.172}$$

Substituting equation (4.172) into equation (4.171) produces

$$S = M\left(E_{dc}^2 + 2E_{dc}E_{ac} + E_{ac}^2\right). \tag{4.173}$$

The first term in the expansion on the right-hand side represents a constant strain induced by the dc electric field. Denoting this as $ME_{dc}^2 = S_{dc}$ yields

$$S - S_{dc} = 2ME_{dc}E_{ac} + ME_{ac}^2. \tag{4.174}$$

Assuming that the alternating field is much smaller than the product of the dc and ac fields,

$$E_{ac}^2 \ll E_{dc}E_{ac}, \tag{4.175}$$

the alternating mechanical strain about the dc strain is written as the approximation

$$S - S_{dc} \approx 2ME_{dc}E_{ac}. \tag{4.176}$$

The important fact to note about equation (4.176) is that that the ac strain is a *linear* function of the applied ac field under the assumption that equation (4.175) is valid. The coefficient $2ME_{dc}$ is constant due to the fact that the dc field is constant with time. If this constant is denoted d, equation (4.176) is rewritten

$$S - S_{dc} \approx dE_{ac}, \tag{4.177}$$

which is identical to the expression for a piezoelectric material.

This analysis demonstrates that an electrostrictive material can function as a linear piezoelectric material by applying an electric field that consists of the superposition of a dc bias and an ac field. The coefficient $d = 2ME_{dc}$, called the *effective piezoelectric strain coefficient*, can be compared directly with the strain coefficients of a piezoelectric material.

Example 4.15 Compute the effective piezoelectric strain coefficient for the material studied in Example 4.14 under the application of a dc electric field of 10 kV/cm.

Solution The effective piezoelectric strain coefficient is

$$d = 2(8 \times 10^{-17}\ \text{m}^2/\text{V}^2)(1 \times 10^6\ \text{V/m}) = 160\ \text{pm/V}.$$

The units of the coefficient are equal to m/V, which is equivalent to the units of d for piezoelectric materials.

The effective piezoelectric strain coefficient is proportional to the applied dc field, and increasing the dc field will increase the effective strain coefficient. For many types of electrostrictive materials the effective strain coefficient can be as great if not greater than that of a piezoelectric material.

The effective piezoelectric strain coefficient has a simple mathematical interpretation. Taking the derivative of equation (4.171) with respect to the electric field yields

$$\frac{d\mathrm{S}}{d\mathrm{E}} = 2M\mathrm{E}. \tag{4.178}$$

Evaluating this equation at the bias field E_{dc} produces the same value as the effective piezoelectric strain coefficient. Thus, the effective piezoelectric strain coefficient can be interpreted as the slope of the strain-electric field curve evaluated at the bias.

4.8.2 Polarization-Based Models of Electrostriction

The model proposed in equation (4.170) is a relationship between the mechanical strain and the applied electric field. This model serves as a useful comparison to piezoelectric materials because in previous chapters we have introduced piezoelectricity as a linear coupling between strain and applied field.

An alternative set of constitutive equations are written in terms of the *electric polarization*, P. The strain is written as a quadratic function of the polarization in the manner

$$\mathrm{S}_{ij} = Q_{ijmn}\mathrm{P}_m\mathrm{P}_n. \tag{4.179}$$

The fourth-rank tensor Q_{ijmn} is the electrostrictive coefficients in units of $\mathrm{m}^4/\mathrm{C}^2$. The electric polarization is related to the electric displacement through the expression

$$\mathrm{P}_m = \mathrm{D}_m - \varepsilon_o\mathrm{E}_m \tag{4.180}$$

and has the same units as the electric displacement (charge/area).

Why express the strain as a function of polarization instead of field? If the polarization is a linear function of the electric field,

$$\mathrm{P}_m \propto \mathrm{E}_m, \tag{4.181}$$

the strain is simply proportional to the square of the electric field. This is identical to the model presented in Section 4.8.1, in which the strain was written as a quadratic function of the electric field.

In actuality, though, it is known through experimentation that the electric polarization is not proportional to the electric field. The polarization–electric field relationship exhibits a *saturation* phenomenon as the electric field is increased, which implies that the polarization remains approximately constant as the electric field is increased. Figure 4.33*a* illustrates this phenomenon and compares it to a perfectly linear relationship between P and E.

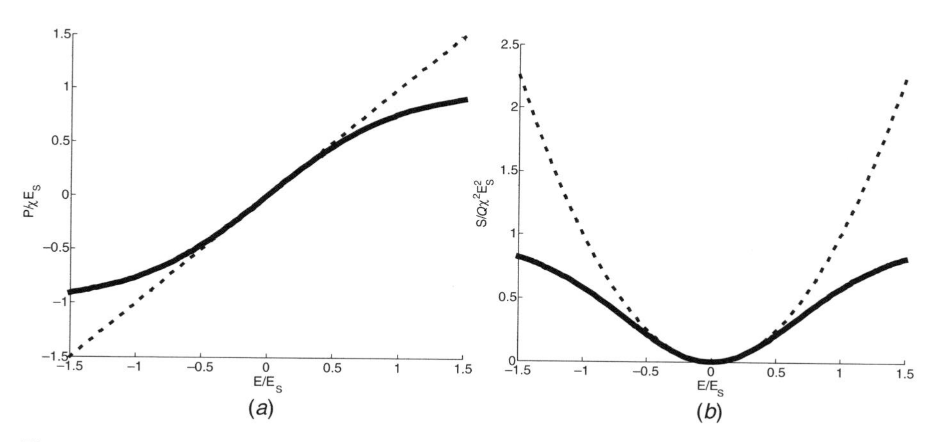

Figure 4.33 (*a*) Polarization-to-electric field relationship for an electrostrictive material: with saturation (solid), without saturation (dashed); (*b*) strain-to-electric field relationship for an electrostrictive material: with saturation (solid), without saturation (dashed).

To understand the effect of polarization saturation on the strain-to-field relationship, let us once again reduce the general constitutive relationship, equation (4.179), to a single dimension:

$$\mathrm{S} = Q\mathrm{P}^2. \tag{4.182}$$

A model of the relationship between polarization and electric field that accounts for saturation is

$$\mathrm{P} = \chi \mathrm{E_S} \tanh \frac{\mathrm{E}}{\mathrm{E_S}}, \tag{4.183}$$

where χ is the *pseudosusceptibility* and $\mathrm{E_S}$ is the *saturation electric field*. Substituting equation (4.183) into equation (4.182) produces a relationship between strain and electric field (Figure 4.33*b*) that incorporates a model of saturation:

$$\mathrm{S} = Q\chi^2 \mathrm{E_S^2} \tanh^2 \frac{\mathrm{E}}{\mathrm{E_S}}. \tag{4.184}$$

This model is analogous to the model developed earlier in the chapter for low electric fields. The hyperbolic tangent has the property that $\tanh x \approx x$ when $x \ll 1$. Thus, when $\mathrm{E} \ll \mathrm{E_S}$, $\tanh^2(\mathrm{E}/\mathrm{E_S}) \approx (\mathrm{E}/\mathrm{E_S})^2$ and

$$\mathrm{S} \approx Q\chi^2 \mathrm{E}^2, \tag{4.185}$$

which is identical to the original model proposed for electrostriction. This analysis demonstrates that the strain in an electrostrictive material is approximately

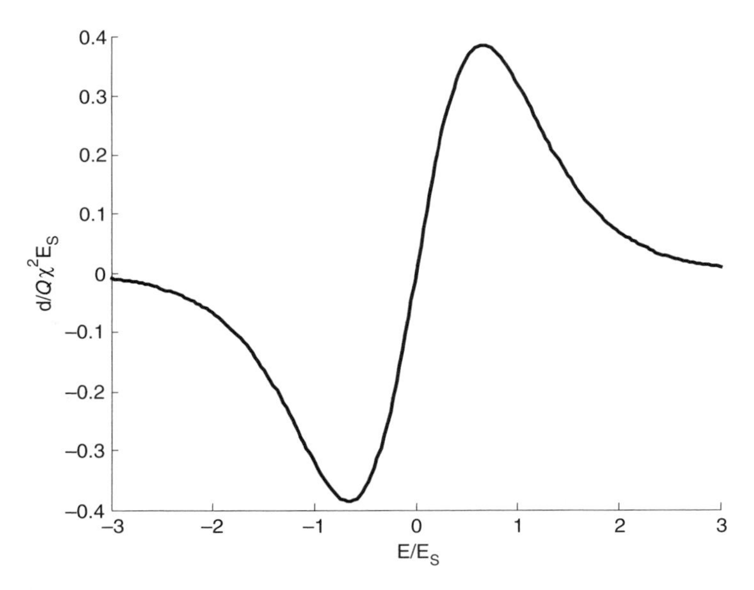

Figure 4.34 Variation in the effective strain coefficient for an electrostrictive material that exhibits polarization saturation as a function of a normalized electric field.

proportional to the square of the electric field at field values much less than those of the saturation electric field.

In Section 4.8.1 it was demonstrated that an electrostrictive material can be operated similar to a piezoelectric material by applying a dc bias in addition to an ac electric field. As long as the ac field is "small" compared to the dc field, the material exhibits linear electromechanical coupling. The effective piezoelectric strain coefficient of the material is obtained by computing the derivative of the strain-to-electric field response. For an electrostrictive material that exhibits saturation, the effective strain coefficient is derived by taking the derivative of equation (4.184):

$$\frac{dS}{dE} = d = Q\chi^2 E_S \tanh\frac{E}{E_S}\left[1 - \tanh\left(\frac{E}{E_S}\right)^2\right]. \tag{4.186}$$

Plotting the function $d/Q\chi^2 E_S$ as a function of E/E_S as shown in Figure 4.34 illustrates that the effective piezoelectric strain coefficient exhibits a maximum value of $\approx 0.38 Q\chi^2 E_S$ when $E/E_S \approx \frac{2}{3}$. A negative bias field of $E/E_S \approx -\frac{2}{3}$ will yield an effective strain coefficient of the same value but opposite in sign.

Example 4.16 Measurements of the electric field–polarization properties of an electrostrictive material indicate that the pseudosusceptibility χ is equal to 2.5×10^{-10} F/mm, and the saturation electric field E_S is equal to 800 V/mm. Using the model developed in Section 4.8.1, determine (a) the bias electric field that maximizes the effective piezoelectric strain coefficient, and (b) the electrostriction coefficient Q that is required to achieve an effective piezoelectric strain coefficient of 300 pm/V.

Solution (a) The electric field that maximizes the effective piezoelectric strain coefficient is equal to $\frac{2}{3}\mathrm{E_S}$:

$$\mathrm{E} = \frac{2}{3}(800\ \mathrm{V/mm}) = 533\ \mathrm{V/mm}.$$

(b) From the discussion preceding this example we know that the maximum effective strain coefficient is approximately equal to $0.38Q\chi^2\mathrm{E_S}$ at the optimal bias field. Thus, the electrostrictive coefficient required is computed from

$$\begin{aligned} Q = \frac{d}{0.38\chi^2\mathrm{E_S}} &= \frac{300 \times 10^{-12}\ \mathrm{m/V}}{0.38\left(2.5 \times 10^{-7}\ \mathrm{F/m}\right)^2\left(8 \times 10^5\ \mathrm{V/m}\right)} \\ &= 0.0158\ \mathrm{m^4/C^2}. \end{aligned}$$

Example 4.17 An electrostrictive ceramic with a pseudosusceptibility of 2×10^{-7} F/m and an electrostrictive coefficient $Q = 0.012\ \mathrm{m^4/C^2}$ is being excited by an electric field. If the saturation electric field is 750 V/mm, compute the strain obtained at the application of an electric field of 1500 V/mm.

Solution The electrostrictive strain is computed from equation (4.184). The coefficient in the expression is

$$\begin{aligned} Q\chi^2\mathrm{E_S^2} &= (0.012\ \mathrm{m^4/C^2})(2 \times 10^{-7}\ \mathrm{F/m})^2(7.5 \times 10^5\ \mathrm{V/m})^2 \\ &= 2.7 \times 10^{-4}\ \mathrm{m/m}. \end{aligned}$$

This coefficient is modified by the hyperbolic tangent function in equation (4.184)

$$\tanh\frac{\mathrm{E}}{\mathrm{E_S}} = \tanh\frac{1500\ \mathrm{V/mm}}{750\ \mathrm{V/mm}} = 0.964.$$

The strain is computed from

$$\mathrm{S} = (2.7 \times 10^{-4}\ \mathrm{m/m})(0.964) = 260\ \mu\mathrm{strain}.$$

From Figure 4.33 we note that the hyperbolic tangent function asymptotes at 1 for large values of $\mathrm{E/E_S}$; therefore, computing the coefficient $Q\chi^2\mathrm{E_S^2}$ allows an estimation of the maximum strain achievable with an electrostrictive material. The hyperbolic tangent function can then be computed at the specified field to determine the strain that can be achieved.

4.8.3 Constitutive Modeling

Thus far in this chapter we have introduced the basic properties of electrostrictive materials and concentrated the analysis on systems that can be modeled in one dimension. As we did with piezoelectric materials, we now extend the analysis to the full constitutive expressions for electrostrictive materials. A set of constitutive expressions will allow us to analyze the use of electrostrictive materials in multiple dimensions.

The constitutive relationships for an electrostrictive material, written in indicial notation, are

$$\mathrm{T}_{ij} = C_{ijkl}\mathrm{S}_{kl} - C_{ijkl}Q_{klmn}\mathrm{P}_m\mathrm{P}_n \tag{4.187}$$

$$\mathrm{E}_m = -2C_{ijkl}Q_{klmn}\mathrm{S}_{ij}\mathrm{P}_n + \chi_m^{-1}\mathrm{P}_m^{\mathrm{S}} \operatorname{arctanh}\frac{\mathrm{P}_m}{\mathrm{P}_m^{\mathrm{S}}} + 2C_{ijkl}Q_{klmn}Q_{ijpq}\mathrm{P}_n\mathrm{P}_p\mathrm{P}_q. \tag{4.188}$$

Voigt notation can be introduced into these two expressions to yield a less cumbersome version of the constitutive relationships. Using the Voigt notation, equations (4.187) and (4.188) are rewritten

$$\mathrm{T}_i = c_{ij}^{\mathrm{P}}\mathrm{S}_j - c_{ij}Q_{jkl}\mathrm{P}_k\mathrm{P}_l \tag{4.189}$$

$$\mathrm{E}_k = -2c_{ij}^{\mathrm{P}}Q_{jkl}\mathrm{S}_i\mathrm{P}_l + \chi_k^{-1}\mathrm{P}_k^{\mathrm{S}} \operatorname{arctanh}\frac{\mathrm{P}_k}{\mathrm{P}_k^{\mathrm{S}}} + 2c_{ij}^{\mathrm{P}}Q_{jkl}Q_{ipq}\mathrm{P}_l\mathrm{P}_p\mathrm{P}_q. \tag{4.190}$$

For isotropic materials the matrix of elastic coefficients is expressed as

$$c^{\mathrm{P}} = \frac{Y}{(1+\nu)(1-2\nu)}\begin{bmatrix} 1-\nu & \nu & \nu & 0 & 0 & 0 \\ \nu & 1-\nu & \nu & 0 & 0 & 0 \\ \nu & \nu & 1-\nu & 0 & 0 & 0 \\ 0 & 0 & 0 & \dfrac{1-2\nu}{2} & 0 & 0 \\ 0 & 0 & 0 & 0 & \dfrac{1-2\nu}{2} & 0 \\ 0 & 0 & 0 & 0 & 0 & \dfrac{1-2\nu}{2} \end{bmatrix}, \tag{4.191}$$

and the electrostrictive coefficients are reduced to

$$\begin{aligned} Q_{111} &= Q_{222} = Q_{333} \\ Q_{122} &= Q_{133} = Q_{211} = Q_{233} = Q_{311} = Q_{322} \\ 2(Q_{111} - Q_{222}) &= Q_{412} = Q_{523} = Q_{613}. \end{aligned} \tag{4.192}$$

All other electrostrictive coefficients are zero. Thus, for isotropic materials the mechanical properties are specified by the elastic modulus Y and Poisson's ratio ν and the electrostrictive properties are specified by two coefficients, Q_{111} and Q_{122}.

Consider the case in which we have an electrostrictive material that is being operated in the 33 mode. For this analysis we assume that

$$\mathrm{P}_1 = \mathrm{P}_2 = 0. \tag{4.193}$$

Substituting the expressions in equation (4.193) into the constitutive expressions reduces the first three constitutive relationships to

$$\begin{aligned}
\mathrm{T}_1 &= c_{11}^{\mathrm{P}}\mathrm{S}_1 + c_{12}^{\mathrm{P}}\mathrm{S}_2 + c_{13}^{\mathrm{P}}\mathrm{S}_3 - \left(c_{11}^{\mathrm{P}}Q_{133} + c_{12}^{\mathrm{P}}Q_{233} + c_{13}^{\mathrm{P}}Q_{333}\right)\mathrm{P}_3^2 \\
\mathrm{T}_2 &= c_{21}^{\mathrm{P}}\mathrm{S}_1 + c_{22}^{\mathrm{P}}\mathrm{S}_2 + c_{23}^{\mathrm{P}}\mathrm{S}_3 - \left(c_{21}^{\mathrm{P}}Q_{133} + c_{22}^{\mathrm{P}}Q_{233} + c_{23}^{\mathrm{P}}Q_{333}\right)\mathrm{P}_3^2 \\
\mathrm{T}_3 &= c_{31}^{\mathrm{P}}\mathrm{S}_1 + c_{32}^{\mathrm{P}}\mathrm{S}_2 + c_{33}^{\mathrm{P}}\mathrm{S}_3 - \left(c_{31}^{\mathrm{P}}Q_{133} + c_{32}^{\mathrm{P}}Q_{233} + c_{33}^{\mathrm{P}}Q_{333}\right)\mathrm{P}_3^2.
\end{aligned} \tag{4.194}$$

Expanding the expression for E_3 produces

$$\begin{aligned}
\mathrm{E}_3 = &-2\left[\left(c_{11}^{\mathrm{P}}Q_{133} + c_{12}^{\mathrm{P}}Q_{233} + c_{13}^{\mathrm{P}}Q_{333}\right)\mathrm{S}_1\right. \\
&+ \left(c_{21}^{\mathrm{P}}Q_{133} + c_{22}^{\mathrm{P}}Q_{233} + c_{23}^{\mathrm{P}}Q_{333}\right)\mathrm{S}_2 \\
&+ \left.\left(c_{31}^{\mathrm{P}}Q_{133} + c_{32}^{\mathrm{P}}Q_{233} + c_{33}^{\mathrm{P}}Q_{333}\right)\mathrm{S}_3\right]\mathrm{P}_3 + \chi_3^{-1}\mathrm{P}_3^{\mathrm{S}}\,\mathrm{arctanh}\,\frac{\mathrm{P}_3}{\mathrm{P}_3^{\mathrm{S}}} \\
&+2\left[\left(c_{11}^{\mathrm{P}}Q_{133} + c_{21}^{\mathrm{P}}Q_{233} + c_{31}^{\mathrm{P}}Q_{333}\right)Q_{133}\right. \\
&+ \left(c_{12}^{\mathrm{P}}Q_{133} + c_{22}^{\mathrm{P}}Q_{233} + c_{32}^{\mathrm{P}}Q_{333}\right)Q_{233} \\
&+ \left.\left(c_{13}^{\mathrm{P}}Q_{133} + c_{23}^{\mathrm{P}}Q_{233} + c_{33}^{\mathrm{P}}Q_{333}\right)Q_{333}\right]\mathrm{P}_3^2.
\end{aligned} \tag{4.195}$$

The expression for E_3 can be placed in terms of the applied stress by rewriting the expressions in equation (4.194) and subsituting them into equation (4.195). The result is a much simpler expression,

$$\mathrm{E}_3 = -2\left(Q_{133}\mathrm{T}_1 + Q_{233}\mathrm{T}_2 + Q_{333}\mathrm{T}_3\right)\mathrm{P}_3 + \chi_3^{-1}\mathrm{P}_3^{\mathrm{S}}\,\mathrm{arctanh}\,\frac{\mathrm{P}_3}{\mathrm{P}_3^{\mathrm{S}}}, \tag{4.196}$$

which, as expected, reduces to the stress-free polarization response of the electrostrictive material when $\mathrm{T}_1 = \mathrm{T}_2 = \mathrm{T}_3 = 0$. Introducing the assumption of isotropy into equation (4.194) produces the expressions

$$\begin{aligned}
\mathrm{T}_1 &= \frac{Y}{(1+\nu)(1-2\nu)}\left[(1-\nu)\,\mathrm{S}_1 + \nu\mathrm{S}_2 + \nu\mathrm{S}_3 - \left(Q_{122} + \nu Q_{111}\right)\mathrm{P}_3^2\right] \\
\mathrm{T}_2 &= \frac{Y}{(1+\nu)(1-2\nu)}\left[\nu\mathrm{S}_1 + (1-\nu)\,\mathrm{S}_2 + \nu\mathrm{S}_3 - \left(Q_{122} + \nu Q_{111}\right)\mathrm{P}_3^2\right] \\
\mathrm{T}_3 &= \frac{Y}{(1+\nu)(1-2\nu)}\left[\nu\mathrm{S}_1 + \nu\mathrm{S}_2 + (1-\nu)\,\mathrm{S}_3 - \left(2\nu Q_{122} + (1-\nu)\,Q_{111}\right)\mathrm{P}_3^2\right].
\end{aligned} \tag{4.197}$$

In many applications it will be necessary to solve for the strain given the state of stress and the input polarization or electric field. In this case equation (4.197) is solved for

Table 4.4 Representative properties for an electrostrictive material

Y (GPa)	ν	Q_{111} (m^4/C^2)	Q_{122} (m^4/C^2)	P_3^S (C/m^2)	χ_3 (F/m)
95	0.33	1×10^{-2}	-6×10^{-3}	0.26	3.12×10^{-7}

the strain components by writing the system of equations as a matrix and then solving for S_1, S_2, and S_3. The result is

$$\begin{aligned} S_1 &= \frac{1}{Y}(T_1 - \nu T_2 - \nu T_3) + Q_{122}P_3^2 \\ S_2 &= \frac{1}{Y}(-\nu T_1 + T_2 - \nu T_3) + Q_{122}P_3^2 \\ S_3 &= \frac{1}{Y}(-\nu T_1 - \nu T_2 + T_3) + Q_{111}P_3^2. \end{aligned} \tag{4.198}$$

It is expected that the expressions for strain would reduce to a quadratic function of the polarization in the stress-free state.

Example 4.18 An electrostrictive ceramic has the properties listed in Table 4.4. Plot the relationship between electric field and polarization in the 3 direction for values of $0 \leq P_3 \leq 0.25$ C/m^2 when a compressive stress of 40 MPa is applied in the 3 direction and no stress is applied in the 1 and 2 directions.

Solution The electric field is computed using equation (4.198). Substituting the values $T_1 = T_2 = 0$ and $T_3 = -40$ MPa into the expression along with the material properties yields

$$\begin{aligned} E_3 &= -2[(1\times 10^{-2}\ m^4/C^2)(-40\times 10^6\ N/m^2)]P_3 + \frac{0.26\ C/m^2}{3.12\times 10^{-7}\ F/m}\operatorname{arctanh}\frac{P_3}{P_3^S} \\ &= (8\times 10^5)P_3 + (8.33\times 10^5)\operatorname{arctanh}\frac{P_3}{P_3^S}. \end{aligned}$$

This expression allows us to plot the relationship between E_3 and P_3. Before plotting the result, note that the relationship is a function of a linear term in the polarization and the inverse hyperbolic tangent. At low values of the polarization the inverse hyperbolic tangent is approximately linear; therefore, we expect that the field-to-polarization curve will be approximately linear when $P_3 \ll P_3^S$. As the polarization input approaches the saturation polarization, we will see more pronounced nonlinearity due to the inverse hyperbolic term. These attributes are exhibited in the field-to-polarization curve shown in Figure 4.35. The relationship is approximately linear up to a polarization of approximately 0.15 C/m^2, or $P_3/P_3^S = 0.58$, and then exhibits a nonlinearity due to the saturation phenomena. It is important to realize that the

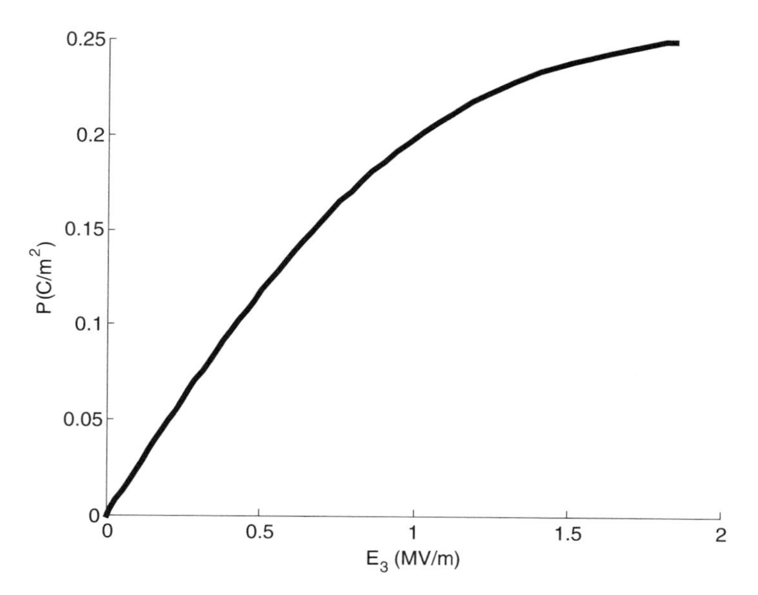

Figure 4.35 Field-to-polarization curve for Example 4.18.

nonlinearity in the field-to-polarization response becomes pronounced at a polarization value much lower than the saturation polarization of the material.

Example 4.19 For the material with properties listed in Table 4.4, plot the strain in the polarization direction as a function of the electric field for polarization values between $-0.25 \leq P_3 \leq 0.25$.

Solution The relationship between electric field and polarization is given in Example 4.18 for the material properties listed in Table 4.4:

$$E_3 = (8 \times 10^5)P_3 + (8.33 \times 10^5)\text{arctanh}\frac{P_3}{0.26\ \text{C/m}^2}.$$

Equation (4.198) is used to compute the relationship between polarization and strain:

$$\begin{aligned} S_3 &= \frac{-40 \times 10^6\ \text{N/m}^2}{95 \times 10^9\ \text{N/m}^2} + (1 \times 10^{-2}\ \text{m}^4/\text{C}^2)P_3^2 \\ &= -421 \times 10^{-6} + (1 \times 10^{-2}\ \text{m}^4/\text{C}^2)P_3^2 \qquad \text{m/m}. \end{aligned}$$

Solving for the strain as a function of electric field directly is complicated by the fact that the polarization is an argument of the inverse hyperbolic tangent function. The relationship between strain and electric field can be plotted by computing the

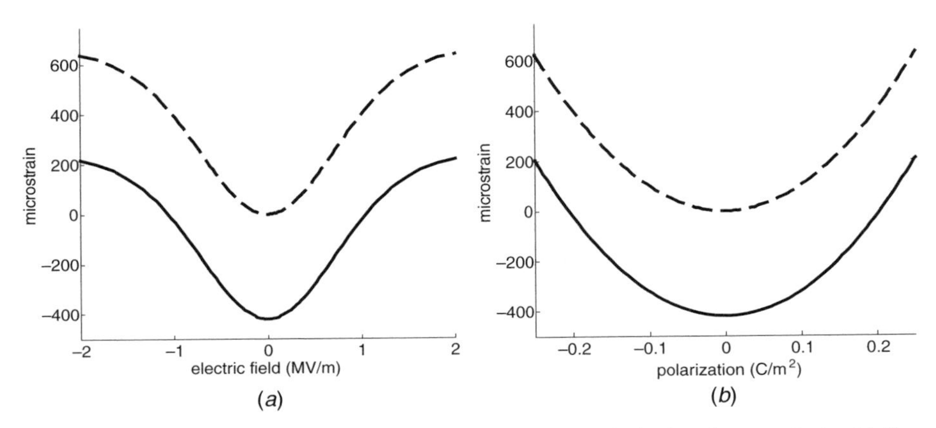

Figure 4.36 (*a*) Strain-to-electric field and (*b*) strain-to-polarization for an electrostrictive material.

strain and electric field numerically for the polarization values specified and plotting S_3 versus E_3. The relationship between strain and electric field exhibits a saturation at electric fields that causes polarization saturation (Figure 4.36*a*). The offset in the strain response is associated with the constant strain applied by the external stress. Often, the strain is plotted with a zero offset to illustrate the amount of strain caused by the applied field. This is shown by the dashed line in Figure 4.36*a*. As illustrated by equation (4.198), the strain-to-polarization response is a quadratic function, as shown in Figure 4.36*b*.

4.8.4 Harmonic Response of Electrostrictive Materials

The primary difference between piezoelectric materials and electrostrictive materials is the nature of the electromechanical coupling. Piezoelectric materials exhibit a linear relationship between mechanical response (strain and stress) and electrical input (polarization and electric field). In addition, electrostrictive materials have been shown to exhibit polarization saturation, which results in a nonlinear relationship between the applied electric field and the polarization of the material. In Section 4.8.3 we demonstrated that these two properties produce electromechanical coupling properties that are substantially different than piezoelectric materials.

The linear coupling exhibited by piezoelectric material properties also has important ramifications when the mechanical or electrical variables are time varying. For example, we know that when the electric field varies as a harmonic function of time (e.g., sine or cosine), the steady-state response of linear system will also consist of a harmonic response at the same frequency but with different amplitude and phase.

Harmonic excitation of electrostrictive materials does not exhibit the same type of behavior as that of a piezoelectric material. The nonlinear coupling between polarization and strain and the saturation phenomena associated with the field response will produce nonlinear behavior in the harmonic response of an electrostrictive material. To examine this property, consider the case in which an electrostrictive material is excited by a polarization input that is aligned with the polarization direction (i.e., the 3 direction), and the polarization is a harmonic function with frequency ω,

$$\mathrm{P}_3 = P \sin\omega t. \tag{4.199}$$

Consider the electric field response when the external stress components are zero

$$\mathrm{E}_3 = \chi_3^{-1}\mathrm{P}_3^{\mathrm{S}} \,\mathrm{arctanh}\left(\frac{P}{\mathrm{P}_3^{\mathrm{S}}}\sin\omega t\right). \tag{4.200}$$

Recall that the inverse hyperbolic tangent function is approximately linear when the argument is less than 0.5. We would expect that the electric field response would be approximately a pure harmonic function when the amplitude of the polarization is less than half the saturation polarization. As the amplitude of the polarization approaches the saturation polarization value we would expect that the electric field response would become more distorted. This result is plotted in Figure 4.37. When $P/\mathrm{P}_3^{\mathrm{S}} = 0.5$, the electric field response is approximately a sine wave. Increasing the amplitude of the polarization closer to the saturation value produces an increasingly distorted electric field, until when $P/\mathrm{P}_3^{\mathrm{S}} = 0.95$, the electric field exhibits pronounced nonlinearity due to the saturation of the electrical response.

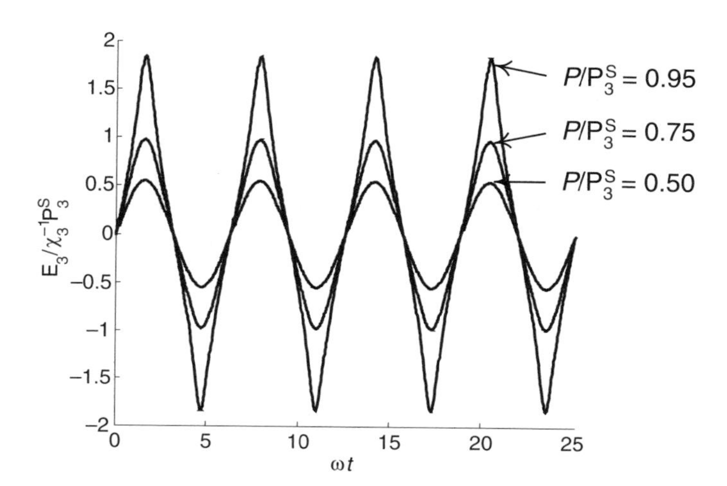

Figure 4.37 Effect of polarization saturation on the harmonic response of an electrostrictive material.

Consider the strain response in which the external stress components are all zero, $T_1 = T_2 = T_3 = 0$,

$$\begin{aligned} S_1 &= Q_{122} P^2 \sin^2 \omega t \\ S_2 &= Q_{122} P^2 \sin^2 \omega t \\ S_3 &= Q_{111} P^2 \sin^2 \omega t. \end{aligned} \tag{4.201}$$

Applying the trigonometric identity for $\sin^2 \omega t$, the strain response is written

$$\begin{aligned} S_1 &= \frac{Q_{122} P^2}{2} (1 - \cos 2\omega t) \\ S_2 &= \frac{Q_{122} P^2}{2} (1 - \cos 2\omega t) \\ S_3 &= \frac{Q_{111} P^2}{2} (1 - \cos 2\omega t). \end{aligned} \tag{4.202}$$

The quadratic relationship between polarization and strain produces a dc bias in the strain response to a harmonic excitation. The dc bias is proportional to the square of the polarization amplitude. The normalized polarization input and normalized strain response are shown in Figure 4.38 as a function of ωt. In addition to creating a bias strain output, the quadratic relationship between polarization and strain produces a response that oscillates at twice the input frequency with a peak-to-peak amplitude that is equivalent to S/QP^2, where S is the strain in either the 1, 2, or 3 direction and Q represents either Q_{111} or Q_{122}.

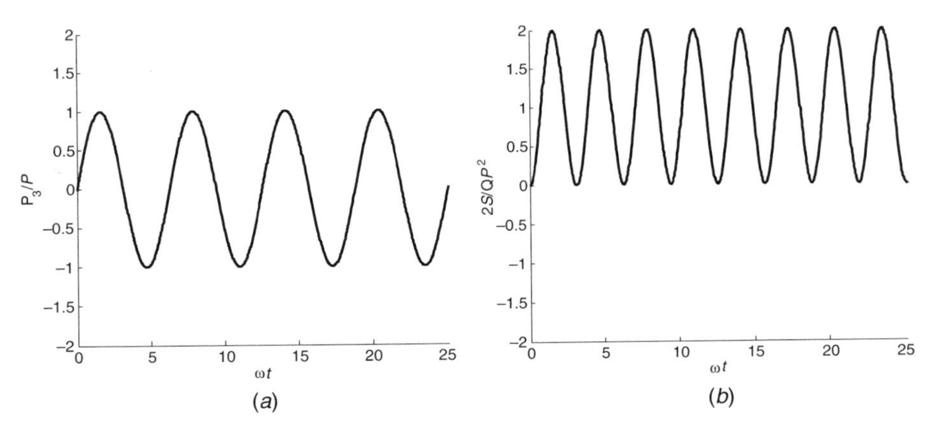

Figure 4.38 (*a*) Harmonic polarization input normalized with respect to the amplitude; (*b*) normalized strain response.

4.9 CHAPTER SUMMARY

The fundamental relationships for a linear piezoelectric material were introduced in this chapter and related to the equations for transducers that utilize piezoeletric material. A linear piezoelectric material was characterized by two pairs of field variables. The mechanical field variables are stress and strain and the electrical field variables are electric displacement and electric field. In the full three-dimensional constitutive equations the stress and strain are related through the compliance matrix, the electric displacement and electric field are related through the dielectric permittivity, and the coupling term is represented by the matrix of piezoelectric strain coefficients.

The equations for piezoelectric transducers were derived directly from the constitutive relationships. Expressions for piezoelectric stack actuators and sensors were obtained using a reduced form of the constitutive equations that contained the terms associated with the 33 direction of the material. Expressions for piezoelectric bimorphs were obtained using a reduced form that contained terms associated with the 31 direction. In both cases the transducer equations were obtained by introducing the geometry of the transducer into the constitutive relationships. One of the most important results in these analyses is the derivation of the force–deflection expressions for piezoelectric actuators and the sensitivity expressions for piezoelectric sensors. The derivations demonstrated that the force–deflection characteristics of piezoelectric transducers can be substantially different, depending on the mode of operation. Generally speaking, piezoelectric stack actuators produce higher force and smaller deflection than a piezoelectric bimorph of similar size. Piezoelectric bimorph actuators are useful in applications that require larger motion but smaller forces than can be obtained with a piezoelectric stack.

Direct comparisons between the various actuators were obtained in the chapter. The time response of stacks and bimorphs are compared by noting that both sets of transducer equations can be analyzed as a second-order dynamic system. A useful basis of comparison is the static energy density that is achieved by an actuator. This analysis demonstrated that piezoelectric stacks typically have energy densities that are five to 10 times greater than that of piezoelectric bimorphs. This advantage is due to more efficient use of the piezoelectric phenomenon and the fact that the 33 mode of operation generally has higher piezoelectric strain coefficients than the 31 operating mode.

In the final section of the chapter we focused on an analysis of electrostrictive material behavior. Electrostrictive behavior is characterized by a quadratic relationship between applied field and strain. The basic properties of electrostrictive materials were introduced and compared to the properties of linear piezoelectric behavior. Methods for linearizing the behavior of electrostrictive materials are based on the application of a constant bias field and operation in a small regime around the bias point. This method allows an electrostrictive material to be treated as a linear piezoelectric material for the purposes of analyzing applications such as actuators.

PROBLEMS

4.1. Compute the stress required to produce 100 microstrain in APC 856 when the applied electric field is held constant at zero. Compute the stress required to produce 100 microstrain when the electric displacement is held equal to zero.

4.2. A new composition of piezoelectric material is found to have a compliance at zero electric field of 18.2 $\mu m^2/N$, a piezoelectric strain coefficient of 330 pm/V, and a relative permittivity of 1500.
(a) Write the one-dimensional constitutive relationship for the material with strain and electric displacement as the dependent variables.
(b) Write the one-dimensional constitutive relationship with stress and electric field as the dependent variables.

4.3. Derive equation (4.20) from the one-dimensional constitutive relationships for a piezoelectric material.

4.4. The short-circuit mechanical compliance of a piezoelectric material has been measured to be 20 $\mu m^2/N$ and the open-circuit mechanical compliance has been measured to be 16.2 $\mu m^2/N$. If the stress-free relative permittivity is equal to 2800, compute the relative permittivity of the material when the strain is constrained to be zero.

4.5. Compute the sensitivity between strain and electric displacement for a piezoelectric material whose material parameters are s^E 16 $\mu m^2/N$ and d 220 pm/V, with a relative permittivity of 1800. Assume that the signal conditioning circuit for the sensor maintains a zero electric field.

4.6. Beginning with equation (4.55), write the constitutive relationships for a piezoelectric material with an electric field applied in the direction of polarization (the 3 direction). Assume that the applied stress is zero except for the 1 and 2 material directions.

4.7. **(a)** Compute the strain vector for the system studied in Problem 4.6 when the applied field is equal to 0.5 MV/m and the applied stress in the 1 and 2 directions is 5 MPa. Assume the material properties of PZT-5H.
(b) Compute the stress vector for this material if we assume that the strain in the 1 and 2 directions is constrained to be zero.
(c) Compute the strain in the 3 direction when the strain in the 1 and 2 directions are zero.

4.8. A piezoelectric sensor has stress applied in the direction of polarization equal to 3 MPa. Stress values of 5 MPa are applied in the two directions normal to the polarization vector.
(a) Compute the electric field vector produced by the applied stress assuming that the electric displacement is held equal to zero. Assume the material properties of PZT-5H.

Table 4.5 Material properties for piezoelectric ceramics

		PZT-A	PZT-B	PZT-C	PZT-D	PZT-E
ε_{33}		1400	1400	1100	5440	1800
ε_{11}		1350	1300	1400	5000	2000
s_{11}^E	(μm^2/N)	12.7	13.1	11.2	14.8	16.5
s_{33}^E	(μm^2/N)	15.4	15.6	15.2	18.1	19.9
d_{13}	(pm/V)	−133	−132	−99	−287	−198
d_{33}	(pm/V)	302	296	226	635	417

(b) Compute the electric displacement in the polarization direction assuming that the electric field is held equal to zero.

4.9. A piezoelectric material operating in the 33 mode has the material properties $d_{33} = 450$ pm/V and $Y_3^E = 63$ GPa.

(a) Compute the blocked stress and free strain under the application of an electric field of 0.75 MV/m.

(b) Compute the voltage required to achieve this blocked stress or free displacement for a wafer that is 250 μm in thick.

4.10. Material properties for several different types of piezoelectric ceramics are shown in Table 4.5.

(a) Plot the blocked stress and free strain for all of these materials on a single plot in a manner similar to Figure (4.39).

(b) Compute the volumetric energy density of each type of material for operation in the 33 mode for an electric field of 1 MV/m.

4.11. The geometry of a piezoelectric stack with annular cross section is shown in Figure (4.39). Compute the stiffness, free displacement, and blocked force of the stack assuming that $r_1 = 5$ mm, $r_2 = 15$ mm, $t = 0.25$ mm, and the applied voltage is 150 V. Assume the material properties of PZT-5H.

4.12. The geometry of a piezoelectric stack with square cross section is shown in Figure 4.40. Compute the stiffness, free displacement, and blocked force of

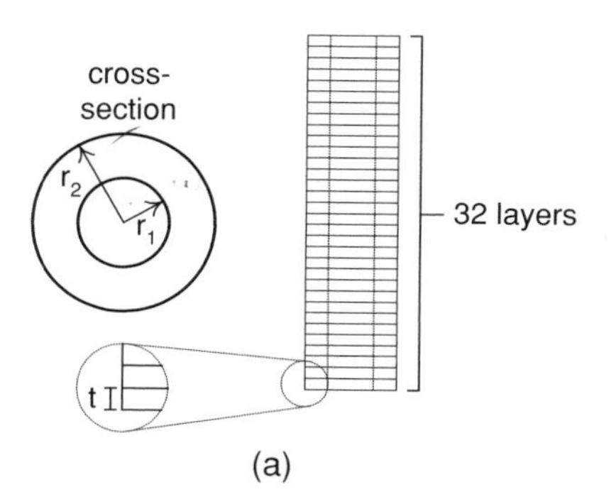

Figure 4.39 Geometry of a piezoelectric stack with an annular cross section.

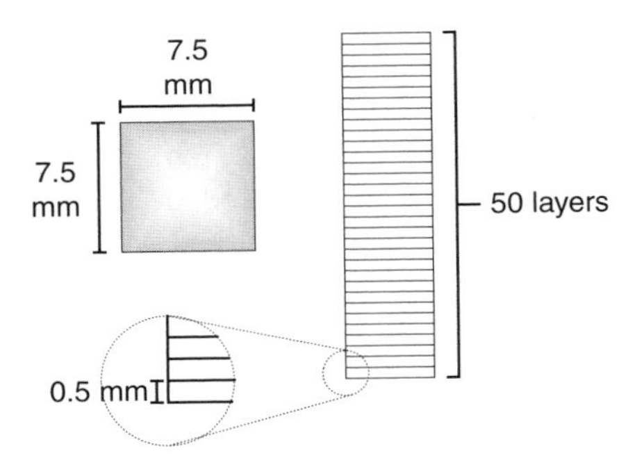

Figure 4.40 Geometry of a piezoelectric stack with a square cross section.

the stack assuming that the applied electric field across each layer is 1 MV/m. Assume the material properties of APC 856.

4.13. Piezoelectric stack actuators are often *preloaded* to reduce the risk of placing the brittle ceramic in tension. A piezoelectric stack with a free displacement of 30 μm and a stiffness of 40 N/μm is placed in a housing as shown in Figure 4.41.

(a) Compute the stiffness of the preload spring such that the output displacement is 20 μm.

(b) Compute the output force of the preloaded stack.

4.14. A piezoelectric stack with a free displacement of 50 μm and a blocked force of 750 N is being used for a static positioning application. The load is modeled as a linear elastic spring.

(a) Determine the load stiffness that will maximize the work performed by the stack.

(b) Compute the maximum work output of the stack.

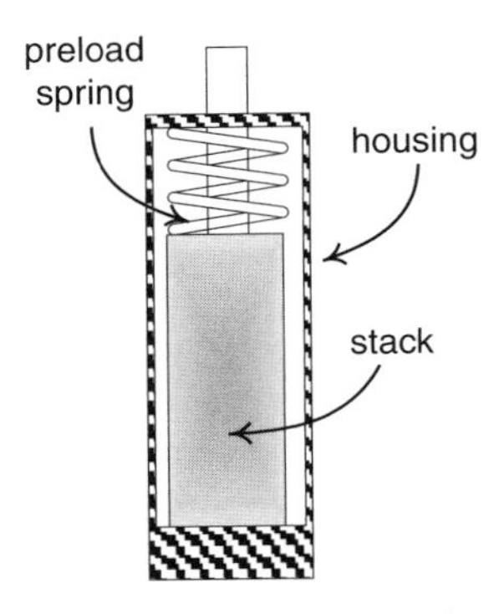

Figure 4.41 Schematic of a preloaded piezoelectric stack.

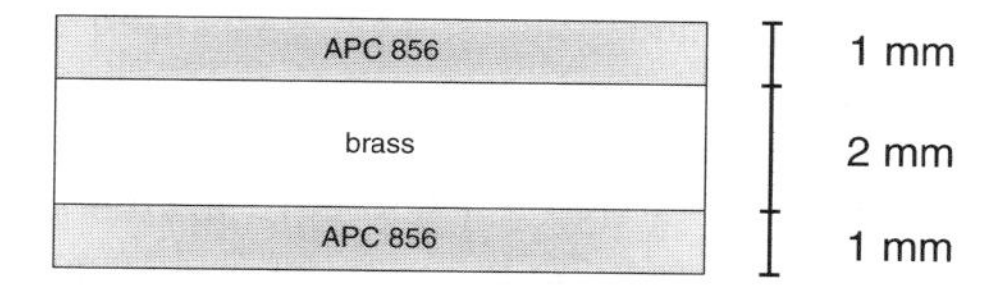

Figure 4.42 Cross section of a piezoelectric bimorph.

4.15. A piezoelectric stack with a short-circuit stiffness of 5 N/μm is driving a load with a mass of 500 grams and a stiffness of 3.5 N/μm. The free displacement of the stack at 100 V is 40 μm. Plot and label the frequency response of the stack displacement per unit voltage over the frequency range 1 to 10,000 Hz.

4.16. A piezoelectric accelerometer is being designed using PZT-5H as the material. The moving mass of the piezoelectric is 5 g and the dimensions of the piezoelectric sensor are 3 mm $\times$ 3 mm by 0.25 mm thick.

(a) Compute the low-frequency sensitivity of the accelerometer in pC/g.

(b) Compute the natural frequency of the accelerometer.

(c) Recompute the low-frequency sensitivity and the natural frequency if the moving mass is increased to 25 g.

4.17. A piezoelectric extender operating in the 13 mode has the cross-sectional geometry shown in Figure 4.42. Compute the strain in the 1 direction for an applied field of 0.75 MV/m. Assume that brass has a modulus of 117 GPa and that the piezoelectric material is APC 856.

4.18. A piezoelectric bimorph operating as a bender actuator has the cross-sectional geometry shown in Figure 4.42.

(a) Compute the nondimensional curvature as shown in equation (4.134) for this geometry.

(b) Compute the curvature of the beam. Assume that brass has a modulus of 117 GPa and that the piezoelectric material is APC 856. The applied field in the polarization direction is 0.75 MV/m.

NOTES

The nomenclature for this chapter (and the remainder of the book) is based on the conventions stated in the *IEEE Standard on Piezoelectricity* [6]. The discussion of transducer relationships for 33 and 31 actuators and sensors was not taken from any set of references, but derived from the general expressions. Additional analyses on the work associated with induced-strain piezoelectric actuators can be found in an article by Giurgiutiu and Rogers [24]. In this reference the basic principles of stiffness matching are discussed as they relate to static and dynamic induced strain actuation. The work of Crawley and Anderson [25] was used as a basis for the discussion of piezoelectric beams. This reference was one of the first detailed analyses of surface-bonded piezoelectric actuation for beam structures. An analyses of plate structures can be found in a paper by Crawley and Lazarus [26]. The analysis of thickness relationships in beams may be found in an article by Leo et al. [27]. Another body of work in the modeling of induced-strain actuated systems is the research of Liang and Rogers on impedance-based methods [28–30]. These methods are an effective way to understand the coupling between a piezoelectric element and the host structure, but it was felt that the energy-based methods discussed in Chapter 5 are a more generalized approach for system modeling; therefore, these methods were not discussed in detail. One of the most highly cited papers in recent years is that of Giurgiutiu et al. [31]. on the comparison of energy characteristics of induced-strain actuators. The section in this chapter is based on that work, and the reader is referred to that paper as a more complete discussion of piezoelectric actuator materials through the mid-1990s. Additional references on electrostrictive materials may be found in an article by Damjanovic and Newnham [32]. Derivations of the constitutive relationships in this book are based on work by Hom and Shankar [33] and Pablo and Petitjean [34].

5

PIEZOELECTRIC MATERIAL SYSTEMS

The constitutive equations for piezoelectric materials were stated in Chapter 4 and used to derive equations that modeled the static and dynamic response of piezoelectric devices such as actuators and sensors. In this chapter we extend these results to engineering systems that incorporate piezoelectric active materials. In Chapter 2 we introduced the variational approach to solving for systems of equations for static and dynamic systems. We will find this approach to be particularly useful for the analysis of systems that incorporate piezoelectric devices. At the close of this chapter we describe a general method for developing equations of motion for such systems based on use of the variational principle.

5.1 DERIVATION OF THE PIEZOELECTRIC CONSTITUTIVE RELATIONSHIPS

In Chapter 4 the constitutive relationships for a linear piezoelectric material were stated without derivation as a relationship between stress, strain, electric displacement, and electric field. We began by defining a material that had elastic properties, which led us to define mechanical compliance terms that related stress to strain, and then we defined the electrical properties in terms of the dielectric permittivity, which is an electrical compliance that relates the charge flow to an electric field. The piezoelectric properties were defined in terms of a linear coupling term that related stress to electric field and charge density to applied stress.

In this chapter we explain in more detail how these constitutive equations are derived using the concept of energy functions and basic thermodynamic principles for reversible systems. In addition to providing a more rigorous derivation of the relationships, the energy approach provides insight into the symmetry exhibited by the constitutive equations derived in Chapter 4. It will also form the basis for modeling techniques that will enable the analysis of structural systems that incorporate piezoelectric materials as actuators and sensors.

A rigorous derivation of the constitutive properties of a piezoelectric material begins by considering the first law of thermodynamics,

$$dU = dQ + dW, \tag{5.1}$$

which states that the change in internal energy, dU, of a system is equivalent to the heat added, dQ, and the work done on the system, dW. For reversible systems, the second law of thermodynamics states that the infinitesimal change in the heat can be written in terms of the absolute temperature, θ, and the infinitesimal change in the entropy, $d\sigma$,

$$dQ = \theta \, d\sigma. \tag{5.2}$$

The external work can be written as the summation of the mechanical work and electrical work,

$$dW = \mathrm{T}_i \, d\mathrm{S}_i + \mathrm{E}_k \, d\mathrm{D}_k, \tag{5.3}$$

where the compact (or Voigt) notation is used to express the summation of the mechanical work and electrical work. In equation (5.3), the index i takes values 1 through 6 and k takes values 1 through 3. Combining equations (5.2), (5.3), and (5.1) yields an expression for the infinitesimal change in internal energy as a function of the thermodynamic state variables entropy, strain, and electric displacement,

$$dU = \theta \, d\sigma + \mathrm{T}_i \, d\mathrm{S}_i + \mathrm{E}_k \, d\mathrm{D}_k. \tag{5.4}$$

Equation (5.4) is derived by combining the first and second laws of thermodynamics with the expression for the work associated with a piezoelectric material. Consider writing the internal energy as a general function of the thermodynamic state variables entropy, strain, and electric displacement,

$$U = U(\sigma, \mathbf{S}, \mathbf{D}), \tag{5.5}$$

and write the total derivative of this function

$$dU = \left(\frac{\partial U}{\partial \sigma}\right) d\sigma + \left(\frac{\partial U}{\partial \mathrm{S}_i}\right) d\mathrm{S}_i + \left(\frac{\partial U}{\partial \mathrm{D}_k}\right) d\mathrm{D}_k. \tag{5.6}$$

In the case in which all of the strains and electric displacements of the material system are held constant such that $d\mathrm{S}_i = d\mathrm{D}_i = 0$, we can write

$$dU = \left(\frac{\partial U}{\partial \sigma}\right) d\sigma = \theta \, d\sigma, \tag{5.7}$$

which leads to the result

$$\theta = \left.\left(\frac{\partial U}{\partial \sigma}\right)\right|_{\mathbf{S},\mathbf{D}}. \tag{5.8}$$

Equation (5.8) states that the temperature is equal to the change in internal energy with respect to entropy when all other thermodynamic state variables are held constant. In the same manner the stress and electric field are written as

$$\begin{aligned} \mathrm{T}_i &= \left.\left(\frac{\partial U}{\partial \mathrm{S}_i}\right)\right|_{\theta,\mathbf{D}} \\ \mathrm{E}_k &= \left.\left(\frac{\partial U}{\partial \mathrm{D}_k}\right)\right|_{\theta,\mathbf{T}}. \end{aligned} \tag{5.9}$$

The functions of temperature, stress, and electric field are, in general, also functions of the thermodynamic state variables:

$$\begin{aligned} \theta &= \theta\,(\sigma, \mathbf{S}, \mathbf{D}) \\ \mathrm{T}_i &= \mathrm{T}_i\,(\sigma, \mathbf{S}, \mathbf{D}) \\ \mathrm{E}_i &= \mathrm{E}_i\,(\sigma, \mathbf{S}, \mathbf{D})\,. \end{aligned} \tag{5.10}$$

Small changes in the temperature, stress, and electric field can be written as total derivatives,

$$\begin{aligned} d\theta &= \left(\frac{\partial \theta}{\partial \sigma}\right) d\sigma + \left(\frac{\partial \theta}{\partial \mathrm{S}_j}\right) d\mathrm{S}_j + \left(\frac{\partial \theta}{\partial \mathrm{D}_l}\right) d\mathrm{D}_l \\ d\mathrm{T}_i &= \left(\frac{\partial \mathrm{T}_i}{\partial \sigma}\right) d\sigma + \left(\frac{\partial \mathrm{T}_i}{\partial \mathrm{S}_j}\right) d\mathrm{S}_j + \left(\frac{\partial \mathrm{T}_i}{\partial \mathrm{D}_l}\right) d\mathrm{D}_l \\ d\mathrm{E}_k &= \left(\frac{\partial \mathrm{E}_k}{\partial \sigma}\right) d\sigma + \left(\frac{\partial \mathrm{E}_k}{\partial \mathrm{S}_j}\right) d\mathrm{S}_j + \left(\frac{\partial \mathrm{E}_k}{\partial \mathrm{D}_l}\right) d\mathrm{D}_l, \end{aligned} \tag{5.11}$$

where the index j ranges from 1 to 6 and l ranges from 1 to 3. Substituting equations (5.8) and (5.9) into these expressions yields

$$\begin{aligned} d\theta &= \left(\frac{\partial^2 U}{\partial \sigma^2}\right) d\sigma + \left(\frac{\partial^2 U}{\partial \sigma \mathrm{S}_j}\right) d\mathrm{S}_j + \left(\frac{\partial^2 U}{\partial \sigma \mathrm{D}_l}\right) d\mathrm{D}_l \\ d\mathrm{T}_i &= \left(\frac{\partial^2 U}{\partial \mathrm{S}_i \sigma}\right) d\sigma + \left(\frac{\partial^2 U}{\partial \mathrm{S}_i \mathrm{S}_j}\right) d\mathrm{S}_j + \left(\frac{\partial^2 U}{\partial \mathrm{S}_i \mathrm{D}_l}\right) d\mathrm{D}_l \\ d\mathrm{E}_k &= \left(\frac{\partial^2 U}{\partial \mathrm{D}_k \sigma}\right) d\sigma + \left(\frac{\partial^2 U}{\partial \mathrm{D}_k \mathrm{S}_j}\right) d\mathrm{S}_j + \left(\frac{\partial^2 U}{\partial \mathrm{D}_k \mathrm{D}_l}\right) d\mathrm{D}_l. \end{aligned} \tag{5.12}$$

Evaluating equation (5.12) at the values of the thermodynamic state variables produces the linear equations of state for the material. Strictly speaking, these equations are valid only in a small neighborhood around the specified state, although under certain assumptions they will produce expressions that are valid over a larger operating range.

5.1.1 Alternative Energy Forms and Transformation of the Energy Functions

Up to this point the discussion of the piezoelectric constitutive equations has been expressed in terms of the internal energy U of the piezoelectric material, and the thermodynamic state of the internal energy is represented in terms of the entropy, strain, and electric displacement. As discussed in Chapter 4, though, the temperature, stress, and electric field are functions of the thermodynamic state variables and thus are not independent variables. The fact that these variables are dependent on the thermodynamic state allows us to write the energy formulation as a function of any combination of three pairs of conjugate state variables: temperature–entropy, stress–strain, and electric field–electric displacement.

Transforming the state variables from one set to another requires that we write an alternative energy expression for the piezoelectric material. The internal energy of the material, U, is represented as a function of three pairs of conjugate variables with entropy, strain, and electric displacement as the independent state variables. These variables are denoted the *principal state variables*. Due to the fact that three pairs of independent variables can be written in one of eight ways, it is possible to define an additional seven energy functions in addition to the internal energy U. Each additional energy expression is defined as a relationship between the new energy function and the internal energy. The seven additional transformations are listed in Table 5.1.

It is important to note that **none of these additional energy expressions contain any more or less information than the expression for internal energy**. Application of these energy transformations simply interchanges the dependent and independent variables in the constitutive relationships. To illustrate how the variables interchange upon application of the energy transformation, take the total derivative of the Helmholtz free energy $A = U - \theta\sigma$,

$$dA = dU - \theta d\sigma - \sigma d\theta. \tag{5.13}$$

Table 5.1 Transformations between internal energy and the alternative energy forms

Helmholtz free energy	$A = U - \theta\sigma$
Enthalpy	$H = U - \mathrm{T}_i\mathrm{S}_i - \mathrm{E}_k\mathrm{D}_k$
Elastic enthalpy	$H_1 = U - \mathrm{T}_i\mathrm{S}_i$
Electric enthalpy	$H_2 = U - \mathrm{E}_k\mathrm{D}_k$
Gibbs free energy	$G = U - \theta\sigma - \mathrm{T}_i\mathrm{S}_i - \mathrm{E}_k\mathrm{D}_k$
Elastic Gibbs energy	$G_1 = U - \theta\sigma - \mathrm{T}_i\mathrm{S}_i$
Electric Gibbs energy	$G_2 = U - \theta\sigma - \mathrm{E}_k\mathrm{D}_k$

Introducing equation (5.4) into equation (5.13) produces

$$dA = \theta d\sigma + \mathrm{T}_i d\mathrm{S}_i + \mathrm{E}_k d\mathrm{D}_k - \theta d\sigma - \sigma d\theta. \tag{5.14}$$

The term $\theta d\sigma$ cancels out of the expression and equation (5.14) is reduced to

$$dA = -\sigma d\theta + \mathrm{T}_i d\mathrm{S}_i + \mathrm{E}_k d\mathrm{D}_k. \tag{5.15}$$

Comparing equation (5.15) with the total derivative of $A(\theta, \mathbf{S}, \mathbf{D})$, we can write the relationships

$$\begin{aligned} \sigma &= -\left(\frac{\partial A}{\partial \theta}\right)\bigg|_{\mathbf{S},\mathbf{D}} \\ \mathrm{T}_i &= \left(\frac{\partial A}{\partial \mathrm{S}_i}\right)\bigg|_{\theta,\mathbf{D}} \\ \mathrm{E}_i &= \left(\frac{\partial A}{\partial \mathrm{D}_i}\right)\bigg|_{\theta,\mathbf{S}}. \end{aligned} \tag{5.16}$$

The first expression provides a new intepretation of the entropy as the negative of the partial derivative of the Helmholtz free energy when holding strain and electric displacement constant.

To visualize this energy transformation, consider a system in which the strain and electric displacement are being held constant such that $d\mathrm{S}_i = d\mathrm{D}_i = 0$. Under these assumptions the differential change in internal energy is written as

$$dU = \theta\, d\sigma. \tag{5.17}$$

Figure 5.1a is a plot of an arbitrary relationship between temperature and entropy. We see from the figure that the differential element of internal energy is depicted as the area

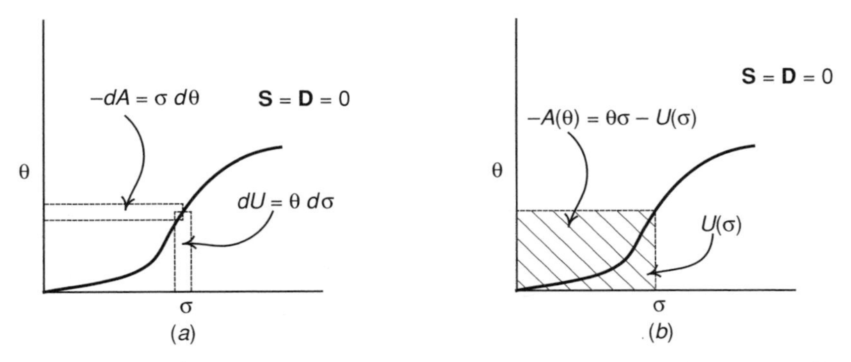

Figure 5.1 (a) Arbitrary temperature–entropy diagram illustrating differential elements of internal energy and Helmholtz free energy; (b) temperature–entropy diagram illustrating the relationship between internal energy and Helmholtz free energy.

Table 5.2 Independent variables associated with each of the energy expressions

Helmholtz free energy	A	S, D, θ
Enthalpy	H	T, E, σ
Elastic enthalpy	H_1	T, D, σ
Electric enthalpy	H_2	S, E, σ
Gibbs free energy	G	T, E, θ
Elastic Gibbs energy	G_1	T, D, θ
Electric Gibbs energy	G_2	S, E, θ

of a differential change in entropy multiplied by the current value of the temperature. Figure 5.1*a* also illustrates a differential change in the Helmholtz free energy, dA, as the negative of the differential change in temperature multiplied by the current value of entropy. It is clear from the figure that for arbitrary temperature–entropy functions, a differential change in the internal energy is not equal to a differential change in the Helmholtz free energy.

Integrating as a function of σ yields the internal energy $U(\sigma)$ for constant strain and constant electric displacement. For any values of temperature and entropy, as shown in Figure 5.1*b*, the area under the curve represents the internal energy. Integrating with respect to θ produces the Helmholtz free energy, and this is represented by the area above the curve in the hatched region. It is clear from the figure that the Helmholtz free energy is the difference between the hatched region and the internal energy $U(\sigma)$. It is also clear from the figure that, in general, $-A(\theta) \neq U(\sigma)$.

Expressions analogous to equations (5.13) to (5.16) can be derived for any of the remaining energy functions. The result of the derivation will be an expression of the dependent variables in terms of the independent variables chosen for the analysis. The transformations listed in Table 5.1 will exchange the dependent and independent variables associated with the energy expression. Table 5.2 lists the independent variables of each form of the energy.

5.1.2 Development of the Energy Functions

The discussion in Section 5.1.1 examined how to derive the constitutive relationships from an energy expression for the piezoelectric material. Often it is necessary to perform the opposite analysis, that is, to derive a form for the energy expression from measured or assumed data about the electromechanical coupling. Determining which energy expression to use in the analysis depends on the type of information available for the analysis. In the case of using experimental data to determine the energy expression, it is often the case that the form of the energy expression will depend on the variables that can be controlled directly in the measurements.

Consider the case in which experimental data are used to determine the energy expression for a material that is assumed to exhibit linear electromechanical coupling. If it is assumed that the parameters that are controlled in the experiment are stress, electric field, and temperature, Table 5.2 states that the Gibbs energy expression is appropriate for this analysis. Electromechanical tests on a sample of material at low fields and low stress levels result in the relationships

$$\begin{aligned} \mathrm{S}_i &= s_{ij}^{\mathrm{E}}\mathrm{T}_j + d_{ik}\mathrm{E}_k \\ \mathrm{D}_k &= d_{ik}\mathrm{T}_i + \varepsilon_{kl}^{\mathrm{T}}\mathrm{E}_k. \end{aligned} \tag{5.18}$$

Note that equation (5.18) implies that no temperature dependence was measured and the energy expression will be independent of temperature. The expression for Gibbs free energy is listed in Table 5.1, $G = U - \theta\sigma - \mathrm{T}_i\mathrm{S}_i - \mathrm{E}_k\mathrm{D}_k$, and the total derivative is

$$dG = dU - \theta d\sigma - \sigma d\theta - \mathrm{T}_i d\mathrm{S}_i - \mathrm{S}_i d\mathrm{T}_i - \mathrm{E}_k d\mathrm{D}_k - \mathrm{D}_k d\mathrm{E}_k. \tag{5.19}$$

Substituting equation (5.4) into equation (5.19) and canceling terms yields

$$dG = -\sigma d\theta - \mathrm{S}_i d\mathrm{T}_i - \mathrm{D}_k d\mathrm{E}_k. \tag{5.20}$$

Comparing equation (5.20) with the total derivative of the function $G(\theta, \mathrm{T}_i, \mathrm{E}_k)$ produces the relationships

$$\begin{aligned} -\sigma &= \frac{\partial G}{\partial \theta} \\ -\mathrm{S}_i &= \frac{\partial G}{\partial \mathrm{T}_i} \\ -\mathrm{D}_k &= \frac{\partial G}{\partial \mathrm{E}_k}. \end{aligned} \tag{5.21}$$

Substituting the first expression of equation (5.18) into the second expression of equation (5.21) yields

$$\frac{\partial G}{\partial \mathrm{T}_i} = -s_{ij}^{\mathrm{E}}\mathrm{T}_j - d_{ik}\mathrm{E}_k. \tag{5.22}$$

Integrating equation (5.22) with respect to stress produces

$$G = -\frac{1}{2}s_{ij}^{\mathrm{E}}\mathrm{T}_i\mathrm{T}_j - d_{ik}\mathrm{T}_i\mathrm{E}_k + g(\mathrm{E}_k). \tag{5.23}$$

The function $g(\mathrm{E}_k)$ represents the constant of integration that is only a function of the electric field terms. Taking the derivative of equation (5.23) with respect to E_k and setting it equal to equation (5.18) yields

$$-d_{ik}\mathrm{T}_i + \frac{\partial g(\mathrm{E}_k)}{\partial \mathrm{E}_k} = -d_{ik}\mathrm{T}_i - \varepsilon_{kl}^{\mathrm{T}}\mathrm{E}_k. \tag{5.24}$$

Canceling like terms and integrating to find $g(\mathrm{E}_k)$ produces

$$g(\mathrm{E}_k) = -\frac{1}{2}\varepsilon^{\mathrm{T}}_{kl}\mathrm{E}_k\mathrm{E}_l. \tag{5.25}$$

Adding equation (5.25) to equation (5.23) produces the the Gibbs free energy expression when the constitutive relationships satisfy equation (5.18):

$$G = -\frac{1}{2}s^{\mathrm{E}}_{ij}\mathrm{T}_i\mathrm{T}_j - d_{ik}\mathrm{T}_i\mathrm{E}_k - \frac{1}{2}\varepsilon^{\mathrm{T}}_{kl}\mathrm{E}_k\mathrm{E}_l. \tag{5.26}$$

This derivation demonstrates that *linear* constitutive properties will produce a *quadratic* energy expression which is a function of the independent variables associated with the energy form. The various components of the energy expression are readily identifiable in relation to the mechanical and electrical energy of the piezoelectric material. The first term on the right-hand side of equation (5.26) is the Gibbs free energy associated with elastic deformation. Similarly, the third term on the right-hand side of the expression is the Gibbs free energy due to the stored electrical energy. The middle term is the coupling term that produces the piezoelectric strain coefficients in the linear constitutive equations.

5.1.3 Transformation of the Linear Constitutive Relationships

The Gibbs free energy expression in equation (5.26) is obtained by assuming linear constitutive relationship and integrating the constitutive relationships according to the relationships between Gibbs free energy and the independent variables stress and electric field. Similar analyses could be performed for each of the energy expressions to yield energy functions with respect to any of the other state variables, such as strain and electric displacement. This series of analyses would yield relationships between the piezoelectric coefficients in the various energy forms.

An alternative method for obtaining the relationships between piezoelectric coefficients in the linear constitutive equations is to transform the dependent and independent variables directly. Consider writing the Gibbs free energy expression in matrix notation as opposed to compact indicial notation,

$$G = -\frac{1}{2}\,\mathbf{T}'\,\mathrm{s}^{\mathrm{E}}\,\mathbf{T} - \mathbf{T}'\,\mathrm{d}\mathbf{E} - \frac{1}{2}\mathbf{E}'\varepsilon^{\mathrm{T}}\mathbf{E}. \tag{5.27}$$

Taking the derivative of the general expression for the Gibbs free energy and combining with equation (5.6) results in

$$dG = \left(\frac{\partial U}{\partial \mathrm{S}_i} - \mathrm{T}_i\right)d\mathrm{S}_i + \left(\frac{\partial U}{\partial \mathrm{D}_i} - \mathrm{E}_i\right)d\mathrm{D}_i - \mathrm{S}_i d\mathrm{T}_i - \mathrm{D}_i d\mathrm{E}_i. \tag{5.28}$$

The terms in parentheses are equal to zero by definition of the stress and electric field in terms of the energy; therefore, the differential change in the Gibbs free energy is equivalent to

$$dG = -\mathrm{S}_i d\mathrm{T}_i - \mathrm{D}_i d\mathrm{E}_i. \tag{5.29}$$

Taking the derivative of equation (5.27) produces

$$dG = -(\mathrm{s}^{\mathrm{E}}\,\mathbf{T} + \mathrm{d}\mathbf{E})d\mathbf{S} - (\mathrm{d}'\mathbf{T} + \varepsilon^{\mathrm{T}}\mathbf{E})d\mathbf{E}. \tag{5.30}$$

Compared to equation (5.30), this produces the constitutive relationships

$$\mathbf{S} = \mathrm{s}^{\mathrm{E}}\,\mathbf{T} + \mathrm{d}\mathbf{E} \tag{5.31}$$

$$\mathbf{D} = \mathrm{d}'\mathbf{T} + \varepsilon^{\mathrm{T}}\mathbf{E}, \tag{5.32}$$

which, as expected, are simply matrix forms of the constitutive equations expressed in equation (5.18).

The various energy formulations provide a means for deriving the constitutive relationships in terms of different sets of independent variables. It is important to emphasize once again that both sets of constitutive relationship are equivalent. Since the original energy terms contained no more or no less information than did every other term, the resulting constitutive equations must be equivalent except for the fact that they are expressed in terms of different independent variables. Thus, it makes sense that sets of constitutive equations can be obtained from one another to determine the relationships between the various parameters in the analyses.

To understand how to transform the constitutive relationships from one form to another, write equations (5.31) and (5.32) in terms of the independent variables stress and electric displacement by premultiplying equation (5.31) by c^{E} and rewriting equation (5.31) as

$$\mathrm{c}^{\mathrm{E}}\mathbf{S} = \mathbf{T} + \mathrm{c}^{\mathrm{E}}\mathrm{d}\mathbf{E}. \tag{5.33}$$

Solving equation (5.33) for stress yields

$$\mathbf{T} = \mathrm{c}^{\mathrm{E}}\mathbf{S} - \mathrm{c}^{\mathrm{E}}\mathrm{d}\mathbf{E}. \tag{5.34}$$

Substituting equation (5.34) into equation (5.32) and combining terms yields

$$\mathbf{D} = \mathrm{d}'\mathrm{c}^{\mathrm{E}}\mathbf{S} + (\varepsilon^{\mathrm{T}} - \mathrm{d}'\mathrm{c}^{\mathrm{E}}\mathrm{d})\mathbf{E}. \tag{5.35}$$

Combining equation (5.34) and (5.35) yields the constitutive relationships

$$\begin{aligned} \mathbf{T} &= \mathrm{c}^{\mathrm{E}}\mathbf{S} - \mathrm{e}\mathbf{E} \\ \mathbf{D} &= \mathrm{e}'\mathbf{S} + \varepsilon^{\mathrm{S}}\mathbf{E}, \end{aligned} \tag{5.36}$$

where

$$\begin{aligned} \mathrm{e} &= \mathrm{c}^{\mathrm{E}}\mathrm{d} \\ \varepsilon^{\mathrm{S}} &= \varepsilon^{\mathrm{T}} - \mathrm{d}'\mathrm{c}^{\mathrm{E}}\mathrm{d}. \end{aligned} \tag{5.37}$$

The constitutive relationships shown in equation (5.36) are an example of a *mixed-variable* form due to the fact that the independent variables are strain and electric field. The upper right and lower left partitions of a mixed-variable form are the transpose of one another but opposite in sign.

Another mixed-variable form is one in which the independent variables are stress and electric displacement. This form of the constitutive equations can be obtained by premultiplying equation (5.32) by β^{T} and rewriting as

$$\mathbf{E} = -\beta^{\mathrm{T}}\mathrm{d}'\mathbf{T} + \beta^{\mathrm{T}}\mathbf{D}. \tag{5.38}$$

Substituting equation (5.38) into equation (5.31) and combining terms produces the expression

$$\mathbf{S} = (\mathrm{s}^{\mathrm{E}} - \mathrm{d}\beta^{\mathrm{T}}\mathrm{d}')\mathbf{T} + \mathrm{d}\beta^{\mathrm{T}}\mathbf{D}. \tag{5.39}$$

Combining equation (5.39) with equation (5.38) yields the constitutive relationships

$$\begin{aligned} \mathbf{S} &= \mathrm{s}^{\mathrm{D}}\mathbf{T} + \mathrm{g}\mathbf{D} \\ \mathbf{E} &= -\mathrm{g}'\mathrm{T} + \beta^{\mathrm{T}}\mathbf{D}, \end{aligned} \tag{5.40}$$

where

$$\begin{aligned} \mathrm{g} &= \mathrm{d}\beta^{\mathrm{T}} \\ \mathrm{s}^{\mathrm{D}} &= \mathrm{s}^{\mathrm{E}} - \mathrm{d}\beta^{\mathrm{T}}\mathrm{d}'. \end{aligned} \tag{5.41}$$

The final constitutive form of the piezoelectric equations is a form in which strain and electric displacement are the independent variables. This form will be particularly important in Section 5.2, where we utilize energy methods to derive equations of piezoelectric material systems. This form can be derived by premultiplying equation (5.40) by c^{D} and solving for the stress:

$$\mathbf{T} = \mathrm{c}^{\mathrm{D}}\mathbf{S} - \mathrm{c}^{\mathrm{D}}\mathrm{g}\mathbf{D}. \tag{5.42}$$

Substituting this result into equation (5.40) and combining terms produces

$$\mathbf{E} = -\mathrm{g}'\mathrm{c}^{\mathrm{D}}\mathrm{S} + (\beta^{\mathrm{T}} + \mathrm{g}'\mathrm{c}^{\mathrm{D}}\mathrm{g})\mathbf{D}. \tag{5.43}$$

Equations (5.42) and (5.43) can be written as a set of constitutive equations:

$$\begin{aligned} \mathbf{T} &= c^D \mathbf{S} - h\mathbf{D} \\ \mathbf{E} &= -h'\mathbf{S} + \beta^S D, \end{aligned} \tag{5.44}$$

where

$$\begin{aligned} h &= c^D g \\ \beta^S &= \beta^T + g' c^D g. \end{aligned} \tag{5.45}$$

The four forms of the constitutive relationships for a linear piezoelectric material are summarized in Table 5.3 in terms of the constitutive parameters and the independent variables. The Gibbs free energy form is a good starting point for the analysis because in most cases the piezoelectric material properties are specified in terms of the mechanical properties, the permittivity, and the *d* coefficients of the material. From these parameters the remaining three forms of the constitutive properties are obtained by applying the matrix transformations listed in Table 5.3.

Example 5.1 A piezoelectric material has the following constitutive properties:

$$s^E = \begin{bmatrix} 12.0 & -4.0 & -5.0 & 0.0 & 0.0 & 0.0 \\ -4.0 & 12.0 & -5.0 & 0.0 & 0.0 & 0.0 \\ -5.0 & -5.0 & 15.0 & 0.0 & 0.0 & 0.0 \\ 0.0 & 0.0 & 0.0 & 39.0 & 0.0 & 0.0 \\ 0.0 & 0.0 & 0.0 & 0.0 & 39.0 & 0.0 \\ 0.0 & 0.0 & 0.0 & 0.0 & 0.0 & 33.0 \end{bmatrix} \mu m^2/N$$

Table 5.3 Summary of the constitutive forms for a linear piezoelectric material neglecting thermal effects

Independent Variables	Constitutive Relationships	Transformations
T, E	$\mathbf{S} = s^E\,\mathbf{T} + d\mathbf{E}$ $\mathbf{D} = d'\mathbf{T} + \varepsilon^T\mathbf{E}$	
S, E	$\mathbf{T} = c^E\mathbf{S} - e\mathbf{E}$ $\mathbf{D} = e'\mathbf{S} + \varepsilon^S\mathbf{E}$	$e = c^E d$ $\varepsilon^S = \varepsilon^T - d'c^E d$
T, D	$\mathbf{S} = s^D\mathbf{T} + g\mathbf{D}$ $\mathbf{E} = -g'T + \beta^T\mathbf{D}$	$g = d\beta^T$ $s^D = s^E - d\beta^T d'$
S, D	$\mathbf{T} = c^D\mathbf{S} - h\mathbf{D}$ $\mathbf{E} = -h'\mathbf{S} + \beta^S\mathbf{D}$	$h = c^D g$ $\beta^S = \beta^T + g'c^D g$

$$\mathrm{d} = \begin{bmatrix} 0.0 & 0.0 & -120.0 \\ 0.0 & 0.0 & -120.0 \\ 0.0 & 0.0 & 290.0 \\ 0.0 & 500.0 & 0.0 \\ 500.0 & 0.0 & 0.0 \\ 0.0 & 0.0 & 0.0 \end{bmatrix} \mathrm{pm/V}$$

$$\varepsilon^{\mathrm{T}} = \begin{bmatrix} 13.3 & 0.0 & 0.0 \\ 0.0 & 13.3 & 0.0 \\ 0.0 & 0.0 & 11.5 \end{bmatrix} \mathrm{nF/m}.$$

Transform the constitutive properties into a form in which strain and electric field are the independent variables.

Solution The transformations required are listed in Table 5.3. The first computation is to invert the short-circuit mechanical compliance matrix,

$$\mathrm{c}^{\mathrm{E}} = \mathrm{s}^{\mathrm{E}^{-1}} = \begin{bmatrix} 138.4 & 75.9 & 71.4 & 0.0 & 0.0 & 0.0 \\ 75.9 & 138.4 & 71.4 & 0.0 & 0.0 & 0.0 \\ 71.4 & 71.4 & 114.3 & 0.0 & 0.0 & 0.0 \\ 0.0 & 0.0 & 0.0 & 25.6 & 0.0 & 0.0 \\ 0.0 & 0.0 & 0.0 & 0.0 & 25.6 & 0.0 \\ 0.0 & 0.0 & 0.0 & 0.0 & 0.0 & 30.3 \end{bmatrix} \times 10^9\ \mathrm{N/m^2}$$

The piezoelectric coefficients in this constitutive form are

$$\mathrm{e} = \mathrm{c}^{\mathrm{E}}\mathrm{d} = \begin{bmatrix} 0.0 & 0.0 & -5.0 \\ 0.0 & 0.0 & -5.0 \\ 0.0 & 0.0 & 16.0 \\ 0.0 & 12.8 & 0.0 \\ 12.8 & 0.0 & 0.0 \\ 0.0 & 0.0 & 0.0 \end{bmatrix} \mathrm{C/m^2}$$

The permittivity matrix at constant strain is

$$\varepsilon^{\mathrm{S}} = \varepsilon^{\mathrm{S}} - \mathrm{d}'\mathrm{c}^{\mathrm{E}}\mathrm{d} = \begin{bmatrix} 6.9 & 0.0 & 0.0 \\ 0.0 & 6.9 & 0.0 \\ 0.0 & 0.0 & 5.7 \end{bmatrix} \mathrm{nF/m}.$$

The linear constitutive equations are now written according to the expressions listed in Table 5.3.

5.2 APPROXIMATION METHODS FOR STATIC ANALYSIS OF PIEZOLECTRIC MATERIAL SYSTEMS

One of the great strengths of the energy formulation is that it can easily be extended to *systems* that incorporate active materials. As discussed earlier, the energy method is a means of developing the constitutive relationships for piezoelectric materials. The combination of thermodynamic principles with an assumed representation of the energy function produces a set of relationships between the variables that represent the state of the material. The assumption of a quadratic energy function produces a linear constitutive relationship among stress, strain, electric field, and electric displacement.

In Chapter 2 we highlighted the fact that energy methods are effective and efficient means of studying the static or dynamic response of a system. The significant advantage is the fact that the energy is conveniently separated into potential, or stored energy, the nonconservative or external work, and the kinetic energy of a system. Once the first two terms are specified, the static response of a system can be computed through use of the variational principle for static systems. Specifying the third term allows us to apply the variational principle to dynamic systems.

We study the development of equations for piezoelectric material systems by studying static systems first and then moving on to dynamic systems. Although we will see that the equations for a static system are simply a subset of those derived for the dynamic response, it is instructional to study systems in equilibrium before generalizing the result. In addition, we will see that there are a number of applications in which the static response is of interest.

Begin by considering an arbitrarily shaped structure that has piezoelectric elements attached at various points within the volume (Figure 5.2). The elastic, electrical, and electromechanical coupling properties of the piezoelectric elements are assumed to be linear. We denote a set of axes that represent the global coordinates as x_1, x_2, and x_3, and assume that each piezoelectric element has coordinates defined by $\tilde{x}_1^i$, $\tilde{x}_2^i$, and

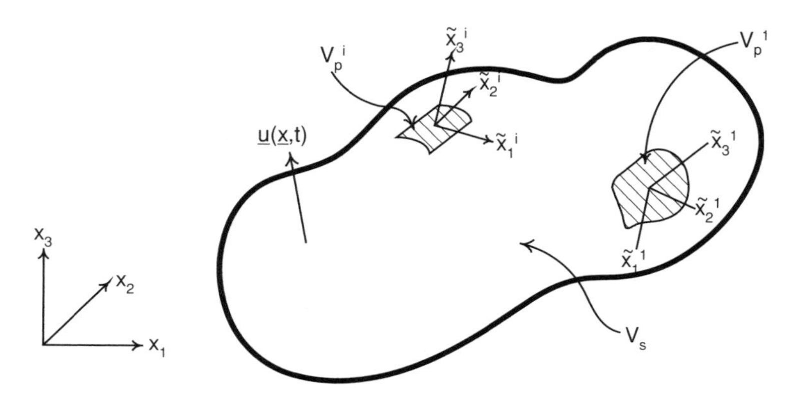

Figure 5.2 Volume with piezoelectric elements.

$\tilde{x}_3^i$. The volume occupied by the structure is denoted V_s, and the volume occupied by each piezoelectric element is denoted V_p^i.

The total potential energy of the structure is expressed as an integral of the strain energy over the volume,

$$V_{\text{struc}} = \int_{V_s} \frac{1}{2}\, \mathbf{S}'\mathrm{c_s}\mathbf{S}\, dV_s. \tag{5.46}$$

The potential energy of the piezoelectric elements is expressed as an integration of the energy function associated with the linear piezoelectric material:

$$V_{\text{piezo}} = \sum_{i=1}^{N_p} \int_{V_p^i} \left(\frac{1}{2}\, \tilde{\mathbf{S}}'\mathrm{c}^{\mathrm{D}}\tilde{\mathbf{S}} - \tilde{\mathbf{S}}'\mathrm{h}\tilde{\mathbf{D}} + \frac{1}{2}\tilde{\mathbf{D}}'\beta^{\mathrm{S}}\tilde{\mathbf{D}} \right) dV_p^i. \tag{5.47}$$

The tilde over a symbol denotes the fact that the strain and electric displacement are expressed in terms of the local piezoelectric coordinates and not the global coordinates. The local coordinate system for each piezoelectric element can be transformed to the global coordinates through the transformations

$$\begin{aligned} \tilde{\mathbf{S}} &= \mathrm{R_S^i}\mathbf{S} \\ \tilde{\mathbf{D}} &= \mathrm{R_D^i}\mathbf{D} \end{aligned} \tag{5.48}$$

and substituted into equation (5.47) to yield the expression

$$V_{\text{piezo}} = \sum_{i=1}^{N_p} \int_{V_p^i} \left(\frac{1}{2}\, \mathbf{S}'\mathrm{R_S^i}'\mathrm{c}\mathrm{R_S^i}\mathbf{S} - \mathbf{S}'\mathrm{R_S^i}'\mathrm{h}\mathrm{R_D^i}\mathbf{D} + \frac{1}{2}\mathbf{D}'\mathrm{R_D^i}'\beta^{\mathrm{S}}\mathrm{R_D^i}\mathbf{D} \right) dV_p^i. \tag{5.49}$$

The total potential energy is the sum of equations (5.46) and (5.49):

$$\begin{aligned} V &= V_{\text{struc}} + V_{\text{piezo}} \\ &= \int_{V_s} \frac{1}{2}\, \mathbf{S}'\mathrm{c_s}\mathbf{S}\, dV_s \\ &\quad + \sum_{i=1}^{N_p} \int_{V_p^i} \left(\frac{1}{2}\, \mathbf{S}'\mathrm{R_S^i}'\mathrm{c}^{\mathrm{D}}\mathrm{R_S^i}\mathbf{S} - \mathbf{S}'\mathrm{R_S^i}'\mathrm{h}\mathrm{R_D^i}\mathbf{D} + \frac{1}{2}\mathbf{D}'\mathrm{R_D^i}'\beta^{\mathrm{S}}\mathrm{R_D^i}\mathbf{D} \right) dV_p^i. \end{aligned} \tag{5.50}$$

The variational work associated with the system is written as

$$\delta W^{\text{ext}} = \sum_{j=1}^{N_f} \boldsymbol{f}_j \cdot \delta\boldsymbol{u}(\mathbf{x}_j) + \sum_{i=1}^{N_p} \boldsymbol{v}^i \cdot \delta\boldsymbol{q}^i. \tag{5.51}$$

In this book the *Ritz method* is used to solve for the generalized coordinates of the piezoelectric material system. The Ritz method is a common technique in structural mechanics for solving problems that are expressed by a variational principle. In this method, solution of the variational principle is reduced to the solution of a set of linearly independent equations by substituting an assumed form for the generalized coordinates into the energy terms. To represent the potential energy as a function of the generalized coordinates, first apply the differential operator associated with the elasticity problem. The differential operator defines a relationship between strain and displacement and is represented by the differential operator matrix $\mathrm{L_u}$. Applying the differential operator to the displacement vector yields the expression

$$\mathbf{S}(\mathbf{x}) = \mathrm{L_u}\boldsymbol{u}(\mathbf{x}). \tag{5.52}$$

The displacements are written as a linear combination of the generalized coordinates $\boldsymbol{r}$:

$$\boldsymbol{u}(\mathbf{x}) = \mathrm{N_r}(\mathbf{x})\boldsymbol{r}, \tag{5.53}$$

where $\mathrm{N_r}(\mathbf{x})$ is a set of *admissible shape functions* associated with the analysis. Admissible shape functions must satisfy the following conditions:

- The kinematic boundary conditions of the problem must be satisfied.
- The shape functions must form a linearly independent set.
- The shape functions must be differentiable $m - 1$ times, where m is the highest-order derivative of the differential operator associated with the strain energy function.

Additionally, the accuracy of the solution is increased by forming the shape functions from a *complete set*.

Combining equation (5.52) and (5.53) defines the relationship between the strain vector and the generalized coordinates:

$$\mathbf{S}(\mathbf{x}) = \mathrm{L_u}\mathrm{N_r}(\mathbf{x})\boldsymbol{r} = \mathrm{B_r}(\mathbf{x})\boldsymbol{r}. \tag{5.54}$$

Let us define a similar relationship for the relationship between the electric displacements of the ith piezoelectric, $\mathbf{D}^i$, and the generalized coordinates $\boldsymbol{q}$:

$$\mathrm{D}^i(\mathbf{x}) = \mathrm{B}_{\mathrm{q}}^{\mathrm{i}}(\mathbf{x})\boldsymbol{q}. \tag{5.55}$$

Substituting equations (5.54) and (5.55) into the expression for the potential energy, equation (5.50), will produce an expression for the potential energy as a function of generalized coordinates $\boldsymbol{r}$ and $\boldsymbol{q}$. The result can be written as the matrix expression

$$V = \frac{1}{2}\boldsymbol{r}'\mathrm{K_s}\boldsymbol{r} + \frac{1}{2}\boldsymbol{r}'\mathrm{K}_{\mathrm{p}}^{\mathrm{D}}\boldsymbol{r} - \boldsymbol{r}'\Theta\boldsymbol{q} + \frac{1}{2}\boldsymbol{q}'\mathrm{C}_{\mathrm{p}}^{\mathrm{S}^{-1}}\boldsymbol{q}, \tag{5.56}$$

where the matrices are defined by the integrals

$$\mathrm{K_s} = \int_{V_s} \mathrm{B_r}(\mathbf{x})'\mathrm{c_s}\mathrm{B_r}(\mathbf{x})\, dV_s \tag{5.57}$$

$$\mathrm{K_p^D} = \sum_{i=1}^{N_p} \int_{V_p^i} \mathrm{B_r}(\mathbf{x})'\mathrm{R_S^i}{}'\mathrm{c^D}\mathrm{R_S^i}\mathrm{B_r}(\mathbf{x})\, dV_p^i \tag{5.58}$$

$$\Theta = \sum_{i=1}^{N_p} \int_{V_p^i} \mathrm{B_r}(\mathbf{x})'\mathrm{R_S^i}{}'\mathrm{h}\mathrm{R_D^i}\mathrm{B_q^i}(\mathbf{x})\, dV_p^i \tag{5.59}$$

$$\mathrm{C_p^{S^{-1}}} = \sum_{i=1}^{N_p} \int_{V_p^i} \mathrm{B_q^i}(\mathbf{x})'\mathrm{R_D^i}{}'\beta^{\mathrm{S}}\mathrm{R_D^i}\mathrm{B_q^i}(\mathbf{x})\, dV_p^i. \tag{5.60}$$

The variational principle for a static system states that the variation in potential energy is equal to the variation in external work. The external work terms in equation (5.51) must be expressed in terms of the generalized coordinates so that they can be added to the potential energy function. First consider the external mechanical work. Using the fact that the order of the terms in equation (5.51) can be reversed and rewriting the dot product as a vector multiplication gives us

$$\sum_{j=1}^{N_f} \delta \boldsymbol{u}(\mathbf{x}_j)\cdot \boldsymbol{f}_j = \sum_{j=1}^{N_f} \delta \boldsymbol{u}(\mathbf{x}_\mathrm{j})' \boldsymbol{f}_j. \tag{5.61}$$

Substituting equation (5.53) into equation (5.61) produces

$$\sum_{j=1}^{N_f} \delta \boldsymbol{u}(\mathbf{x}_\mathrm{j})' \boldsymbol{f}_j = \delta \boldsymbol{r}' \left[\sum_{j=1}^{N_f} \mathrm{N_r}(\mathrm{x_j})'.\boldsymbol{f}_j \right]. \tag{5.62}$$

The term in brackets is expanded:

$$\sum_{j=1}^{N_f} \mathrm{N_r}(\mathbf{x}_\mathrm{j})'\boldsymbol{f}_j = \begin{bmatrix} \mathrm{N_r}(\mathbf{x}_1)' & \mathrm{N_r}(\mathbf{x}_2)' & \dots & \mathrm{N_r}(\mathbf{x}_{\mathrm{N_f}})' \end{bmatrix} \begin{bmatrix} \boldsymbol{f}_1 \\ \boldsymbol{f}_2 \\ \vdots \\ \boldsymbol{f}_{N_f} \end{bmatrix}. \tag{5.63}$$

Letting

$$\mathrm{B_f} = \begin{bmatrix} \mathrm{N_r}(\mathbf{x}_1)' & \mathrm{N_r}(\mathbf{x}_2)' & \dots & \mathrm{N_r}(\mathbf{x}_{\mathrm{N_f}})' \end{bmatrix} \tag{5.64}$$

and

$$\boldsymbol{f} = \begin{bmatrix} f_1 \\ f_2 \\ \vdots \\ f_{N_f} \end{bmatrix}, \tag{5.65}$$

equation (5.62) is rewritten as

$$\delta \boldsymbol{r}' \left[\sum_{j=1}^{N_f} \mathrm{N_r(x_j)}' \boldsymbol{f}_j \right] = \delta \boldsymbol{r}' \mathrm{B_f} \boldsymbol{f}. \tag{5.66}$$

In the same manner, the work due to external electrical loads is written

$$\sum_{i=1}^{N_p} \boldsymbol{v}^i \cdot \delta \boldsymbol{q}^i = \delta \boldsymbol{q}' \mathrm{B_v} \boldsymbol{v}. \tag{5.67}$$

The variation of the external work is written as

$$\delta W^{\text{ext}} = \delta \boldsymbol{r}' \mathrm{B_f} \boldsymbol{f} + \delta \boldsymbol{q}' \mathrm{B_v} \boldsymbol{v} \tag{5.68}$$

by combining equation (5.66) and (5.67). Taking the variation of equation (5.56) and combining with equation (5.68) yields

$$\delta \boldsymbol{r}' \left(\mathrm{B_f} \boldsymbol{f} - \mathrm{K_s} \boldsymbol{r} - \mathrm{K_p^D} \boldsymbol{r} + \Theta \boldsymbol{q} \right) + \delta \boldsymbol{q}' \left(\mathrm{B_v} \boldsymbol{v} + \Theta' \boldsymbol{r} - \mathrm{C_p^{S^{-1}}} \boldsymbol{q} \right) = 0. \tag{5.69}$$

Since the variational displacements are arbitrary, the terms in parentheses must both be equal to zero for equilibrium to be satisfied. The result is a set of matrix equations of the form

$$\begin{aligned} \left(\mathrm{K_s} + \mathrm{K_p^D} \right) \boldsymbol{r} - \Theta \boldsymbol{q} &= \mathrm{B_f} \boldsymbol{f} \\ -\Theta' \boldsymbol{r} + \mathrm{C_p^{S^{-1}}} \boldsymbol{q} &= \mathrm{B_v} \boldsymbol{v}. \end{aligned} \tag{5.70}$$

The expressions in equation (5.70) are the matrix expressions for a linear piezoelectric material system that is in static equilibrium. The form of the equations is independent of the type of problem that is being analyzed.

5.2.1 General Solution for Free Deflection and Blocked Force

Expressions for the blocked force and free deflection of a system that incorporates linear piezoelectric materials are obtained from equation (5.70). In Chapter 4, these two quantities often characterized the performance of an actuator and could be used for sizing a device. The same two quantities can be derived from equation (5.70) by proper manipulation of the matrix expressions.

Consider the deflection first. The output deflection at any point in the system, $\mathbf{x}_o$, can be written from equation (5.53),

$$\boldsymbol{u}(\mathbf{x}_o) = \mathrm{N_r}(\mathbf{x}_o)\boldsymbol{r}. \tag{5.71}$$

To be consistent with the notation introduced in Chapter 3, denote

$$\mathrm{H_d} = \mathrm{N_r}(\mathbf{x}_o) \tag{5.72}$$

as the *output matrix* for the system. Solving equation (5.70) for $\boldsymbol{q}$, we obtain

$$\boldsymbol{q} = \mathrm{C_p^S B_v}\boldsymbol{v} + \mathrm{C_p^S}\Theta'\boldsymbol{r}. \tag{5.73}$$

Substituting equation (5.73) into equation (5.70) produces

$$\left(\mathrm{K_s} + \mathrm{K_p^D} - \Theta\mathrm{C_p^S}\Theta'\right)\boldsymbol{r} = \Theta\mathrm{C_p^S B_v}\boldsymbol{v} + \mathrm{B_f}\boldsymbol{f}. \tag{5.74}$$

Solving the previous expression for $\boldsymbol{r}$ and substituting into equation (5.71) produces

$$\boldsymbol{u}(\mathbf{x}_o) = \mathrm{H_d}\mathcal{K}^{-1}\Theta\mathrm{C_p^S B_v}\boldsymbol{v} + \mathrm{H_d}\mathcal{K}^{-1}\mathrm{B_f}\boldsymbol{f}, \tag{5.75}$$

where

$$\mathcal{K} = \mathrm{K_s} + \mathrm{K_p^D} - \Theta\mathrm{C_p^S}\Theta'. \tag{5.76}$$

Equation (5.75) is the expression for the displacement of the system as a function of the input voltage to the piezoelectric material and the applied force. This expression can be used to compute the output deflection at any point due to electrical and mechanical excitation.

Equation (5.75) can also be used to obtain expressions for the free displacement and blocked force of the piezoelectric material system. The free deflection is obtained by setting $\boldsymbol{f} = 0$ in the expression and solving for the displacement,

$$\boldsymbol{u}(\mathbf{x}_o)|_{\boldsymbol{f}=0} = \mathrm{H_d}\mathcal{K}^{-1}\Theta\mathrm{C_p^S B_v}\boldsymbol{v}. \tag{5.77}$$

The blocked force is obtained by setting the left-hand side of equation (5.75) to zero and solving for the force:

$$\boldsymbol{f}|_{\boldsymbol{u}=0} = \left(\mathrm{H_d}\mathcal{K}^{-1}\mathrm{B_f}\right)^{-1}\mathrm{H_d}\mathcal{K}^{-1}\Theta\mathrm{C_p^S B_v}\boldsymbol{v}. \tag{5.78}$$

Equation (5.78) is a rather complicated matrix expression, but it is instructive to examine the solution to see if it makes sense from a physical perspective. Substituting equation (5.77) into equation (5.78) yields,

$$f|_{u=0} = \left(\mathrm{H_d}\mathcal{K}^{-1}\mathrm{B_f}\right)^{-1} \boldsymbol{u}(\mathbf{x}_o)|_{f=0} \tag{5.79}$$

and demonstrates that the blocked force is related directly to the free deflection. The remaining matrix term in the expression is analyzed by setting $\boldsymbol{v} = 0$ in equation (5.75) and solving for the force. In this manner we interpret the matrix expression $(\mathrm{H_d}\mathcal{K}^{-1}\mathrm{B_f})^{-1}$ as the *stiffness* elements associated with the system. Thus, the blocked force for a system that incorporates piezoelectric elements is interpreted as the product of the stiffness and the free deflection, which is the same interpretation that was obtained in the case of a piezoelectric actuator derived in Chapter 4.

5.3 PIEZOELECTRIC BEAMS

The energy method can be applied to determine the static response of structures with piezoelectric elements. It is instructive to first analyze the response of geometrically simple systems such as beams to understand the fundamental properties of the system response.

Beams were introduced in Chapter 2 as mechanical elements that supported a load perpendicular to their length whose geometry was such that their width and thickness were much smaller than their length. In the case of a beam that satisfied the Euler–Bernoulli assumptions that plane sections remain plane, displacement in the direction along the length of the beam could be written as the product of the local beam rotation and the distance from the neutral axis. These assumptions are assumed to be valid in the following analysis.

5.3.1 Cantilevered Bimorphs

Consider the case of a cantilever bimorph beam with two separate layers of piezoelectric material as shown in Figure 5.3. The coordinate systems are chosen such that the poling direction of the piezoelectric is in the same direction as the beam deflection. Under this assumption we can write

$$\mathrm{R_s} = \mathrm{I} \qquad \mathrm{R_D} = \mathrm{I}, \tag{5.80}$$

which simplifies the expressions for the potential energy. Assuming that the Euler–Bernoulli beam assumptions are valid and that the beam is in a state of uniaxial stress,

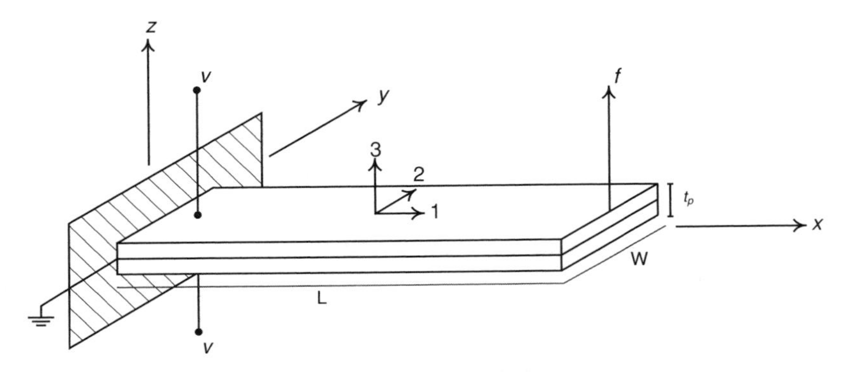

Figure 5.3 Cantilevered piezoelectric bimorph.

the strain vector is written

$$\begin{Bmatrix} S_1(\mathbf{x}) \\ S_2(\mathbf{x}) \\ S_3(\mathbf{x}) \\ S_4(\mathbf{x}) \\ S_5(\mathbf{x}) \\ S_6(\mathbf{x}) \end{Bmatrix} = \begin{bmatrix} 0 & 0 & z\dfrac{\partial^2}{\partial x^2} \\ 0 & 0 & -\nu_{12}z\dfrac{\partial^2}{\partial x^2} \\ 0 & 0 & -\nu_{13}z\dfrac{\partial^2}{\partial x^2} \\ 0 & 0 & 0 \\ 0 & 0 & 0 \\ 0 & 0 & 0 \end{bmatrix} \begin{Bmatrix} u_1(\mathbf{x}) \\ u_2(\mathbf{x}) \\ u_3(\mathbf{x}) \end{Bmatrix}. \tag{5.81}$$

A general expression for the displacement vector is written as a power series in x:

$$\begin{Bmatrix} u_1(\mathbf{x}) \\ u_2(\mathbf{x}) \\ u_3(\mathbf{x}) \end{Bmatrix} = \begin{bmatrix} 0 & 0 & 0 & 0 \\ 0 & 0 & 0 & 0 \\ 1 & x & x^2 & x^3 \end{bmatrix} \begin{Bmatrix} r_o \\ r_1 \\ r_2 \\ r_3 \end{Bmatrix}. \tag{5.82}$$

Admissible shape functions are determined by satisfying the geometric boundary conditions $u_3(0) = 0$ and $du_3/dx|_{x=0} = 0$:

$$\begin{aligned} 0 &= r_o \\ 0 &= r_1. \end{aligned} \tag{5.83}$$

Substituting these constraints into equation (5.82) produces the shape functions

$$\mathrm{N_r}(\mathbf{x}) = \begin{bmatrix} 0 & 0 \\ 0 & 0 \\ x^2 & x^3 \end{bmatrix}. \tag{5.84}$$

The shape functions that represent the strain as a function of the generalized coordinates are

$$\mathrm{B_r}(\mathbf{x}) = \mathrm{L_u N_r}(\mathbf{x})\boldsymbol{r} = \begin{bmatrix} 2z & 6zx \\ -\nu_{12}2z & -\nu_{12}6zx \\ -\nu_{13}2z & -\nu_{13}6zx \\ 0 & 0 \\ 0 & 0 \\ 0 & 0 \end{bmatrix}. \tag{5.85}$$

The generalized coordinates for the electric displacement are chosen as the charge associated with the top and bottom piezoelectric layers normalized by the surface area. Denoting the charge on the top layer as q_1 and the charge on the bottom layer as q_2 results in the shape functions

$$\mathrm{B_q^1}(\mathbf{x}) = \begin{bmatrix} 0 & 0 \\ 0 & 0 \\ \dfrac{1}{wL} & 0 \end{bmatrix} \tag{5.86}$$

$$\mathrm{B_q^2}(\mathbf{x}) = \begin{bmatrix} 0 & 0 \\ 0 & 0 \\ 0 & \dfrac{1}{wL} \end{bmatrix}. \tag{5.87}$$

The matrices on the left-hand side of equation (5.70) are determined once the matrices $\mathrm{B_r}(\mathbf{x})$ and $\mathrm{B_q}(\mathbf{x})$ are defined. Assuming that the coordinate rotation matrices are both identity matrices as shown in equation (5.80), we have

$$\mathrm{B_r}(\mathbf{x})'\mathrm{c^D B_r}(\mathbf{x}) = \hat{c}^D z^2 \begin{bmatrix} 4 & 12x \\ 12x & 36x^2 \end{bmatrix} \tag{5.88}$$

$$\mathrm{B_r}(\mathbf{x})'\mathrm{h B_q^1}(\mathbf{x}) = \hat{h}\frac{z}{wL} \begin{bmatrix} 2 & 0 \\ 6x & 0 \end{bmatrix} \tag{5.89}$$

$$\mathrm{B_r}(\mathbf{x})'\mathrm{h B_q^2}(\mathbf{x}) = \hat{h}\frac{z}{wL} \begin{bmatrix} 0 & 2 \\ 0 & 6x \end{bmatrix} \tag{5.90}$$

$$\mathrm{B_q^1}(\mathbf{x})'\beta^S \mathrm{B_q^1}(\mathbf{x}) = \beta_{33}^S \frac{1}{w^2L^2} \begin{bmatrix} 1 & 0 \\ 0 & 0 \end{bmatrix} \tag{5.91}$$

$$\mathrm{B_q^2}(\mathbf{x})'\beta^S \mathrm{B_q^2}(\mathbf{x}) = \beta_{33}^S \frac{1}{w^2L^2} \begin{bmatrix} 0 & 0 \\ 0 & 1 \end{bmatrix}. \tag{5.92}$$

The coefficients $\hat{c}^{\mathrm{D}}$ and $\hat{h}$ are defined as

$$\begin{aligned} \hat{c}^{\mathrm{D}} &= c_{11}^{\mathrm{D}} - 2\nu_{12}c_{12}^{\mathrm{D}} - 2\nu_{13}c_{13}^{\mathrm{D}} + \nu_{12}^2 c_{22}^{\mathrm{D}} + 2\nu_{12}\nu_{13}c_{23}^{\mathrm{D}} + \nu_{13}^2 c_{33}^{\mathrm{D}} \\ \hat{h} &= h_{13} - \nu_{12}h_{23} - \nu_{13}h_{33}. \end{aligned} \tag{5.93}$$

The matrices in equations (5.58) to (5.60) can now be determined from the integration over the volume:

$$\begin{aligned} \mathrm{K}_{\mathrm{p}}^{\mathrm{D}} &= \hat{c}^{\mathrm{D}} \int_{-w/2}^{w/2} \int_0^L \int_{-t_p/2}^{t_p/2} \left\{ z^2 \begin{bmatrix} 4 & 12x \\ 12x & 36x^2 \end{bmatrix} \right\} dz\, dx\, dy \\ &= \hat{c}^{\mathrm{D}} w t_p^3 \begin{bmatrix} L/3 & L^2/2 \\ L^2/2 & L^3 \end{bmatrix} \end{aligned} \tag{5.94}$$

$$\begin{aligned} \Theta &= \frac{\hat{h}}{wL} \int_{-w/2}^{w/2} \int_0^L \left\{ \int_0^{t_p/2} z \begin{bmatrix} 2 & 0 \\ 6x & 0 \end{bmatrix} dz + \int_{-t_p/2}^{0} z \begin{bmatrix} 0 & 2 \\ 0 & 6x \end{bmatrix} dz \right\} dx\, dy \\ &= \hat{h} t_p^2 \begin{bmatrix} 1/4 & -1/4 \\ 3L/8 & -3L/8 \end{bmatrix} \end{aligned} \tag{5.95}$$

$$\begin{aligned} \mathrm{C}_{\mathrm{p}}^{\mathrm{S}^{-1}} &= \frac{\beta_{33}^{\mathrm{S}}}{w^2 L^2} \int_{-w/2}^{w/2} \int_0^L \left\{ \int_0^{t_p/2} \begin{bmatrix} 1 & 0 \\ 0 & 0 \end{bmatrix} dz + \int_{-t_p/2}^{0} \begin{bmatrix} 0 & 0 \\ 0 & 1 \end{bmatrix} dz \right\} dx\, dy \\ &= \beta_{33}^{\mathrm{S}} \frac{t_p}{wL} \begin{bmatrix} 1/2 & 0 \\ 0 & 1/2 \end{bmatrix}. \end{aligned} \tag{5.96}$$

Combining the matrices according to equation (5.70) yields the matrix expression

$$\begin{bmatrix} \hat{c}^{\mathrm{D}} w t_p^3 L/3 & \hat{c}^{\mathrm{D}} w t_p^3 L^2/2 & \hat{h} t_p^2/4 & -\hat{h} t_p^2/4 \\ \hat{c}^{\mathrm{D}} w t_p^3 L^2/2 & \hat{c}^{\mathrm{D}} w t_p^3 L^3 & 3\hat{h} t_p^2 L/8 & -3\hat{h} t_p^2 L/8 \\ \hat{h} t_p^2/4 & 3\hat{h} t_p^2 L/8 & \beta_{33}^{\mathrm{S}} t/2wL & 0 \\ -\hat{h} t_p^2/4 & -3\hat{h} t_p^2 L/8 & 0 & \beta_{33}^{\mathrm{S}} t_p/2wL \end{bmatrix} \begin{Bmatrix} r_2 \\ r_3 \\ q_1 \\ q_2 \end{Bmatrix} = \begin{Bmatrix} L^2 \\ L^3 \\ 0 \\ 0 \end{Bmatrix} f + \begin{Bmatrix} 0 \\ 0 \\ 1 \\ -1 \end{Bmatrix} v. \tag{5.97}$$

Equation (5.97) represents the equilibrium expressions for the ideal piezoelectric bimorph. The transducer expressions can be determined by solving for the displacement and charge using the expressions

$$\begin{Bmatrix} u_3(L) \\ q \end{Bmatrix} = \begin{bmatrix} L^2 & L^3 & 0 & 0 \\ 0 & 0 & 1 & -1 \end{bmatrix} \begin{Bmatrix} r_2 \\ r_3 \\ q_1 \\ q_2 \end{Bmatrix}. \tag{5.98}$$

The solution can be obtained by solving for the generalized coordinates from equation (5.97) and substituting into equation (5.98). The result is a set of transducer equations of the form

$$u_3(L) = \frac{4}{\hat{c}^{\mathrm{D}}}\frac{L^2}{wt_p^3}\frac{1-3\hat{k}^2/16}{1-3\hat{k}^2/4}f - 3\frac{\hat{h}}{\hat{c}^{\mathrm{D}}\beta_{33}^{\mathrm{S}}}\frac{L^2}{t_p^2}\frac{1}{1-3\hat{k}^2/4}v \tag{5.99}$$

$$q = -3\frac{\hat{h}}{\hat{c}^{\mathrm{D}}\beta_{33}^{\mathrm{S}}}\frac{L^2}{t_p^2}\frac{1}{1-3\hat{k}^2/4}f + \frac{4}{\beta_{33}^{\mathrm{S}}}\frac{wL}{t_p}\frac{1}{1-3\hat{k}^2/4}v. \tag{5.100}$$

The free deflection and blocked force can be obtained from equation (5.99):

$$\begin{aligned} u_3(L)|_{f=0} &= -3\frac{\hat{h}}{\hat{c}^{\mathrm{D}}\beta_{33}^{\mathrm{S}}}\frac{L^2}{t_p^2}\frac{1}{1-3\hat{k}^2/4}v \\ f|_{u_3(L)=0} &= \frac{3}{4}\frac{\hat{h}}{\beta_{33}^{\mathrm{S}}}\frac{wt_p}{L}\frac{1}{1-3\hat{k}^2/16}v, \end{aligned} \tag{5.101}$$

where

$$\hat{k}^2 = \frac{\hat{h}^2}{\hat{c}^{\mathrm{D}}\beta_{33}^{\mathrm{S}}} \tag{5.102}$$

can be thought of as the square of a *generalized coupling coefficient*, $\hat{k}$. The coefficients in equation (5.99) and (5.100) can be inverted to obtain a set of transducer relationships that have deflection and charge as the independent variables:

$$\begin{aligned} f &= \frac{\hat{c}^{\mathrm{D}}}{4}\frac{wt_p^3}{L^3}u_3(L) + \frac{3}{16}\hat{h}\frac{t_p^2}{L^2}q \\ v &= \frac{3}{16}\hat{h}\frac{t_p^2}{L^2}u_3(L) + \frac{\beta_{33}^{\mathrm{S}}}{4}\frac{t_p}{wL}\left(\frac{1-3\hat{k}^2}{16}\right)q. \end{aligned} \tag{5.103}$$

In this form we recognize that the coefficient of $u_3(L)$ in the equations is the open-circuit stiffness of the bimorph, and the coefficient of q is the inverse of the strain-free capacitance.

5.3.2 Pinned–Pinned Bimorphs

One of the strengths of the energy method is that the different geometries can be studied with exactly the same methodology simply by changing the shape functions used in the analysis. The shape functions utilized for a cantilever beam produce displacement, slope, and strain functions that match those obtained from an elementary analysis using strength of materials. The displacement is a cubic function when a point load is applied at the tip and a quadratic function for a distributed moment is applied to the surface. The result is that the stiffness coefficients in K match those obtained from an elementary analysis.

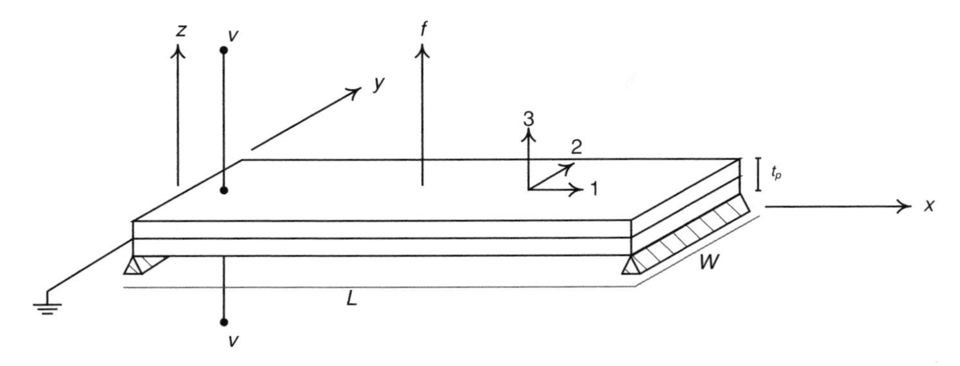

Figure 5.4 Pinned–pinned piezoelectric bimorph.

More care must be taken when discontinuities exist in the solution of the strength of materials analysis. A good example of this situation is for a pinned–pinned beam with a point load at the center span. The "exact" solution to the analysis yields a moment diagram that exhibits a discontinuity at the center span. This discontinuity must be approximated by the shape functions chosen for the analysis. If the shape functions have smooth derivatives, this leads to an approximation of the exact solution that will become more accurate as the number of shape functions is added to the energy analysis.

To illustrate this concept, consider a pinned–pinned beam with the geometry shown in Figure 5.4. Let us write the displacement as a cubic function, as shown in equation (5.82). A candidate set of shape functions are found by solving equation (5.82) for the boundary conditions

$$u_3(0) = u_3(L) = 0. \tag{5.104}$$

Eliminating r_0 and r_1 from the general solution of the displacement produces the shape functions for a pinned–pinned beam

$$\mathrm{N_r}(\mathbf{x}) = \begin{bmatrix} 0 & 0 \\ 0 & 0 \\ x\,(x-L) & x(x^2-L^2) \end{bmatrix}. \tag{5.105}$$

Proceeding with the energy analysis we see that the function $\mathrm{B_r}(\mathbf{x})'$ is identical for the cantilevered and pinned–pinned beam, indicating that the matrices $\mathrm{K_p^D}$, θ, and $\mathrm{C_p^S}$ are the same for both cases. The only difference in the analysis is the input vector, which for a pinned–pinned beam is equal to

$$\mathrm{B_f} = \begin{bmatrix} -L^2/4 \\ -3L^3/8 \end{bmatrix}. \tag{5.106}$$

The equilibrium equations can be written as

$$\begin{bmatrix} \hat{c}^{\mathrm{D}} w t_p^3 L/3 & \hat{c}^{\mathrm{D}} w t_p^3 L^2/2 & \hat{h} t_p^2/4 & -\hat{h} t_p^2/4 \\ \hat{c}^{\mathrm{D}} w t_p^3 L^2/2 & \hat{c}^{\mathrm{D}} w t_p^3 L^3 & 3\hat{h} t_p^2 L/8 & -3\hat{h} t_p^2 L/8 \\ \hat{h} t_p^2/4 & 3\hat{h} t_p^2 L/8 & \beta_{33}^{\mathrm{S}} t_p/2wL & 0 \\ -\hat{h} t_p^2/4 & -3\hat{h} t_p^2 L/8 & 0 & \beta_{33}^{\mathrm{S}} t_p/2wL \end{bmatrix} \begin{Bmatrix} r_2 \\ r_3 \\ q_1 \\ q_2 \end{Bmatrix}$$

$$= \begin{Bmatrix} -L^2/4 \\ -3L^3/8 \\ 0 \\ 0 \end{Bmatrix} f + \begin{Bmatrix} 0 \\ 0 \\ 1 \\ -1 \end{Bmatrix} v \tag{5.107}$$

and the transducer equations are obtained from the solution of

$$\begin{Bmatrix} u_3(L/2) \\ q \end{Bmatrix} = \begin{bmatrix} -L^2/4 & -3L^3/8 & 0 & 0 \\ 0 & 0 & 1 & -1 \end{bmatrix} \begin{Bmatrix} r_2 \\ r_3 \\ q_1 \\ q_2 \end{Bmatrix}, \tag{5.108}$$

which results in

$$u_3(L/2) = \frac{3}{16\hat{c}^{\mathrm{D}}} \frac{L^2}{w t_p^3} \frac{1}{1 - 3\hat{k}^2/4} f + \frac{3}{4} \frac{\hat{h}}{\hat{c}^{\mathrm{D}} \beta_{33}^{\mathrm{S}}} \frac{L^2}{t_p^2} \frac{1}{1 - 3\hat{k}^2/4} v \tag{5.109}$$

$$q = \frac{3}{4} \frac{\hat{h}}{\hat{c}^{\mathrm{D}} \beta_{33}^{\mathrm{S}}} \frac{L^2}{t_p^2} \frac{1}{1 - 3\hat{k}^2/4} f + \frac{4}{\beta_{33}^{\mathrm{S}}} \frac{wL}{t_p} \frac{1}{1 - 3\hat{k}^2/4} v. \tag{5.110}$$

The blocked force and free displacement for a pinned–pinned beam can be computed from equations (5.109) and (5.110):

$$u_3(L/2)|_{f=0} = \frac{3}{4} \frac{\hat{h}}{\hat{c}^{\mathrm{D}} \beta_{33}^{\mathrm{S}}} \frac{L^2}{t_p^2} \frac{1}{1 - 3\hat{k}^2/4} v \tag{5.111}$$

$$f|_{u_3(L)=0} = -4 \frac{\hat{h}}{\beta_{33}^{\mathrm{S}}} \frac{w t_p}{L} v. \tag{5.112}$$

Inverting equations (5.109) and (5.110), we can solve for the transducer expressions in terms of charge and displacement:

$$f = \frac{16\hat{c}^{\mathrm{D}}}{3} \frac{w t_p^3}{L^3} u_3(L/2) - \hat{h} \frac{t_p^2}{L^2} q \tag{5.113}$$

$$v = -\hat{h} \frac{t_p^2}{L^2} u_3(L/2) + \frac{\beta_{33}^{\mathrm{S}}}{4} \frac{t_p}{wL} q. \tag{5.114}$$

The coefficient in front of $u_3(L/2)$ in equation (5.113) is the stiffness of the pinned–pinned beam. Substituting the expression $I = \frac{1}{12} w t_p^3$ for the moment of inertia into the expression, we note that the stiffness of the pinned–pinned beam from this analysis is equal to $64\hat{c}^{\mathrm{D}}I/L^3$, which is larger than the value that would be obtained from an elementary strength of materials analysis, $48\hat{c}^{\mathrm{D}}I/L^3$.

The discrepancy arises from the fact that the shape functions that we have chosen for the analysis cannot represent the moment caused by a point load located at the center span. The second derivative of the shape function is equal to

$$\frac{d^2u_3}{dx^2} = 2r_2 + 6xr_3, \tag{5.115}$$

which is not a good approximation to the second derivative of the displacement obtained from an elementary analysis.

This discrepancy is alleviated if we increase the order of our shape functions used in the analysis. For example, when we increase the order of the shape functions to

$$\begin{Bmatrix} u_1 \\ u_2 \\ u_3 \end{Bmatrix} = \begin{bmatrix} 0 & 0 & 0 & 0 & 0 \\ 0 & 0 & 0 & 0 & 0 \\ 1 & x & x^2 & x^3 & x^4 \end{bmatrix} \begin{Bmatrix} r_o \\ r_1 \\ r_2 \\ r_3 \\ r_4 \end{Bmatrix} \tag{5.116}$$

and repeat the energy analysis, we obtain the transducer equations

$$f = 48.76\frac{\hat{c}^{\mathrm{D}}I}{L^3}u_3(L/2) - \frac{16}{21}\hat{h}\frac{t_p^2}{L^2}q \tag{5.117}$$

$$v = -\frac{16}{21}\hat{h}\frac{t_p^2}{L^2}u_3(L/2) + \frac{\beta_{33}^{\mathrm{S}}}{4}\frac{t_p}{wL}\left(\frac{1-5\hat{k}^2}{28}\right)q. \tag{5.118}$$

We see from equation (5.117) that the stiffness term is approaching the value obtained from an elementary strength of materials analysis. The reason for this increase in accuracy is that the second derivative of the shape functions yields a second-order polynomial, which can more accurately represent the exact solution.

Varying the shape functions also has an effect on the remaining transducer coefficients. The coupling term between force and charge changes from -1 to $-16/21$, and an additional term due to the piezoelectric coupling coefficient is introduced into the relationship between voltage and charge. If we write the transducer expressions as

$$f = T_{11}\frac{\hat{c}^{\mathrm{D}}I}{L^3}u_3(L/2) - T_{12}\hat{h}\frac{t_p^2}{L^2}q \tag{5.119}$$

$$v = -T_{12}\hat{h}\frac{t_p^2}{L^2}u_3(L/2) + \frac{\beta_{33}^{\mathrm{S}}}{4}\frac{t_p}{wL}\left(1 - T_{22}\hat{k}^2\right)q, \tag{5.120}$$

we can study the convergence of the transducer relationships as the shape functions change. A general polynomial expression for the displacement is

$$\left\{\begin{array}{l} u_1 \\ u_2 \\ u_3 \end{array}\right\} = \left[\begin{array}{c} 0 \\ 0 \\ \sum_{i=0}^{N} x^i r_i \end{array}\right]. \tag{5.121}$$

Introducing the geometric constraints for a pinned–pinned beam produces the shape functions

$$\mathrm{N_r}(\mathbf{x}) = \left[\begin{array}{c} 0 \\ 0 \\ \sum_{i=1}^{N-1} x(x^i - L^i) r_{i+1} \end{array}\right]. \tag{5.122}$$

Performing the energy analysis for an increasing number of shape functions illustrates that the transducer coefficients converge. Increasing the number of shape functions from three to four produces a large change in the coefficients. Increasing the number of shape functions produces small changes in the transducer coefficients. As Figure 5.5 illustrates, the stiffness coefficient (T_{11}) is converging to 48 while the other two terms in the transducer expressions are converging to 3/4 and ≈ 0.19, respectively.

Comparing the analysis of the pinned–pinned beam to that of a cantilevered beam, we see that the energy method is an approximate method. The accuracy of the coefficient terms is increased as the shape functions become more accurate representations of the moment, slope, and displacement functions. Increasing the number of shape

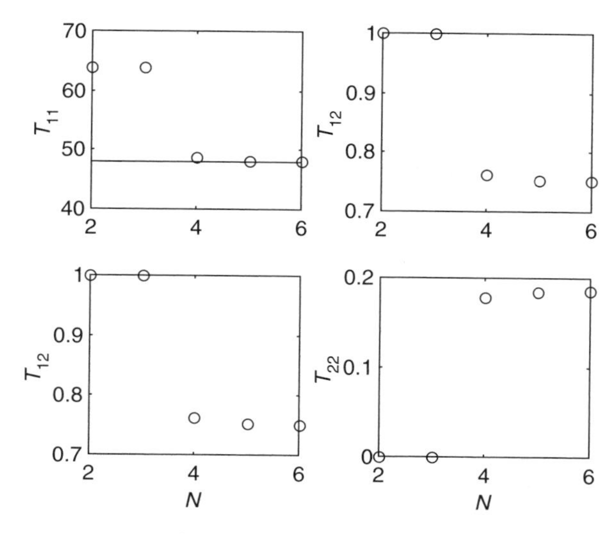

Figure 5.5 Convergence properties of the transducer coefficients of a pinned–pinned beam as a function of the number of polynomial terms in the displacement expression.

functions allows a study of the convergence properties of the solution obtained using the energy method.

The solution for a pinned–pinned beam with sufficient number of terms also allows us to compare the free deflection and blocked force of the bimorph geometry to that of a cantilever beam. Using $N = 8$ in equation (5.121) and solving for the blocked force and free deflection yields

$$u_3(L/2)|_{f=0} \rightarrow \frac{3}{4}\frac{\hat{h}}{\hat{c}^D \beta_{33}^S}\frac{L^2}{t_p^2}\frac{1}{1-3\hat{k}^2/4}v \tag{5.123}$$

$$f|_{u_3(L/2)=0} \rightarrow -3\frac{\hat{h}}{\beta_{33}^S}\frac{wt_p}{L}v. \tag{5.124}$$

The blocked force for a pinned–pinned bimorph is four times that of a cantilevered beam, while the free deflection is four times less. Thus, we see that changing the boundary conditions of the bimorph is another method of trading the blocked force and the free deflection of a piezoelectric transducer.

5.4 PIEZOELECTRIC MATERIAL SYSTEMS: DYNAMIC ANALYSIS

Dynamic analysis of piezoelectric material systems is performed by applying the variational principle for dynamic systems. The two differences between static and dynamic analysis are the addition of the kinetic energy term to the analysis and integration with respect to time of the system Lagrangian.

The displacement function for a dynamic analysis is a function of both space and time; therefore,

$$\boldsymbol{u}(\mathbf{x},t) = \mathrm{N_r}(\mathbf{x})\boldsymbol{r}(t), \tag{5.125}$$

where $\mathrm{N_r}(\mathbf{x})$ is a set of admissible shape functions and $r(t)$ is the time-dependent generalized coordinate.

The total kinetic energy of the system is written as

$$T = \frac{1}{2}\int_{V_s} \rho_s \dot{\boldsymbol{u}}'(\mathbf{x},t)\boldsymbol{u}(\mathbf{x},t)\,dV_s + \sum_{i=1}^{N_p}\frac{1}{2}\int_{V_p} \rho_p^i \dot{\boldsymbol{u}}'(\mathbf{x},t)\boldsymbol{u}(\mathbf{x},t)\,dV_p, \tag{5.126}$$

where ρ_s is the density of the substrate and ρ_p is the density of the piezoelectric elements. Substituting equation (5.125) into equation (5.126) allows the kinetic energy to be rewritten in matrix form:

$$T = \frac{1}{2}\dot{\boldsymbol{r}}(t)'\mathrm{M_s}\dot{\boldsymbol{r}}(t) + \frac{1}{2}\dot{\boldsymbol{r}}'(t)\mathrm{M_p}\dot{\boldsymbol{r}}(t), \tag{5.127}$$

where

$$\mathrm{M_s} = \int_{V_s} \rho_s \mathrm{N_r}(\mathbf{x})' \mathrm{N_r}(\mathbf{x}) \, dV_s \tag{5.128}$$

$$\mathrm{M_p} = \sum_{i=1}^{N_p} \int_{V_p} \rho_p^i \mathrm{N_r}(\mathbf{x})' \mathrm{N_r}(\mathbf{x}) \, dV_p. \tag{5.129}$$

The system Lagrangian is obtained by combining equation (5.127) with equation (5.56):

$$\begin{aligned} L = T - V = {} & \frac{1}{2}\dot{\boldsymbol{r}}'(t)\mathrm{M_s}\dot{\boldsymbol{r}}(t) + \frac{1}{2}\dot{\boldsymbol{r}}'(t)\mathrm{M_p}\dot{\boldsymbol{r}}(t) - \frac{1}{2}\boldsymbol{r}'\mathrm{K_s}\boldsymbol{r} - \frac{1}{2}\boldsymbol{r}'\mathrm{K_p^D}\boldsymbol{r}(t) \\ & + \boldsymbol{r}'(t)\Theta\boldsymbol{q}(t) - \frac{1}{2}\boldsymbol{q}'(t)\mathrm{C_p^{S^{-1}}}\boldsymbol{q}(t). \end{aligned} \tag{5.130}$$

The variation of the Lagrangrian produces the expression

$$\begin{aligned} \delta L = {} & \delta\dot{\boldsymbol{r}}'(t)\{\mathrm{M_s}\dot{\boldsymbol{r}}(t) + \mathrm{M_p}\dot{\boldsymbol{r}}(t)\} + \delta\boldsymbol{r}'(t)\left[-\mathrm{K_s}\boldsymbol{r}(t) - \mathrm{K_p^D}\boldsymbol{r}(t) + \Theta\boldsymbol{q}(t)\right] \\ & + \delta\boldsymbol{q}'(t)\left[\Theta'\boldsymbol{r}(t) - \mathrm{C_p^{S^{-1}}}\boldsymbol{q}(t)\right]. \end{aligned} \tag{5.131}$$

Integrating the variation of the Lagrangian with respect to time allows us to write

$$\begin{aligned} \int_{t_1}^{t_2} \delta L \, dt = {} & \int_{t_1}^{t_2} \{\delta\boldsymbol{r}'(t)\left[-(\mathrm{M_s} + \mathrm{M_p})\ddot{\boldsymbol{r}}(t) - (\mathrm{K_s} + \mathrm{K_p^D})\boldsymbol{r}(t) + \Theta\boldsymbol{q}(t)\right] \\ & + \delta\boldsymbol{q}'\left[\Theta'\boldsymbol{r}(t) - \mathrm{C_p^{S^{-1}}}\boldsymbol{q}(t)\right]\}dt. \end{aligned} \tag{5.132}$$

The external work terms, equation (5.68), are added to equation (5.132). Since the integral must be zero for arbitrary variational displacements, we obtain the matrix set of equations for dynamic analysis of a piezoelectric material system:

$$(\mathrm{M_s} + \mathrm{M_p})\ddot{\boldsymbol{r}}(t) + (\mathrm{K_s} + \mathrm{K_p^D})\boldsymbol{r}(t) - \Theta\boldsymbol{q}(t) = \mathrm{B_f}\boldsymbol{f}(t) \tag{5.133}$$

$$-\Theta'\boldsymbol{r}(t) + \mathrm{C_p^{S^{-1}}}\boldsymbol{q}(t) = \mathrm{B_v}\boldsymbol{v}(t). \tag{5.134}$$

The equations of motion for the dynamic system are a second-order matrix set of matrix equations that reduce to the static equations of equilibrium if the mass terms are set equal to zero.

5.4.1 General Solution

The second-order matrix expressions in equation (5.133) and (5.134) can be solved to yield a general solution for the dynamic electromechanical response of a piezoelectric material system. The lack of inertial terms associated with the charge coordinates allows us to eliminate $\boldsymbol{q}$ from equation (5.134) and substitute it into equation (5.133), with the following result

$$(\mathrm{M_s} + \mathrm{M_p})\ddot{\boldsymbol{r}}(t) + \left(\mathrm{K_s} + \mathrm{K_p^D} - \Theta \mathrm{C_p^S} \Theta'\right)\boldsymbol{r}(t) = \mathrm{B_f}\boldsymbol{f}(t) + \Theta \mathrm{C_p^S} \mathrm{B_v} \boldsymbol{v}(t). \tag{5.135}$$

Transforming into the frequency domain produces

$$\left[\mathrm{K_s} + \mathrm{K_p^D} - \Theta \mathrm{C_p^S} \Theta' - (\mathrm{M_s} + \mathrm{M_p})\omega^2\right]\boldsymbol{R}(\omega) = \mathrm{B_f}\boldsymbol{F}(\omega) + \Theta \mathrm{C_p^S} \mathrm{B_v} \boldsymbol{V}(\omega). \tag{5.136}$$

Solving for $\boldsymbol{R}(\omega)$ yields

$$\boldsymbol{R}(\omega) = \Delta^{-1}(\omega)\mathrm{B_f}\boldsymbol{F}(\omega) + \Delta^{-1}(\omega)\Theta \mathrm{C_p^S} \mathrm{B_v} \boldsymbol{V}(\omega), \tag{5.137}$$

where $\Delta(\omega)$ is the *dynamic stiffness matrix* expressed as

$$\Delta(\omega) = \mathrm{K_s} + \mathrm{K_p^D} - \Theta \mathrm{C_p^S} \Theta' - (\mathrm{M_s} + \mathrm{M_p})\omega^2. \tag{5.138}$$

Defining the output at the location $\mathbf{x}_o$ as

$$\boldsymbol{u}(\mathbf{x}_o, t) = \mathrm{H_d}\boldsymbol{r}(t), \tag{5.139}$$

we can transform equation (5.139) into the frequency domain and express the displacement as the frequency-dependent function

$$\boldsymbol{U}(\mathbf{x}_o, \omega) = \mathrm{H_d}\Delta^{-1}(\omega)\mathrm{B_f}\boldsymbol{F}(\omega) + \mathrm{H_d}\Delta^{-1}(\omega)\Theta \mathrm{C_p^S} \mathrm{B_v} \boldsymbol{V}(\omega). \tag{5.140}$$

Equation (5.140) is the expression for displacement as a function of the force and voltage inputs. The electrical output of the system can also be expressed as a frequency-dependent expression by transforming equation (5.134) into the frequency domain and solving for the charge:

$$\boldsymbol{Q}(\omega) = \mathrm{C_p^S} \mathrm{B_v} \boldsymbol{V}(\omega) + \mathrm{C_p^S} \Theta' \boldsymbol{R}(\omega). \tag{5.141}$$

Substituting equation (5.137) into equation (5.141) we have

$$\boldsymbol{Q}(\omega) = \mathrm{C_p^S} \Theta' \Delta^{-1}(\omega)\mathrm{B_f}\boldsymbol{F}(\omega) + \left[\mathrm{C_p^S} + \mathrm{C_p^S} \Theta' \Delta^{-1}(\omega)\Theta \mathrm{C_p^S}\right]\mathrm{B_v} \boldsymbol{V}(\omega). \tag{5.142}$$

The first term on the right-hand side of equation (5.142) is the relationship between applied force and charge in the piezoelectric material. This expression could be used to analyze the force-sensing response of the piezoelectrics. The second term (in brackets) is interpreted as the *dynamic capacitance* of the material.

Equations (5.140) and (5.142) represent the general solution for displacement and charge in the frequency domain as a function of the force and voltage inputs to the system. The vibration characteristics of the system can be obtained by solving for the natural frequencies using the homogeneous equations. As discussed in Chapter 4, the vibration response of a piezoelectric material system is a function of the electrical boundary conditions. For the case of multi-degree-of-freedom systems, this is analyzed by setting the charge coordinates and the voltage inputs as zero and solving for the system natural frequencies.

In the case in which the charge is specified to be zero, the homogeneous equations are reduced to

$$(\mathrm{M_s} + \mathrm{M_p})\ddot{\boldsymbol{r}} + \left(\mathrm{K_s} + \mathrm{K_p^D}\right)\boldsymbol{r} = 0. \tag{5.143}$$

Denoting the natural frequencies with zero charge as ω_i^{D}, we can solve for the natural frequencies from

$$\left(\mathrm{K_s} + \mathrm{K_p^D}\right)\boldsymbol{v}^{\mathrm{D}} = \omega_i^{\mathrm{D}^2}(\mathrm{M_s} + \mathrm{M_p})\boldsymbol{v}^{\mathrm{D}}, \tag{5.144}$$

which will yield N natural frequencies ω_i^{D} and N eigenvectors $\boldsymbol{v}_i^{\mathrm{D}}$.

The natural frequencies of the system for nonzero charge can be solved from the homogeneous expressions

$$(\mathrm{M_s} + \mathrm{M_p})\ddot{\boldsymbol{r}} + \left(\mathrm{K_s} + \mathrm{K_p^D}\right)\boldsymbol{r} - \Theta \boldsymbol{q} = 0 \tag{5.145}$$

$$-\Theta' \boldsymbol{r} + \mathrm{C_p^{-1}}\boldsymbol{q} = 0. \tag{5.146}$$

Solving equation (5.146) for the charge and substituting into equation (5.145) produces a single expression. The short-circuit (E = 0) natural frequencies, ω_i^{E}, can be solved from the eigenvalue problem

$$\left(\mathrm{K_s} + \mathrm{K_p^D} - \Theta \mathrm{C_p^S}\Theta'\right)\boldsymbol{v}^{\mathrm{E}} = \omega_i^{\mathrm{E}^2}(\mathrm{M_s} + \mathrm{M_p})\boldsymbol{v}^{\mathrm{E}}, \tag{5.147}$$

which will yield N natural frequencies ω_i^{E} and N eigenvectors $\boldsymbol{v}_i^{\mathrm{E}}$.

5.5 SPATIAL FILTERING AND MODAL FILTERS IN PIEZOELECTRIC MATERIAL SYSTEMS

The analysis in Section 5.4 is the general solution for a piezoelectric material system. We see that the assumption of linear elasticity and linear piezoelectric coupling in the energy function produces a set of second-order matrix expressions that can be solved in the time and frequency domains to yield the response of the mechanical and electrical coordinates.

Before applying this analysis to representative problems, let us discuss some general properties of the solution that can provide insight into the physics of using piezoelectric materials as sensors and actuators. It is well known that for linear vibration problems, we can represent the solution as a linear combination of *mode shapes*,

$\phi_i(\mathbf{x})$, as

$$\boldsymbol{u}(\mathbf{x},t) = \sum_{i=1}^{N} \phi_i(\mathbf{x}) r_i(t). \tag{5.148}$$

The mode shapes are unique to a constant; therefore, we can normalize the functions in any manner that we choose. Mass-normalizing the mode shapes implies that we choose $\phi_i(\mathbf{x})$ such that

$$\int_{V_s} \rho_s \phi_i(\mathbf{x}) \phi_j(\mathbf{x}) \, dV_s = \delta_{ij}, \tag{5.149}$$

where $\delta_{ij} = 1$ when $i = j$ and $\delta_{ij} = 0$ when $i \neq j$. Vibration analysis of continuous systems reveals that this condition also normalizes the stiffness terms such that

$$\int_{V_s} [\mathrm{L_u} \phi_i(\mathbf{x})]' \mathrm{c_s} [\mathrm{L_u} \phi_j(\mathbf{x})] \, dV_s = \omega_i^2 \delta_{ij}, \tag{5.150}$$

where ω_i is the natural frequency of the structure without addition of the piezoelectric elements. Using mass-normalized mode shapes in the displacement expression produces diagonal mass and stiffness matrices of the form

$$\mathrm{M_s} = \mathrm{I} \qquad \mathrm{K_s} = \Lambda = \mathrm{diag}\left(\omega_1^2, \omega_2^2, \ldots, \omega_N^2\right). \tag{5.151}$$

Although the mass and stiffness matrix of the structure are diagonal, note that in general the mass and stiffness matrices of the piezoelectric elements are not diagonal, due to the fact that the integration is not over the entire volume of the structure. The off-diagonal terms in the mass and stiffness matrix of the piezoelectric elements introduce coupling between the vibration modes. In general, this will change the mode shape and natural frequencies of the piezoelectric system as compared to the mode shapes and frequencies of the system without the piezoelectric elements.

Another measure of the coupling introduced by the piezoelectric elements is contained within the matrix Θ. Let us assume for a moment that we have a system whose mode shapes are only a function of a single dimension, x, and that the rotation matrices are both identity matrices (indicating that the local axis of the piezoelectric material lines up with the global axes). Under these assumptions, the (i, j)th element of Θ is proportional to

$$\Theta_{ij} \propto \int_{x_1}^{x_2} [\mathrm{L_u} \phi_i(x)] \, \mathrm{B_{q}}_j(x) \, dx. \tag{5.152}$$

To illustrate the concept of spatial filtering, let's assume that the system is a beam in which $\mathrm{L_u} = z d^2/dx^2$; therefore,

$$\Theta_{ij} \propto \int_{x_1}^{x_2} \frac{d^2\phi_i(x)}{dx^2} \mathrm{B}_{\mathrm{q}_j}(x)\, dx. \tag{5.153}$$

In the simplest case the electric displacement will have no spatial variation, in which integration produces the expression

$$\Theta_{ij} \propto \left.\frac{d\phi_i(x)}{dx}\right|_{x_2} - \left.\frac{d\phi_i(x)}{dx}\right|_{x_1}. \tag{5.154}$$

Equation (5.154) indicates that the coupling term associated with the piezoelectric element is proportional to the difference in the slopes at the ends of the element.

This concept is illustrated in Figure 5.6. A representative mode shape and its derivative are shown in the figure. In the leftmost plot, we see that placing a piezoelectric element along the length from $x_1 = 0$ to $x_2 = L/3$ produces a differential in slope of approximately 0.32. Increasing the length of the piezoelectric element to a point at which the slope at x_1 and x_2 are both zero produces a coupling term of approximately zero. This is equivalent to saying that integration of the strain over the length of the piezoelectric element is zero. Finally, a piezoelectric element that spans the length of the beam produces the maximum coupling term but yields a value that is opposite in sign to the case in which $x_2 = L/3$.

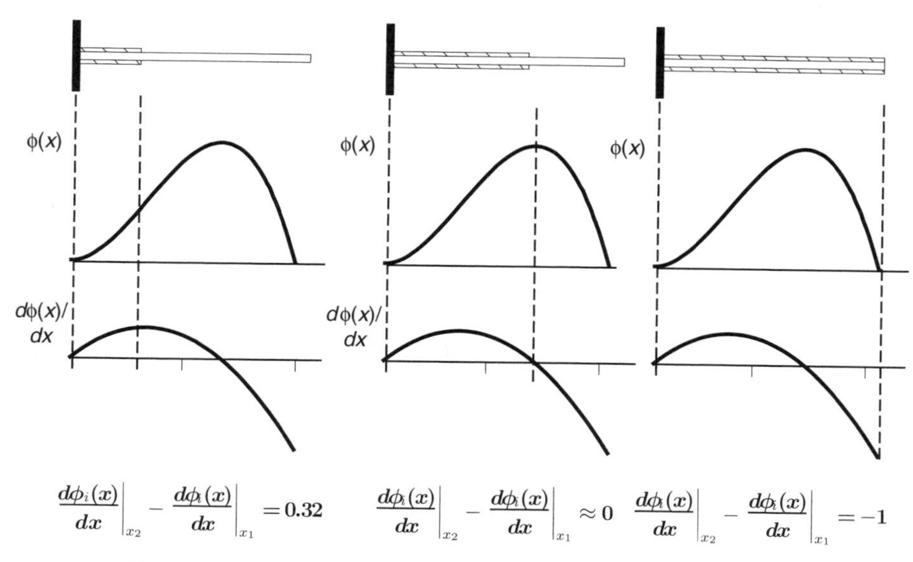

Figure 5.6 Effect of length of a piezoelectric on the coupling term θ_{ij}.

Figure 5.6 illustrates how spatial filtering can be used to tailor the coupling properties of a piezoelectric element. If the wavelength of the mode shape is long compared to the spatial dimension of the piezoelectric element, the strain variation will be small and the slope differential between the ends will also be small. This situation will lead to a low coupling value. Positioning the piezoelectric element in this manner will filter out the contribution of that mode in the coupling matrix.

This technique is especially interesting for dynamic problems when one considers how the coupling will vary when multiple modes are considered in the analysis. From vibration analysis of continuous systems, we know that the standing mode shapes of vibrating structures are harmonic functions in the spatial dimension. To illustrate the concept of spatial filtering for multiple modes, assume that

$$\phi_i(x) = \sin\frac{i\pi x}{L}, \tag{5.155}$$

where L is the length in the x direction and $0 \leq x \leq L$. The first three mode shapes for this expansion are plotted in Figure 5.7 along with the slope functions

$$\frac{d\phi_i(x)}{dx} = \frac{i\pi}{L}\cos\frac{i\pi x}{L}. \tag{5.156}$$

The three parts of Figure 5.7 demonstrate the concept of spatial filtering for multiple modes. In Figure 5.7*a*, the piezoelectric element is centered around the midspan of the beam. For this placement we see that the slope differential for the second mode is zero, due to the fact that the piezoelectric is centered around a node for $\phi_2(x)$.

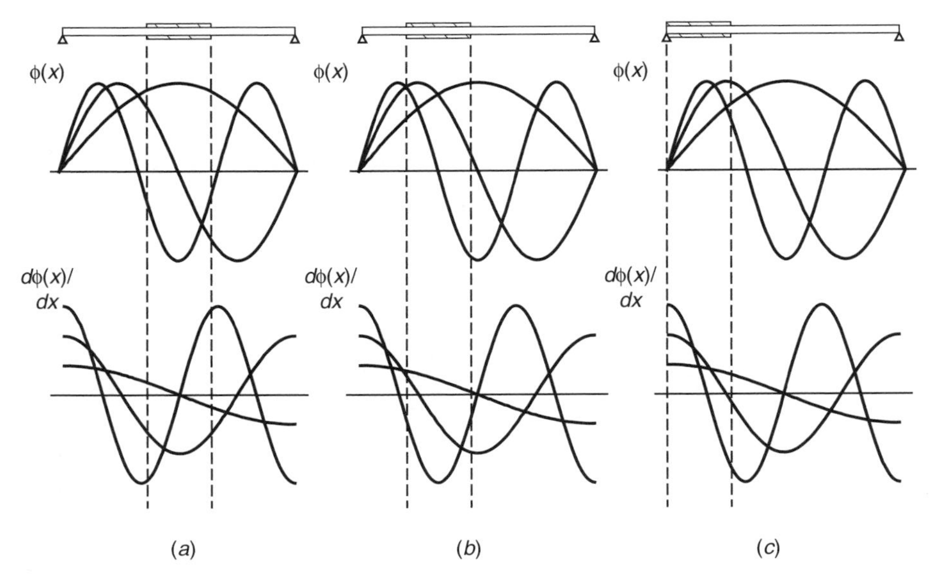

Figure 5.7 Variation in the piezoelectric coupling term for a representative modal expansion.

Table 5.4 Piezoelectric coupling terms for various locations on a structure with harmonic mode shapes

	$L/2$	$L/3$	$L/6$
$\left.\frac{d\phi_1(x)}{dx}\right\rvert_{x_2} - \left.\frac{d\phi_1(x)}{dx}\right\rvert_{x_1}$	$-\pi$	-2.72	$-\pi/2$
$\left.\frac{d\phi_2(x)}{dx}\right\rvert_{x_2} - \left.\frac{d\phi_2(x)}{dx}\right\rvert_{x_1}$	0	-9.42	-3π
$\left.\frac{d\phi_3(x)}{dx}\right\rvert_{x_2} - \left.\frac{d\phi_3(x)}{dx}\right\rvert_{x_1}$	6π	0	-6π

Moving the piezoelectric element to a location in which it is symmetric about the node for $\phi_3(x)$ (Figure 5.7*b*) produces a zero slope differential for the third mode. In both of these cases we have filtered out the contribution of particular modes in the piezoelectric coupling matrix. In Figure 5.7*c*, we have a situation in which none of the modes are filtered out and all three modes make a contribution to the piezoelectric coupling terms.

The relative contribution of the modes to the piezoelectric coupling terms can be evaluated by computing the slope differential for specified locations on the structure. As an example, consider a piezoelectric element with a length of $L/3$ with varying center locations as shown in Figure 5.7. The slope differentials for three cases of center location are shown in Table 5.4. From the table we see that centering the piezoelectric element at $L/2$ or $L/3$ filters out the modal contribution in the second and third modes, respectively. Furthermore, we see that centering the element at $L/2$ produces a sign change between the piezoelectric coupling term for the first and third modes. This sign change indicates that there will be a phase difference between the contribution between these two modes. This contrasts with the contributions when the element is centered at $L/6$. In this location we see that the sign of all the elements is the same, indicating that there will not be a phase change in the piezoelectric coupling term for the first three modes of the structure.

The analysis was performed for a particular type of mode shape, but the concept can be generalized for arbitrary structures once the mode shapes are known. Our discussion of spatial filtering has shown that the relative contribution of the modes in the piezoelectric coupling matrix can be tailored by appropriate choice of location and element size. Modes with wavelengths that are long compared to the element size will tend to be attenuated in the coupling matrix due to the fact that the strain variation over the element length is small. Modes can be filtered completely by choosing the location of the element such that the slope differential is negligible compared to the slope differential of the remaining mode shapes. This is equivalent to choosing a location such that the integral of the strain over the piezoelectric dimension is zero.

5.5.1 Modal Filters

In Section 5.4 we discussed how piezoelectric materials couple to a structural system. We saw that the coupling was related to integration of the strain over the dimensions of the piezoelectric element. By proper choice of the location, we could tailor the coupling properties of the piezoelectric element and in some cases completely eliminate the contribution of certain modes in the piezoelectric response.

We can take this analysis one step further and show that the piezoelectric element can be designed such that it *couples to only one vibration mode*. Examining equation (5.153), we see that if

$$\mathrm{B}_{\mathrm{q}_j}(x) \propto \frac{d^2\phi_i(x)}{dx^2}, \tag{5.157}$$

the piezoelectric coupling term is

$$\Theta_{ij} \propto \int_{x_1}^{x_2} \frac{d^2\phi_i(x)}{dx^2}\frac{d^2\phi_i(x)}{dx^2}dx. \tag{5.158}$$

Extending the limits of integration over the entire length, and recalling that the mode shapes are orthogonal produces the relationship

$$\Theta_{ij} \propto \int_{0}^{L} \frac{d^2\phi_i(x)}{dx^2}\frac{d^2\phi_i(x)}{dx^2}dx \propto \omega_i^2\delta_{ij}. \tag{5.159}$$

Equation (5.159) shows that if we can shape the electrodes of the piezoelectric element such that it is orthogonal to the second derivative of the mode shape, the coupling terms of the piezoelectric material will be zero *except for a single vibration mode of the structure*.

To illustrate the concept of modal filtering, let us plot the electrode shape required to filter the modes shown in Figure 5.7. The second derivative of each of the mode shapes is also proportional to a sine function:

$$\frac{d^2\phi_i(x)}{dx^2} \propto \sin\frac{i\pi x}{L}; \tag{5.160}$$

therefore, the required spatial distribution of the electrode is also a sine function. These results are shown in Figure 5.8. Figure 5.8*a* illustrates the electrode pattern for a modal filter whose only nonzero coupling term is for the first mode of the beam. In Figure 5.8*b* we have an electrode pattern for the second mode shape. The plus and minus signs indicate that we have to change the phase of the signal to the different regions of the electrode, due to the fact that the second derivative of the mode shape changes from positive to negative over the length of the beam. Figure 5.8*c* shows a modal filter for the third structural mode shape.

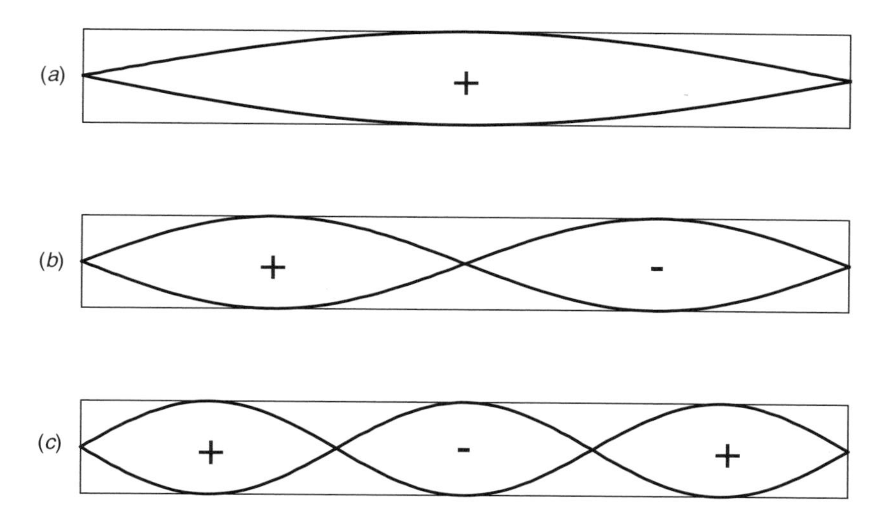

Figure 5.8 Electrode patterns for modal filters for a structure with mode shapes $\sin(i\pi x/L)$.

5.6 DYNAMIC RESPONSE OF PIEZOELECTRIC BEAMS

Beamlike structures offer an opportunity to study the fundamental dynamic response properties of piezoelectric material systems. The procedure is similar to the method described earlier in the chapter for static analysis. The main differences are that we must form the mass matrices of the structure and the piezoelectric materials, and we must choose a set of admissible shape functions that represent the dynamic response of the structure. Unlike the system studied earlier in the chapter, we assume that the inactive substrate is of nonnegligible thickness and that the piezoelectric element does not span the full length of the beam. This will allow us to study the effect of substrate properties and placement on the dynamic response.

The dynamic response of a beam is a well-known problem in vibration analysis. It is known that the motion of the beam can be expressed as a linear combination of functions known as *mode shapes*, $\phi_i(x)$, and that these are a function of the boundary conditions. To use the energy analysis described earlier for static problems, the displacement vector for the system is chosen to be

$$\left\{\begin{array}{c} u_1 \\ u_2 \\ u_3 \end{array}\right\} = \left[\begin{array}{c} 0 \\ 0 \\ \sum_{i=0}^{N} \phi_i(x) r_i(t) \end{array}\right], \tag{5.161}$$

which leads to the shape functions

$$\mathrm{N_r}(x) = \left[\begin{array}{ccc} 0 & \cdots & 0 \\ 0 & \cdots & 0 \\ \phi_1(x) & \cdots & \phi_N(x) \end{array}\right]. \tag{5.162}$$

The differential operator that relates the strain in the piezoelectric material to the displacement is

$$\mathbf{L}_{\mathrm{u_p}} = \begin{bmatrix} 0 & 0 & z\dfrac{\partial^2}{\partial x^2} \\ 0 & 0 & -\nu_{12}z\dfrac{\partial^2}{\partial x^2} \\ 0 & 0 & -\nu_{13}z\dfrac{\partial^2}{\partial x^2} \\ 0 & 0 & 0 \\ 0 & 0 & 0 \\ 0 & 0 & 0 \end{bmatrix}, \tag{5.163}$$

and the differential operator that relates the strain to the displacement in the substrate is

$$\mathbf{L}_{\mathrm{u_s}} = \begin{bmatrix} 0 & 0 & z\dfrac{\partial^2}{\partial x^2} \\ 0 & 0 & -\nu z\dfrac{\partial^2}{\partial x^2} \\ 0 & 0 & -\nu z\dfrac{\partial^2}{\partial x^2} \\ 0 & 0 & 0 \\ 0 & 0 & 0 \\ 0 & 0 & 0 \end{bmatrix}. \tag{5.164}$$

The shape functions for the strain in the piezoelectric are obtained by combining equations (5.161) and (5.163):

$$\mathbf{B}_{\mathrm{r_p}} = \mathbf{L}_{\mathrm{u_p}}\mathbf{N}_{\mathrm{r}}(x) = z\begin{bmatrix} \dfrac{d^2\phi_1(x)}{dx^2} & \cdots & \dfrac{d^2\phi_N(x)}{dx^2} \\ -\nu_{12}\dfrac{d^2\phi_1(x)}{dx^2} & \cdots & -\nu_{12}\dfrac{d^2\phi_N(x)}{dx^2} \\ -\nu_{13}\dfrac{d^2\phi_1(x)}{dx^2} & \cdots & -\nu_{13}\dfrac{d^2\phi_N(x)}{dx^2} \\ 0 & \cdots & 0 \\ 0 & \cdots & 0 \\ 0 & \cdots & 0 \end{bmatrix}. \tag{5.165}$$

The shape functions for the substrate are obtained similarly by combining equations (5.161) and (5.164):

$$\mathrm{B}_{\mathrm{r_s}} = \mathrm{L}_{\mathrm{u_s}}\mathrm{N}_{\mathrm{r}}(x) = z \begin{bmatrix} \dfrac{d^2\phi_1(x)}{dx^2} & \cdots & \dfrac{d^2\phi_N(x)}{dx^2} \\ -\nu\dfrac{d^2\phi_1(x)}{dx^2} & \cdots & -\nu\dfrac{d^2\phi_N(x)}{dx^2} \\ -\nu\dfrac{d^2\phi_1(x)}{dx^2} & \cdots & -\nu\dfrac{d^2\phi_N(x)}{dx^2} \\ 0 & \cdots & 0 \\ 0 & \cdots & 0 \\ 0 & \cdots & 0 \end{bmatrix}. \tag{5.166}$$

The matrices for the charge coordinates are identical to those shown in equations (5.86) and (5.87), except for the fact that the length of the piezoelectric elements is assumed to be equal to $L_2 - L_1$; therefore,

$$\mathrm{B}_{\mathrm{q}}^1(\mathbf{x}) = \begin{bmatrix} 0 & 0 \\ 0 & 0 \\ \dfrac{1}{w\,(L_2 - L_1)} & 0 \end{bmatrix} \tag{5.167}$$

$$\mathrm{B}_{\mathrm{q}}^2(\mathbf{x}) = \begin{bmatrix} 0 & 0 \\ 0 & 0 \\ 0 & \dfrac{1}{w\,(L_2 - L_1)} \end{bmatrix}. \tag{5.168}$$

Defining the shape functions allows us to define the terms associated with the stiffness, coupling, capacitance, and mass matrices. The stiffness matrix of the substrate is obtained from

$$\mathrm{B}_{\mathrm{r_s}}(\mathbf{x})'\mathrm{c_s}\mathrm{B}_{\mathrm{r_s}}(\mathbf{x}) = Y_s z^2 \begin{bmatrix} \dfrac{d^2\phi_1(x)}{dx^2}\dfrac{d^2\phi_1(x)}{dx^2} & \cdots & \dfrac{d^2\phi_1(x)}{dx^2}\dfrac{d^2\phi_N(x)}{dx^2} \\ \vdots & \ddots & \vdots \\ \dfrac{d^2\phi_N(x)}{dx^2}\dfrac{d^2\phi_1(x)}{dx^2} & \cdots & \dfrac{d^2\phi_N(x)}{dx^2}\dfrac{d^2\phi_N(x)}{dx^2} \end{bmatrix}, \tag{5.169}$$

where Y_s is the modulus of the substrate material. The stiffness matrix of the piezoelectric element is obtained from

$$\mathrm{B_{r_p}}(\mathbf{x})'\mathrm{c^D}\mathrm{B_{r_p}}(\mathbf{x}) = \hat{c}^{\mathrm{D}} z^2 \begin{bmatrix} \frac{d^2\phi_1(x)}{dx^2}\frac{d^2\phi_1(x)}{dx^2} & \cdots & \frac{d^2\phi_1(x)}{dx^2}\frac{d^2\phi_N(x)}{dx^2} \\ \vdots & \ddots & \vdots \\ \frac{d^2\phi_N(x)}{dx^2}\frac{d^2\phi_1(x)}{dx^2} & \cdots & \frac{d^2\phi_N(x)}{dx^2}\frac{d^2\phi_N(x)}{dx^2} \end{bmatrix}, \tag{5.170}$$

where

$$\hat{c}^{\mathrm{D}} = c_{11}^{\mathrm{D}} - 2\nu_{12}c_{12}^{\mathrm{D}} - 2\nu_{13}c_{13}^{\mathrm{D}} + \nu_{12}^2 c_{22}^{\mathrm{D}} + 2\nu_{12}\nu_{13}c_{23}^{\mathrm{D}} + \nu_{13}^2 c_{33}^{\mathrm{D}}. \tag{5.171}$$

The remaining matrices are

$$\mathrm{B_r}(\mathbf{x})'\mathrm{h}\mathrm{B_q^1}(\mathbf{x}) = \tilde{h}\frac{z}{w(L_2 - L_1)} \begin{bmatrix} \frac{d^2\phi_1(x)}{dx^2} & 0 \\ \vdots & \vdots \\ \frac{d^2\phi_N(x)}{dx^2} & 0 \end{bmatrix} \tag{5.172}$$

$$\mathrm{B_r}(\mathbf{x})'\mathrm{h}\mathrm{B_q^2}(\mathbf{x}) = \tilde{h}\frac{z}{w(L_2 - L_1)} \begin{bmatrix} 0 & \frac{d^2\phi_1(x)}{dx^2} \\ \vdots & \vdots \\ 0 & \frac{d^2\phi_N(x)}{dx^2} \end{bmatrix} \tag{5.173}$$

$$\mathrm{B_q^1}(\mathbf{x})'\beta^{\mathrm{S}}\mathrm{B_q^1}(\mathbf{x}) = \beta_{33}^{\mathrm{S}}\frac{1}{w^2(L_2 - L_1)^2}\begin{bmatrix} 1 & 0 \\ 0 & 0 \end{bmatrix} \tag{5.174}$$

$$\mathrm{B_q^2}(\mathbf{x})'\beta^{\mathrm{S}}\mathrm{B_q^2}(\mathbf{x}) = \beta_{33}^{\mathrm{S}}\frac{1}{w^2(L_2 - L_1)^2}\begin{bmatrix} 0 & 0 \\ 0 & 1 \end{bmatrix}. \tag{5.175}$$

The mass terms are obtained from the matrices

$$\mathrm{N_r}(\mathbf{x})'\mathrm{N_r}(\mathbf{x}) = \begin{bmatrix} \phi_1(x)\phi_1(x) & \cdots & \phi_1(x)\phi_N(x) \\ \vdots & \ddots & \vdots \\ \phi_N(x)\phi_1(x) & \cdots & \phi_N(x)\phi_N(x) \end{bmatrix}. \tag{5.176}$$

Mass normalizing the mode shapes leads to the expressions

$$\mathrm{M_s} = \rho_s \int_{-t_s/2}^{t_s/2}\int_{-w/2}^{w/2}\int_0^L$$

$$\times \begin{bmatrix} \phi_1(x)\phi_1(x) & \cdots & \phi_1(x)\phi_N(x) \\ \vdots & \ddots & \vdots \\ \phi_N(x)\phi_1(x) & \cdots & \phi_N(x)\phi_N(x) \end{bmatrix} dx\,dy\,dz = \begin{bmatrix} 1 & \cdots & 0 \\ \vdots & \ddots & \vdots \\ 0 & \cdots & 1 \end{bmatrix} \tag{5.177}$$

$$\mathrm{K_s} = Y_s \int_{-t_s/2}^{t_s/2} \int_{-w/2}^{w/2} \int_0^L z^2 \times \begin{bmatrix} \frac{d^2\phi_1(x)}{dx^2}\frac{d^2\phi_1(x)}{dx^2} & \cdots & \frac{d^2\phi_1(x)}{dx^2}\frac{d^2\phi_N(x)}{dx^2} \\ \vdots & \ddots & \vdots \\ \frac{d^2\phi_N(x)}{dx^2}\frac{d^2\phi_1(x)}{dx^2} & \cdots & \frac{d^2\phi_N(x)}{dx^2}\frac{d^2\phi_N(x)}{dx^2} \end{bmatrix} dx\,dy\,dz = \begin{bmatrix} \omega_1^2 & \cdots & 0 \\ \vdots & \ddots & \vdots \\ 0 & \cdots & \omega_N^2 \end{bmatrix}. \tag{5.178}$$

The mass and stiffness matrices for the piezoelectric elements are

$$\mathrm{M_p} = \rho_p \int_{-w/2}^{w/2} \int_{L_1}^{L_2} \left\{ \int_{-t_s/2-t_p/2}^{-t_s/2} \begin{bmatrix} \phi_1(x)\phi_1(x) & \cdots & \phi_1(x)\phi_N(x) \\ \vdots & \ddots & \vdots \\ \phi_N(x)\phi_1(x) & \cdots & \phi_N(x)\phi_N(x) \end{bmatrix} dz + \int_{t_s/2}^{t_s/2+t_p/2} \begin{bmatrix} \phi_1(x)\phi_1(x) & \cdots & \phi_1(x)\phi_N(x) \\ \vdots & \ddots & \vdots \\ \phi_N(x)\phi_1(x) & \cdots & \phi_N(x)\phi_N(x) \end{bmatrix} dz \right\} dx\,dy \tag{5.179}$$

$$\mathrm{K_p^D} = \hat{c}^{\mathrm{D}} \int_{-w/2}^{w/2} \int_{L_1}^{L_2} \left\{ \int_{-t_s/2-t_p/2}^{-t_s/2} z^2 \begin{bmatrix} \frac{d^2\phi_1(x)}{dx^2}\frac{d^2\phi_1(x)}{dx^2} & \cdots & \frac{d^2\phi_1(x)}{dx^2}\frac{d^2\phi_N(x)}{dx^2} \\ \vdots & \ddots & \vdots \\ \frac{d^2\phi_N(x)}{dx^2}\frac{d^2\phi_1(x)}{dx^2} & \cdots & \frac{d^2\phi_N(x)}{dx^2}\frac{d^2\phi_N(x)}{dx^2} \end{bmatrix} dz + \int_{t_s/2}^{t_s/2+t_p/2} z^2 \begin{bmatrix} \frac{d^2\phi_1(x)}{dx^2}\frac{d^2\phi_1(x)}{dx^2} & \cdots & \frac{d^2\phi_1(x)}{dx^2}\frac{d^2\phi_N(x)}{dx^2} \\ \vdots & \ddots & \vdots \\ \frac{d^2\phi_N(x)}{dx^2}\frac{d^2\phi_1(x)}{dx^2} & \cdots & \frac{d^2\phi_N(x)}{dx^2}\frac{d^2\phi_N(x)}{dx^2} \end{bmatrix} dz \right\} dx\,dy. \tag{5.180}$$

The coupling matrix is obtained from

$$\Theta = \frac{\tilde{h}}{w(L_2 - L_1)} \int_{-w/2}^{w/2} \int_{L_1}^{L_2} \times \left\{ \int_{t_s/2}^{t_s/2+t_p/2} z \begin{bmatrix} \frac{d^2\phi_1(x)}{dx^2} & 0 \\ \vdots & \vdots \\ \frac{d^2\phi_N(x)}{dx^2} & 0 \end{bmatrix} dz + \int_{-t_s/2-t_p/2}^{-t_s/2} z \begin{bmatrix} 0 & \frac{d^2\phi_1(x)}{dx^2} \\ \vdots & \vdots \\ 0 & \frac{d^2\phi_N(x)}{dx^2} \end{bmatrix} dz \right\} dx\, dy \tag{5.181}$$

and the capacitive terms are

$$C_p^{S^{-1}} = \frac{\beta_{33}^S}{w^2(L_2 - L_1)^2} \int_{-w/2}^{w/2} \int_{L_1}^{L_2} \times \left\{ \int_{t_s/2}^{t_s/2+t_p/2} \begin{bmatrix} 1 & 0 \\ 0 & 0 \end{bmatrix} dz + \int_{-t_s/2-t_p/2}^{-t_s/2} \begin{bmatrix} 0 & 0 \\ 0 & 1 \end{bmatrix} dz \right\} dx\, dy. \tag{5.182}$$

The integrations associated with the mass, stiffness, and coupling matrices can be simplified by making the following substitutions into the expressions:

$$z = t_p \zeta \tag{5.183}$$

$$x = L\xi \tag{5.184}$$

$$\tau = t_s/t_p. \tag{5.185}$$

Substituting these expressions into the mass and stiffness matrices for the substrate yields

$$M_s = \rho_s w t_s L \int_0^1 \begin{bmatrix} \phi_1(\xi)\phi_1(\xi) & \cdots & \phi_1(\xi)\phi_N(\xi) \\ \vdots & \ddots & \vdots \\ \phi_N(\xi)\phi_1(\xi) & \cdots & \phi_N(\xi)\phi_N(\xi) \end{bmatrix} d\xi \tag{5.186}$$

$$K_s = \frac{Y_s I_s}{L^3} \int_0^1 \begin{bmatrix} \frac{d^2\phi_1(\xi)}{d\xi^2}\frac{d^2\phi_1(\xi)}{d\xi^2} & \cdots & \frac{d^2\phi_1(\xi)}{d\xi^2}\frac{d^2\phi_N(\xi)}{d\xi^2} \\ \vdots & \ddots & \vdots \\ \frac{d^2\phi_N(\xi)}{d\xi^2}\frac{d^2\phi_1(\xi)}{d\xi^2} & \cdots & \frac{d^2\phi_N(\xi)}{dx^2}\frac{d^2\phi_N(xi)}{d\xi^2} \end{bmatrix} d\xi, \tag{5.187}$$

where $I_s = \frac{1}{12} w t_s^3$.

The piezoelectric mass matrix can be written

$$\mathrm{M_p} = \rho_p w t_p L \int_{\xi_1}^{\xi_2} \begin{bmatrix} \phi_1(\xi)\phi_1(\xi) & \cdots & \phi_1(\xi)\phi_N(\xi) \\ \vdots & \ddots & \vdots \\ \phi_N(\xi)\phi_1(\xi) & \cdots & \phi_N(\xi)\phi_N(\xi) \end{bmatrix} d\xi \tag{5.188}$$

and the piezoelectric stiffness matrix can be written as the integration

$$\mathrm{K_p^D} = \frac{\hat{c}^{\mathrm{D}} I_p}{L^3}(1 + 3\tau + 3\tau^2) \int_{\xi_1}^{\xi_2} \begin{bmatrix} \dfrac{d^2\phi_1(\xi)}{d\xi^2}\dfrac{d^2\phi_1(\xi)}{d\xi^2} & \cdots & \dfrac{d^2\phi_1(\xi)}{d\xi^2}\dfrac{d^2\phi_N(\xi)}{d\xi^2} \\ \vdots & \ddots & \vdots \\ \dfrac{d^2\phi_N(\xi)}{d\xi^2}\dfrac{d^2\phi_1(\xi)}{d\xi^2} & \cdots & \dfrac{d^2\phi_N(\xi)}{dx^2}\dfrac{d^2\phi_N(xi)}{d\xi^2} \end{bmatrix} d\xi, \tag{5.189}$$

where $I_p = \frac{1}{12} w t_p^3$.

The integration in the definition of the coupling matrix can be performed to yield the expression

$$\Theta = \frac{\tilde{h}}{8(\xi_2 - \xi_1)} \frac{t_p^2}{L^2}(2\tau + 1) \times \begin{bmatrix} \left.\dfrac{d\phi_1(\xi)}{d\xi}\right|_{\xi_2} - \left.\dfrac{d\phi_1(\xi)}{d\xi}\right|_{\xi_1} & -\left(\left.\dfrac{d\phi_1(\xi)}{d\xi}\right|_{\xi_2} - \left.\dfrac{d\phi_1(\xi)}{d\xi}\right|_{\xi_1}\right) \\ \vdots & \vdots \\ \left.\dfrac{d\phi_N(\xi)}{d\xi}\right|_{\xi_2} - \left.\dfrac{d\phi_N(\xi)}{d\xi}\right|_{\xi_1} & -\left(\left.\dfrac{d\phi_N(\xi)}{d\xi}\right|_{\xi_2} - \left.\dfrac{d\phi_N(\xi)}{d\xi}\right|_{\xi_1}\right) \end{bmatrix}. \tag{5.190}$$

The capactive term can be integrated to yield

$$\mathrm{C_p^{S}}^{-1} = \frac{\beta_{33}^{\mathrm{S}} t_p}{wL(\xi_2 - \xi_1)} \begin{bmatrix} 1/2 & 0 \\ 0 & 1/2 \end{bmatrix}. \tag{5.191}$$

Recall from basic vibration theory that the mode shapes are not unique and can be normalized arbitrarily. A common normalization is to define $\phi_i(x)$ such that

$$\int_0^1 \phi_i(\xi)\phi_j(\xi)\, d\xi = \delta_{ij}, \tag{5.192}$$

which automatically defines the integration

$$\int_0^1 \frac{d^2\phi_i(\xi)}{d\xi^2}\frac{d^2\phi_j(\xi)}{d\xi^2}d\xi = \lambda_i^4\delta_{ij}. \tag{5.193}$$

where λ_i is the ith eigenvalue.

Substituting the normalizations into the matrix expressions simplifies their form. It is instructive to examine the mass and stiffness matrices to understand how the addition of the piezoelectric element changes the equations of motion. The sum of the substrate mass and piezoelectric mass can be written

$$\begin{aligned} \mathrm{M_s} + \mathrm{M_p} &= \rho_s w t_s L \\ &\times \left\{ \begin{bmatrix} 1 & \cdots & 0 \\ \vdots & \ddots & \vdots \\ 0 & \cdots & 1 \end{bmatrix} + \frac{\rho_p}{\rho_s}\frac{1}{\tau}\int_{\xi_1}^{\xi_2} \begin{bmatrix} \phi_1(\xi)\phi_1(\xi) & \cdots & \phi_1(\xi)\phi_N(\xi) \\ \vdots & \ddots & \vdots \\ \phi_N(\xi)\phi_1(\xi) & \cdots & \phi_N(\xi)\phi_N(\xi) \end{bmatrix} d\xi \right\}. \end{aligned} \tag{5.194}$$

It is clear from equation (5.194) that as the ratio $\rho_p/\tau\rho_s$ approaches zero, the piezoelectric mass has little effect on the mass matrix of the system. Physically, this quantity represents the ratio of the mass per unit area of the piezoelectric to the mass per unit area of the substrate. This quantity can be small if the piezoelectric element has a much lower density than the substrate or if the thickness of the piezoelectric element is much smaller than the thickness of the substrate ($\tau \gg 1$).

A similar analysis can be performed for the stiffness matrices. The sum of the stiffness matrix of the piezoelectric element and the substrate can be written

$$\begin{aligned} \mathrm{K_s} + \mathrm{K_p^D} &= \frac{Y_s I_s}{L^3} \left\{ \begin{bmatrix} \lambda_1^4 & \cdots & 0 \\ \vdots & \ddots & \vdots \\ 0 & \cdots & \lambda_N^4 \end{bmatrix} + \frac{\hat{c}^D}{Y_s} g(\tau) \int_{\xi_1}^{\xi_2} \right. \\ &\times \left. \begin{bmatrix} \frac{d^2\phi_1(\xi)}{d\xi^2}\frac{d^2\phi_1(\xi)}{d\xi^2} & \cdots & \frac{d^2\phi_1(\xi)}{d\xi^2}\frac{d^2\phi_N(\xi)}{d\xi^2} \\ \vdots & \ddots & \vdots \\ \frac{d^2\phi_N(\xi)}{d\xi^2}\frac{d^2\phi_1(\xi)}{d\xi^2} & \cdots & \frac{d^2\phi_N(\xi)}{dx^2}\frac{d^2\phi_N(xi)}{d\xi^2} \end{bmatrix} d\xi \right\}, \end{aligned} \tag{5.195}$$

where

$$g(\tau) = \frac{1 + 3\tau + 3\tau^2}{\tau^3}. \tag{5.196}$$

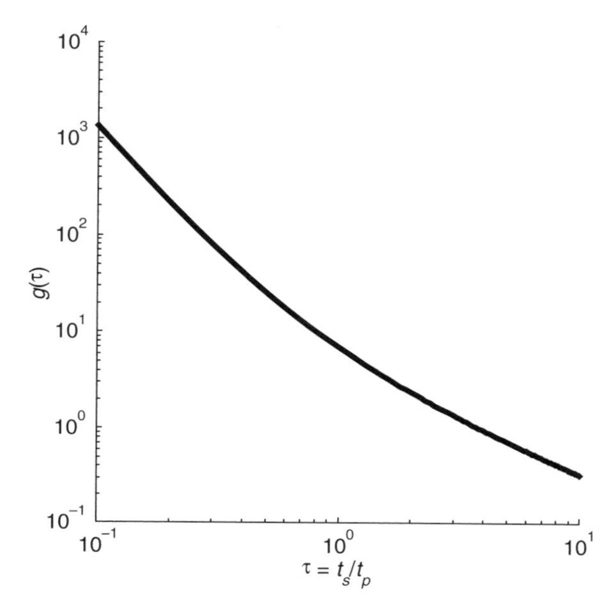

Figure 5.9 Variation in the function $g(\tau)$ with respect to the thickness ratio.

Equation (5.195) illustrates that the beam stiffness matrix is modified by the addition of the piezoelectric element. The additional stiffness terms are a function of the ratio of the material moduli. If the piezoelectric element has a modulus that is much smaller than that of the beam, $\hat{c}^D \ll Y_s$, and it is possible that the additional stiffness terms will be small compared to the stiffness matrix of the beam.

The other factor that determines the magnitude of the additional stiffness terms is the thickness ratio between the substrate and the piezoelectric. The effect of the thickness ratio is quantified by the function $g(\tau)$, which is shown in equation (5.196). Figure 5.9 is a plot of the function $g(\tau)$ with respect to the thickness ratio t_s/t_p. We see that even for a thickness ratio of 1, the stiffness of the piezoelectric element can have a substantial effect on the stiffness elements of the beam, due to the fact that $g(\tau) = 10$. Reducing the thickness ratio produces a large increase in the function, whereas very thin piezoelectric elements [i.e., $g(\tau) \gg 1$] reduce the function to much less than 1.

One of the most important aspects of this analysis is that it highlights the relationship between the mode shapes of the structure and the mass, stiffness, and coupling terms of the system matrices. We see from the expressions for the mass, stiffness, and coupling terms that the mode shapes determine the individual elements of the matrices.

5.6.1 Cantilevered Piezoelectric Beam

To examine this relationship for a particular type of system, let's use the example of a cantilever beam to illustrate how the mass, stiffness, and coupling matrices depend

Table 5.5 Mode shape parameters for a cantilever beam

i	λ_i	σ_i
1	1.875	0.7341
2	4.694	1.0185
3	7.855	0.9992
4	10.996	1.0000
5	14.137	1.0000
>5	$\frac{(2i-1)\pi}{2}$	1

on the mode shapes. The mode shapes of a cantilever beam are expressed as

$$\phi_i(\xi) = \cosh\lambda_i\xi - \cos\lambda_i\xi - \sigma_i(\sinh\lambda_i\xi - \sin\lambda_i\xi), \tag{5.197}$$

where the values for λ_i and σ_i are shown in Table 5.5. These mode shapes are normalized such that equations (5.192) and (5.193) are satisfied.

A plot of $\phi_i(\xi)\phi_j(\xi)$ for the first two modes of a cantilever beam are shown in Figure 5.10. The elements of the piezoelectric mass matrix will be related to integration under these curves between the endpoints of the element, ξ_2 and ξ_1. We see from the figure that the integration will be small if the element is placed close

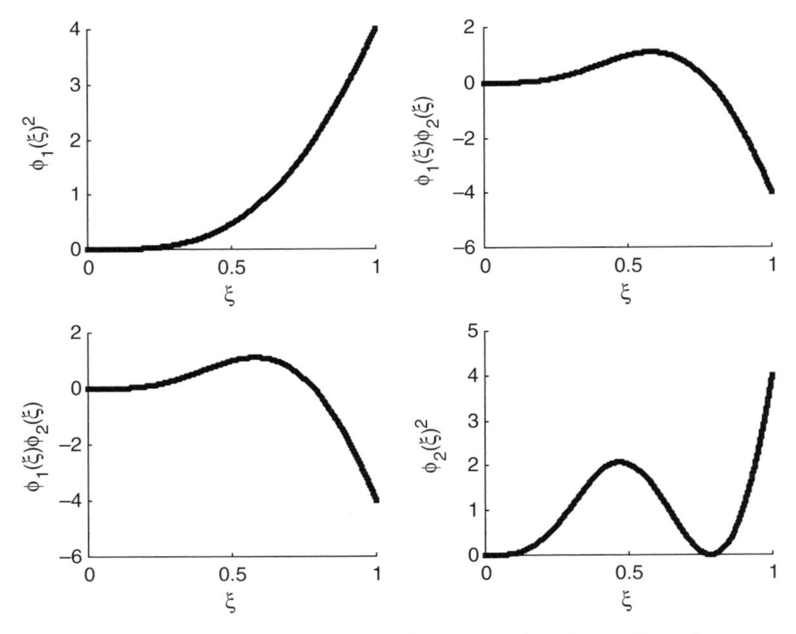

Figure 5.10 Plot of $\phi_i(\xi)\phi_j(\xi)$ for the first two modes of a cantilever beam.

to the clamped end of the beam ($\xi = 0$). Moving the element along the beam will affect the individual terms of M_p differently. For example, an element placed near the midspan of the beam with a length less than the full length of the beam will have a small effect on the (1,1) element of the piezoelectric mass matrix while having a correspondingly large effect on the (2,2) element of the matrix. Figure 5.10 also demonstrates that the off-diagonal terms of the mass matrix can be negative, while the diagonal terms are always positive.

The expression for the piezoelectric coupling term is related directly to the slope of the mode shapes at the ends of the piezoelectric element. The first derivative of the mode shape is

$$\frac{d\phi_i(\xi)}{d\xi} = \lambda_i[\sinh\lambda_i\xi + \sin\lambda_i\xi - \sigma_i(\cosh\lambda_i\xi - \cos\lambda_i\xi)]. \tag{5.198}$$

Figure 5.11 is a plot of the slope of the first two mode shapes of a cantilever beam. The endpoints of the piezo element define the entries in the coupling matrix Θ.

The elements of the stiffness matrix are related to the shape of the functions

$$\frac{d^2\phi_i(\xi)}{d^2\xi} = \lambda_i^2[\cosh\lambda_i\xi + \cos\lambda_i\xi - \sigma_i(\sinh\lambda_i\xi + \sin\lambda_i\xi)]. \tag{5.199}$$

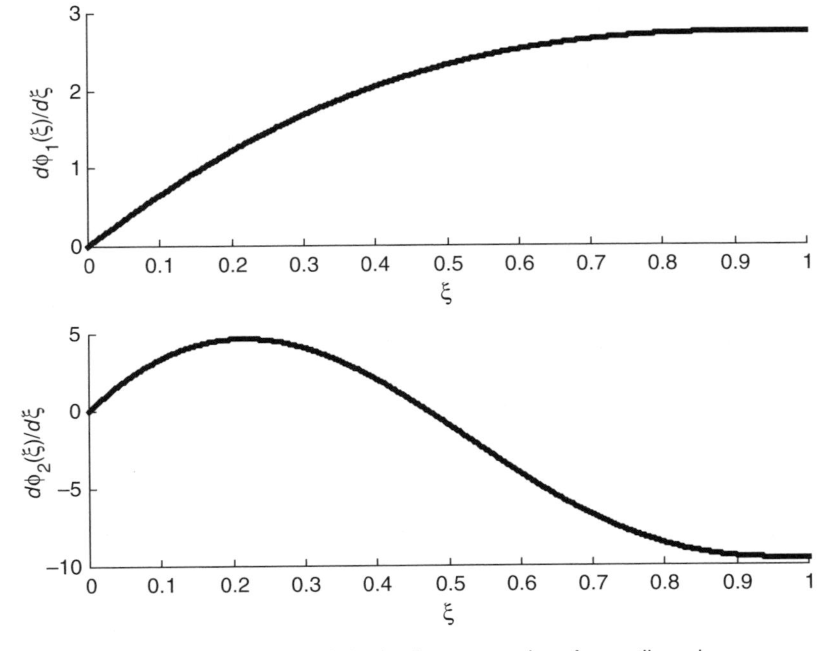

Figure 5.11 Plot of $d\phi_i/d\xi$ for the first two modes of a cantilever beam.

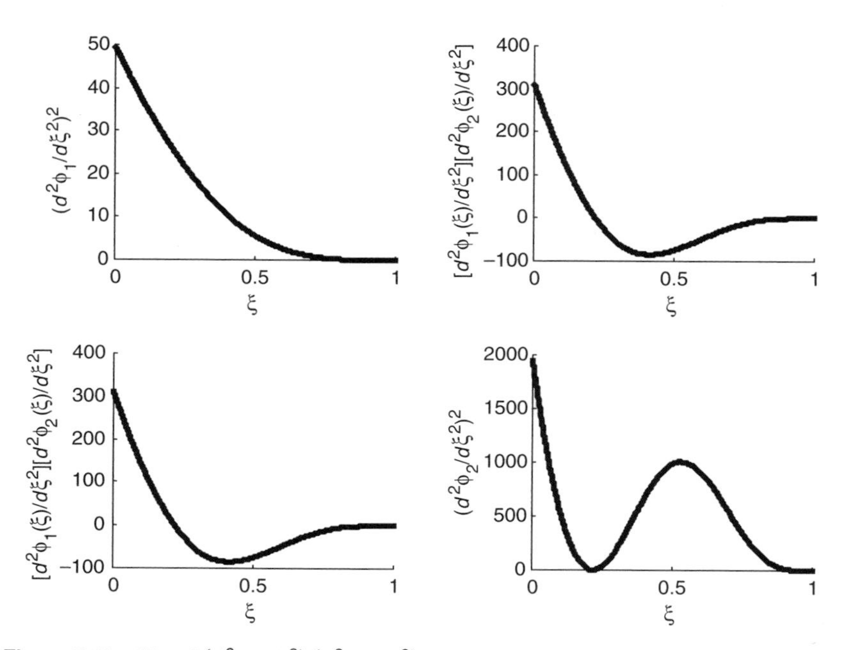

Figure 5.12 Plot of $(d^2\phi_i/d\xi^2)\,(d^2\phi_j/d\xi^2)$ for the first two modes of a cantilever beam.

Figure 5.12 is a plot of $(d^2\phi_i/d\xi^2)\,(d^2\phi_j/d\xi^2)$ for the first two modes of a cantilever beam. The elements of the piezoelectric stiffness matrix are related to the integration under these curves over the length of the piezoeletric element. We see from the figures that the elements of $\mathrm{K}_\mathrm{p}^\mathrm{D}$ will be greatest when the piezoelectric element is placed near the root of the beam. This contrasts with the effect of the piezoelectric element on the mass matrix of the system. Once again, we note that placing the element near the midspace will produce a small entry for the (1,1) element of the piezoelectric stiffness matrix while producing a correspondingly greater value for the (2,2) element. Also, Figure 5.12 illustrates that the off-diagonal terms of the matrix can be negative while the on-diagonal terms are always positive.

Example 5.2 Determine expressions for the stiffness, mass, and coupling matrix for a cantilever beam for two separate locations of the piezoelectric element. The first element is located between 0 and $0.25L$, where L is the total length of the beam, and the second element is located between $0.375L$ and $0.625L$. Assume that the model incorporates the first three modes of the beam.

Solution First consider the piezoelectric element located between 0 and $0.25L$. The nondimensional coordinates associated with this element are $\xi_1 = 0$ and $\xi_2 = 0.25$. Equation (5.194) is the expression for the system mass matrix separated into the mass matrix of the substrate and the piezoelectric element. We see from the expression that that (i, j)th element of the mass matrix is related to the integration of the mode shapes according to the expression

$$\mathrm{M}_{\mathrm{p}_{ij}} = \frac{\rho_p}{\rho_s}\frac{1}{\tau}\int_{\xi_1}^{\xi_2} \phi_i(\xi)\phi_j(\xi)\,d\xi.$$

The expression for the mode shapes of a cantilever beam are shown in equation (5.197). Integration of the mode shapes produces

$$\mathrm{M}_{\mathrm{p}} = \frac{\rho_p}{\rho_s}\frac{1}{\tau}\begin{bmatrix} 0.0020 & 0.0092 & 0.0182 \\ 0.0092 & 0.0429 & 0.0853 \\ 0.0182 & 0.0853 & 0.1711 \end{bmatrix}.$$

As expected, these values are much less than 1, due to the fact that the area under the curve of the functions shown in Figure 5.10 is small over the range 0 to 0.25.

The individual elements of the stiffness matrix are computed from the expression

$$\mathrm{K}^{\mathrm{D}}_{\mathrm{p}\,ij} = \frac{\hat{c}^{\mathrm{D}}}{Y_s} g(\tau)\int_{\xi_1}^{\xi_2} \frac{d^2\phi_i(\xi)}{d\xi^2}\frac{d^2\phi_j(\xi)}{d\xi^2}d\xi.$$

Integration of the second derivative of the mode shapes produces

$$\mathrm{K}^{\mathrm{D}}_{\mathrm{p}} = \frac{\hat{c}^{\mathrm{D}}}{Y_s} g(\tau)\begin{bmatrix} 8.6011 & 29.1651 & 28.6165 \\ 29.1651 & 136.7469 & 271.0647 \\ 28.6165 & 271.0647 & 900.2436 \end{bmatrix}.$$

The coupling matrix is obtained from the first derivative of the mode shapes according to equation (5.190):

$$\Theta = \frac{\tilde{h}}{8}\frac{t_p^2}{L^2}(2\tau + 1)\begin{bmatrix} 5.8242 & -5.8242 \\ 18.2905 & -18.2905 \\ 12.6857 & -12.6857 \end{bmatrix}.$$

Repeating the analysis for the case when $\xi_1 = 0.375$ and $\xi_2 = 0.625$ yields the solution

$$\mathrm{M}_{\mathrm{p}} = \frac{\rho_p}{\rho_s}\frac{1}{\tau}\begin{bmatrix} 0.1244 & 0.2278 & -0.0223 \\ 0.2278 & 0.4558 & 0.0279 \\ -0.0223 & 0.0279 & 0.1327 \end{bmatrix}$$

$$\mathrm{K}^{\mathrm{D}}_{\mathrm{p}} = \frac{\hat{c}^{\mathrm{D}}}{Y_s} g(\tau)\begin{bmatrix} 1.5375 & -17.6507 & -4.8399 \\ -17.6507 & 221.3999 & -38.0437 \\ -4.8399 & -38.0437 & 506.2961 \end{bmatrix}$$

$$\Theta = \frac{\tilde{h}}{8}\frac{t_p^2}{L^2}(2\tau + 1)\begin{bmatrix} 2.4094 & -2.4094 \\ -29.6798 & 29.6798 \\ 2.8920 & -2.8920 \end{bmatrix}.$$

Table 5.6 Geometric and material parameters for the beam example

Geometric	Substrate	Piezoelectric
$L = 20$ cm	$t_s = 1$ mm	$t_p = 0.25$ mm
$w = 10$ mm	$\rho_s = 2700$ kg/m^3	$\rho_p = 7800$ kg/m^3
	$Y_s = 69$ GPa	$\hat{c}^D = 130$ GPa
		$\tilde{h} = -2.7 \times 10^9$ N/C
		$\beta_{33}^S = 1.5 \times 10^8$ m/F

Comparing the two solutions, we see several changes in the matrices when the piezoelectric element is moved. In the mass matrix we note that the magnitude of the elements has increased. This result is consistent with the figures of the mode shape functions (Figure 5.10). In the stiffness and coupling matrices we see that the size of the elements associated with the first mode have decreased while those of the second mode have increased. In addition, we now have negative terms in the off-diagonal of the stiffness matrix. Finally, we note that there is a sign difference in the coupling matrix, which will become important when we discuss the frequency response of piezoelectric material systems.

In Example 5.2 we saw that we can compute the nondimensional elements of the system matrices simply by knowing the mode shapes of the beam. Determining the system matrices then requires knowledge of the geometric and material parameters of the substrate and piezoelectric element. This is analyzed in the following example.

Example 5.3 Compute the short- and open-circuit natural frequencies of a beam with a piezoelectric element located between $x = 0$ and $x = L/4$. The geometric and material parameters for the beam and substrate are shown in Table 5.6.

Solution The definition of the geometric and material parameters allows us to compute the nondimensional parameters of the equations. The thickness ratio is

$$\tau = \frac{t_s}{t_p} = \frac{1 \text{ mm}}{0.25 \text{ mm}} = 4.$$

The term associated with the mass matrices is

$$\frac{\rho_p}{\rho_s \tau} = \frac{7800 \text{ kg/m}^3}{(2700 \text{ kg/m}^3)(4)} = 0.722$$

and the term associated with the stiffness matrix is

$$\frac{\hat{c}^D}{Y_s} g(\tau) = \frac{130 \times 10^9 \text{ Pa}}{69 \times 10^9 \text{ Pa}} \left[\frac{1 + 3(4) + 3(4)^2}{4^3} \right] = 1.796.$$

The expressions for the mass, stiffness, and coupling matrices have leading terms that define the units of the matrix. These computations can also be performed:

$$
\begin{aligned}
\rho_s w L t_s &= (2700\,\text{kg/m}^3)(10 \times 10^{-3}\,\text{m})(20 \times 10^{-2}\,\text{m})(1 \times 10^{-3}\,\text{m}) \\
&= 0.0054\,\text{kg} \\
\frac{Y_s I_s}{L^3} &= \frac{(69 \times 10^9\,\text{N/m}^2)\frac{1}{12}(10 \times 10^{-3}\,\text{m})(1 \times 10^{-3}\,\text{m})^3}{(20 \times 10^{-2}\,\text{m})^3} \\
&= 7.1875\,\text{N/m} \\
\frac{\tilde{h}}{8}\frac{t_p^2}{L^2}(2\tau + 1) &= \frac{(-2.7 \times 10^9\,\text{N/C})(0.25 \times 10^{-3}\,\text{m})^2}{(8)(20 \times 10^{-2}\,\text{m})^2}[(2)(4) + 1] \\
&= -4.7461 \times 10^3\,\text{N/C}
\end{aligned}
$$

The numerical results from Example 5.2 can be combined with these computations to determine the mass and stiffness matrices. The mass matrix is

$$
\begin{aligned}
\text{M} &= (0.0054\,\text{kg}) \left\{ \begin{bmatrix} 1 & 0 & 0 \\ 0 & 1 & 0 \\ 0 & 0 & 1 \end{bmatrix} + (0.722) \begin{bmatrix} 0.0020 & 0.0092 & 0.0182 \\ 0.0092 & 0.0429 & 0.0853 \\ 0.0182 & 0.0853 & 0.1711 \end{bmatrix} \right\} \\
&= \begin{bmatrix} 5.4077 \times 10^{-3} & 3.5835 \times 10^{-5} & 7.0787 \times 10^{-5} \\ 3.5835 \times 10^{-5} & 5.5674 \times 10^{-3} & 3.3262 \times 10^{-4} \\ 7.0787 \times 10^{-5} & 3.3262 \times 10^{-4} & 6.0671 \times 10^{-3} \end{bmatrix} \text{kg.}
\end{aligned}
$$

The stiffness matrix is computed from

$$
\begin{aligned}
\text{K} = (7.1875\,\text{N/m}) &\left\{ \begin{bmatrix} 1.8750^4 & 0 & 0 \\ 0 & 4.6940^4 & 0 \\ 0 & 0 & 7.8550^4 \end{bmatrix} \right. \\
&\left. + (1.7696) \begin{bmatrix} 8.6011 & 29.1651 & 28.6165 \\ 29.1651 & 136.7469 & 271.0647 \\ 28.6165 & 271.0647 & 900.2436 \end{bmatrix} \right\} \\
= &\begin{bmatrix} 0.0200 & 0.0376 & 0.0369 \\ 0.0376 & 0.5254 & 0.3499 \\ 0.0369 & 0.3499 & 3.8982 \end{bmatrix} \times 10^4\,\text{N/m.}
\end{aligned}
$$

The coupling matrix is computed from

$$
\begin{aligned}
\Theta &= -4.7461 \times 10^3\,\text{N/C} \begin{bmatrix} 5.8242 & -5.8242 \\ 18.2905 & -18.2905 \\ 12.6857 & -12.6857 \end{bmatrix} \\
&= \begin{bmatrix} -2.7642 & 2.7642 \\ -8.6809 & 8.6809 \\ -6.0207 & 6.0207 \end{bmatrix} \times 10^4\,\text{N/C.}
\end{aligned}
$$

The capacitive term is

$$\mathbf{C}_{\mathrm{p}}^{\mathrm{S}^{-1}} = \frac{(1.5 \times 10^8 \text{ m/F})(0.25 \times 10^{-3} \text{ m})}{(10 \times 10^{-3} \text{ m})(20 \times 10^{-2} \text{ m})(0.25 - 0)} \begin{bmatrix} 1/2 & 0 \\ 0 & 1/2 \end{bmatrix}$$
$$= \begin{bmatrix} 37.5 & 0 \\ 0 & 37.5 \end{bmatrix} \times 10^6 \text{ F}^{-1}.$$

All the matrices necesary to compute the natural frequencies have been determined. The open-circuit natural frequencies are determined from the eigenvalue problem stated in equation (5.144):

$$\left\{ \begin{bmatrix} 0.0200 & 0.0376 & 0.0369 \\ 0.0376 & 0.5254 & 0.3499 \\ 0.0369 & 0.3499 & 3.8982 \end{bmatrix} \times 10^3 \right\} \mathbf{v}^{\mathrm{D}}$$
$$= \omega^{\mathrm{D}^2} \left\{ \begin{bmatrix} 5.4077 & 0.0358 & 0.0708 \\ 0.0358 & 5.5674 & 0.3326 \\ 0.0708 & 0.3326 & 6.0671 \end{bmatrix} \times 10^{-3} \right\} \mathbf{v}^{\mathrm{D}},$$

which can be solved to yield the natural frequencies

$$\omega_1^{\mathrm{D}} = 178.2 \text{ rad/s}$$
$$\omega_2^{\mathrm{D}} = 944.9 \text{ rad/s}$$
$$\omega_3^{\mathrm{D}} = 2536.8 \text{ rad/s}.$$

The eigenvalue problem stated in equation (5.147) can be used to compute the short-circuit natural frequencies:

$$\left\{ \begin{bmatrix} 0.0159 & 0.0248 & 0.0281 \\ 0.0248 & 0.4852 & 0.3220 \\ 0.0281 & 0.3220 & 3.8789 \end{bmatrix} \times 10^3 \right\} \mathbf{v}^{\mathrm{E}}$$
$$= \omega^{\mathrm{E}^2} \left\{ \begin{bmatrix} 5.4077 & 0.0358 & 0.0708 \\ 0.0358 & 5.5674 & 0.3326 \\ 0.0708 & 0.3326 & 6.0671 \end{bmatrix} \times 10^{-3} \right\} \mathbf{v}^{\mathrm{E}},$$

which can be solved to yield the natural frequencies

$$\omega_1^{\mathrm{E}} = 164.2 \text{ rad/s}$$
$$\omega_2^{\mathrm{E}} = 909.5 \text{ rad/s}$$
$$\omega_3^{\mathrm{E}} = 2529.8 \text{ rad/s}.$$

The results for both computations are summarized in Table 5.7 along with the percentage change in the natural frequencies from open-circuit to short-circuit boundary conditions. The results are consistent with the expectations since the natural frequencies

Table 5.7 Short- and open-circuit natural frequencies computed in Example 5.3

Open-Circuit f^{D} (Hz)	Short-Circuit f^{E} (Hz)	% Change
28.3662	26.1280	−8.5665
150.3811	144.7459	−3.8932
403.7383	402.6287	−0.2756

decrease when changing the boundary condition from an open circuit (zero charge) to a short circuit (zero field). Examining the results, we see that the percentage change in the natural frequencies are not the same for each mode. In this example the first natural frequency changed by almost 10%, while the third changed by less than 1%. This variation is due directly to placement of the piezoelectric element along the beam. Changing the location of the element would change the relative effects on each mode.

Examples 5.2 and 5.3 demonstrated the development of a dynamic model of a piezoelectric beam with material that did not span the length of the beam. The result of the model is a set of matrices that define the mass, stiffness, coupling, and capacitive properties of the system. Defining the equations of motion in this manner allows the analysis of the frequency response of the system that incorporates the piezoelectric material using the methods described earlier in the chapter. As described by equations (5.140) (5.142), the frequency response of the displacement and charge can be obtained from a frequency-by-frequency analysis of the equations of motion.

Equations (5.140) and (5.142) are general expressions that relate the input force and voltage to the displacement at any location and the charge response of the piezoelectric material. These equations can be simplified by assuming that the displacement measurement is at the location of the forcing input and that the charge measurement has the same form as the application of the voltage to the piezoelectric material. To illustrate, consider the case in which the displacement is measured at the location of the force application, $\mathbf{x}_f$; therefore,

$$\boldsymbol{U}(\mathbf{x}_f, \omega) = \mathrm{B}_\mathrm{f}' \boldsymbol{r}(t). \tag{5.200}$$

and thus $C_o = \mathrm{B}_\mathrm{f}'$ in equation (5.140). If the charge is measured in the same manner as the voltage is applied to the individual piezoelectric layers, the measured charge, $\boldsymbol{Q}_m(\omega)$, is equal to

$$\boldsymbol{Q}_m(\omega) = \mathrm{B}_\mathrm{v}' \boldsymbol{Q}(\omega). \tag{5.201}$$

Substituting equations into equations (5.140) and (5.142) produces

$$\boldsymbol{U}(\mathbf{x}_f, \omega) = \mathrm{B}_\mathrm{f}' \Delta^{-1}(\omega) \mathrm{B}_\mathrm{f} \boldsymbol{F}(\omega) + \mathrm{B}_\mathrm{f}' \Delta^{-1}(\omega) \Theta \mathrm{C}_\mathrm{p}^\mathrm{S} \mathrm{B}_\mathrm{v} \boldsymbol{V}(\omega) \tag{5.202}$$

$$\boldsymbol{Q}_m(\omega) = \mathrm{B}_\mathrm{v}' \mathrm{C}_\mathrm{p}^\mathrm{S} \Theta' \Delta^{-1}(\omega) \mathrm{B}_\mathrm{f} \boldsymbol{F}(\omega) + \mathrm{B}_\mathrm{v}' \left[\mathrm{C}_\mathrm{p}^\mathrm{S} + \mathrm{C}_\mathrm{p}^\mathrm{S} \Theta' \Delta^{-1}(\omega) \Theta \mathrm{C}_\mathrm{p}^\mathrm{S}\right] \mathrm{B}_\mathrm{v} \boldsymbol{V}(\omega). \tag{5.203}$$

Equations (5.202) and (5.203) enable plotting of the sensing and actuation frequency response for structures with integrated piezoelectric materials. Before using an example to illustrate the frequency response functions, there are several aspects of the equations that provide insight into the physics of the piezoelectric material systems. First consider the case in which the voltage applied to the piezoelectric materials is zero. Using the notation introduced earlier in the chapter, the relationship between the deflection and the force is written as

$$T_{uf}(\omega) = \mathrm{B}_{\mathrm{f}}' \Delta^{-1}(\omega) \mathrm{B}_{\mathrm{f}}, \tag{5.204}$$

where the matrix $\Delta(\omega)$ is defined in equation (5.138). Equation (5.204) is the frequency-dependent compliance at the point of force application. This matrix is a symmetric matrix at every frequency ω due to the symmetry of the constituent matrices. Since the inverse of a symmetric matrix is also symmetric, the matrix $T_{uf}(\omega)$ is also symmetric at each frequency.

The symmetry of the compliance matrix introduces bounds on the frequency response function between the applied force and the deflection at the same location. In the general case in which there are multiple forcing locations and multiple measurement locations, the frequency response function between each input–output pair becomes large at the resonances of the structure due to the singularity of the dynamic stiffness matrix $\Delta(\omega)$. Another important feature of the dynamic response is the existence of *transmission zeros*, or simply zeros, between the input force and output deflection. Zeros represent a frequency at which a nonzero input will produce zero output. In the case of the matrix $T_{uf}(\omega)$, this represents a frequency at which the deflection at the measurement points will be identically equal to zero for a nonzero input force.

The symmetry of the matrix $T_{uf}(\omega)$ introduces constraints on the location of the transmission zeros. Denoting the frequency of the ith transmission zero as z_i, the transmission zeros are bounded by the expressions

$$\begin{aligned} \omega_1 &\leq z_1 \leq \omega_{m+1} \\ \omega_2 &\leq z_2 \leq \omega_{m+2} \\ &\vdots \\ \omega_{n-m} &\leq z_{n-m} \leq \omega_n, \end{aligned} \tag{5.205}$$

where m is the number of forcing inputs and therefore the number of measurement points. In the simple case in which there is only a single force and measurement location, $m = 1$, and equation (5.205) shows that the ith transmission zero z_i will be bounded between ω_i and ω_{i+1}. This property is called *pole–zero interlacing*.

The relationship between charge and voltage exhibits similar characteristics as the compliance matrix. The matrix $T_{qv}(\omega)$ can be separated into two components,

$$T_{qv}(\omega) = \mathrm{B}_{\mathrm{v}}' \mathrm{C}_{\mathrm{p}}^{\mathrm{S}} \mathrm{B}_{\mathrm{v}} + \mathrm{B}_{\mathrm{v}}' \mathrm{C}_{\mathrm{p}}^{\mathrm{S}} \Theta' \Delta^{-1}(\omega) \Theta \mathrm{C}_{\mathrm{p}}^{\mathrm{S}} \mathrm{B}_{\mathrm{v}}. \tag{5.206}$$

The first term in the expression is the strain-free capacitance. It is clear from the expression that this term is independent of frequency, which is sensible since a piezoelectric material has an inherent capacitance due to the dielectric properties of the material. The second term in equation (5.205) is a function of frequency and represents the variation in capacitance due to electromechanical coupling. As in the case of the compliance matrix, this matrix is symmetric at each frequency and therefore will exhibit pole–zero interlacing due to the constraints on the transmission zeros. The transmission zeros of $T_{qv}(\omega)$ will be different, though, due to the addition of strain-free capacitance associated with the dielectric properties of the piezoelectric. In the simplest case in which there is only a single voltage input and a single charge measurement, the transmission zeros will be at the frequencies in which

$$\mathrm{B}_{\mathrm{v}}'\mathrm{C}_{\mathrm{p}}^{\mathrm{S}}\mathrm{B}_{\mathrm{v}} = -\mathrm{B}_{\mathrm{v}}'\mathrm{C}_{\mathrm{p}}^{\mathrm{S}}\Theta'\Delta^{-1}(\omega)\Theta\mathrm{C}_{\mathrm{p}}^{\mathrm{S}}\mathrm{B}_{\mathrm{v}}. \tag{5.207}$$

which is associated with the frequency at which the strain-free capacitance is equal in magnitude but opposite in phase to the capacitance generated by the electromechanical coupling.

The final attribute of equations (5.202) and (5.203) worth noting is related to the symmetry associated with sensing and actuation. The remaining frequency response functions $T_{uv}(\omega)$ and $T_{qf}(\omega)$ represent the free displacement and charge-to-force relationship for the piezoelectric material system, respectively. These two frequency response functions quantify the free displacement of the system in actuation and the amount of charge produced by an applied force in sensing. From equations (5.202) and (5.203) we note that these matrices are the transpose of one another in the case in which the displacement measurement is at the same location as the force application and the charge is measured in the same manner as the voltage is applied. Under these conditions, $T_{qf}(\omega) = T_{uv}'(\omega)$, and the sensing and actuation of the system are reciprocal relationships.

The concept of frequency response will be illustrated with the equations derived in Example 5.3. The dynamic stiffness matrix is computed as a function of frequency,

$$\Delta(\omega) = \begin{bmatrix} 159.1 - 0.005408\,\omega^2 & 248.5 - 0.00003583\,\omega^2 & 280.6 - 0.00007079\,\omega^2 \\ 248.5 - 0.00003583\,\omega^2 & 4,852.0 - 0.005567\,\omega^2 & 3,220.0 - 0.0003326\,\omega^2 \\ 280.6 - 0.00007079\,\omega^2 & 3,220.0 - 0.0003326\,\omega^2 & 38,790.0 - 0.006067\,\omega^2 \end{bmatrix}. \tag{5.208}$$

The input matrix associated with a force acting at the tip is equal to the value of the mode shapes at the tip of the beam,

$$B_f = \begin{bmatrix} 1.9998 \\ -2.0016 \\ 2.0320 \end{bmatrix}. \tag{5.209}$$

Equation (5.204) is used to compute the frequency response function between the deflection and the force. Figure 5.13a is a plot of the magnitude of the frequency

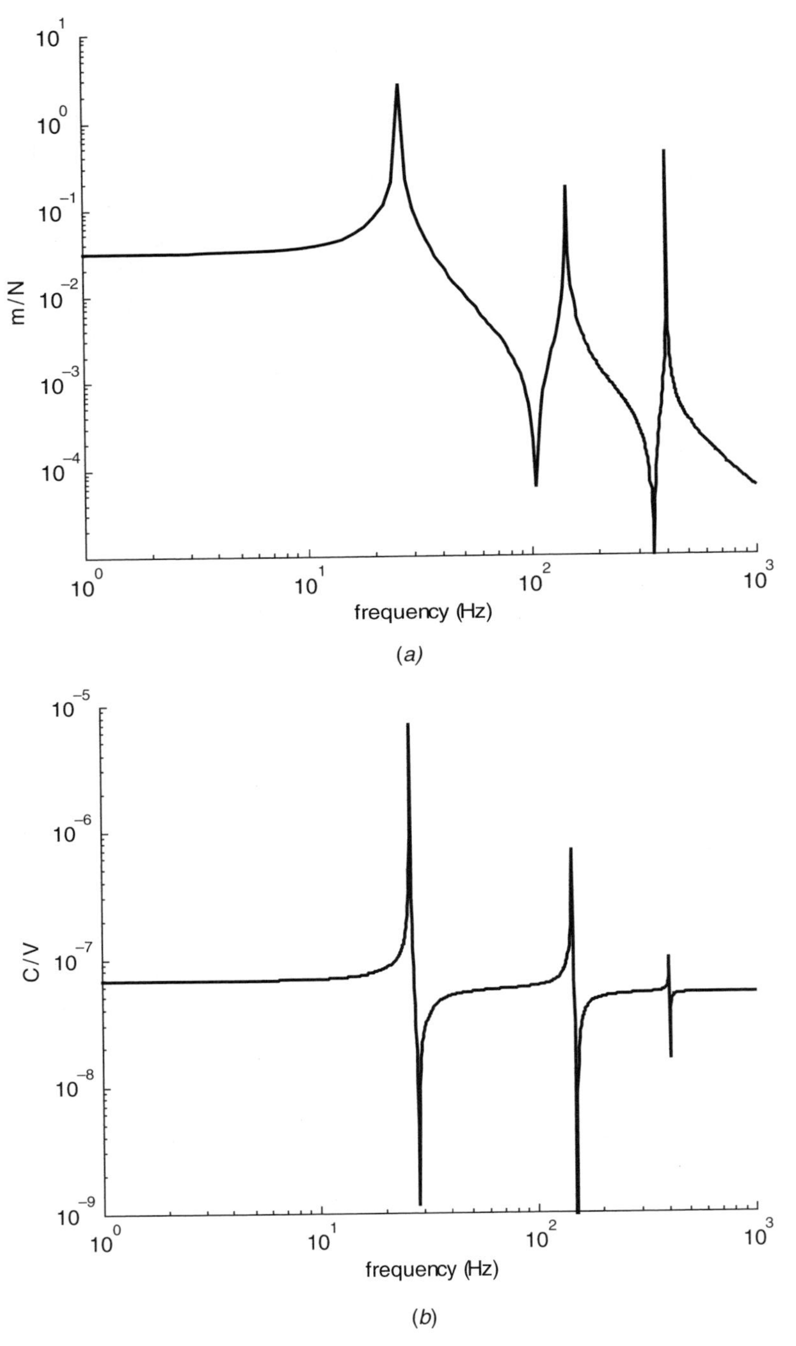

Figure 5.13 Frequency response functions (*a*) $T_{uf}(\omega)$ and (*b*) $T_{qv}(\omega)$ for the values obtained in Example 5.3.

response function $T_{uf}(\omega)$. The sharp peaks in the frequency response function are the resonant frequencies. At these frequencies the response becomes unbounded for a bounded input, due to the singularity in the dynamic stiffness matrix. The height of the response at these three frequencies is not relevant, due to the fact that it is simply an artifact of the frequency spacing utilized in the simulation. The sharp notches in the frequency response function are the transmission zeros of the frequency response. As expected, the resonances and zeros are interlaced due to the form of the matrices that make up the frequency response function. The frequency region that is much lower than the first resonance is often characterized as a "quasistatic" response, due to the fact that the magnitude of the frequency response is flat over this range. The asymptotic value of the compliance response at low frequencies is the inverse of the structural stiffness. The inertial terms begin to dominate at frequencies above the first resonance, and the displacement frequency response exhibits a roll-off which is characterized by a decreasing average magnitude.

The relationship between charge output and applied voltage is computed using equation (5.206). The input vector B_v is defined as

$$B_v = \begin{bmatrix} 1 \\ -1 \end{bmatrix}. \tag{5.210}$$

The plot of the charge-to-voltage frequency response is shown in Figure 5.13*b*. The resonant peaks are the same as those shown for the deflection-to-force frequency response, but the zeros are at different frequencies and the frequency response does not exhibit the roll-off at higher frequencies that is characteristic of $T_{uf}(\omega)$. These two attributes of the charge-to-voltage response are understood by once again examining the two components of equation (5.206).

This decomposition of $T_{qv}(\omega)$ is illustrated in Figure 5.14*a*. The first component is a constant positive value that represents the strain-free capacitance of the piezoelectric material. The electromechanical coupling adds an additional frequency-dependent term to the capacitance function. At frequencies well below the first resonance, the strain-free capacitance and the capacitance due to electromechanical coupling are in phase; therefore, they add, and the total capacitance is larger than the strain-free capacitance. At frequencies between the first resonance and the first transmission zero of the electromechanical coupling capacitance (this is represented by the dotted line in the figure), the strain-free capacitance and the capacitance due to electromechanical coupling are out of phase; therefore, addition of the two terms creates a zero in the total capacitance at the frequency at which the magnitudes are equal. At frequencies above the first transmission zero and below the second resonance, the two capacitance terms are once again in phase.

This analysis could be carried out at frequencies near each resonance of the system. As shown in Figure 5.13*b*, the spacing between the resonance and transmission zero changes at each resonance frequency. The spacing between the peak and zero is largest near the first resonance and the spacing decreases for the second and third natural frequencies of the system. In the limit as the capacitance due to electromechanical coupling becomes small compared to the strain-free capacitance, the frequency

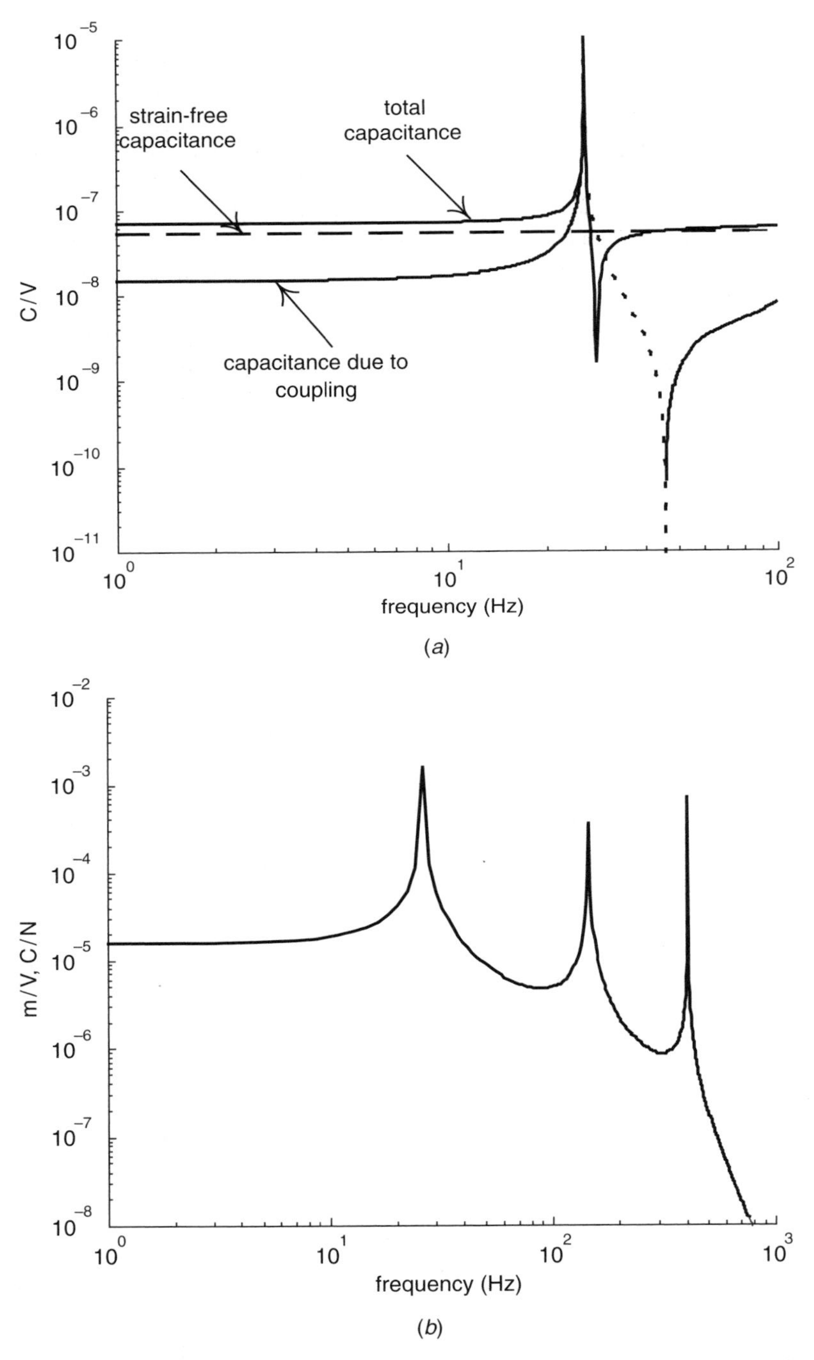

Figure 5.14 (*a*) Decomposition of the charge-to-voltage frequency response, illustrating the effects of electromechanical coupling; (*b*) displacement to voltage response.

response function $T_{qv}(\omega)$ will exhibit very small separation between the resonance and transmission zero. This situation indicates that there is very small electromechanical coupling at that frequency, and the charge-to-voltage frequency response is dominated by the strain-free capacitance at those frequencies. This issue will become important when we study passive and active control methods using piezoelectric materials.

The final frequency response of interest is the relationship between deflection and voltage. As discussed previously, this relationship is identical to the relationship between charge and force due to the symmetry of the frequency response functions. The frequency response for $T_{uv}(\omega)$ is shown in Figure 5.14*b*. The resonance peaks are at the same frequency as the peaks in the other two frequency response functions, but the frequency response does not exhibit any antiresonances, or notches, due to transmission zeros. This example highlights that fact that the resonance frequencies are not a function of the input–output relationship being examined, but the transmission zeros vary as a function of the input–output pair.

5.6.2 Generalized Coupling Coefficients

The discussion in Section 5.6.1 highlights basic relationships in the dynamic response of piezoelectric material systems. One of the important factors that arises is the relationship between the properties and placement of the piezoelectric material and the modal response of the system. The discussion and examples in Section 5.6.1 indicate that the piezoelectric material couples differently to different modes of vibration in the structure. The coupling is affected by the location, shape, and geometry of the piezoelectric material as it relates to the host structure.

In Chapter 4 several metrics were introduced to quantify the coupling in the piezoelectric material. An important parameter that was introduced was the piezoelectric coupling coefficient, k_{ij}, where i and j define the directions associated with the mechanical and electrical fields. As discussed previously, k_{ij} quantified the cyclic electromechanical energy conversion in the material. Originally, this parameter was defined by the material properties of the piezoelectric: strain coefficient, material compliance, and dielectric permittivity. A higher value of k_{ij} indicated larger energy conversion, and hence, better coupling in the material.

Analyzing the frequency response of piezoelectric material systems, we see that a piezoelectric transducer will couple differently to different modes of a structure. This fact has led to the development of *generalized coupling coefficients* which define the relative coupling of a piezoelectric transducer to the vibration modes of a structure. In Chapter 4 a relationship was derived that related the coupling coefficient of a single-mode system to the change in natural frequency between the short- and open-circuit conditions,

$$k_{ij}^2 = \frac{\omega^{\mathrm{D}^2} - \omega^{\mathrm{E}^2}}{\omega^{\mathrm{E}^2}}. \tag{5.211}$$

This relationship illustrates that if there is no change in the natural frequency between short- and open-circuit conditions, the coupling coefficient is zero. The larger the

Table 5.8 Generalized piezoelectric coupling coefficients for Example 5.2

Open-Circuit f^D (Hz)	Short-Circuit f^E (Hz)	k_{13}
28.3662	26.1280	0.42
150.3811	144.7459	0.28
403.7383	402.6287	0.07

variation in the natural frequency between short- and open-circuit conditions, the larger the coupling coefficient.

This concept can be generalized quite easily to a structure with multiple vibration modes. The examples in Section 5.6.1 indicate that the change in natural frequency as a function of the electrical boundary condition varies with each vibration mode. A simple extension of the concept of piezoelectric coupling is to apply equation (5.211) to each vibration mode separately and compute a generalized coupling coefficient that represents the amount of coupling between the piezoelectric material and the individual mode. As is the case with the original definition, smaller changes in the natural frequency from short- to open-circuit conditions will produce a smaller generalized coupling coefficient.

The concept is illustrated using the values obtained in the examples from Section 5.6.1. Table 8 lists the variations in natural frequencies for Example 5.3. Equation (5.211) is applied to the computation of the generalized coupling coefficients, and these values are listed in Table 5.8. The results demonstrate that there can be a wide variation in coupling for the separate vibration modes. For the values used in Example 5.3, the generalized coupling coefficients vary from 0.42 for the first mode to only 0.07 for the third mode. The numerical values correlate with the frequency response plot shown in Figure 5.13*b*, which illustrates that the pole–zero spacing decreases from the first to the third modes.

Quantifying the coupling between the piezoelectric material is useful for determining the location and properties of the piezoelectric transducer. For example, equation (5.211) is useful for studying the design of a system in which particular modes are targeted for study or, for example, for vibration suppression. This result will become important in future chapters when we study the use of piezoelectric materials for vibration control and structural damping, when it will be shown that coupling to the individual modes plays an important role in determining the effectiveness of control or damping using piezoelectric materials.

5.6.3 Structural Damping

The equations for piezoelectric material systems derived thus far do not include any energy dissipation mechanisms and have resulted in the development of a set of *undamped* equations of motion. In actuality, though, dynamic systems generally have some measure of energy dissipation that is introduced by physical mechanisms such as sliding of components, inherent damping in materials, or damping introduced by mechanical or electronic means to suppress vibration. Modeling of damping is an active field of research and a priori estimates of damping are often difficult to incorporate in structural models. In certain instances, though, there are straightforward models of damping that can be introduced into the equations of motion for the purpose of designing structural systems that incorporate piezoelectric materials.

A common method of introducing the effects of energy dissipation is to incorporate a *structural damping matrix* into the equations of motion. The structural damping matrix is assumed to be a linear function of the generalized velocities, $\dot{\boldsymbol{r}}$, and is denoted $\mathrm{D_s}$. The addition of a damping matrix into equations (5.133) and (5.134) yields the expressions

$$(\mathrm{M_s} + \mathrm{M_p})\ddot{\boldsymbol{r}} + \mathrm{D_s}\dot{\boldsymbol{r}} + \left(\mathrm{K_s} + \mathrm{K_p^D}\right)\boldsymbol{r} - \Theta \boldsymbol{q} = \mathrm{B_f}\boldsymbol{f} \qquad (5.212)$$

$$-\Theta' \boldsymbol{r} + \mathrm{C_p}^{-1}\boldsymbol{q} = \mathrm{B_v}\boldsymbol{v}. \qquad (5.213)$$

The structural damping matrix represents the loss inherent in the host structure. The damping matrix can be derived from first principles analysis, or it is often added after the undamped equations of motion have been derived. In an engineering design problem, the damping is often added to the model afterward to account for energy dissipation that has been measured or estimated from experimental data. One method of adding structural damping is to assume that the damping matrix is *proportionally damped* and of the form

$$\mathrm{D_s} = \alpha \mathrm{M_s} + \beta \mathrm{K_s}, \qquad (5.214)$$

where α and β are scalar constants chosen to match the damping in the model to some prediction or measurement of the system damping. This model of damping has the advantage that the eigenvectors of the structural mass and stiffness matrices will diagonalize the damping matrix, thus simplifying the solution of the vibration problem.

Another damping model that is commonly added to structures is *modal damping*. Modal damping is especially convenient when the structural mass and stiffness matrices are diagonalized through proper choice of the structural mode shapes. A modal damping matrix is a diagonal matrix of the form

$$\mathrm{D_s} = \begin{bmatrix} c_i & 0 & 0 \\ 0 & \ddots & 0 \\ 0 & 0 & c_N \end{bmatrix}, \qquad (5.215)$$

where c_i is the modal damping coefficient. The fact that the mass and stiffness matrices are diagonal results in a set of uncoupled equations of motion where m_i represents the *modal mass* and k_i represents the *modal stiffness*. With these definitions the modal damping coefficient can be written as a function of the modal mass, stiffness, and damping ratio ζ_i using an expression that is identical to that of a single-degree-of-freedom oscillator:

$$c_i = 2\zeta_i \sqrt{m_i k_i}. \tag{5.216}$$

The modal damping ratio ζ_i is the variable that is typically chosen to match that of experimental data or the results of separate damping analysis. Equation (5.216) is solved for each mode to form the damping matrix $\mathrm{D_s}$.

A structural damping matrix of the form $\mathrm{D_s}\dot{\boldsymbol{r}}$ is advantageous from the standpoint of the analysis techniques presented thus far in the book. The frequency response functions $T_{uf}(\omega)$, $T_{uv}(\omega)$, $T_{qf}(\omega)$, and $T_{qv}(\omega)$ have all been derived as a function of the dynamic stiffness matrix $\Delta(\omega)$. Rederiving these expressions for the damped equations of motion, equation (5.212), it is shown that expressions for the frequency response functions are identical to those derived previously except for the fact that the dynamic stiffness matrix has the form

$$\Delta(j\omega) = \mathrm{K_s} + \mathrm{K_p^D} - \Theta \mathrm{C_p^S} \Theta' - (\mathrm{M_s} + \mathrm{M_p})\omega^2 + j\mathrm{D_s}\omega, \tag{5.217}$$

where j is the imaginary number. The primary difference in the analysis is that the dynamic stiffness matrix is a complex-valued symmetric matrix at each frequency. Thus, the inverse of the dynamic stiffness matrix is also complex, which can complicate the analysis for larger-order dynamic systems.

The addition of structural damping will affect both the time- and frequency-domain responses of the piezoelectric material system. In the time domain the response will exhibit fewer oscillations when excited with impulse inputs and will exhibit a smaller steady-state response when excited near resonance. The steady-state response can be analyzed by plotting the frequency response functions for various values of damping. Figure 5.15*a* is a plot of the deflection-to-force frequency response for the undamped system and damped system with increasing values of the modal damping ratio, ζ_i. Increasing the modal damping ratio affects the frequency response, and hence the steady-state mangitude to harmonic excitations, most strongly near the system resonances and transmission zeros. The peak resonance near resonance decreases as the modal damping ratio increases and the depth of the notch associated with a transmission zero increases with increasing ζ_i. Structural damping has only negligible effects at other frequencies, as illustrated by the fact that the three curves in Figure 5.15*a* overlay at frequencies away from the system resonances and transmission zeros.

Electromechanical coupling in the piezoelectric materials will result in a change in the charge-to-voltage response when structural damping is introduced into the system. Figure 5.15*b* is a plot of $T_{qv}(\omega)$ for three values of modal damping. The plot illustrates that modal damping affects the frequency response most strongly near the resonance peaks and the transmission zeros in the same manner that $T_{uf}(\omega)$ is affected. The reduction in charge response due to mechanical response produces a reduction in the spacing between the resonance and transmission zero of $T_{qv}(\omega)$. For the example studied, the variation in capacitance from the strain-free capacitance

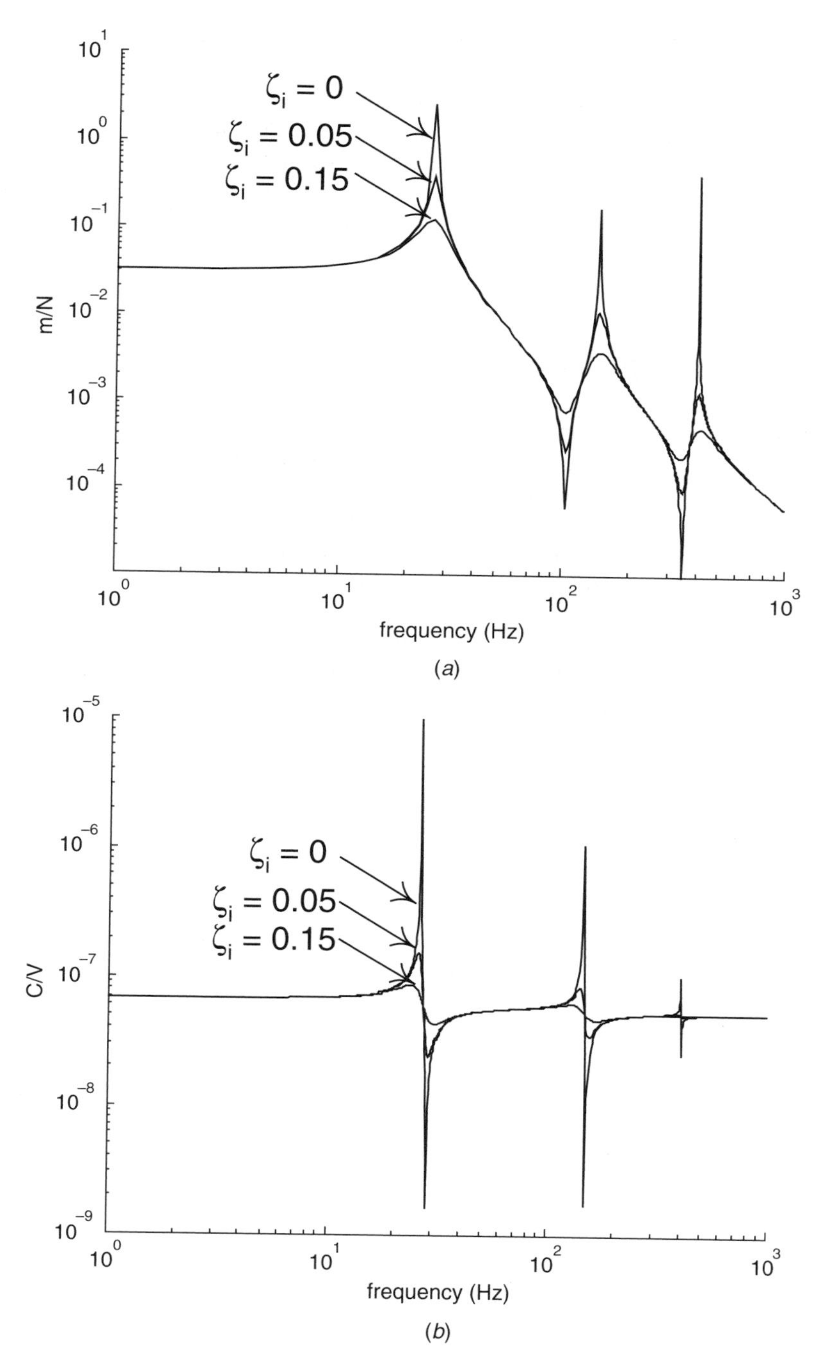

Figure 5.15 Frequency response functions for (*a*) deflection-to-force and (*b*) charge-to-voltage for three values of modal damping.

practically disappears when the modal damping ratio is increased to 0.05 and 0.15. Similar reductions in the charge response occur at the other resonances.

This example illustrates that the addition of structural damping tends to mask, or decrease, the electromechanical coupling in a piezoelectric material system. The reduction in coupling produces an electrical response that is dominated by the strain-free capacitance of the piezoelectric material. Once again, this physical process will have ramifications in our use of piezoelectric materials as passive and active control elements for structures.

5.7 PIEZOELECTRIC PLATES

The Ritz method that was applied to beams in Section 5.6 is extended to plates that incoporate piezoelectric properties. As discussed in Section 5.6, although all problems are strictly three-dimensional, the solution of certain problems can be reduced to a lower dimension and still yield accurate results. This is the case for thin, slender members such as beams, where only a single dimension is required to express the displacement functions and hence the solution to the static and dynamic analysis.

Plates are solid bodies bounded by two parallel flat surfaces whose lateral dimensions are large compared to the distance between the flat surfaces. Typical plate geometries are rectangular plates, as shown in Figure 5.16, or plates with circular geometries. In this discussion we restrict ourselves to a discussion of rectangular plates. Thin plates are those in which the ratio of the thickness to the length of the smaller span length is less than approximately $\frac{1}{20}$. The value of $\frac{1}{20}$ is a commonly accepted ratio, which, when satisfied, allows the full three-dimensional problem to be reduced to a problem in two dimensions.

The fundamental assumptions of classical plate theory are:

1. The displacements of the midsurface are small compared with the plate thickness, and the slope of the deflected surface is small compared to unity.

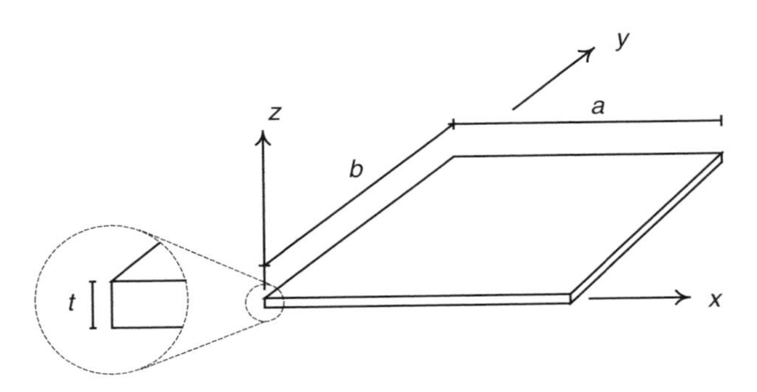

Figure 5.16 Rectangular thin plate.

2. The midplane of the plate remains unstrained and therefore can be considered the neutral plane after bending.
3. Plane sections normal to the surface before bending are normal to the midsurface after bending. This implies that the shear strains S_{xy} and S_{yz} are zero. The deflection of the plate is principally associated with bending, which also implies that the normal strain in the z direction, S_{zz}, can be neglected.
4. The normal stress in the z direction, T_{zz}, is small compared to the other stresses and can be neglected.

Assumptions 3 and 4 imply that the thin plate is analyzed as a plane stress problem. Under this assumption the strain in the thin plate is

$$\begin{aligned}
S_{xx} = S_1 &= \frac{\partial u_1(x,y)}{\partial x} - z\frac{\partial^2 u_3(x,y)}{\partial x^2} \\
S_{yy} = S_2 &= \frac{\partial u_2(x,y)}{\partial x} - z\frac{\partial^2 u_3(x,y)}{\partial y^2} \\
S_{xz} = S_6 &= \frac{\partial u_1(x,y)}{\partial y} + \frac{\partial u_2(x,y)}{\partial x} - 2z\frac{\partial^2 u_3(x,y)}{\partial x y}.
\end{aligned} \tag{5.218}$$

To utilize the Ritz method in conjunction with the variational approach, the strain is written as the matrix relationship

$$\begin{Bmatrix} S_1(\mathbf{x}) \\ S_2(\mathbf{x}) \\ S_3(\mathbf{x}) \\ S_4(\mathbf{x}) \\ S_5(\mathbf{x}) \\ S_6(\mathbf{x}) \end{Bmatrix} = \begin{bmatrix} \frac{\partial}{\partial x} & 0 & -z\frac{\partial^2}{\partial x^2} \\ 0 & \frac{\partial}{\partial x} & -z\frac{\partial^2}{\partial y^2} \\ 0 & 0 & 0 \\ 0 & 0 & 0 \\ 0 & 0 & 0 \\ \frac{\partial}{\partial y} & \frac{\partial}{\partial x} & -2z\frac{\partial^2}{\partial x y} \end{bmatrix} \begin{Bmatrix} u_1(\mathbf{x}) \\ u_2(\mathbf{x}) \\ u_3(\mathbf{x}) \end{Bmatrix}. \tag{5.219}$$

Equation (5.219) represents the strain–displacement relationship for a thin plate undergoing deformation in all three directions.

5.7.1 Static Analysis of Piezoelectric Plates

The general strain–displacement relationships stated in equation (5.219) are simplified under the assumption that the deflection is only in the transverse direction. With this assumption, equation (5.219) is reduced to

$$\left\{\begin{array}{c} S_1(\mathbf{x}) \\ S_2(\mathbf{x}) \\ S_3(\mathbf{x}) \\ S_4(\mathbf{x}) \\ S_5(\mathbf{x}) \\ S_6(\mathbf{x}) \end{array}\right\} = \left\{\begin{array}{c} -z\dfrac{\partial^2 u_3}{\partial x^2} \\ -z\dfrac{\partial^2 u_3}{\partial y^2} \\ 0 \\ 0 \\ 0 \\ -2z\dfrac{\partial^2 u_3}{\partial xy} \end{array}\right\}. \tag{5.220}$$

A polynomial expansion of the displacement function is

$$u_3(x,y) = \sum_{i=0}^{m}\sum_{j=0}^{n} r_{mn}x^m y^n. \tag{5.221}$$

Consider the case of a plate that is pinned at $x = 0$ and $x = a$. The kinematic boundary conditions are

$$\begin{aligned} u_3(0,\, y) &= 0 \\ u_3(a,\, y) &= 0, \end{aligned} \tag{5.222}$$

which introduces the constraints

$$\begin{aligned} \sum_{j=0}^{n} r_{0n} y^n &= 0 \\ \sum_{i=0}^{m}\sum_{j=0}^{n} r_{mn}a^m y^n &= 0. \end{aligned} \tag{5.223}$$

A two-term expansion yields the displacement function

$$u_3(x,y) = r_{20}\,(x-a)\,x + r_{21}\,(x-a)\,xy + r_{22}\,(x-a)\,xy^2. \tag{5.224}$$

The associated shape function matrix is

$$\mathrm{N_r}(\mathbf{x}) = \left[\,(x-a)\,x \;\; (x-a)\,xy \;\; (x-a)\,xy^2\,\right]. \tag{5.225}$$

The differential operator for the strain–displacement relationships is applied to the shape function matrix to produce the shape functions for the strain in the plate,

$$\mathrm{B_r}(\mathbf{x}) = \begin{bmatrix} 2z & 2zy & 2zy^2 \\ 0 & 0 & 2z(x-a)x \\ 0 & 0 & 0 \\ 0 & 0 & 0 \\ 0 & 0 & 0 \\ 0 & z(2x-a) & z(2xy + 2(x-a)y) \end{bmatrix}. \tag{5.226}$$

Similarly, the shape functions for the electric displacement are defined as

$$\mathrm{B_q}(\mathbf{x}) = \begin{bmatrix} 0 & 0 \\ 0 & 0 \\ \dfrac{1}{ab} & \dfrac{1}{ab} \end{bmatrix}. \tag{5.227}$$

The stiffness matrix is computed using equation (5.57):

$$\mathrm{K^D} = \begin{bmatrix} \dfrac{c_{11}^{\mathrm{D}}\, abt^3}{3} & \dfrac{c_{11}^{\mathrm{D}}\, ab^2t^3}{6} & \dfrac{c_{11}^{\mathrm{D}} a^3bt^3}{18} K_{13} \\ \dfrac{c_{11}^{\mathrm{D}}\, ab^2t^3}{6} & \dfrac{c_{11}^{\mathrm{D}} a^3bt^3}{36} K_{22} & \dfrac{c_{11}^{\mathrm{D}} b^2a^3t^3}{36} K_{23} \\ \dfrac{c_{11}^{\mathrm{D}} a^3bt^3}{18} K_{13} & \dfrac{c_{11}^{\mathrm{D}} b^2a^3t^3}{36} K_{23} & \dfrac{c_{11}^{\mathrm{D}} ba^5t^3}{270} K_{33} \end{bmatrix}, \tag{5.228}$$

where

$$K_{22} = 4\frac{b^2}{a^2} + \frac{c_{66}^{\mathrm{D}}}{c_{11}^{\mathrm{D}}}$$

$$K_{13} = 2\frac{b^2}{a^2} - \frac{c_{12}^{\mathrm{D}}}{c_{11}^{\mathrm{D}}}$$

$$K_{23} = 3\,\frac{b^2}{a^2} - \frac{c_{12}^{\mathrm{D}}}{c_{11}^{\mathrm{D}}} + \frac{c_{66}^{\mathrm{D}}}{c_{11}^{\mathrm{D}}}$$

$$K_{33} = 18\,\frac{b^4}{a^4} - 10\,\frac{b^2}{a^2}\frac{c_{12}^{\mathrm{D}}}{c_{11}^{\mathrm{D}}} + 10\,\frac{b^2}{a^2}\frac{c_{66}^{\mathrm{D}}}{c_{11}^{\mathrm{D}}} + 3\,\frac{c_{22}^{\mathrm{D}}}{c_{11}^{\mathrm{D}}}.$$

The coupling matrix is computed using equation (5.58):

$$\Theta = \begin{bmatrix} \frac{1}{4}\, h_{13}\, t^2 & -\frac{1}{4}\, h_{13}\, t^2 \\ \frac{1}{8}\, h_{13}\, bt^2 & -\frac{1}{8}\, h_{13}\, bt^2 \\ -\frac{1}{24}\left(h_{23}\, a^2 - 2\, h_{13}\, b^2\right) t^2 & \frac{1}{24}\left(h_{23}\, a^2 - 2\, h_{13}\, b^2\right) t^2 \end{bmatrix}. \tag{5.229}$$

Table 5.9 Material properties for Example 5.4

Mechanical	Electrical	Coupling
$c_{11}^{D} = 131.6$ GPa	$\beta_{33}^{S} = 1.48 \times 10^{8}$ m/F	$h_{13} = -2.72 \times 10^{9}$ N/C
$c_{22}^{D} = 131.6$ GPa		$h_{23} = -2.72 \times 10^{9}$ N/C
$c_{12}^{D} = 84.2$ GPa		
$c_{66}^{D} = 3.0$ GPa		

The inverse of the capacitance matrix is computed using equation (5.59):

$$\mathrm{C}_{\mathrm{p}}^{\mathrm{S}^{-1}} = \begin{bmatrix} \dfrac{1}{2}\dfrac{\beta_{33}^{S} t}{ab} & 0 \\ 0 & \dfrac{1}{2}\dfrac{\beta_{33}^{S} t}{ab} \end{bmatrix}. \tag{5.230}$$

The equations of equilibrium for the piezoelectric plate are formed by combining equations (5.228), (5.229), and (5.230) with the appropriate input matrices for the forcing function and the applied voltage. This procedure is discussed in the following example.

Example 5.4 Compute the free displacement per unit voltage and blocked force per unit voltage for a 10 cm $\times$ 10 cm piezoelectric bimorph plate with a thickness of 0.5 mm. The material properties for the plate are listed in Table 5.9.

Solution From the problem statement the geometric parameters are $a = b = 0.1$ m and $t = 0.5 \times 10^{-3}$ m. Substituting these values and the values listed in Table 5.9 into equation (5.228) yields

$$\mathrm{K}^{\mathrm{D}} = \begin{bmatrix} 54.84 & 2.742 & 0.1243 \\ 2.742 & 0.1839 & 0.01089 \\ 0.1243 & 0.01089 & 0.0009038 \end{bmatrix} \times 10^{-3} \text{ N/m}.$$

The coupling matrix is computed using equation (5.229):

$$\Theta = \begin{bmatrix} -169.9 & 169.9 \\ -8.496 & 8.496 \\ -0.2832 & 0.2832 \end{bmatrix} \text{ N/C}.$$

The inverse of the capacitance matrix is computed from equation (5.230):

$$\mathrm{C}_{\mathrm{p}}^{\mathrm{S}^{-1}} = \begin{bmatrix} 3.711 & 0 \\ 0 & 3.711 \end{bmatrix} \times 10^{6} \text{ F}^{-1}.$$

The generalized stiffness matrix expressed in equation (5.75) is now computed:

$$\mathcal{K} = \begin{bmatrix} 39.29 & 1.964 & 0.09839 \\ 1.964 & 0.1450 & 0.009595 \\ 0.09839 & 0.009595 & 0.0008605 \end{bmatrix} \times 10^{-3}\ \text{N/m}.$$

The input vectors of the system are defined as

$$\mathrm{B_f} = \mathrm{N_r}'(a/2, b/2) = \begin{bmatrix} -2.500 \\ -0.125 \\ -0.006 \end{bmatrix}$$

and

$$\mathrm{B_v} = \begin{bmatrix} 1 \\ -1 \end{bmatrix}.$$

The free deflection per unit volt is now obtained using the expression in equation (5.77):

$$u_3(a/2, b/2)|_{f=0} = 5.83\ \mu\text{m/V}, \tag{5.231}$$

and the blocked force per unit volt is obtained using equation (5.78):

$$f|_{u_3=0} = 36.7\ \text{mN/V}. \tag{5.232}$$

The solution presented in Example 5.4 is obtained using a three-term expansion for the plate deflection. As discussed earlier in the chapter, the Ritz method offers an effective means of approximating the solution to the governing equations and often requires additional terms to obtain sufficient convergence in the solution. For a pinned–pinned plate with the boundary conditions stated in equation (5.222), a general form of the displacement expansion is the series

$$u_3(x, y) = \sum_{i=2}^{m} \sum_{j=0}^{n} r_{ij}(x^{i-1} - a^{i-1})xy^j, \tag{5.233}$$

where $m = n = 2$ in Example 5.4. Note that the kinematic boundary conditions are satisfied by this assumed deflection shape.

A study of the solution as a function of the number of terms in the expansion, N, illustrates that the free strain and blocked force results converge after a small number of terms. Figure 5.17 demonstrates that the use of four terms in the expansion produces sufficient convergence in the computation of the free strain per unit voltage. Small gains are achieved in the blocked force computation by increasing the number of terms in the expansion to six.

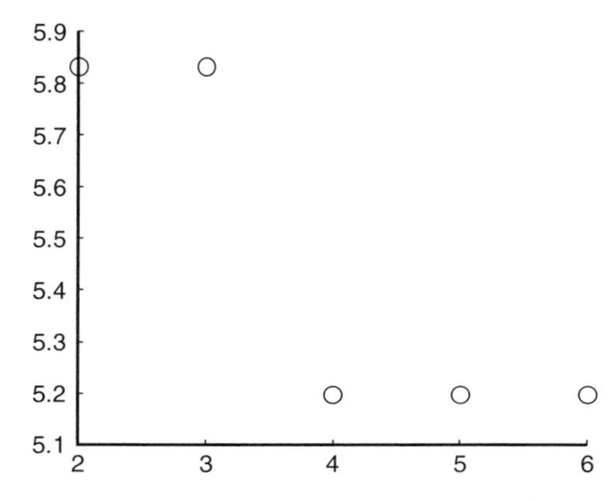

Figure 5.17 Convergence properties of the plate example as a function of terms in the displacement expansion.

The variational approach is also valid for plates in which the piezoelectric elements do not cover the entire surface. Consider the case of a thin plate with two piezoelectric layers attached as shown in Figure 5.18. The lower left corner of the piezoelectric layers are located at $x = x_p$, $y = y_p$, and each layer has dimension $a_p \times b_p$. Assume that the piezoelectric layers are wired to produce bending when the same voltage is applied, as in all cases studied previously.

Expanding the deflection in terms of the displacement functions $\phi_i(x, y)$ that are consistent with the kinematic boundary conditions,

$$u_3(x, y) = \sum_{k=1}^{N} \phi_k(x, y) r_k, \tag{5.234}$$

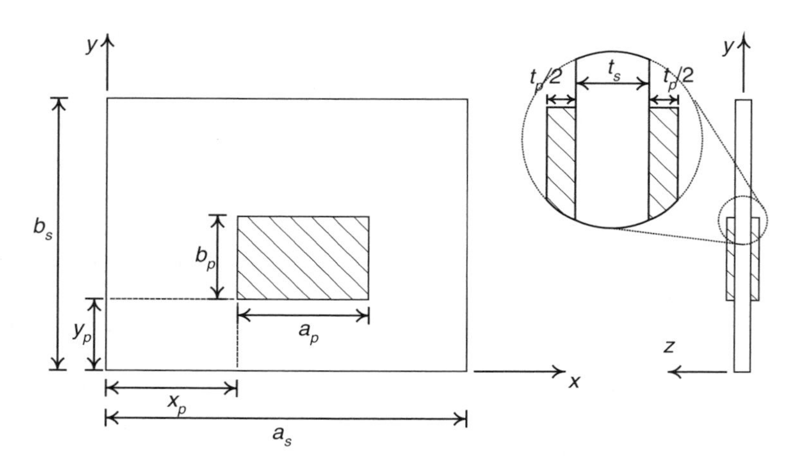

Figure 5.18 Plate with a piezoelectric bimorph.

produces the following displacement function matrix:

$$\mathrm{N_r}(\mathbf{x}) = [\,\phi_1(x,y)\ \phi_2(x,y)\ \ldots\ \phi_N(x,y)\,]. \tag{5.235}$$

The shape function matrix for the strain is

$$\mathrm{B_r}(\mathbf{x}) = \begin{bmatrix} -z\dfrac{\partial^2\phi_1}{\partial x^2} & \cdots & -z\dfrac{\partial^2\phi_N}{\partial x^2} \\ -z\dfrac{\partial^2\phi_1}{\partial y^2} & \cdots & -z\dfrac{\partial^2\phi_N}{\partial y^2} \\ 0 & \cdots & 0 \\ 0 & \cdots & 0 \\ 0 & \cdots & 0 \\ -2z\dfrac{\partial^2\phi_1}{\partial xy} & \cdots & -2z\dfrac{\partial^2\phi_N}{\partial xy} \end{bmatrix}. \tag{5.236}$$

The shape functions for the upper and lower piezoelectric elements are defined in the same manner as for previous systems:

$$\mathrm{B_q^1}(\mathbf{x}) = \begin{bmatrix} 0 & 0 \\ 0 & 0 \\ \dfrac{1}{a_p b_p} & 0 \end{bmatrix} \tag{5.237}$$

$$\mathrm{B_q^2}(\mathbf{x}) = \begin{bmatrix} 0 & 0 \\ 0 & 0 \\ 0 & \dfrac{1}{a_p b_p} \end{bmatrix}. \tag{5.238}$$

The stiffness matrix for the substrate is computed from the shape functions using equation (5.57). After integration in the thickness direction, the individual elements of the stiffness matrix are

$$\mathrm{K_{s_{ij}}} = \frac{t_s^3}{12}\int_0^{a_s}\int_0^{b_s}\left[\frac{\partial^2\phi_i}{\partial x^2}\left(c_{11}\frac{\partial^2\phi_j}{\partial x^2} + c_{12}\frac{\partial^2\phi_j}{\partial y^2}\right) + \frac{\partial^2\phi_i}{\partial y^2}\left(c_{12}\frac{\partial^2\phi_j}{\partial x^2} + c_{22}\frac{\partial^2\phi_j}{\partial y^2}\right) + 4c_{66}\frac{\partial^2\phi_i}{\partial xy}\left(\frac{\partial^2\phi_j}{\partial xy}\right)\right] dy\, dx. \tag{5.239}$$

The matrix equations that describe the piezoelectric material are obtained using equations (5.58) to (5.60). Combining the strain shape functions with equation (5.58) and integrating over the thickness yields

$$\mathrm{K}_{\mathrm{P_{ij}}} = \left(\frac{1}{4\,t_p{}^2 t_s} + \frac{1}{4\,t_p\,t_s{}^2} + \frac{1}{12\,t_s{}^3}\right) \int_{x_p}^{x_p+a_p} \int_{y_p}^{y_p+b_p} \left[\frac{\partial^2 \phi_i}{\partial x^2}\left(c_{11}^{\mathrm{D}}\frac{\partial^2 \phi_j}{\partial x^2} + c_{12}^{\mathrm{D}}\frac{\partial^2 \phi_j}{\partial y^2}\right) + \frac{\partial^2 \phi_i}{\partial y^2}\left(c_{12}^{\mathrm{D}}\frac{\partial^2 \phi_j}{\partial x^2} + c_{22}^{\mathrm{D}}\frac{\partial^2 \phi_j}{\partial y^2}\right) + 4c_{66}^{\mathrm{D}}\frac{\partial^2 \phi_i}{\partial xy}\left(\frac{\partial^2 \phi_j}{\partial xy}\right)\right] dy\, dx. \tag{5.240}$$

Integrating the shape functions and the piezoelectric strain matrix h according to equation (5.59) produces the following expression for the ith row of the coupling matrix:

$$\Theta_i = \frac{2t_s t_p + t_p^2}{8a_p b_p} \int_{x_p}^{x_p+a_p} \int_{y_p}^{y_p+b_p} \left(h_{13}\frac{\partial^2 \phi_i}{\partial x^2} + h_{23}\frac{\partial^2 \phi_i}{\partial y^2}\right) dy\, dx \times [\,1 \quad -1\,]. \tag{5.241}$$

The inverse of the capacitance matrix for each piezoelectric element is

$$\mathrm{C}_{\mathrm{p}}^{\mathrm{S}^{-1}} = \begin{bmatrix} \dfrac{1}{2}\dfrac{\beta_{33}^{\mathrm{S}} t_p}{a_p b_p} & 0 \\ 0 & \dfrac{1}{2}\dfrac{\beta_{33}^{\mathrm{S}} t_p}{a_p b_p} \end{bmatrix}. \tag{5.242}$$

Example 5.5 A 10 cm × 10 cm plate pinned along $x = 0$ and $x = a_s$ and free along the other two edges has two piezoelectric layers bonded at the position $x_p = y_p = 0.045$. The piezoelectric layers have dimensions $a_p = b_p = 1$ cm. The material properties for the piezoelectric materials are shown in Table 5.9, and the material properties for the substrate are shown in Table 5.10. The substrate thickness is 0.5 mm and the thickness of both piezoelectric layers is 0.5 mm. (a) Compute the deflection field $u_3(x, y)$ to a 1-V input to the piezoelectric layers using the displacement approximation shown in equation (5.233) for $m = n = 2$. No force is applied to the plate. (b) Plot the displacement field and compute the displacement at the center of the plate: $x = 1/20$, $y = 1/20$.

Solution (a) The shape functions for the displacement expansion have the same form as equation (5.225) when $m = n = 2$. Using equation (5.239) to compute the

Table 5.10 Material properties for the substrate in Example 5.5

$c_{11} = 83.5$ GPa
$c_{22} = 83.5$ GPa
$c_{12} = 35.6$ GPa
$c_{66} = 23.8$ GPa

individual components of the substrate stiffness matrix, we obtain

$$\mathrm{K_s} = \begin{bmatrix} 34.78 & 1.739 & 0.09108 \\ 1.739 & 0.1490 & 0.01076 \\ 0.09108 & 0.01076 & 0.001087 \end{bmatrix} \times 10^{-3}\ \mathrm{N/m}.$$

Similarly, computing the stiffness terms for the piezoelectric layers using equation (5.240) results in

$$\mathrm{K_p^D} = \begin{bmatrix} 3.84 & 0.192 & 0.00351 \\ 0.192 & 0.00963 & 0.000179 \\ 0.00351 & 0.000179 & 0.0000176 \end{bmatrix} \times 10^{-3}\ \mathrm{N/m}.$$

The coupling terms are computed using equation (5.241):

$$\Theta = \begin{bmatrix} 510.0 & -510.0 \\ 25.5 & -25.5 \\ 0.00850 & -0.00850 \end{bmatrix} \mathrm{N/C}.$$

The inverse of the capacitance matrix is computed from equation (5.242):

$$\mathrm{C_p^S} = \begin{bmatrix} 0.371 & 0 \\ 0 & 0.371 \end{bmatrix} \times 10^{9}\ \mathrm{F}^{-1}.$$

The coefficients of the displacement expansion, r_i, are obtained by solving the matrix expression in equation (5.74). Substituting the matrices into this expression produces the equation

$$\begin{bmatrix} 37.2 & 1.86 & 0.0946 \\ 1.86 & 0.155 & 0.0109 \\ 0.0946 & 0.0109 & 0.00111 \end{bmatrix} \begin{pmatrix} r_0 \\ r_1 \\ r_2 \end{pmatrix} = \begin{bmatrix} 0.00275 \\ 0.000137 \\ 0.0000000458 \end{bmatrix} (1).$$

Note that a factor of $\times 10^{-3}$ has been removed from both sides of the expression. The solution of the matrix expression yields the coefficients

$$\boldsymbol{r} = \begin{bmatrix} -0.377 \\ -285.0 \\ 2850.0 \end{bmatrix} \times 10^{-5}.$$

The equation for the displacement field in the 3 direction is obtained by substituting the coefficients into the displacement expansion, equation (5.233):

$$\begin{aligned} u_3(x,y) = [&-0.377\,(x - 1/10)\,x - 285\,(x - 1/10)\,xy \\ &+ 2850\,(x - 1/10)\,xy^2] \times 10^{-5}\ \mathrm{m}. \end{aligned}$$

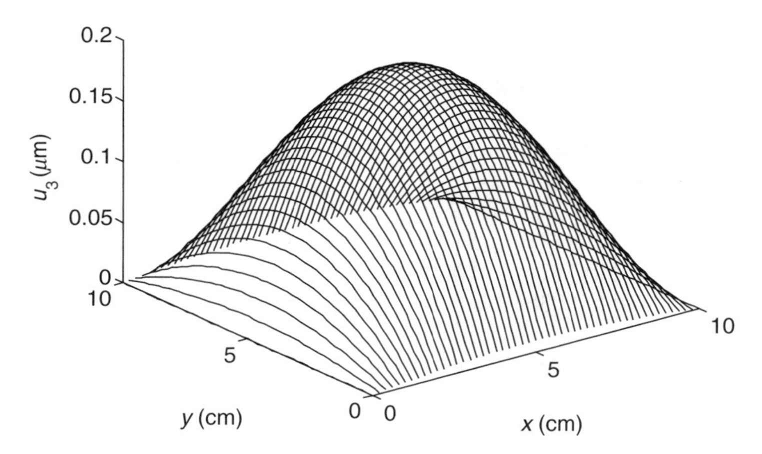

Figure 5.19 Displacement field for the plate of Example 5.5 with $m = n = 2$.

(b) The plot of the displacement field over the range is shown in Figure 5.19. The shape functions constrain the displacement response to be quadratic in x and y; therefore, cutting the response at any value of constant x or y would yield a parabolic shape.

The response at the center of the plate is computed from

$$\begin{aligned} u_3\left(\frac{1}{20}, \frac{1}{20}\right) &= [-0.377\,(-1/10)\,(1/20) - 285\,(-1/10)\,(1/20)^2 \\ &\quad + 2850\,(-1/10)\,(1/20)^3] \times 10^{-5}\ \text{m} \\ &= 0.1875\ \mu\text{m}. \end{aligned}$$

for a 1-V input to the piezoelectric layers. In Example 5.4 the piezoelectric layers covered the entire plate and the substrate was assumed to be negligible. For these conditions the free response per volt was on the order of 5.2 μm; therefore, it is sensible that the free displacement should be substantially less (by a factor of 20) in the case in which the piezoelectric layers cover only a small portion of the plate and the substrate has nonnegligible thickness.

Example 5.5 illustrates the computation of the coefficients of the displacement field for a low-order approximation of the solution. As pointed out in the example, the approximation constrains the displacement field to be quadratic in both directions. Increasing the order of the displacement approximation will produce superior approximations to the solution for the displacement field and yield improved predictions of the free response of the plate.

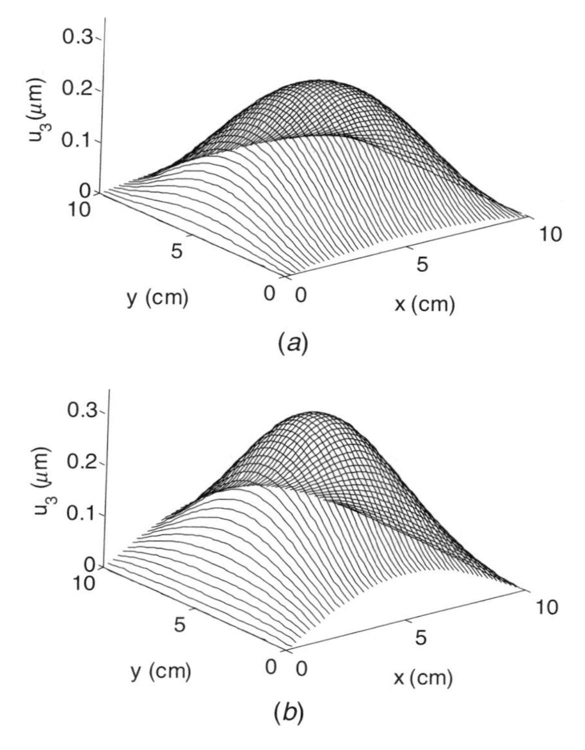

Figure 5.20 Displacement field for the plate of Example 5.5 with (*a*) $m = 2$, $n = 4$; (*b*) $m = 4$, $n = 4$.

Performing a series of computations for increasing orders of approximation will yield information on the convergence of the solution. Repeating the analysis illustrated in Example 5.5 for $m = 2, n = 4$, $m = 4, n = 4$ yields the coefficients shown in Table 5.11. Figure 5.20 is a plot of the displacement fields for the two additional computations. The figure illustrates that the higher-order terms in the displacement expansion produce additional curvature near the location of the piezoelectric patch. This result is consistent with intuition since the piezoelectric layers are producing a localized bending moment on the plate.

The additional curvature induced by the piezoelectric layers is better illustrated by viewing the displacement field along the x and y directions (see Figure 5.21). When viewed along the y axis (Figure 5.21a) the additional curvature appears as an increase in the slope of the displacement field near the free edges of the plate. The induced curvature due the piezoelectrics is not as visible along the x axis, due to the difference in boundary conditions. The view shown in Figure 5.21b illustrates that the displacement field along the x axis is approximately parabolic for any value of y.

Table 5.11 Coefficients of the displacement expansion for Example 5.5 for increasing orders of the approximation

m	n	r_{20} $\times 10^{-5}$	r_{21} $\times 10^{-2}$	r_{22}	r_{23} $\times 10^{-1}$	r_{24}	r_{30} $\times 10^{-3}$	r_{31}	r_{32}	r_{33}	r_{34} $\times 10^{4}$	r_{40} $\times 10^{-2}$	r_{41}	r_{42}	r_{43} $\times 10^{4}$	r_{44} $\times 10^{4}$
2	2	−0.377	−0.285	0.285												
2	4	−0.883	−0.130	−0.725	1.71	−8.55										
4	4	0.894	−1.18	9.61	−16.9	84.4	−0.730	0.176	−17.1	306	−0.153	0.365	−0.878	85.4	−0.153	0.766

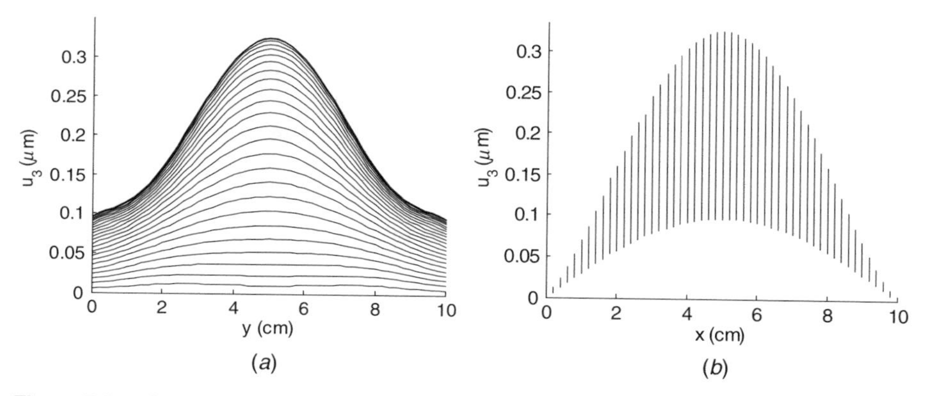

Figure 5.21 Side views of the displacement field for the pinned–pinned beam studied in Example 5.5 for $m = 4$, $n = 6$.

5.7.2 Dynamic Analysis of Piezoelectric Plates

Dynamic analysis of plates with piezoelectric materials proceeds in a manner similar to that for piezoelectric beams. The variational method applied to linear elastic materials with linear piezoelectric materials produces a set of matrix equations. The additional terms required for dynamic analysis over those for static analysis discussed in Section 5.7.1 are the mass matrices associated with the substrate and piezoelectric layers.

The mass matrices for the substrate and piezoelectric elements are obtained from equations (5.128) and (5.129). Applying these expressions to the geometry illustrated in Figure 5.18, and assuming that the materials are homogeneous, the individual elements of the mass and stiffness matrices are

$$\mathrm{M}_{\mathrm{s}_{\mathrm{ij}}} = \rho_s t_s \int_0^{a_s} \int_0^{b_s} \phi_i(x,y)\phi_j(x,y)\, dy\, dx \tag{5.243}$$

$$\mathrm{M}_{\mathrm{p}_{\mathrm{ij}}} = \rho_p t_p \int_{x_p}^{x_p+a_p} \int_{y_p}^{y_p+b_p} \phi_i(x,y)\phi_j(x,y)\, dy\, dx. \tag{5.244}$$

Solutions for the natural frequencies of plates with a variety of boundary conditions are available. As we have seen throughout the discussion of numerical methods for computing the solution of piezoelectric material systems, choice of the shape functions for the displacement expansion is critical to the convergence of the solution and the accuracy of the computed response. In the case of dynamic analysis of beams with piezoelectric elements, the mode shapes of the beams with the appropriate boundary conditions were used as the shape functions for the displacement approximation. In addition to providing reasonable convergence properties, the use of the mode shapes derived from the closed-form solution had the added benefit that the mass and stiffness matrices of the substrate were diagonal, leading to a nondimensional approach to analyzing the system response.

Closed-form solutions for the mode shapes of plates with common boundary conditions on the four edges do not exist. However, numerical analysis of the response of a thin plate can be performed assuming that the shape functions can be decomposed into the elementary mode shapes of a beam with the appropriate boundary conditions on the opposite two edges. Under this assumption the mode shapes are decomposed according to

$$\phi_i(x,y) = \Phi_i(x)\Psi_i(y), \tag{5.245}$$

where $\Phi_i(x)$ are the mode shapes of a beam with edge conditions specified by the plate boundaries $x = 0$ and $x = a_s$ and $\Psi_i(y)$ are mode shapes of a beam with boundary conditions defined by the edge conditions of the plate at $y = 0$ and $y = b_s$. Substituting equation (5.245) into equations (5.243) and (5.244) reduces the integrations to

$$\mathrm{M}_{\mathrm{s_{ij}}} = \rho_s t_s \left[\int_0^{a_s} \Phi_i(x)\Phi_j(x)\,dx\right]\left[\int_0^{b_s} \Psi_i(y)\Psi_j(y)\,dy\right] \tag{5.246}$$

$$\mathrm{M}_{\mathrm{p_{ij}}} = \rho_p t_p \left[\int_{x_p}^{x_p+a_p} \Phi_i(x)\Phi_j(x)\,dx\right]\left[\int_{y_p}^{y_p+b_p} \Psi_i(y)\Psi_j(y)\,dy\right]. \tag{5.247}$$

Equations (5.246) and (5.247) illustrate that separating the mode shapes into the x and y directions allows the integration of the mass matrices to be separated into the product of two integrations. Judicious choice of the mode shape functions allows the mass matrices of the substrate to be diagonalized in the same manner as for the beam analysis presented earlier in the chapter. If the mode shapes are chosen such that

$$\int_0^{a_s} \Phi_i(x)\Phi_j(x)\,dx = \delta_{ij} \tag{5.248}$$

$$\int_0^{b_s} \Psi_i(y)\Psi_j(y)\,dy = \delta_{ij}, \tag{5.249}$$

the mass matrix of the substrate is

$$\mathrm{M}_{\mathrm{s_{ij}}} = \rho_s t_s \delta_{ij}. \tag{5.250}$$

The mass matrix of the piezoelectric layers will only be diagonal if the integration in equation (5.247) is performed over the surface of the entire plate. In general, though, $\mathrm{M_p}$ will not be diagonal but the product of the integrations will be less than 1 if the mode shapes are chosen to satisfy equations (5.248) and (5.249).

Example 5.6 Determine the expression for the mass matrices of a thin plate that is pinned on all four sides with piezoelectric layers placed at an arbitrary location. Use the following displacement expansion for the problem:

$$u_3(x,y) = \sum_{i=1}^{m}\sum_{j=1}^{n} r_{ij}\, 2\sin\frac{i\pi x}{a_s}\sin\frac{j\pi y}{b_s}.$$

Solution Using the displacement expansion specified in the problem, the shape functions for the plate can be written

$$\mathrm{N_r}(\mathbf{x}) = \Bigg[\begin{array}{ccc} 2\sin\frac{\pi x}{a_s}\sin\frac{\pi y}{b_s} & \dots & 2\sin\frac{\pi x}{a_s}\sin\frac{n\pi y}{b_s} \\ 2\sin\frac{2\pi x}{a_s}\sin\frac{\pi y}{b_s} & \dots & 2\sin\frac{2\pi x}{a_s}\sin\frac{n\pi y}{b_s} \\ \vdots & & \vdots \\ 2\sin\frac{m\pi x}{a_s}\sin\frac{\pi y}{b_s} & \dots & 2\sin\frac{m\pi x}{a_s}\sin\frac{n\pi y}{b_s} \end{array}\Bigg].$$

Specifying the shape functions in this manner shows that the matrix

$$\mathrm{N_r}(\mathbf{x})'\mathrm{N_r}(\mathbf{x}) = \begin{bmatrix} 4\sin^2\frac{\pi x}{a_s}\sin^2\frac{\pi y}{b_s} & 4\sin^2\frac{\pi x}{a_s}\sin\frac{\pi y}{b_s}\sin\frac{2\pi y}{b_s} & \dots \\ 4\sin^2\frac{\pi x}{a_s}\sin\frac{\pi y}{b_s}\sin\frac{2\pi y}{b_s} & 4\sin^2\frac{2\pi x}{a_s}\sin^2\frac{2\pi y}{b_s} & \dots \\ \vdots & \vdots & \ddots \end{bmatrix}$$

only has a product of squared terms along the diagonal. All other entries have at least a single term in which a sine function is multiplied by a sine function with a different period. Due to the orthogonality property of the sine (and cosine) function,

$$\int_0^{\pi} \sin mx \sin nx \, dx = \frac{\pi}{2}\delta_{mn},$$

the mass matrix for the substrate will be a diagonal matrix of the form

$$\mathrm{M_s} = \rho_s t_s a_s b_s \mathrm{I}.$$

The expression for the mass matrix of the piezoelectric elements on the plate is given by equation (5.247). Due to the fact that the piezoelectric element is placed at an arbitrary location, the orthogonality of the shape functions will not necessarily reduce the number of elements in the mass matrix. A general expression would be

$$\mathrm{M_s} = \rho_p t_p \int_{x_p}^{x_p+a_p} \int_{y_p}^{y_p+b_p} \times \begin{bmatrix} 4\sin^2\frac{\pi x}{a_s}\sin^2\frac{\pi y}{b_s} & 4\sin^2\frac{\pi x}{a_s}\sin\frac{\pi y}{b_s}\sin\frac{2\pi y}{b_s} & \cdots \\ 4\sin^2\frac{\pi x}{a_s}\sin\frac{\pi y}{b_s}\sin\frac{2\pi y}{b_s} & 4\sin^2\frac{2\pi x}{a_s}\sin^2\frac{2\pi y}{b_s} & \cdots \\ \vdots & \vdots & \ddots \end{bmatrix} dy\,dx.$$

The integral in the preceding equation would have to be solved as a function of the piezoelectric element location. More insight can be derived for the form of the solution by realizing that the integrations in the mass matrix of the piezoelectric take one of two forms. The first form is

$$\int \sin^2\frac{m\pi\xi}{a}d\xi = \frac{1}{2}\frac{m\pi\xi - a\cos(m\pi\xi/a)\sin(m\pi\xi/a)}{m\pi},$$

where m is an integer and a is a constant. The other form in the integrand is

$$\int \sin m\pi\xi \sin n\pi\xi d\xi = \frac{1}{2}\frac{\sin[(m\pi - n\pi)\xi]}{m\pi - n\pi} - \frac{1}{2}\frac{\sin[(m\pi + n\pi)\xi]}{m\pi + n\pi},$$

assuming that $m \neq n$. It is interesting to note that the elements of the mass matrices are themselves periodic. This periodicity leads to the effect of spatial filtering that we discussed for piezoelectric beams earlier in the chapter.

Example 5.6 illustrates the manner in which the mass matrix is computed for dynamic analysis of a plate pinned on all edges. In the next example the full system matrices are computed for a model that includes four displacement expansion functions.

Example 5.7 A plate pinned on all four edges has two piezoelectric layers bonded symmetrically across the midplane and connected in a bimorph configuration. The material properties for the substrate and piezoelectric layers are the same as those used in Examples 5.5 and 5.6 (see Tables 5.9 and 5.10), with the exception that $a_p = b_p = 0.25$ and the piezoelectric layers are located at $x_p = y_p = 0$. Use the displacement approximation

$$u_3(x,y) = \sum_{i=1}^{2}\sum_{j=1}^{2} r_{ij}\, 2\sin\frac{i\pi x}{a_s}\sin\frac{j\pi y}{b_s}$$

for this problem. Compute (a) the mass matrices for the substrate and piezoelectric layers, (b) the stiffness matrices for the substrate and piezoelectric layers, and (c) the coupling matrix and the inverse of the capacitance matrix for the piezoelectric layers.

Solution (a) The mass matrices for the substrate are computed using equation (5.243). Using the shape functions defined in the displacement expansion, the result is

$$\mathbf{M}_{\mathrm{s}} = \begin{bmatrix} 13.5 & 0 & 0 & 0 \\ 0 & 13.5 & 0 & 0 \\ 0 & 0 & 13.5 & 0 \\ 0 & 0 & 0 & 13.5 \end{bmatrix} \times 10^{-3} \text{ kg}.$$

As a check, we note that the values along the diagonal of $\mathbf{M}_{\mathrm{s}}$ are equivalent to $\rho_s t_s a_s b_s$. This is due to the choice of shape functions for the displacement expansion.

The mass matrix for the piezoelectric layers is computed using equation (5.244). The result is

$$\mathbf{M}_{\mathrm{p}} = \begin{bmatrix} 0.3219 & 0.5316 & 0.5316 & 0.8781 \\ 0.5316 & 0.8857 & 0.8781 & 1.4630 \\ 0.5316 & 0.8781 & 0.8857 & 1.4630 \\ 0.8781 & 1.4630 & 1.4630 & 2.4375 \end{bmatrix} \times 10^{-3} \text{ kg}.$$

(b) The stiffness matrix for the substrate is computed using equation (5.239):

$$\mathbf{K}_{\mathrm{s}} = \begin{bmatrix} 33.8746 & 0 & 0 & 0 \\ 0 & 211.7165 & 0 & 0 \\ 0 & 0 & 211.7165 & 0 \\ 0 & 0 & 0 & 541.9942 \end{bmatrix} \times 10^{3} \text{ N/m}.$$

Once again we note that the stiffness matrix for the substrate is diagonal, due to the choice of the shape functions for the displacement. The stifffness matrix for the piezoelectric layers is computed from equation (5.240):

$$\mathbf{K}_{\mathrm{p}}^{\mathrm{D}} = \begin{bmatrix} 3.9719 & 12.5643 & 12.5643 & 30.7188 \\ 12.5643 & 50.4894 & 42.8428 & 120.1989 \\ 12.5643 & 42.8428 & 50.4894 & 120.1989 \\ 30.7188 & 120.1989 & 120.1989 & 315.2521 \end{bmatrix} \times 10^{3} \text{ N/m}.$$

(c) The computation for the coupling matrix Θ is shown in equation (5.241):

$$\Theta = \begin{bmatrix} -0.1399 & 0.1399 \\ -0.5972 & 0.5972 \\ -0.5972 & 0.5972 \\ -1.6313 & 1.6313 \end{bmatrix} \times 10^{6} \text{ N/C},$$

and the inverse of the capacitance matrix is computed using equation (5.242):

$$\mathrm{C_p^{S^{-1}}} = \begin{bmatrix} 5.9382 & 0 \\ 0 & 5.9382 \end{bmatrix} \times 10^7\ \mathrm{F}^{-1}.$$

Inverting the matrix yields the capacitance matrix

$$\mathrm{C_p^S} = \begin{bmatrix} 16.8403 & 0 \\ 0 & 16.8403 \end{bmatrix} \times 10^{-9}\ \mathrm{F}$$

for the piezoelectric layers.

Example 5.7 illustrates how the system matrices are computed for a plate with piezoelectric layers. This example illustrates that, in general, the mass, stiffness, and coupling matrices associated with the piezoelectric layers are fully populated. The relationship between these matrices and the natural frequencies of the system will be studied shortly. Before studying these relationships, though, the next example focuses on the computation of the system matrices for piezoelectric layers that are located such that the coupling to certain vibration modes of the plate are minimized.

Example 5.8 Repeat Example 5.7 with $x_p = y_p = 0.0375$ m. This location places the piezoelectric in the center of the pinned plate. Compare the mass, stiffness, and coupling matrices for this location of the piezoelectric layers to the matrices obtained in Example 5.7.

Solution The mass and stiffness matrices for the substrate are not a function of the location of the piezoelectric layers; therefore, they are the same as those computed in Example 5.7. The computation of the mass, stiffness, and coupling matrices for the piezoelectric layers yields

$$\mathrm{M_p} = \begin{bmatrix} 8.8023 & 0 & 0 & 0 \\ 0 & 1.6832 & 0 & 0 \\ 0 & 0 & 1.6832 & 0 \\ 0 & 0 & 0 & 0.3219 \end{bmatrix} \times 10^{-3}\ \mathrm{kg}$$

$$\mathrm{K_p^D} = \begin{bmatrix} 69.2147 & 0 & 0 & 0 \\ 0 & 89.6030 & 0 & 0 \\ 0 & 0 & 89.6030 & 0 \\ 0 & 0 & 0 & 63.5512 \end{bmatrix} \times 10^3\ \mathrm{N/m}$$

$$\Theta = \begin{bmatrix} -0.9556 & 0.9556 \\ 0 & 0 \\ 0 & 0 \\ 0 & 0 \end{bmatrix} \times 10^6 \text{ N/C}.$$

The obvious difference between these three matrices and the results of Example 5.7 is the sparseness of the mass, stiffness, and coupling matrices when the piezoelectric layers are moved to the center of the plate. As discussed previously, the location of the piezoelectric layers strongly affects the coupling between the piezoelectric material and the vibration modes of the plate. This coupling is quantified by $\mathrm{M_p}$, $\mathrm{K_p^D}$, and Θ.

Examples 5.7 and 5.8 illustrate that, as expected, the form of the system matrices for the piezoelectric elements are strongly affected by the location of the piezoelectric layers. This, in turn, will affect the coupling that the layers exhibit to the vibration modes of the plate. Earlier in the chapter the coupling between a piezoelectric element and the vibration modes of the structure was quantified by the generalized coupling coefficient, equation (5.211). The coupling coefficient is an indicator of the energy transfer between the vibration modes of the substrate and the piezoelectric element. In Example 5.9 the generalized coupling coefficient is computed for the systems analyzed in Examples 5.7 and 5.8.

Example 5.9 Compute the generalized coupling coefficients for the systems analyzed in Examples 5.7 and 5.8. Compare the coupling coefficients when the piezoelectric layers are located in the corner of the plate to the case in which they are located in the middle of the plate.

Solution Computation of the generalized coupling coefficient requires first computing the open-circuit and short-circuit natural frequencies of the system. With the system matrices computed in Example 5.7, the natural frequencies are computed using the eigenvalue problem stated in equation (5.144):

$$\mathrm{K_s} + \mathrm{K_p^D} = \begin{bmatrix} 37.8466 & 12.5643 & 12.5643 & 30.7188 \\ 12.5643 & 262.2059 & 42.8428 & 120.1989 \\ 12.5643 & 42.8428 & 262.2059 & 120.1989 \\ 30.7188 & 120.1989 & 120.1989 & 857.2463 \end{bmatrix} \times 10^3 \text{ N/m}.$$

The mass matrix for the eigenvalue computation is

$$\mathrm{M_s} + \mathrm{M_p} = \begin{bmatrix} 13.8219 & 0.5316 & 0.5316 & 0.8781 \\ 0.5316 & 14.3857 & 0.8781 & 1.4630 \\ 0.5316 & 0.8781 & 14.3857 & 1.4630 \\ 0.8781 & 1.4630 & 1.4630 & 15.9375 \end{bmatrix} \times 10^{-3} \text{ kg}.$$

The open-circuit natural frequencies (in rad/s) are computed from the eigenvalue problem:

$$\left(\begin{bmatrix} 37.8466 & 12.5643 & 12.5643 & 30.7188 \\ 12.5643 & 262.2059 & 42.8428 & 120.1989 \\ 12.5643 & 42.8428 & 262.2059 & 120.1989 \\ 30.7188 & 120.1989 & 120.1989 & 857.2463 \end{bmatrix} \times 10^3\right) \boldsymbol{v}_{\mathrm{i}}^{\mathrm{D}}$$
$$= \left(\begin{bmatrix} 13.8219 & 0.5316 & 0.5316 & 0.8781 \\ 0.5316 & 14.3857 & 0.8781 & 1.4630 \\ 0.5316 & 0.8781 & 14.3857 & 1.4630 \\ 0.8781 & 1.4630 & 1.4630 & 15.9375 \end{bmatrix} \times 10^{-3}\right) \omega_i^{\mathrm{D}^2} \boldsymbol{v}_{\mathrm{i}}^{\mathrm{D}}.$$

Solving for the eigenvalues and transforming them into hertz (cycles/s) produces

$$\begin{aligned} f_1^{\mathrm{D}} &= 258.4 \text{ Hz} \\ f_2^{\mathrm{D}} &= 641.4 \text{ Hz} \\ f_3^{\mathrm{D}} &= 674.5 \text{ Hz} \\ f_4^{\mathrm{D}} &= 1172.0 \text{ Hz}. \end{aligned}$$

The short-circuit natural frequencies are computed from the eigenvalue problem stated in equation (5.147). The short-circuit stiffness matrix of the system is

$$\mathrm{K}_s + \mathrm{K}_{\mathrm{p}}^{\mathrm{D}} - \Theta \mathrm{C}_{\mathrm{p}}^{\mathrm{S}} \Theta' = \begin{bmatrix} 37.1870 & 9.7494 & 9.7494 & 23.0302 \\ 9.7494 & 250.1925 & 30.8294 & 87.3858 \\ 9.7494 & 30.8294 & 250.1925 & 87.3858 \\ 23.0302 & 87.3858 & 87.3858 & 767.6216 \end{bmatrix} \times 10^3 \text{ N/m},$$

and the eigenvalue problem for the short-circuit system is

$$\left(\begin{bmatrix} 37.1870 & 9.7494 & 9.7494 & 23.0302 \\ 9.7494 & 250.1925 & 30.8294 & 87.3858 \\ 9.7494 & 30.8294 & 250.1925 & 87.3858 \\ 23.0302 & 87.3858 & 87.3858 & 767.6216 \end{bmatrix} \times 10^3\right) \boldsymbol{v}_{\mathrm{i}}^{\mathrm{E}}$$
$$= \left(\begin{bmatrix} 13.8219 & 0.5316 & 0.5316 & 0.8781 \\ 0.5316 & 14.3857 & 0.8781 & 1.4630 \\ 0.5316 & 0.8781 & 14.3857 & 1.4630 \\ 0.8781 & 1.4630 & 1.4630 & 15.9375 \end{bmatrix} \times 10^{-3}\right) \omega_i^{\mathrm{E}^2} \boldsymbol{v}_{\mathrm{i}}^{\mathrm{E}}.$$

The short-circuit natural frequencies are

$$\begin{aligned} f_1^{\mathrm{E}} &= 257.8 \\ f_2^{\mathrm{E}} &= 641.4 \\ f_3^{\mathrm{E}} &= 663.6 \\ f_4^{\mathrm{E}} &= 1105.9. \end{aligned}$$

Table 5.12 Generalized coupling coefficients for the two systems studied in Examples 5.7 and 5.8

	$x_p = y_p = 0$			$x_p = y_p = 0.0375$ m		
Mode	f_i^{D} (Hz)	f_i^{E} (Hz)	k_{ij}	f_i^{D} (Hz)	f_i^{E} (Hz)	k_{ij}
1	258.4	257.8	0.0636	342.2	286.6	0.6520
2	641.4	641.4	0	709.0	709.0	0
3	674.5	663.6	0.1819	709.0	709.0	0
4	1172.0	1105.9	0.3507	1053.4	1053.4	0

Equation (5.211) is used to computed the generalized coupling coefficients, and the results are summarized in Table 5.12. The analysis is repeated for the case in which the piezoelectric layer is located in the center of the plate. These results are also listed in Table 5.12.

The results obtained from the analysis of the generalized coupling coefficients are consistent with an examination of the coupling terms computed for the system matrices. In the case when the piezoelectric elements are placed in the middle of the plate, the coupling terms for the second, third, and fourth modes are equal to zero. Once again this result is due to the spatial filtering effect caused by the fact that the piezoelectric materials couple to the integral of the strain over the element.

In addition to the spatial filtering caused by relocating the piezoelectric elements, comparing the results from the two examples demonstrates how the coupling is strongly affected by the placement of the piezoelectric materials. Placing the piezoelectric material in the corner of the plate produces weak coupling to the first mode of the plate, while placing it in the center significantly increases coupling to the first mode, in addition to filtering out the contributions of the remaining shape functions incorporated into the model.

5.8 CHAPTER SUMMARY

The relationship between the constitutive relationships stated in Chapter 4 and the energy formulations of piezoelectric materials were examined at the outset of the chapter. In this discussion it was shown that the constitutive relationships for a piezoelectric material can be derived from one of a number of energy functions that describe the elastic, electric, and coupling energy terms associated with the material. The choice of the energy derivation for a linear material is arbitrary in the sense that the constitutive properties in one form have no more and no less information than do the constitutive properties derived in a different form. The primary difference between the constitutive relationships are the dependent and independent variables, and it was shown that the various forms of the constitutive properties for a linear piezoelectric material can be derived from a different form through a series of matrix transformations.

The energy formulation was extended to *systems* that incorporate piezoelectric materials by combining the internal energy form of a piezoelectric material with the Lagrangian derivation introduced in Chapter 2. Combining the internal energy with the Lagrangian formulation enables efficient derivation of the system equations of motion for an elastic structure with incorporated, or embedded, linear piezoelectric materials. The variational principle for static systems was applied to the derivation of equations of equilibrium for beams and plates with surface-bonded piezoelectric elements. The variational principle for dynamic systems yields a set of second-order matrix expressions that can be solved for the natural frequencies, mode shapes, and dynamic response of a system that incorporates piezoelectric elements. The analysis in this chapter focused on understanding the relationship between the placement of piezoelectric materials and their coupling to various modes of vibration in the structure. In numerous examples it was demonstrated that placement of the piezoelectric elements relative to the mode shapes of the structure strongly influences their coupling to the vibration modes. The coupling is quantified by the generalized coupling coefficient, which is related directly to the change in the natural frequencies between short- and open-circuit conditions. Methods for computing the frequency response and for adding structural damping to the equations of motion were also introduced.

PROBLEMS

5.1. The material properties for a piezoelectric are

$$
\mathrm{s}^{\mathrm{E}} = \begin{bmatrix} 12.0 & -4.0 & -5.0 & 0.0 & 0.0 & 0.0 \\ -4.0 & 12.0 & -5.0 & 0.0 & 0.0 & 0.0 \\ -5.0 & -5.0 & 15.0 & 0.0 & 0.0 & 0.0 \\ 0.0 & 0.0 & 0.0 & 39.0 & 0.0 & 0.0 \\ 0.0 & 0.0 & 0.0 & 0.0 & 39.0 & 0.0 \\ 0.0 & 0.0 & 0.0 & 0.0 & 0.0 & 33.0 \end{bmatrix} \mu\mathrm{m}^2/\mathrm{N}
$$

$$
\mathrm{d} = \begin{bmatrix} 0.0 & 0.0 & -110.0 \\ 0.0 & 0.0 & -110.0 \\ 0.0 & 0.0 & 280.0 \\ 0.0 & 450.0 & 0.0 \\ 450.0 & 0.0 & 0.0 \\ 0.0 & 0.0 & 0.0 \end{bmatrix} \mathrm{pm/V}
$$

$$
\varepsilon^{\mathrm{T}} = \begin{bmatrix} 12 & 0.0 & 0.0 \\ 0.0 & 12 & 0.0 \\ 0.0 & 0.0 & 10.5 \end{bmatrix} \mathrm{nF/m}.
$$

(a) Write the constitutive relationships in a form with strain and electric field as the independent variables.

(b) Write the constitutive relationships in a form with stress and electric displacement as the independent variables.

(c) Write the constitutive relationships in a form with strain and electric displacement as the independent variables.

5.2. The static displacement field for a piezoelectric material system is assumed to be

$$\begin{Bmatrix} u_1(x) \\ u_2(x) \\ u_3(x) \end{Bmatrix} = \begin{Bmatrix} 0 \\ 0 \\ r_0 + r_1 x + r_2 x^2 + r_3 x^3 \end{Bmatrix}.$$

(a) Write the matrix of shape functions $N_r(x)$ for this problem assuming that the generalized coordinates are r_i, $i = 0, 1, 2, 3$.

(b) Assuming that the differential operator for the problem is

$$\begin{bmatrix} 0 & 0 & z\dfrac{\partial^2}{\partial x^2} \\ 0 & 0 & 0 \\ 0 & 0 & 0 \\ 0 & 0 & 0 \\ 0 & 0 & 0 \\ 0 & 0 & 0 \end{bmatrix},$$

determine the matrix $B_r(x)$.

(c) Assuming that the differential operator for the problem is

$$\begin{bmatrix} \dfrac{\partial}{\partial x} & 0 & -z\dfrac{\partial^2}{\partial x^2} \\ 0 & \dfrac{\partial}{\partial x} & -z\dfrac{\partial^2}{\partial y^2} \\ 0 & 0 & 0 \\ 0 & 0 & 0 \\ 0 & 0 & 0 \\ \dfrac{\partial}{\partial y} & \dfrac{\partial}{\partial x} & -2z\dfrac{\partial^2}{\partial xy} \end{bmatrix},$$

determine the matrix $B_r(x)$. Compare your result to that of part (b).

5.3. The static displacement field for a piezoelectric material system is assumed to be

$$\begin{Bmatrix} u_1(x) \\ u_2(x) \\ u_3(x) \end{Bmatrix} = \begin{Bmatrix} 0 \\ 0 \\ r_0 + r_1 xy + r_2 x^2 y + r_3 x^3 y \end{Bmatrix}.$$

(a) Write the matrix of shape functions $N_r(x)$ for this problem assuming that the generalized coordinates are r_i, $i = 0, 1, 2, 3$.

(b) Assuming that the differential operator for the problem is

$$\begin{bmatrix} 0 & 0 & z\frac{\partial^2}{\partial x^2} \\ 0 & 0 & 0 \\ 0 & 0 & 0 \\ 0 & 0 & 0 \\ 0 & 0 & 0 \\ 0 & 0 & 0 \end{bmatrix},$$

determine the matrix $B_r(x)$.

(c) Assuming that the differential operator for the problem is

$$\begin{bmatrix} \frac{\partial}{\partial x} & 0 & -z\frac{\partial^2}{\partial x^2} \\ 0 & \frac{\partial}{\partial x} & -z\frac{\partial^2}{\partial y^2} \\ 0 & 0 & 0 \\ 0 & 0 & 0 \\ 0 & 0 & 0 \\ \frac{\partial}{\partial y} & \frac{\partial}{\partial x} & -2z\frac{\partial^2}{\partial xy} \end{bmatrix},$$

determine the matrix $B_r(x)$. Compare your result to that of part (b).

5.4. Use the material properties shown in Table 5.13 for this problem. Assuming a width of 10 mm, thickness of 0.25 mm, and length of 30 mm:

(a) Compute the equations for static equilibrium of a cantilever beam as listed in equation (5.97).

(b) Solve the equations for the generalized coordinates when $f = 0$ and $v =$ 50 V.

(c) Solve the equations for the generalized coordinates when $f = 1$ N and $v = 0$.

Table 5.13 Geometric and material parameters

Piezoelectric
$\rho = 7800$ kg/m^3
$c_{11}^{D} = 130$ GPa
$h_{13} = -2.7 \times 10^9$ N/C
$\beta_{33}^{S} = 1.5 \times 10^8$ m/F

5.5. Using the values listed in Table 5.13 as material properties for a cantilevered bimorph of length 25 mm, width 5 mm, and total thickness 0.25 mm:

(a) Determine the transducer equations for this bimorph using displacement and charge as the independent variables.

(b) Compute the free displacement and blocked force of this transducer.

(c) Determine the voltage produced by a displacement at the free end when the charge in the bimorph is held equal to zero.

5.6. Derive equation (5.122) from equation (5.121).

5.7. A pinned–pinned ideal piezoelectric bimorph has a length of 25 mm, a width of 5 mm, and a total thickness of 0.5 mm. Use Table 5.13 for this problem.

(a) Determine the transducer equations assuming a third-order expansion for the displacement shape functions.

(b) Determine the transducer equations assuming a fourth-order expansion for the displacement shape functions.

(c) Compare the predicted blocked force and free deflection of the bimorph for each of the expansions analyzed in parts (a) and (b). Compare the results.

5.8. **(a)** Plot the first three mode shapes of a cantilevered beam.

(b) Plot the derivative of the mode shapes for the first three modes of a cantilever beam.

5.9. Fill out Table 5.4 for the first three modes of a cantilevered beam. Determine if any of the locations of the piezoelectric produce a zero in the coefficient matrix for a vibration mode.

5.10. **(a)** Determine the locations of a piezoelectric element on a cantilever beam that will produce a zero in the coupling matrix for the second vibration mode.

(b) Repeat part (a) for the third vibration mode.

5.11. Draw the electrode patterns required for obtaining modal filters for the first three modes on a cantilevered beam. Compare the patterns to those shown in Figure 5.8.

5.12. A cantilever beam has piezoelectric elements bonded from the clamped end to one-fourth of the free length. Using the values listed in Table 5.14:

Table 5.14 Geometric and material parameters for the beam example

Substrate	Piezoelectric
$t_s = 0.75$ mm	$t_p = 0.25$ mm
$\rho_s = 2700$ kg/m^3	$\rho_p = 7800$ kg/m^3
$Y_s = 74$ GPa	$c_{11}^D = 125$ GPa
	$h_{13} = -2.5 \times 10^9$ N/C
	$\beta_{33}^S = 1.7 \times 10^8$ m/F

(a) Determine the nondimensional mass matrix $\mathrm{M_s} + \mathrm{M_p}/\rho_s w t_s L$ from equation (5.194) for a cantilever beam with normalized mode shapes listed in equation (5.197) for a thickness ratio of 1.

(b) Determine the nondimensional stiffness matrix $\mathrm{K_s} + \mathrm{K_p^D}/(Y_s I_s/L^3)$ from equation (5.195) for a cantilever beam with normalized mode shapes listed in equation (5.197) for a thickness ratio of 1.

(c) Repeat parts (a) and (b) for thickness ratios of $\frac{1}{10}$ and 10.

5.13. A cantilever beam has piezoelectric elements bonded from a location at one-half of the free length to three-fourth of the free length. Using the values listed in Table 5.14:

(a) Determine the nondimensional mass matrix $\mathrm{M_s} + \mathrm{M_p}/\rho_s w t_s L$ from equation (5.194) for a cantilever beam with normalized mode shapes listed in equation (5.197) for a thickness ratio of 1.

(b) Determine the nondimensional stiffness matrix $\mathrm{K_s} + \mathrm{K_p^D}/(Y_s I_s/L^3)$ from equation (5.195) for a cantilever beam with normalized mode shapes listed in equation (5.197) for a thickness ratio of 1.

(c) Repeat parts (a) and (b) for thickness ratios of 1/10 and 10.

5.14. A 50-cm-long cantilever beam with a width of 3 cm has piezoelectric elements bonded from the clamped end to one-fifth of the free length. Using the material properties listed in Table 5.14:

(a) Compute the first three natural frequencies of the beam without the piezoelectric elements.

(b) Compute the first three open-circuit natural frequencies of the beam.

(c) Compute the first three short-circuit natural frequencies of the beam. Tabulate all three results.

5.15. A 10-cm-long cantilever beam with a width of 1 cm has piezoelectric elements bonded from the clamped end to one-third of the free length. Using the material properties listed in Table 5.14:

(a) Compute the magnitude of the frequency response function $T_{uf}(\omega)$ for a measurement of the displacement at the free end of the beam.

(b) Repeat part (a) for a measurement location at $2L/3$.

(c) Repeat part (a) for a measurement location at $L/3$.

5.16. A 10-cm-long cantilever beam with a width of 1 cm has piezoelectric elements bonded from the clamped end to one-third of the free length. Using the material properties listed in Table 5.14, compute the frequency response function $T_{qv}(\omega)$.

5.17. A 10-cm-long cantilever beam with a width of 1 cm has piezoelectric elements bonded from the clamped end to one-third of the free length. Using the material properties listed in Table 5.14:

(a) Compute the frequency response function $T_{uv}(\omega)$ for a measurement location at the free end of the beam.

(b) Repeat part (a) for a measurement location at $L/2$ on the beam.

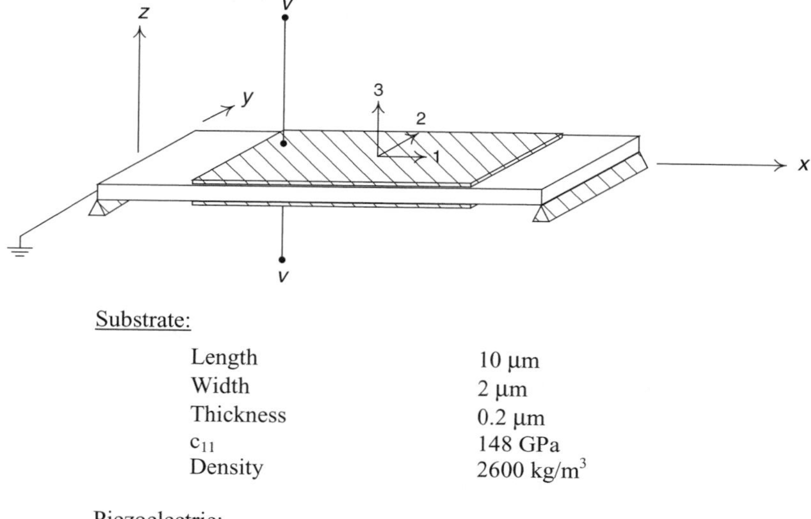

Substrate:

Length	10 μm
Width	2 μm
Thickness	0.2 μm
c_{11}	148 GPa
Density	2600 kg/m^3

Piezoelectric:

L1	2.5 μm
L2	7.5 μm
Thickness of **1 layer**	0.05 μm
c_{11}^D	143 GPa
h_{13}	-0.9×10^9 N/C
β^S_3	147×10^6 m/F
Density	7800 kg/m^3

Figure 5.22 Microresonator modeled as a pinned–pinned piezoelectric beam.

5.18. The natural frequencies of a structure with integrated piezoelectric elements have been measured to be 24, 34.5, and 56 Hz when the piezoelectric elements are open-circuit. The short-circuit frequencies have been measured to be 22.4, 34.1, and 51 Hz. Compute the generalized coupling coefficients of the piezoelectric elements for these three modes.

5.19. Microresonators are mechanical devices used as filters for communication systems. A schematic of a microresonator modeled as a pinned–pinned beam with surface-bonded piezoelectric elements is shown in Figure 5.22. The material properties are also shown in the figure.

(a) Compute the first three short-circuit natural frequencies of the microresonator.

(b) Compute the first three open-circuit natural frequencies of the resonator.

(c) Compute the generalized coupling coefficients of the first three modes.

5.20. Repeat Problem 5.15 assuming that a modal damping matrix has been incorporated into the model:

Table 5.15 Material properties for Problems 5.22 to 5.24

Mechanical	Electrical	Coupling
$c_{11}^{D} = 128$ GPa	$\beta_{33}^{S} = 1.3 \times 10^{8}$ m/F	$h_{13} = -2.72 \times 10^{9}$ N/C
$c_{22}^{D} = 128$ GPa		$h_{23} = -2.72 \times 10^{9}$ N/C
$c_{12}^{D} = 82$ GPa		
$c_{66}^{D} = 2.5$ GPa		

(a) Assume that each mode has a modal damping ratio of 0.01.
(b) Assume that each mode has a modal damping ratio of 0.05.
(c) Assume that each mode has a modal damping ratio of 0.5.

5.21. Repeat Problem 5.16 assuming that a modal damping matrix has been incorporated into the model:
(a) Assume that each mode has a modal damping ratio of 0.01.
(b) Assume that each mode has a modal damping ratio of 0.05.
(c) Assume that each mode has a modal damping ratio of 0.5.

5.22. Compute the open-circuit stiffness matrix using equation (5.228) for the properties listed in Table 5.15.

5.23. Compute the piezoelectric coupling matrix using equation (5.229) for the properties listed in Table 5.15.

5.24. Repeat Example 5.5 for the properties listed in Table 5.15 and the 9 country shown in Figrue 5.23.

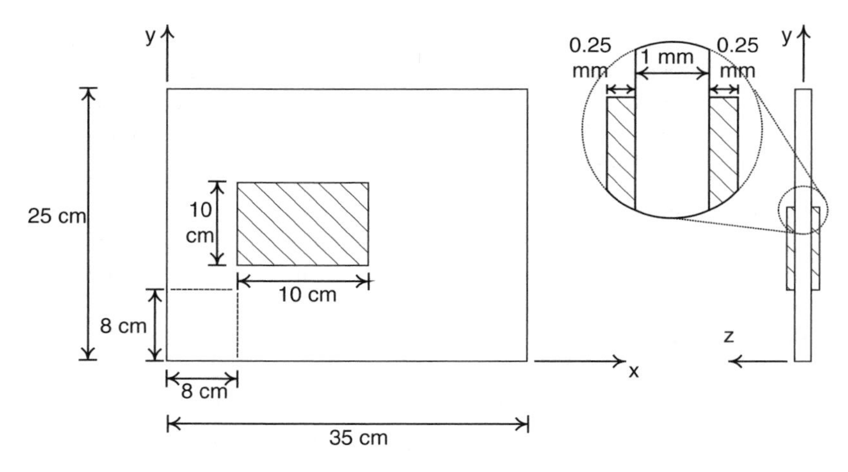

Figure 5.23 Plate dimensions for Homework XXX.

NOTES

The material on transformation of the constitutive relationships for a piezoelectric material is based on the *IEEE Standard on Piezoelectricity* [6] as well as the text by Ikeda [4]. The information in this book is much more focused on the matrix manipulations than these references, but these two references provide additional detail on the equations for transforming the constitutive properties. The derivations of the expressions for a piezoelectric material system are based on the work of Hagood et al. [35], with one important exception. In the work of Hagood et al., the equations are derived using a different set of variables, or equivalently, a different energy formulation. In their work the formulation is derived in terms of strain and electric field. This allows them to write the constitutive equations in terms of the piezoelectric strain coefficients, thus eliminating the need to transform the constitutive relationships before writing the equations of motion. In this book it was decided to express the energy formulation in terms of strain and electric displacement to make the analogy to the work and energy methods described earlier in the book. Although this necessitates transformation of the constitutive relationships, it is a more direct application of the variational methods derived in earlier chapters. Seminal references on the theory of modal sensors and actuators are those of Burke et al. [36,37], Lee et al. [38–40], and Sullivan et al. [41]. Additional material on the use of modal sensors and actuators for other applications such as noise control may be found in the work of Clark et al. [42]. Damping models for structural systems are an active area of research, but in engineering analysis damping is generally added to a structural model by estimating the engineering dissipation or through system identification. Linear viscous damping models are discussed in this chapter. Other types of damping models are discussed by Inman [43]. More detail on active and semiactive damping introduced by smart materials is given later. An excellent reference text on natural frequency and mode shapes of common structural elements is that of Blevins [44].

6

SHAPE MEMORY ALLOYS

In Chapters 4 and 5 we analyzed the properties of piezoelectric materials and their use as actuators, sensors, and integrated components of structural systems. As emphasized in Chapters 4 and 5, many applications of piezoelectric materials are based on linear constitutive properties that can be derived from energy principles. At the close of Chapter 4 the nonlinear behavior of piezoelectric materials was studied in relation to electrostrictive behavior and saturation of the electronic polarization.

In this chapter we study a fundamentally different type of smart material. Shape memory materials exhibit the ability to induce large mechanical strains upon heating and cooling. Many shape memory materials are metal alloys; therefore, they can also produce large mechanical stress when thermally activated. These properties make them well suited for applications in controllable shape change, vibration control, and active and semiactive damping. Other types of shape memory materials are also being studied, including shape memory polymers and magnetically activated shape memory materials.

In this chapter we focus on analyzing the fundamental behavior of thermally activated shape memory materials and present mathematical models for their thermomechanical behavior. In keeping with the spirit of the chapters on piezoelectric materials, we first present an overview of the basic properties of shape memory materials before discussing the material properties that give rise to this behavior. Once the basic properties of shape memory phenomena have been presented, we present mechanics models that can be used in the analysis and design of actuators that incorporate shape memory materials.

6.1 PROPERTIES OF THERMALLY ACTIVATED SHAPE MEMORY MATERIALS

The stress–strain properties of linear elastic materials were described earlier. The fundamental property of a linear elastic material is that the elastic properties are modeled as a proportionality between stress and strain. In a single dimension, this

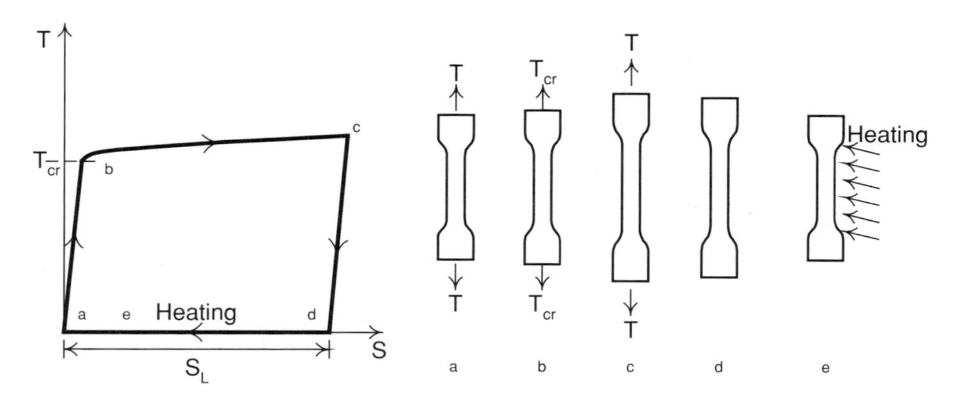

Figure 6.1 Shape memory effect in shape memory materials.

proportionality is modeled as a constant, whereas in multiple dimensions the proportionality is modeled as a tensor or matrix of elastic coefficients.

The stress–strain behavior of shape memory materials exhibits two interesting nonlinear phenomena, the shape memory and pseudoelastic effects. The *shape memory effect* is a property by which very large mechanical strains can be recovered by heating the material above a critical temperature. This strain recovery property produces large contractions in the shape memory materials and enables their use as thermomechanical actuators. The second property, the *pseudoelastic effect*, is a property by which the material exhibits a very large strain upon loading that is recovered fully when the material is unloaded. A shape memory material exhibiting the pseudoelastic effect exhibits a very large hysteresis loop in the stress–strain curve.

The shape memory and pseudoelastic behavior of shape memory materials can be visualized by considering a material that is under uniaxial loading. Loading the material from a zero stress–strain state produces a linear elastic response up to a critical stress, denoted T_{cr} in Figure 6.1. Increasing the load beyond this critical stress produces very large, apparently plastic, strain in the material accompanied by a slight increase in the load. Physically, it would seem that the material is very soft during this portion of the stress–strain curve. Unloading the material would produce a linear elastic response that would result in a residual strain S_L. So far in the discussion there would be nothing to distinguish the shape memory material from any material that has been loaded to the point of plastic deformation. The defining characteristic of a shape memory material is that the residual strain can be *fully recovered* by heating the material beyond a critical temperature. As shown in Figure 6.1, heating the material produces a recovery in the strain and returns the material to the zero stress–strain state. Points a through e represent the critical transitions in the stress–strain behavior of the shape memory material. These transitions are defined by the material properties, as discussed in detail in an upcoming section.

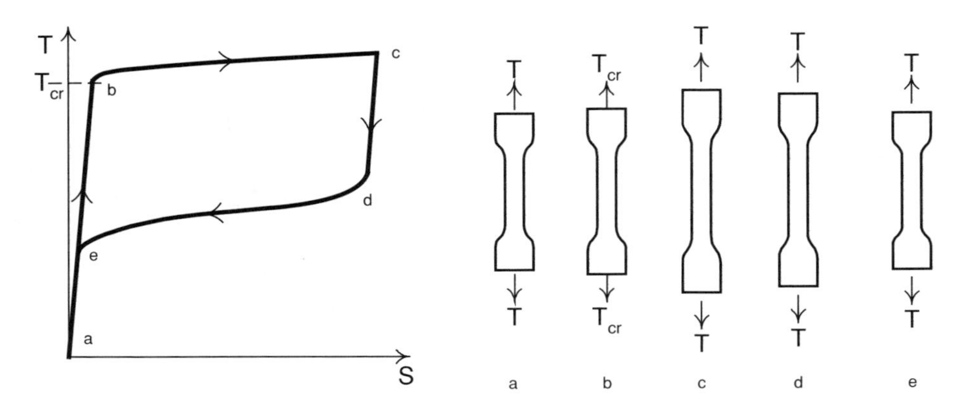

Figure 6.2 Pseudoelastic effect in shape memory materials.

The pseudoelastic effect produces a distinctively different stress–strain diagram than the shape memory effect. Loading a shape memory material that exhibits the pseudoelastic response produces a stress–strain curve that also exhibits linearly elastic behavior. As with the shape memory effect, loading the material beyond the critical stress produces an apparently plastic response in which the material undergoes large strain for a small increase in stress. During unloading, though, a material exhibiting the pseudoelastic effect will once again reach a critical stress in which there will be a large change in the strain for a small change in stress. Effectively, the material will exhibit a reversible plastic response that will result in a stress–strain state that lies on the curve produced by the initial linear elastic response during loading (Figure 6.2). Continued unloading of the material will produce a linear elastic behavior that eventually returns the material to the zero stress–strain state. As in the case of the shape memory effect, the pseudoelastic effect produces a hysteresis curve that represents the mechanical work performed during the process.

6.2 PHYSICAL BASIS FOR SHAPE MEMORY PROPERTIES

Why does a shape memory alloy exhibit the shape memory effect and pseudoelasticity? As with the case for understanding the piezoelectric effect, understanding the physical basis for shape memory behavior requires a discussion of the science associated with the structure of shape memory materials. Most common shape memory materials are metal alloys that exhibit the shape memory effect due to heating and cooling. Some of the earliest shape memory materials were alloys of gold–cadmium and brass; the shape memory effect in these materials was discovered in the 1930s. The most common type of shape memory alloy available today is a mixture of nickel and titanium and is generally referred to by the acronym Nitinol, which stands for nickel–tinanium (NiTi) with the *nol* standing for the Naval Ordnance Laboratory, where the material was developed.

The ability for shape memory alloys such as Nitinol to fully recover large strains is a result of a phase transformation that occurs due to the application of stress and heat. At high temperatures in a stress-free state, shape memory alloys exist in the *austenitic* phase. When the temperature of the material is decreased, the material phase transforms into *martensite*. The phase transformation between the martensitic and austenitic phases induces large mechanical strains in the shape memory alloy and gives rise to both the shape memory effect and the pseudoelastic effect.

In a stress-free state the transformation between the martensitic and austenitic phases is characterized by four transition temperatures. The transformation from martensite to austenite is characterized by A_s and A_f, which are the temperatures at which the phase transformation starts and finishes, respectively. Similarly, the transition from austenite to martensite is characterized by the start and finish temperatures M_s and M_f. The materials used most commonly fall into the category of Type I materials, in which the transition temperatures follow the relationship

$$M_f < M_s < A_s < A_f. \tag{6.1}$$

An important parameter in modeling the behavior of shape memory alloy materials is the fraction of martensite and austenite within the material. At any value of stress or temperature, the material can be in one of three states: fully martensitic, fully austenitic, or a mixture of austenite and martensite. The martensitic fraction of the material can exist in multiple *variants* which have identical crystallography but differ in orientation. These *twin-related* martensite variants are evenly distributed throughout the material when the shape memory alloy is in a stress-free state and is fully martensitic.

The shape memory effect can be explained in terms of the transformation between martensite and austenite as a function of stress and temperature. Consider a shape memory alloy in a stress-free state which is in the fully austenitic phase. Increasing the loading will produce a linear elastic response until a critical stress is reached, as shown in Figure 6.1. At this critical stress the austenitic phase in the material will begin transformation to martensite, resulting in a large change in strain for a small increase in stress. If the stress is increased to the point in which full martensitic transformation occurs, unloading the material will produce no reverse phase change to austenite and the material will exhibit linear elastic behavior. If the temperature of the sample is between the martensitic and austenitic start temperatures ($M_s < \theta < A_s$), unloading the specimen completely will produce a residual strain that can be recovered during heating (Figure 6.1). If the temperature is greater than the austenitic finish temperature ($\theta > A_f$), the instability of the martensitic variants at high temperature will produce a phase transition back to austenite as the specimen is unloaded. This situation will produce the pseudoelastic effect shown in Figure 6.2.

Other transformations are possible depending on the initial state of the shape memory material. If the material has been cooled from austenite to martensite in a stress-free state ($\theta < M_f$), the material will exist in a fully martensitic phase except for the fact that the martensite will exist in multiple variants, or "twins." Upon loading, a critical stress will be reached at which the multiple twins will begin converting to a single preferred variant that is aligned with the axis of loading. The transformation to a single variant of martensite will also yield nonlinear elastic behavior and result in a large mechanical strain for a small increase in load. Since the single-variant form of martensite is stable for temperatures less then A_s, unloading the specimen completely will yield residual strain in a material that is still in a fully martensitic condition. Heating the specimen will recover the strain by transforming the material to a fully austenitic phase. Upon cooling in the stress-free state the material will revert to a twinned martensitic condition.

6.3 CONSTITUTIVE MODELING

In the remainder of this chapter we concentrate on the development of a one-dimensional model of thermally activated shape memory alloys. These constitutive models will allow us to quantify the stress–strain behavior as a function of temperature and time. Constitutive models of shape memory alloys will enable analysis of the stress–strain behavior as a function of load and temperature. In keeping with the discussion of piezoelectric materials, first the fundamental constitutive model is introduced for the purpose of illustrating the basic concepts of modeling shape memory alloys. This model is then coupled to a model of heat transfer in the material to enable time-dependent analysis of shape memory alloys.

6.3.1 One-Dimensional Constitutive Model

Consider a shape memory alloy specimen that is subjected to tensile loads along its 1 axis as shown in Figure 6.1. Neglecting all of the strains except for those in the 1 direction and neglecting thermal expansion of the material, the mechanical constitutive relationship for stress in the 1 direction is written

$$\mathrm{T}_1 - \mathrm{T}_1^0 = Y_1(\mathrm{S}_1 - \mathrm{S}_1^0) + \Omega(\xi - \xi^0), \tag{6.2}$$

where ξ is the *martensitic fraction* of the material and Ω is the *transformation coefficient* of the material. The superscript 0 represents initial quantities of strain and the martensitic fraction. As discussed in Section 6.2, the martensitic fraction ξ is bounded between 0 (fully austenite) and 1 (fully martensite). For the moment we do not distinguish between the single- and multivariant forms of martensite, which will limit our model to temperatures above M_s when only the single-variant form of martensite is present. Later in the chapter this assumption is relaxed to increase the utility of the constitutive model.

The constitutive equation is defined by two material parameters. The parameter Y_1 is the elastic modulus of the material and is generally obtained from stress–strain tests of the shape memory material over the linear elastic regime. The second material parameter, the transformation coefficient Ω, is obtained by considering the hysteresis loop caused by the shape memory effect shown in Figure 6.1. Consider an initial state of zero stress and zero strain such that $S_1^0 = T_1^0 = 0$ when the material is initially in a state of fully austenitic ($\xi_0 = 0$). Referring to the points defined in Figure 6.1, consider an isothermal loading history such that the material is loaded to induce martensitic phase transformation (a–b), loaded to induce full phase transformation (b–c), and then unloaded to zero stress (c–d). As shown in the figure, this will result in a residual strain, which is denoted S_L. Since the final state is at zero stress and a martensitic fraction of $\xi = 1$, equation (6.2) is written

$$0 = Y_1 S_L + \Omega(1). \tag{6.3}$$

Equation (6.3) is solved to obtain an expression for the tranformation coefficient in terms of the elastic modulus and the residual strain,

$$\Omega = -S_L Y_1, \tag{6.4}$$

which can be substituted into equation (6.2) to yield

$$T_1 - T_1^0 = Y_1 \left(S_1 - S_1^0 \right) - S_L Y_1 (\xi - \xi^0). \tag{6.5}$$

Continuing with the discussion of the constitutive model, equation (6.2) must be augmented with a *kinetic law* that governs the transformation behavior of the material. The kinetic law is best discussed by visualizing the relationship between the martensitic fraction and temperature. Four transition temperatures characterize the relationship between ξ and temperature, and type I materials follow the relationship described in equation (6.1). To visualize the relationship between the martensitic fraction and temperature, plot the variable ξ versus temperature and label the four transition temperatures as shown in Figure 6.3. When the material is fully martensite,

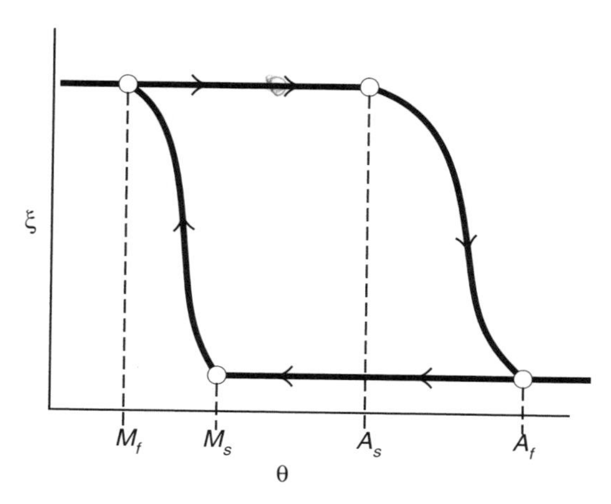

Figure 6.3 Transformation of the martensitic fraction as a function of temperature.

$\xi = 1$ and an increase in the temperature will not induce a phase transformation until A_s. For this reason a straight line is drawn from the left to the right from below M_s to A_s. Similarly, a straight line is drawn from $\theta > A_f$ to M_s, due to the fact that a phase change will not be induced unless the material is cooled from above the martensitic start temperature.

A critical feature of the martensitic fraction as a function of temperature is the expression for the austenitic-to-martensitic and martensitic-to-austenitic phase transformations. Early models of shape memory alloy behavior assumed an exponential transformation process, but more recent models of shape memory alloys have modeled the transformation with a cosine function that traces the curves shown in Figure 6.3. For a stress-free condition, the transformation from martensite to austenite is modeled with the function

$$\xi_{M \to A} = \frac{1}{2} \left\{ \cos \left[a_A \left(\theta - A_s \right) \right] + 1 \right\}, \tag{6.6}$$

where

$$a_A = \frac{\pi}{A_f - A_s} \tag{6.7}$$

is a material constant defined by the austenite start and finish temperatures. Similarly, the transformation from austenite to martensite is expressed by

$$\xi_{A \to M} = \frac{1}{2} \{ \cos[a_M(\theta - M_f)] + 1 \}, \tag{6.8}$$

where

$$a_M = \frac{\pi}{M_s - M_f}. \tag{6.9}$$

Equations (6.6) and (6.8) are for zero-stress conditions in the shape memory alloy. Loading also induces shape memory behavior, due to the fact that the transition temperatures are a function of the applied stress. The ideal relationships between the transition temperatures and applied tensile stress are shown in Figure 6.4. The relationship is idealized as a straight line as a function of stress, where A_s, A_f, M_s, and M_f represent the transformation temperatures at zero stress. Assuming that the angles α and β are known (typically, from experimental data), the transformation temperatures at any nonzero value of stress are

$$\begin{aligned} M_f^* &= M_f + \frac{\mathrm{T}}{C_M} \\ M_s^* &= M_s + \frac{\mathrm{T}}{C_M} \\ A_s^* &= A_s + \frac{\mathrm{T}}{C_A} \end{aligned} \tag{6.10}$$

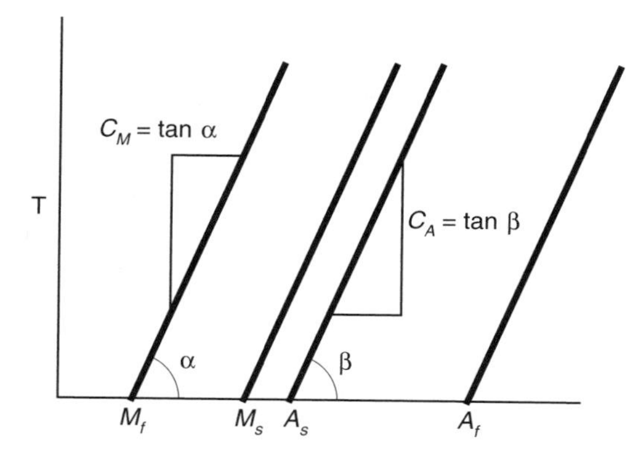

Figure 6.4 Relationship between stress and transformation temperatures in a shape memory alloy.

$$A_f^* = A_f + \frac{\mathrm{T}}{C_A}.$$

The transformation laws can now be modified to include the effects of stress on the variation in martensitic fraction during phase transformation. Consider equation (6.6) for the martensitic-to-austenitic phase transformation. Substituting A_s^* from equation (6.11) into equation (6.6) results in the expression

$$\xi_{M\to A} = \frac{1}{2}\left\{\cos\left[a_A\left(\theta - A_s - \frac{\mathrm{T}}{C_A}\right)\right] + 1\right\}, \tag{6.11}$$

which is rewritten as

$$\xi_{M\to A} = \frac{1}{2}\left\{\cos\left[a_A\left(\theta - A_s\right) - \frac{a_A}{C_A}\mathrm{T}\right] + 1\right\}. \tag{6.12}$$

Equation (6.12) demonstrates that *an increase in stress is equivalent to a reduction in the temperature of the material*. Equation (6.8) is rewritten in a similar manner as

$$\xi_{A\to M} = \frac{1}{2}\left\{\cos\left[a_M(\theta - M_f) - \frac{a_M}{C_M}\mathrm{T}\right] + 1\right\}. \tag{6.13}$$

The change in the kinetic law due to applied stress is illustrated in Figure 6.5. The applied stress is equivalent to a reduction in the temperature and is visualized as a shift to the right for all of the tranformation temperatures. Figure 6.5 also illustrates how the applied stress can induce a phase transformation in the material. Assuming that the specimen is in full austenite phase ($\xi = 0$) and is held at a constant temperature, θ_0, near the martensitic start temperature, applying stress such that the martensitic start temperature increases beyond θ_0 produces an increase in the martensitic fraction

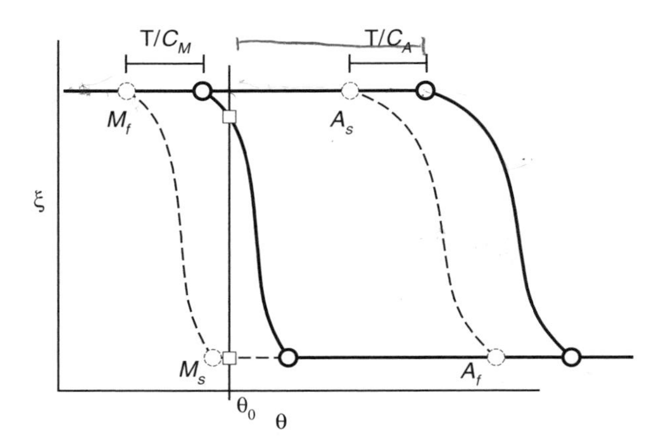

Figure 6.5 Change in the kinetic law as a function of applied stress.

of the material. This is indicated by the squares in Figure 6.5. Increasing the stress enough will induce a complete transformation from austenite to martensite in the material.

Example 6.1 A Nitinol shape memory alloy wire with the properties listed in Table 6.1 is in a zero-stress state at a temperature of 23°C. (a) Compute the martensitic fraction if the material is cooled to a temperature of 15°C in the stress-free state. (b) Assuming that the temperature is held constant, compute the martensitic fraction if a tensile stress of 90 MPa is applied to the specimen.

Solution (a) The martensitic fraction for the austenitic-to-martensitic phase transformation is obtained from equation (6.13). In a stress-free state, T = 0. The material parameter a_M is computed from equation (6.9):

$$a_M = \frac{\pi}{23^\circ\text{C} - 5^\circ\text{C}} = 0.175^\circ\text{C}^{-1}.$$

Table 6.1 Representative properties of the shape memory alloy Nitinol

M_s	23°C
M_f	5°C
A_s	29°C
A_f	51°C
C_A	4.5MPa/°C
C_M	11.3MPa/°C

Substituting this value and T = 0 into equation (6.13) yields

$$\xi_{A\rightarrow M} = \frac{1}{2}\{\cos[(0.175^\circ\mathrm{C}^{-1})(15^\circ\mathrm{C} - 5^\circ\mathrm{C})] + 1\} = 0.411.$$

(b) Equation (6.13) is also used to compute the martensitic fraction under an applied load. For a temperature of $\theta = 23^\circ$C and T = 90 MPa,

$$\begin{aligned}\xi_{A\rightarrow M} &= \frac{1}{2}\left\{\cos\left[(0.175^\circ\mathrm{C}^{-1})\,(23^\circ\mathrm{C} - 5^\circ\mathrm{C}) - \frac{0.175^\circ\mathrm{C}^{-1}}{11.3\mathrm{MPa}/^\circ\mathrm{C}}\,(90\ \mathrm{MPa})\right] + 1\right\}\\ &= 0.412.\end{aligned}$$

These examples illustrate that an applied load under isothermal conditions can induce the same phase transformtion as a change in the temperature with zero stress.

Modeling the shape memory effect and the pseudoelastic effect requires combining the kinetic transformation law with the constitutive behavior expressed in equation (6.2). Models for these two processes are developed in the following sections.

6.3.2 Modeling the Shape Memory Effect

There are four processes associated with the model of the shape memory effect shown in Figure 6.1. The initial conditions of the process are assumed to be 0% martensite in the material and zero stress and strain. The initial temperature, θ_0, is assumed to be $M_s < \theta < A_f$. The four processes that lead to the hysteresis loop shown in Figure 6.1 are:

1. The material is loaded in an isothermal state until a critical stress is reached which induces the start of martensitic phase transformation. The material is assumed to be linear elastic in this regime.
2. The material is loaded in an isothermal state until the martensitic phase transformation is complete. The stress–strain behavior of the material in this regime is assumed to be governed by a combination of the constitutive law, equation (6.5), and the kinetic law for $A \rightarrow M$ transformation.
3. The material is unloaded to zero stress with the resulting strain equal to the residual strain in the shape memory alloy. The material is assumed to be linear elastic in this regime.
4. The material is heated in a zero-stress state above A_f and cooled to the initial temperature to return the material to full austenite.

The model of the first process is obtained by substituting $\mathrm{T}_0 = \mathrm{S}_0 = 0$ into equation (6.5), where the subscripts have been dropped for convenience. The initial and final martensitic fractions of this process are zero; therefore, $\xi_o = \xi = 0$. The result is

$$\mathrm{T}^{\mathrm{a}\to\mathrm{b}} = Y\mathrm{S}, \tag{6.14}$$

which illustrates that the stress–strain behavior is linear elastic in this regime. The stress that induces the start of martensitic phase transformation, T^{b}, is obtained from equation (6.11) with the substitution $M_f^* = \theta$:

$$\mathrm{T}^{\mathrm{b}} = C_M(\theta_0 - M_s)\,. \tag{6.15}$$

The strain at the critical stress is obtained by combining equations (6.15) and (6.14):

$$\mathrm{S}^{\mathrm{b}} = \frac{\mathrm{T}^{\mathrm{b}}}{Y} = \frac{C_M(\theta_0 - M_s)}{Y}. \tag{6.16}$$

The second step in the process is modeled by combining the constitutive equation, equation (6.5), with the kinetic law, equation (6.13). The stress–strain relationship in this regime is most easily solved by first noting that that stress will be bounded by

$$C_M(\theta_0 - M_s) \leq \mathrm{T}^{\mathrm{b}\to\mathrm{c}} \leq C_M(\theta_0 - M_f), \tag{6.17}$$

assuming that the end state is equivalent to full martensitic transformation. Knowing that the stress is bounded by equation (6.17), the martensitic fraction is computed from equation (6.13):

$$\xi^{\mathrm{b}\to\mathrm{c}} = \frac{1}{2}\left\{\cos\left[a_M(\theta_0 - M_f) - \frac{a_M}{C_M}\mathrm{T}^{\mathrm{b}\to\mathrm{c}}\right] + 1\right\} \tag{6.18}$$

for each value of stress between the two bounds. The strain for this regime is computed by solving equation (6.5) for S:

$$\mathrm{S}^{\mathrm{b}\to\mathrm{c}} = \frac{1}{Y}\mathrm{T}^{\mathrm{b}\to\mathrm{c}} + \mathrm{S}_L\xi^{\mathrm{b}\to\mathrm{c}}. \tag{6.19}$$

Note that the initial stress and strain terms cancel one another out since $\mathrm{T}^{\mathrm{a}} = Y\mathrm{S}^{\mathrm{a}}$. The stress and strain state at the completion of the martensitic phase transformation is

$$\begin{aligned} \mathrm{T}^{\mathrm{c}} &= C_M(\theta_0 - M_f) \\ \mathrm{S}^{\mathrm{c}} &= \frac{C_M(\theta_0 - M_f)}{Y} + \mathrm{S}_L. \end{aligned} \tag{6.20}$$

The third step in the the process for the shape memory effect is to compute the strain when the material is unloaded in its full martensitic state. In this step of the process, the martensitic fraction remains at 1; therefore, the constitutive relationship is reduced to

$$\mathrm{T}^{\mathrm{c}\to\mathrm{d}} - C_M(\theta_0 - M_f) = Y\left[\mathrm{S}^{\mathrm{c}\to d} - \frac{C_M(\theta_0 - M_f)}{Y} - \mathrm{S}_L\right]. \tag{6.21}$$

This equation is reduced to

$$\mathrm{T}^{\mathrm{c}\to\mathrm{d}} = Y(\mathrm{S}^{\mathrm{c}\to\mathrm{d}} - \mathrm{S}_L), \tag{6.22}$$

illustrating that the material is linear elastic when it is unloaded in its martensitic state. The state of stress and strain at point d in Figure 6.1 is

$$\begin{aligned} \mathrm{T}^{\mathrm{d}} &= 0 \\ \mathrm{S}^{\mathrm{d}} &= \mathrm{S}_L. \end{aligned} \tag{6.23}$$

This is as expected, since the residual strain S_L was defined as the strain that resulted after unloading in the martensitic phase.

The final step in the process to return the shape memory material to the zero stress–strain state is a heating cycle in which the temperature is raised beyond A_f and then cooled to θ_0. No phase change will occur until $\theta = A_s$ at which the martensitic fraction will decrease and be expressed by the relationship

$$\xi^{\mathrm{d}\to\mathrm{e}} = \frac{1}{2}\left\{\cos\left[a_A\left(\theta - A_s\right)\right] + 1\right\}. \tag{6.24}$$

In this regime the constitutive relationship is reduced to

$$0 = \mathrm{S}^{\mathrm{d}\to\mathrm{e}} - \mathrm{S}_L - \mathrm{S}_L(\xi^{\mathrm{d}\to\mathrm{e}} - 1). \tag{6.25}$$

Equation (6.25) can be solved for the strain to yield

$$\mathrm{S}^{\mathrm{d}\to\mathrm{e}} = \mathrm{S}_L\xi^{\mathrm{d}\to\mathrm{e}}. \tag{6.26}$$

The endpoints and computations associated with the transitions are summarized in Table 6.2 assuming complete phase transformation due to induced stress and applied temperature.

Example 6.2 A Nitinol wire with the material properties shown in Table 6.1, an elastic modulus of 13 GPa, and a recovery strain of 7% is loaded to induce the shape memory effect with full phase transformation (points a to d in Figure 6.1). The initial temperature is 25°C and the material is initially in full austenitic phase. Determine the stress and strain at points a, b, c, and d of the stress–strain diagram.

Table 6.2 Endpoints and computations associated with the shape memory effect, assuming complete phase transformations

	T	S	θ	ξ
a	0	0	θ_0	0
a → b	$\mathrm{T}^{a\to b} = Y\,\mathrm{S}^{a\to b}$		θ_0	0
b	$C_M(\theta_0 - M_s)$	$\dfrac{C_M(\theta_0 - M_s)}{Y}$	θ_0	0
b → c	$C_M(\theta_0 - M_s) \to C_M(\theta_0 - M_f)$	$\dfrac{1}{Y}\mathrm{T}^{b\to c} + \mathrm{S}_L\xi^{b\to c}$	θ_0	$\dfrac{1}{2}\left\{\cos\left[a_M(\theta_0 - M_f) - \dfrac{a_M}{C_M}\mathrm{T}^{b\to c}\right] + 1\right\}$
c	$C_M(\theta_0 - M_f)$	$\dfrac{C_M(\theta_0 - M_f)}{Y} + \mathrm{S}_L$	θ_0	1
c → d	$\mathrm{T}^{c\to d} = Y\left(\mathrm{S}^{c\to d} - \mathrm{S}_L\right)$		θ_0	1
d	0	S_L	θ_0	1
d → e	0	$\mathrm{S}_L\xi^{d\to e}$	$A_s \to A_f$	$\dfrac{1}{2}\{\cos[a_A(\theta - A_s)] + 1\}$

Solution The initial temperature is greater than the martensitic start temperature and less than the austenitic finish temperature. The stress and strain at point b in the diagram are computed from

$$\begin{aligned} T^b &= (11.3\text{MPa}/^\circ\text{C})\,(25^\circ\text{C} - 23^\circ\text{C}) = 22.6\ \text{MPa} \\ S^b &= \frac{22.6\ \text{MPa}}{13,000\ \text{MPa}} = 0.001739 = 1739\ \mu\text{strain}. \end{aligned}$$

The stress–strain relationship during the austenite-to-martensite transition is obtained by computing the martensitic fraction as a function of the applied stress and then computing the strain. The bounds on the stress during the transition are

$$22.6\ \text{MPa} \leq T^{b\to c} \leq (11.3\text{MPa}/^\circ\text{C})\,(25^\circ\text{C} - 5^\circ\text{C}) = 226\ \text{MPa}.$$

As an example of the computation, compute the martensitic fraction at the stress of 100 MPa. First compute the martensitic fraction at this stress level:

$$\begin{aligned} \xi &= \frac{1}{2}\left\{\cos\left[(0.175^\circ\text{C}^{-1})\,(25^\circ\text{C} - 5^\circ\text{C}) - \left(\frac{0.175^\circ\text{C}^{-1}}{11.3\text{MPa}/^\circ\text{C}}\right)(100\ \text{MPa})\right] + 1\right\} \\ &= 0.3143. \end{aligned}$$

The strain is

$$S = \frac{100\ \text{MPa}}{13{,}000\ \text{MPa}} + (0.07)\,(0.3143) = 0.0297 = 2.97\%.$$

The stress and strain at point c are

$$\begin{aligned} T^c &= 226\ \text{MPa} \\ S^c &= \frac{226\ \text{MPa}}{13{,}000\ \text{MPa}} + (0.07)(1) = 8.74\%. \end{aligned}$$

The strain at the completion of unloading is equal to

$$\begin{aligned} T^d &= 0 \\ S^d &= 0.07. \end{aligned}$$

A plot of the stress–strain behavior illustrating the hysteresis loop is shown in Figure 6.6.

6.3.3 Modeling the Pseudoelastic Effect

The second common use of shape memory materials is the pseudoelastic effect illustrated in Figure 6.2. As discussed above, shape memory materials exhibit a

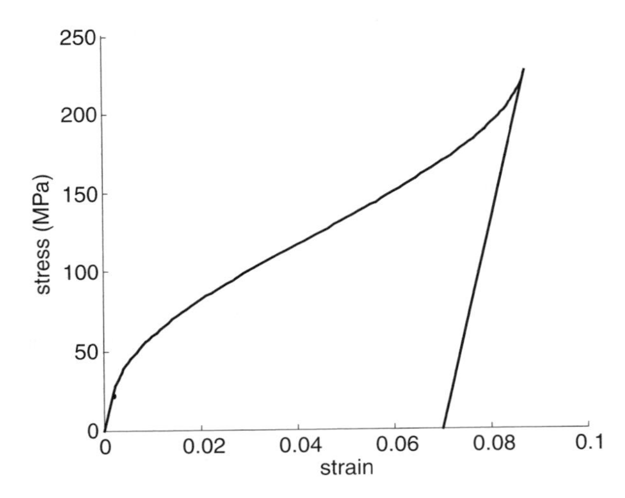

Figure 6.6 Stress–strain plot illustrating the shape memory effect for the material studied in Example 6.2.

pseudoelastic effect when they are loaded and unloaded at a temperature equal to or greater than their austenitic start temperature. Full phase transformation and complete hysteresis loops occur when the temperature is above the austenitic finish temperature.

As with the shape memory effect, the pseudoelastic effect is modeled by combining the constitutive equation with the kinetic law that relates the martensitic fraction to temperature and stress. Moreover, the first two processes (and computations) associated with the shape memory effect are identical to the process associated with computing the pseudoelastic effect. The initial conditions for the pseudoelastic effect are that the material is in full austenitic phase ($\xi = 0$) and that the temperature is equal to or greater than the austenitic finish temperature, $\theta_0 \geq A_f$. The processes associated with the pseudoelastic effect are:

1. The material is loaded in an isothermal state until a critical stress is reached that induces the start of martensitic phase transformation. The material is assumed to be linear elastic in this regime.
2. The material is loaded in an isothermal state until the martensitic phase transformation is complete. The stress–strain behavior of the material in this regime is assumed to be governed by a combination of the constitutive law, equation (6.5), and the kinetic law for austenitic-to-martensitic transformation.
3. The material is unloaded in an isothermal state until the austenitic-to-martensitic phase transformation is induced. The material is linear elastic in this regime.
4. The material is unloaded to zero stress in an isothermal state which completes the austenitic-to-martensitic phase transformation.

Note that in the pseudoelastic effect, the material is never heated; the applied stress induces both martensitic-to-austenitic and austenitic-to-martensitic phase transformations.

Since the first two processes of the pseudoelastic effect are identical to the processes associated with the shape memory effect, equations (6.14) to (6.20) represent the equations needed to compute the stress–strain behavior up until full martensitic phase transformation of the material (points a to c in Figure 6.2). The third step in the process is an unloading of the stress until the austenitic-to-martensitic phase transformation begins to occur. Assuming that point c in the diagram occurs at full Martensitic phase transformation, the stress bounds for the c $\rightarrow$ d transformation are

$$C_M(\theta_0 - M_f) \leq \mathrm{T}^{\mathrm{c}\rightarrow\mathrm{d}} \leq C_A(\theta_0 - A_s). \tag{6.27}$$

The stress–strain relationship in this regime is

$$\mathrm{T}^{\mathrm{c}\rightarrow\mathrm{d}} - C_M(\theta_0 - M_f) = Y\left[\mathrm{S}^{\mathrm{c}\rightarrow\mathrm{d}} - \frac{C_M(\theta_0 - M_f)}{Y} - \mathrm{S}_L\right], \tag{6.28}$$

which can be reduced to

$$\mathrm{T}^{\mathrm{c}\rightarrow\mathrm{d}} = Y(\mathrm{S}^{\mathrm{c}\rightarrow\mathrm{d}} - \mathrm{S}_L). \tag{6.29}$$

The stress–strain state at point d is obtained by combining equations (6.27) and (6.29):

$$\begin{aligned} \mathrm{T}^{\mathrm{d}} &= C_A\,(\theta_0 - A_s) \\ \mathrm{S}^{\mathrm{d}} &= \frac{C_A\,(\theta_0 - A_s)}{Y} + \mathrm{S}_L. \end{aligned} \tag{6.30}$$

The fourth step in the process is an unloading of the stress until the austenitic-to-martensitic phase transformation is completed. The stress in this regime is bounded by

$$C_A\,(\theta_0 - A_s) \leq \mathrm{T}^{\mathrm{d}\rightarrow\mathrm{e}} \leq C_A(\theta_0 - A_f), \tag{6.31}$$

and the martensitic fraction is determined from the expression

$$\xi^{\mathrm{d}\rightarrow\mathrm{e}} = \frac{1}{2}\left\{\cos\left[a_A\,(\theta - A_s) - \frac{a_A}{C_A}\mathrm{T}^{\mathrm{d}\rightarrow\mathrm{e}}\right] + 1\right\}. \tag{6.32}$$

The stress–strain relationship is obtained by substituting the stress and strain at point d into equation (6.5):

$$\mathrm{T}^{\mathrm{d}\rightarrow\mathrm{e}} - C_A\,(\theta_0 - A_s) = Y\left[\mathrm{S}^{\mathrm{d}\rightarrow\mathrm{e}} - \frac{C_A\,(\theta_0 - A_s)}{Y} - \mathrm{S}_L\right] - \mathrm{S}_L Y(\xi^{\mathrm{d}\rightarrow\mathrm{e}} - 1). \tag{6.33}$$

Equation (6.33) is solved for the strain:

$$\mathrm{S}^{\mathrm{d}\to\mathrm{e}} = \frac{\mathrm{T}^{\mathrm{d}\to\mathrm{e}}}{Y} + \mathrm{S}_L \xi^{\mathrm{d}\to\mathrm{e}}. \tag{6.34}$$

The stress and strain at point e are

$$\begin{aligned} \mathrm{T}^{\mathrm{e}} &= C_A (\theta_0 - A_s) \\ \mathrm{S}^{\mathrm{e}} &= \frac{\mathrm{T}^{\mathrm{e}}}{Y}. \end{aligned} \tag{6.35}$$

If the initial temperature is equal to the austenitic finish temperature, point e will have zero stress and zero strain. If $\theta_0 > A_s$, point e will have nonzero stress and strain. Unloading the material will produce linear elastic behavior to a state of zero stress and strain.

The steps associated with computing the stress–strain behavior for the pseudoelastic effect are shown in Table 6.3.

Example 6.3 A Nitinol wire with the material properties shown in Table 6.1, an elastic modulus of 13 GPa, and a recovery strain of 7% is loaded to induce a pseudoelastic effect with full phase transformation (points a to e in Figure 6.2). The initial temperature is 51°C and the material is initially in the full austenitic phase. Determine the stress and strain at points a, b, c, d, and e of the stress–strain diagram.

Solution The initial temperature is equal to the austenitic finish temperature. The stress and strain at point b in the diagram are computed from

$$\begin{aligned} \mathrm{T}^{\mathrm{b}} &= (11.3\mathrm{MPa}/^\circ\mathrm{C})\,(51^\circ\mathrm{C} - 23^\circ\mathrm{C}) = 316.4\ \mathrm{MPa} \\ \mathrm{S}^{\mathrm{b}} &= \frac{316.4\ \mathrm{MPa}}{13{,}000\ \mathrm{MPa}} = 0.0243 = 2.43\%. \end{aligned}$$

The stress–strain relationship during the austenite-to-martensite transition is obtained by computing the martensitic fraction as a function of the applied stress and then computing the strain. The bounds on the stress during the transition are

$$316.4\ \mathrm{MPa} \le \mathrm{T}^{\mathrm{b}\to\mathrm{c}} \le (11.3\ \mathrm{MPa}/^\circ\mathrm{C})\,(51^\circ\mathrm{C} - 5^\circ\mathrm{C}) = 519.8\ \mathrm{MPa}.$$

As an example of the computation, compute the martensitic fraction at a stress of 400 MPa. First compute the martensitic fraction at this stress level:

$$\begin{aligned} \xi &= \frac{1}{2}\left\{\cos\left[(0.175^\circ\mathrm{C}^{-1})\,(51^\circ\mathrm{C} - 5^\circ\mathrm{C}) - \left(\frac{0.175^\circ\mathrm{C}^{-1}}{11.3\ \mathrm{MPa}/^\circ\mathrm{C}}\right)(400\ \mathrm{MPa})\right] + 1\right\} \\ &= 0.3597. \end{aligned}$$

Table 6.3 End points and computations associated with the pseudoelastic effect, assuming complete phase transformations

	T	S	θ	ξ
a	0	0	θ_0	0
a → b	$\mathrm{T}^{a \to b} = Y \mathrm{S}^{a \to b}$		θ_0	0
b	$C_M(\theta_0 - M_s)$	$\dfrac{C_M(\theta_0 - M_s)}{Y}$	θ_0	0
b → c	$C_M(\theta_0 - M_s) \to C_M(\theta_0 - M_f)$	$\dfrac{1}{Y}\mathrm{T}^{b \to c} + \mathrm{S}_L \xi^{b \to c}$	θ_0	$\dfrac{1}{2}\left\{\cos\left[a_M(\theta_0 - M_f) - \dfrac{a_M}{C_M}\mathrm{T}^{b \to c}\right] + 1\right\}$
c	$C_M(\theta_0 - M_f)$	$\dfrac{C_M(\theta_0 - M_f)}{Y} + \mathrm{S}_L$	θ_0	1
c → d	$\mathrm{T}^{c \to d} = Y\left(\mathrm{S}^{c \to d} - \mathrm{S}_L\right)$		θ_0	1
d	$C_A(\theta_0 - A_s)$	$\dfrac{C_A(\theta_0 - A_s)}{Y} + \mathrm{S}_L$	θ_0	1
d → e	$C_A(\theta_0 - A_s) \to C_A(\theta_0 - A_f)$	$\dfrac{\mathrm{T}^{d \to e}}{Y} + \mathrm{S}_L \xi^{d \to e}$	θ_0	$\dfrac{1}{2}\left\{\cos\left[a_A(\theta - A_s) - \dfrac{a_A}{C_A}\mathrm{T}^{d \to e}\right] + 1\right\}$
e	$C_A(\theta_0 - A_s)$	$\dfrac{\mathrm{T}^{e}}{Y}$	θ_0	0

The strain is

$$S = \frac{400 \text{ MPa}}{13{,}000 \text{ MPa}} + (0.07)(0.3597) = 5.59\%.$$

The stress and strain at point c are

$$T^c = 519.8 \text{ MPa}$$

$$S^c = \frac{519.8 \text{ MPa}}{13{,}000 \text{ MPa}} + (0.07)(1) = 11\%.$$

The regime between points c and d is characterized by linear elastic behavior. The stress and strain at point d are

$$\begin{aligned} T^d &= (4.5 \text{ MPa}/^\circ\text{C})(51^\circ\text{C} - 29^\circ\text{C}) = 99 \text{ MPa} \\ S^d &= \frac{99 \text{ MPa}}{13{,}000 \text{ MPa}} + 0.07 = 7.76\%. \end{aligned} \tag{6.36}$$

Since the temperature is equal to the austenitic finish temperature, the final stress–strain state will be equal to 0. The stress–strain behavior at intermediate values is determined by computing the martensitic fraction at a specified stress level and then computing the strain. An example of the computation for T = 50 MPa is

$$\begin{aligned} \xi &= \frac{1}{2}\left\{\cos\left[(0.143^\circ\text{C}^{-1})(51^\circ\text{C} - 29^\circ\text{C}) - \left(\frac{0.143^\circ\text{C}^{-1}}{4.5 \text{ MPa}/^\circ\text{C}}\right)(50 \text{ MPa})\right] + 1\right\} \\ &= 0.5068. \end{aligned}$$

The strain at this value of martensitic fraction is

$$S = \frac{50 \text{ MPa}}{13{,}000 \text{ MPa}} + (0.07)(0.0955) = 3.93\%.$$

The temperature is equal to the austenitic finish temperature, so the state at point e in the diagram will be equal to a stress and strain of zero. The stress–strain diagram for this example is plotted in Figure 6.7.

The analyses so far in Section 6.3 have assumed that the material undergoes full phase transformations due to the application of stress or heat. In certain instances the phase transformation can be incomplete and the stress–strain behavior will be analyzed from an initial condition in which the fraction of martensite is neither zero nor 1. In the case in which the austenitic-to-martensitic transformation begins from the state (ξ_A, θ_A), where $0 < \xi_A < 1$, we assume that no new martensite phase will be formed until the material is cooled to a temperature below M_s. At this point the

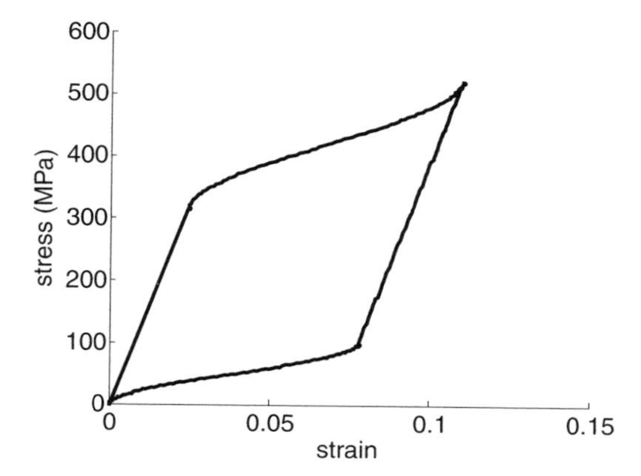

Figure 6.7 Stress–strain plot illustrating the pseudoelastic effect for the material studied in Example 6.3.

kinetic law for transformation is

$$\xi_{A\to M} = \frac{1-\xi_A}{2}\left\{\cos\left[a_M(\theta - M_f) - \frac{a_M}{C_M}\mathrm{T}\right]\right\} + \frac{1+\xi_A}{2}. \tag{6.37}$$

Equation (6.37) reduces to equation (6.13) if the martensitic fraction is initially 0. Similarly, if the $M \to A$ transformation starts from a state of (ξ_M, θ_M), the kinetic law for transformation is

$$\xi_{M\to A} = \frac{\xi_M}{2}\left\{\cos\left[a_A\,(\theta - A_s) - \frac{a_A}{C_A}\mathrm{T}\right] + 1\right\}. \tag{6.38}$$

This kinetic law also reduces to the original definition, equation (6.12), if the martensitic fraction is initially 1.

One of the assumptions of the constitutive law expressed in equation (6.5) is that the material properties are constants. For a number of typical shape memory alloys (e.g., Nitinol) it is known that the material properties also vary as a function of the martensitic fraction. Most notable is the fact that the elastic modulus of the material can vary by a factor of 2 to 3 between the austentic and martensitic phases. Generally, the material has a lower elastic modulus in the martensitic phase than it does in the austenitic phase.

The variation in the material parameters as a function of martensitic fraction is modeled by writing the modulus, Y, and the transformation coefficient, Ω, as functions of ξ in equation (6.2):

$$\mathrm{T} - \mathrm{T}_0 = Y(\xi)\mathrm{S} - Y(\xi_0)\mathrm{S}_0 + \Omega(\xi)\xi - \Omega(\xi_0)\xi_0. \tag{6.39}$$

For the case in which the elastic modulus is assumed to be *a linear function* of the martensitic fraction,

$$Y(\xi) = Y_A + \xi\,(Y_M - Y_A)\,, \tag{6.40}$$

where Y_A is the elastic modulus in the austenitic state and Y_M is the elastic modulus in the martensitic state:

$$\Omega(\xi) = -\mathrm{S}_L Y(\xi). \tag{6.41}$$

Combining equations (6.39) and (6.41) produces

$$\mathrm{T} - \mathrm{T}_0 = Y(\xi)\,(\mathrm{S} - \mathrm{S}_L\xi) - Y(\xi_0)\,(\mathrm{S}_0 - \mathrm{S}_L\xi_0)\,. \tag{6.42}$$

Examining the steps associated with computing the stress–strain behavior of the shape memory material listed in Tables 6.2 and 6.3, it becomes clear that the variation in modulus will affect the computation of the strain for both the linear and nonlinear portions of the mechanical behavior.

Example 6.4 Repeat Example 6.3 assuming that the elastic modulus of the shape memory alloy is 32.5 GPa when the material is in full austenite phase and decreases linearly as a function of ξ to 13 GPa when the material transforms to full martensite.

Solution The methodology for computing the stress and strain at each of the critical points a, b, c, and d is the same as in Example 6.3 except for the fact that the elastic modulus is now a function of the martensitic fraction. Assuming a linear relationship between the elastic modulus and the martensitic fraction, the expression for the elastic modulus is

$$Y(\xi) = 32.5 + \xi\,(13 - 32.5)\ \text{GPa}.$$

The computations for the stress and strain at point b are

$$\begin{aligned}\mathrm{T}^{\mathrm{b}} &= (11.3\text{MPa}/^\circ\text{C})\,(51^\circ\text{C} - 23^\circ\text{C}) = 316.4\ \text{MPa}\\ \mathrm{S}^{\mathrm{b}} &= \frac{316.4\ \text{MPa}}{Y(0)} = \frac{316.4\ \text{MPa}}{32{,}500\ \text{MPa}} = 0.0097 = 0.97\%.\end{aligned}$$

The expression for the elastic modulus does not change the computation of the martensitic fraction at a particular stress, but the strain associated with the computed martensitic fraction does change according to

$$\mathrm{S} = \frac{400\ \text{MPa}}{32{,}500 + (0.3597)\,(-19{,}500)\ \text{MPa}} + (0.07)\,(0.3597) = 4.09\%.$$

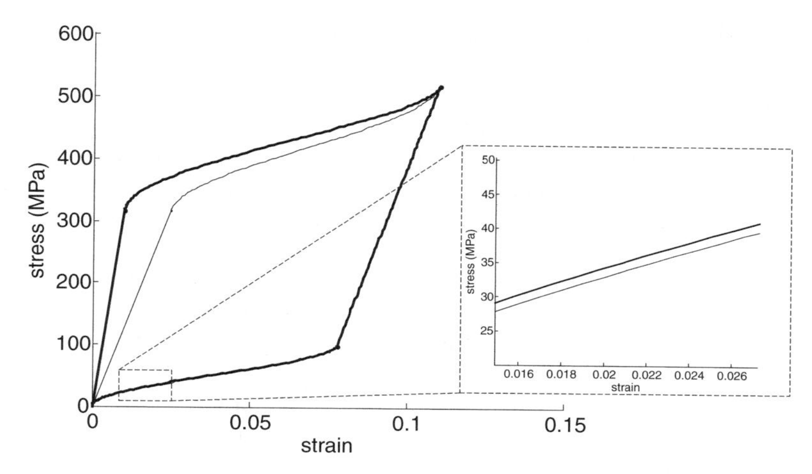

Figure 6.8 Stress–strain behavior for the material studied in Examples 6.3 and 6.4 with nonconstant modulus properties. The darker line is for the material with nonconstant properties, the lighter line is for the material with constant modulus properties.

Note that the strain is lower due to the fact that the material has a higher elastic modulus when it is not in full martensite phase.

The strain computed at points c and d is equal to the values computed in Example 6.3 since the elastic modulus is equal to 13 GPa when the material is in the full Martensitie phase. The computation of the strain does change in the process from points d to e due to the change in modulus. At the value of $\xi = 0.0955$ computed for a value of T = 50 MPa, the strain is

$$S = \frac{50 \text{ MPa}}{32,500 + (0.5068)\,(-19{,}500) \text{ MPa}} + (0.07)\,(0.5068) = 3.77\%.$$

The stress and strain at point e will both be zero, due to the fact the initial temperature is equal to the austenitic finish temperature.

Figure 6.8 is a plot of the stress–strain behavior for material with nonconstant modulus properties compared to the analysis when the elastic modulus is assumed to be constant. The assumption of nonconstant elastic modulus has the most impact in the loading of material in the austenitic state and the nonlinear stress–strain behavior that occurs due to austenitic-to-martensitic phase transformation. There is a slight difference in the stress–strain computations during the martensitic-to-austenitic phase transformation that occurs when the material is unloaded. This difference is highlighted in the insert of Figure 6.8.

6.4 MULTIVARIANT CONSTITUTIVE MODEL

The physical basis of the shape memory effect and the pseudoelastic effect was discussed in Section 6.2. The fundamental process associated with these two effects is conversion of the austenitic phase in the material to the martensitic phase, and vice versa. These phase changes are induced by heating of the material or through the application of stress due to the relationship between the critical temperatures with stress.

In Section 6.2 we introduced that there are two variants of martensite that can exist in the material. The first type is obtained by cooling the material in an unstressed state from the austentic phase to a temperature below M_s. The cooling process causes the transformation to a form of martensite that exhibits *multiple* variants. Cooling the material to below M_f produces a full transformation to martensite in its multiple-variant form. In contrast, stress-induced martensite is a single-variant form that is produced when the material is placed under load. As discussed in Section 6.2, the shape memory effect can be induced through transformation of multiple-variant temperature-induced martensite to single-variant stress-induced martensite in the same manner as the transformation from austenite to martensite.

The model discussed so far does not have the capability of modeling this aspect of the shape memory effect. The parameter that models the phase transformation, ξ, is simply the fraction of martensite in the material and does differentiate between *types* of martensite. This limits the applicability of the model to temperatures that are greater than the martensitic start temperature of the shape memory material, $\theta_0 > M_s$.

To effectively model the transformation from temperature-induced martensite to stress-induced martensite, the martensitic fraction is decomposed further into a summation of two variables,

$$\xi = \xi_S + \xi_T, \tag{6.43}$$

where ξ_S is the fraction of stress-induced (single-variant) martensite in the material and ξ_T is the fraction of temperature-induced (multiple variant) martensite in the material. To capture the effects of temperature-induced martensite, the constitutive law, equation (6.5), is rewritten to include a transformation coefficient associated with ξ_T:

$$\mathrm{T} - \mathrm{T}_0 = Y(\mathrm{S} - \mathrm{S}_0) + \Omega_S(\xi_S - \xi_{S0}) + \Omega_T(\xi_T - \xi_{T0}). \tag{6.44}$$

Expressions for the transformation coefficients are obtained by considering specific cases of loading and unloading. Consider a material specimen that is in the unloaded state at zero strain, $\mathrm{T}_0 = \mathrm{S}_0 = 0$, in full austenite such that $\xi_{S0} = \xi_{T0} = 0$. If the material is loaded through the austenitic-to-martensitic phase transformation and then unloaded to a point of zero stress but residual strain $\mathrm{S} = \mathrm{S}_L$, the martensitic fractions will be $\xi_S = 1$, $\xi_T = 0$, and the constitutive equation is written

$$0 - 0 = Y(\mathrm{S}_L - 0) + \Omega_S(1 - 0) + \Omega_T(0 - 0). \tag{6.45}$$

This expression is solved to yield

$$\Omega_S = -S_L Y. \tag{6.46}$$

Now consider the same final loading conditions except assume that the material is in a state that consists of full temperature-induced martensite, $\xi_{S0} = 0, \xi_{T0} = 1$:

$$0 - 0 = Y(S_L - 0) + \Omega_S(1 - 0) + \Omega_T(0 - 1). \tag{6.47}$$

Combining equations (6.46) and (6.47), it is clear that

$$\Omega_T = 0, \tag{6.48}$$

which results in the constitutive relationship

$$T - T_0 = Y(S - S_0) - S_L Y(\xi_S - \xi_{S0}). \tag{6.49}$$

The significance of this change in the constitutive law is that the shape memory effect can be induced through either transformation of austenite to stress-induced martensite, or *changes in temperature-induced martensite to stress-induced martensite*. The transformation of austenite to stress-induced martensite is identical to the process studied earlier in this section, while the ability to model the transformation of temperature-induced martensite to stress-induced martensite allows the constitutive model to be used for temperatures below the martensitic start temperature.

Decomposition of the martensitic fraction into temperature- and stress-induced martensite also requires modification to the kinetic laws associated with phase transformations. Figure 6.9 illustrates the relationship between the stress and the critical

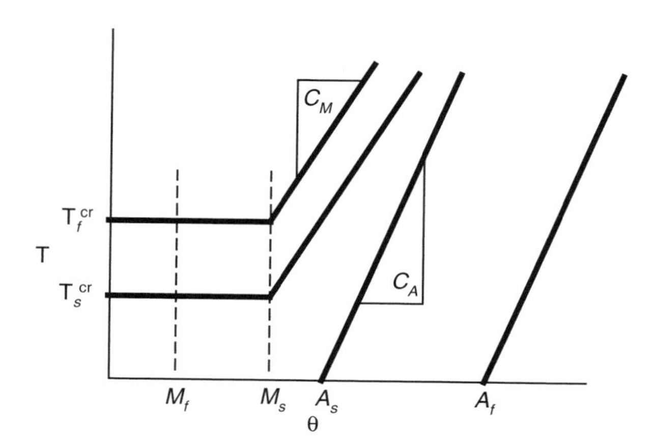

Figure 6.9 Critical temperatures as a function of stress that incorporates constant values below M_S.

temperatures used earlier in the chapter to model stress-induced transformation. Experimental evidence demonstrates that a more accurate description of this relationship is required when the material temperature is initially below the martensitic start temperature. At temperatures below M_s, the critical stresses are assumed to be constant values, as shown in Figure 6.9. Above M_s the critical stresses increase linearly with the slope C_M for M_f and M_s and the slope C_A for A_s and A_f.

The transformation equations also require modification to account for the transformation between the different types of martensite. The kinetic law for conversion from martensite to austenite is

$$\begin{aligned} \theta > A_s \quad &\text{and} \quad C_A(\theta - A_f) < \mathrm{T} < C_A(\theta - A_s) \\ \xi &= \frac{\xi_0}{2}\left\{\cos\left[a_A\left(\theta - A_s - \frac{\mathrm{T}}{C_A}\right)\right] + 1\right\} \\ \xi_S &= \xi_{S0} - \frac{\xi_{S0}}{\xi_0}(\xi_0 - \xi) \\ \xi_T &= \xi_{T0} - \frac{\xi_{T0}}{\xi_0}(\xi_0 - \xi). \end{aligned} \tag{6.50}$$

The kinetic laws of transformation from austenite to martensite become more elaborate, due to the fact that the fraction of stress- and temperature-induced martensite must also be computed during the process. For temperatures above M_s,

$$\begin{aligned} \theta > M_s \quad &\text{and} \quad \mathrm{T}_s^{cr} + C_M(\theta - M_s) < \mathrm{T} < \mathrm{T}_f^{cr} + C_M(\theta - M_s) \\ \xi_S &= \frac{1 - \xi_{S0}}{2}\cos\left\{\frac{\pi}{\mathrm{T}_s^{cr} - \mathrm{T}_f^{cr}}\left[\mathrm{T} - \mathrm{T}_f^{cr} - C_M(\theta - M_s)\right]\right\} + \frac{1 + \xi_{S0}}{2} \\ \xi_T &= \xi_{T0} - \frac{\xi_{T0}}{1 - \xi_{S0}}(\xi_S - \xi_{S0}) \end{aligned} \tag{6.51}$$

and for temperatures below M_s,

$$\begin{aligned} \theta < M_s \quad &\text{and} \quad \mathrm{T}_s^{cr} < \mathrm{T} < \mathrm{T}_f^{cr} \\ \xi_S &= \frac{1 - \xi_{S0}}{2}\cos\left[\frac{\pi}{\mathrm{T}_s^{cr} - \mathrm{T}_f^{cr}}\left(\mathrm{T} - \mathrm{T}_f^{cr}\right)\right] + \frac{1 + \xi_{S0}}{2} \\ \xi_T &= \xi_{T0} - \frac{\xi_{T0}}{1 - \xi_{S0}}(\xi_S - \xi) + \Delta_{T\xi}. \end{aligned} \tag{6.52}$$

The variable $\Delta_{T\xi}$ is defined as

$$\Delta_{T\xi} = \frac{1 - \xi_{T0}}{2}\{\cos[a_M(\theta - M_f)] + 1\} \tag{6.53}$$

Table 6.4 Shape memory alloy material properties

Elastic Properties	Transformation Temperatures	Transformation Constants	Maximum Recoverable Strain
$Y_A = 67$ GPa	$M_f = 9°$C	$C_M = 8$ MPa/°C	$S_L = 0.07$
$Y_M = 26$ GPa	$M_s = 18°$C	$C_A = 14$ MPa/°C	
	$A_s = 35°$C	$T_s^{cr} = 100$ MPa	
	$A_f = 49°$C	$T_f^{cr} = 170$ MPa	

if $M_f < \theta < M_s$ and $\theta < \theta_0$. Otherwise, $\Delta_{T\xi} = 0$.

Equations (6.50) to (6.53) are used to compute the transformation of martensitic fraction as a function of temperature and stress. An example of this computation for a material that is initially in a state of temperature-induced martensite follows.

Example 6.5 A shape memory alloy with material properties listed in Table 6.4 is initially at a temperature of $\theta_0 = 5°$C with $\xi_{T0} = 1$ and $\xi_{S0} = 0$. The material is assumed initially to be at a state of zero stress and zero strain. (a) Compute the strain when the applied stress is 90 MPa. (b) Compute the martensitic fractions when the applied stress is 120 MPa.

Solution (a) The critical temperature for this material is listed as 100 MPa (see Table 6.4); therefore, no phase transformation will occur until the stress becomes greater than 100 MPa. When the stress is below 100 MPa, the material will exhibit linear elastic stress–strain behavior; therefore,

$$\mathrm{S} = \frac{\mathrm{T}}{Y_M} = \frac{100\text{ MPa}}{26{,}000\text{ MPa}} = 0.0038 = 0.38\%.$$

(b) When the applied stress becomes greater than T_s^{cr}, the fractions of stress-induced and temperature martensite will begin to change according to equation (6.52). For an applied stress of 120 MPa, the fraction of stress-induced martensite is

$$\begin{aligned}\xi_S &= \frac{1-0}{2}\cos\left[\frac{\pi}{100\text{ MPa} - 170\text{ MPa}}(120\text{ MPa} - 170\text{ MPa})\right] + \frac{1+0}{2}\\ &= 0.188.\end{aligned}$$

The fraction of temperature-induced martensite is

$$\begin{aligned}\xi_T &= 1 - \frac{1}{1-0}(0.188 - 0) + 0\\ &= 0.812.\end{aligned}$$

Note that $\Delta_{T\xi} = 0$ because $\theta_0 < M_f$.

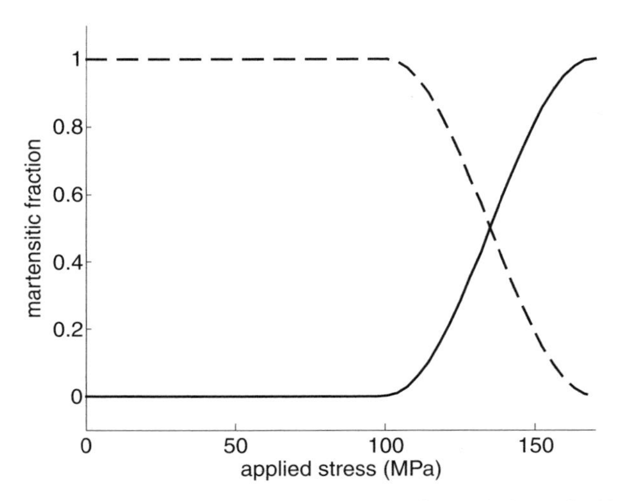

Figure 6.10 Transformation of the martensitic fraction for the analysis studied in Example 6.5: ξ_S (solid), ξ_T (dashed).

Phase transformation will continue in Example 6.5 until the applied stress is equal to T_f^{cr}, at which point all of the temperature-induced martensite will be transformed into stress-induced martensite. This is illustrated in Figure 6.10, which illustrates that no martensitic transformation will occur until the applied stress is greater than 100 MPa. When the applied stress exceeds T_s^{cr}, the temperature-induced martensite will begin transforming to stress-induced martensite according to equation (6.53). When the applied stress is equal to T_f^{cr}, the transformation to stress-induced martensite will be complete.

The stress–strain behavior of the shape memory alloy is computed by coupling the kinetic laws to the constitutive relationships for the material. Equation (6.49) is the constitutive relationship when the material has constant material properties. An analogous constitutive relationship exists when the material properties are a function of the martensitic fraction. Assuming that the elastic modulus exhibits a linear variation as a function of martensitic fraction, as stated in equation (6.40), the constitutive equations are

$$T - T_0 = Y(\xi)S - Y(\xi_0)S_0 + \Omega(\xi)\xi_S - \Omega(\xi_0)\xi_{S0}, \tag{6.54}$$

where $\Omega(\xi) = -S_L Y(\xi)$. Once the martensitic fractions are computed using the transformation laws, equation (6.54) is used to compute the strain for a specified value of stress. This is discussed in the following example.

Example 6.6 The material specimen studied in Example 6.5 is loaded to a stress of 120 MPa. Compute the strain at this value of stress.

Solution The martensitic fractions computed in Example 6.5 for this value of stress are

$$\xi_S = 0.188$$
$$\xi_T = 0.812.$$

Solving equation (6.54) for the strain assuming that $\mathrm{T}_0 = \mathrm{S}_0 = 0$ yields

$$\mathrm{S} = \frac{\mathrm{T} - \Omega(\xi)\xi_S}{Y(\xi)} = \frac{\mathrm{T}}{Y(\xi)} + \mathrm{S}_L \xi_S.$$

Substituting in the values from Table 6.4,

$$\mathrm{S} = \frac{120 \text{ MPa}}{26{,}000 \text{ MPa}} + (0.07)(0.188)$$
$$= 0.0178 = 1.78\%.$$

The elastic modulus used in the computation is equal to the modulus in the full marensite phase because the modulus variation is a function of the *total martensitic fraction*, ξ, not only the fraction of the stress-induced martensite, ξ_S.

The stress–strain behavior is computed by repeating the computation in Example 6.5 over the range of applied stress up to 170 MPa. The resulting stress-strain behavior is shown in Figure 6.11, illustrating the shape memory effect induced by the transformation of temperature-induced martensite to stress-induced martensite.

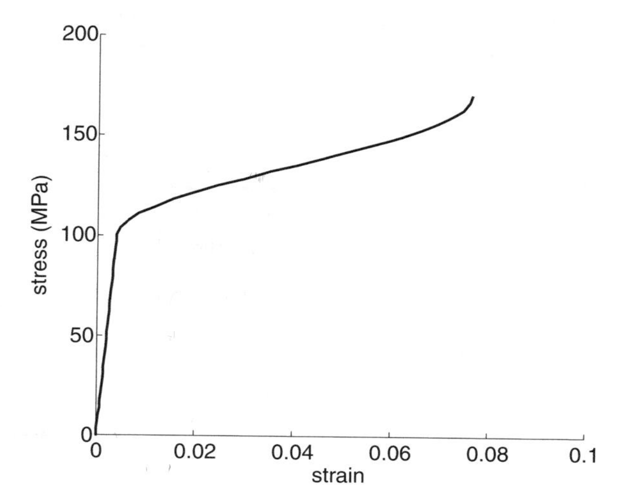

Figure 6.11 Stress–strain behavior for the shape memory response studied in Examples 6.5 and 6.6.

6.5 ACTUATION MODELS OF SHAPE MEMORY ALLOYS

So far in this chapter the discussion has focused on understanding the constitutive behavior of shape memory materials. The critical features of the constitutive behavior include the relationship between temperature, stress, and strain through the transformation of martensite to austenite, and vice versa. The end result of this analysis is a computation of the stress–strain characteristics of the material as a function of initial conditions and loading history.

Understanding the constitutive properties of shape memory alloys enables the analysis of actuation systems based on SMA materials. As with piezoelectric materials, shape memory materials are used for motion control systems. Unlike piezoelectric materials, though, which are generally limited to strains on the order of 0.1 to 0.2% (or $\approx$ 0.5% for single-crystal piezoelectric materials), shape memory materials can generate strains greater than 1%, due to their ability to recover large strains. Furthermore, they can generate large stress upon recovery, due to their high modulus. Thus, they can be used to generate large stress and strain, although, as we shall see later in the chapter, their speed is generally limited, due to the time required to heat the material to cause the strain recovery.

The primary variable of importance for shape memory alloy actuators is the residual strain caused by the shape memory effect. As shown in Figure 6.1 and analyzed in Section 6.3.2, loading the material can cause the shape memory effect, which results in a residual strain. The amount of residual strain is a function of the martensitic fraction of the material after unloading. A martensitic fraction of 1 will cause the full shape memory effect and result in a residual strain equal to the maximum recoverable strain, S_L. Heating the material beyond the austenitic start temperature will cause strain recovery due to the transformation of stress-induced martensite to austenite.

Consider the case of an SMA material that has been loaded and unloaded to produce the shape memory effect shown in Figure 6.12. If the load is high enough to induce full martensitic phase transformation, the residual strain will be equal to the maximum

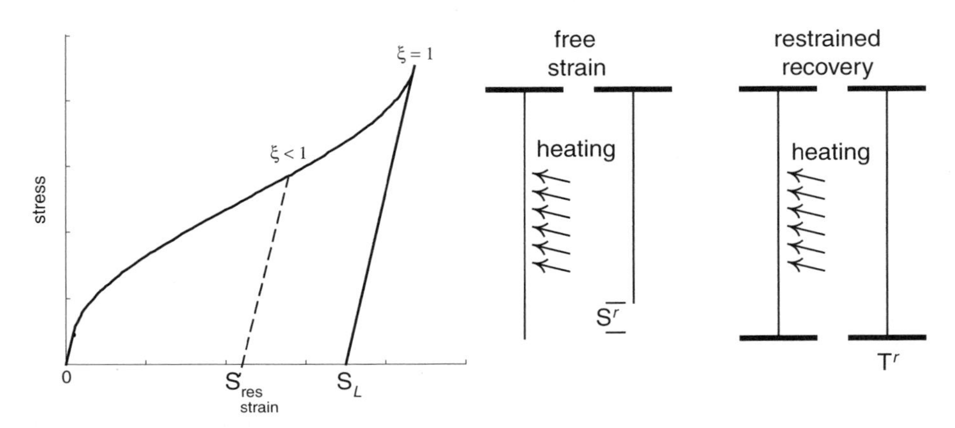

Figure 6.12 Free strain and restrained recovery for a shape memory material.

recoverable strain, S_L, whereas if the load is smaller, only partial transformation will occur and the residual strain will be some fraction of the maximum recoverable strain.

The constitutive relationships for the shape memory material can be solved to show that

$$\xi = \frac{S_{res}}{S_L} \tag{6.55}$$

at the completion of the loading cycle. There are two mechanical boundary conditions of importance when the material is heated. *Free strain recovery* occurs when the mechanical boundary condition is not constrained. In this case the material will contract as a function of the temperature. *Restrained recovery* occurs when the mechanical boundary condition is constrained such that the strain is zero when the material is heated. This will cause the generation of a residual stress, T_{res}, in the material as a function of the temperature. In the next two sections we analyze these cases to determine expressions for the strain as a function of temperature for both the free strain and restrained recovery cases. In both instances we utilize the model developed in Section 6.3.2 and we do not differentiate between temperature- and stress-induced martensite.

6.5.1 Free Strain Recovery

In the case of free strain recovery, we assume that the material has been loaded and unloaded such that the shape memory effect has resulted in a residual strain S_{res} as shown in Figure 6.12. Assuming that the stress is zero, equation (6.42) is reduced to

$$S = S_L \xi \tag{6.56}$$

when solved for the strain. Assuming that the initial temperature is less than A_s, phase transformation, and hence the strain recovery, will not occur until the temperature reaches A_s. For an initial martensitic fraction of $\xi_M = S_{res}/S_L$, equation (6.38) can be substituted into equation (6.56) to produce

$$S^r = \frac{S_{res}}{2}\left\{\cos\left[a_A\left(\theta - A_s\right)\right] + 1\right\}. \tag{6.57}$$

Equation (6.57) is valid for temperatures above the martensitic start temperature, M_s. The variable S^r is used to denote the recovery strain associated with the heating process.

6.5.2 Restrained Recovery

In the case of restrained recovery, it is assumed that the strain remains constant during the heating process and a recovery stress, T^r, is generated due to the mechanical constraint on the shape memory material. Simplifying equation (6.42) for the case of zero change in strain, the resulting expression is

$$\mathrm{T}^r = \frac{Y(\xi)\mathrm{S}_{\mathrm{res}}}{2}\left\{1 - \cos\left[a_A(\theta - A_s) - \frac{a_A}{C_A}\mathrm{T}^r\right]\right\}, \tag{6.58}$$

where $Y(\xi)$ is assumed to be a linear function of the martensitic fraction as defined by equation (6.40). Equation (6.58) is an iterative equation since the recovery stress appears on both the left-hand side and inside the argument of the cosine function. In general, it can be solved by choosing a temperature and iterating to determine the stress that corresponds to the temperature.

Example 6.7 A shape memory material with properties defined in Table 6.4 has been loaded and unloaded to produce a residual strain of 2%, due to the shape memory effect. Compute (a) the free strain recovery and (b) the recovery stress in restrained recovery if the material is heated to 40°C.

Solution (a) The free strain recovery is computed using equation (6.57). The residual strain is stated in the problem, $\mathrm{S}_{\mathrm{res}} = 0.02$, and the temperature is specified to be 40°C. The material property a_A is computed using equation (6.7):

$$a_A = \frac{\pi}{49^\circ\mathrm{C} - 35^\circ\mathrm{C}} = 0.224^\circ\mathrm{C}^{-1}.$$

Substituting this property into equation (6.57) yields

$$\begin{aligned}\mathrm{S}^r &= \frac{0.02}{2}\{\cos[(0.224^\circ\mathrm{C}^{-1})(40^\circ\mathrm{C} - 35^\circ\mathrm{C})] + 1\} \\ &= 0.0144 = 1.44\%.\end{aligned}$$

(b) The recovery stress for restrained recovery is computed using equation (6.58). Substituting the specified temperature into the equation yields

$$\begin{aligned}\mathrm{T}^r &= 0.01Y(\xi)\left\{1 - \cos\left[(0.224^\circ\mathrm{C}^{-1})(40^\circ\mathrm{C} - 35^\circ\mathrm{C}) - \frac{0.224^\circ\mathrm{C}^{-1}}{14\ \mathrm{MPa}/^\circ\mathrm{C}}\mathrm{T}^r\right]\right\} \\ &= 0.01Y(\xi)[1 - \cos(1.12 - 0.016\mathrm{T}^r)].\end{aligned}$$

The expression for the elastic modulus is given by equation (6.40):

$$Y(\xi) = 67{,}000 - 41{,}000\xi,$$

where the martensitic fraction is obtained from equation (6.38):

$$\xi = \frac{0.02/0.07}{2}[\cos(1.12 - 0.016\mathrm{T}^r) + 1].$$

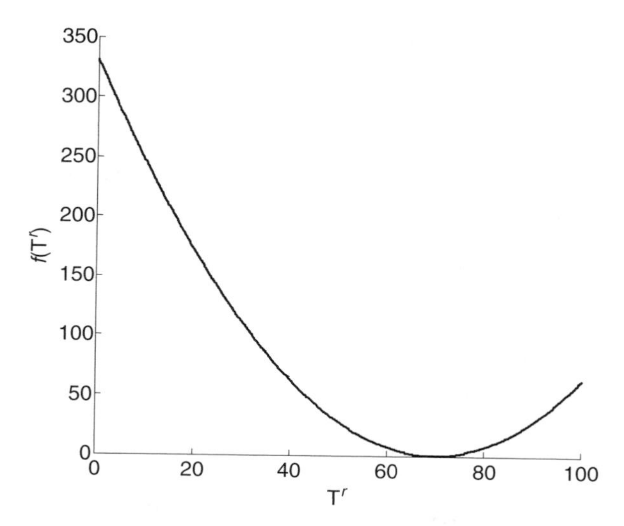

Figure 6.13 Graphical approach to computing the recovery stress for a specified temperature value.

The recovery stress at 40°C is obtained by choosing a value for T^r, computing ξ and $Y(\xi)$, and then substituting into the expression for T^r to determine if the expression is satisfied. This can be done graphically by forming a function

$$f(T^r) = 0.01Y(\xi)[1 - \cos(1.12 - 0.016T^r)] - T^r$$

and then plotting $f(T^r)$ over a range of stress values. The result of this method is shown in Figure 6.13 over the range 0 to 100 MPa. Finding the minimum on the plot yields a recovery stress of 70 MPa for a temperature of 40°C.

The process described in Example 6.7 can be repeated for each value of the temperature to determine the free strain recovery or restrained recovery stress over a range of θ. In a manner similar to the use of free deflection and blocked force for a piezoelectric actuator, this information could be used to determine the applicability of a shape memory alloy for a particular application.

6.5.3 Controlled Recovery

The free strain and restrained recovery studied earlier are analogous to the case of blocked stress and free strain studied for piezoelectric actuators. An intermediate condition (similar to the one studied for piezoelectric actuators) is a mechanical boundary condition that consists of a linear spring of stiffness k. In this case, summing forces produces a relationship between stress and strain,

$$\mathrm{T}^{r} - \mathrm{T}_0 = \frac{kL}{A}(\mathrm{S}_{\mathrm{res}} - \mathrm{S}^{r}), \tag{6.59}$$

where L is the length of the shape memory wire and A is the cross-sectional area. Substituting this relationship into equation (6.42) and solving for the stress yields

$$\left[1 + \frac{AY(\xi)}{kL}\right](\mathrm{T}^{r} - \mathrm{T}_0) = -\mathrm{S}_L Y(\xi)(\xi - \xi_0). \tag{6.60}$$

As in the previous case of restrained recovery, the solution to equation (6.60) is obtained in an interative manner because the stress is on both the left-hand side and in the argument of the martensitic fraction.

6.6 ELECTRICAL ACTIVATION OF SHAPE MEMORY ALLOYS

Until now in our discussion of shape memory materials it has been assumed that temperature and stress are the two parameters that determine the state of the material. In certain instances it is possible to control the temperature of the material directly to induce strain recovery. In other instances, though, the temperature of the material must be controlled indirectly through the application of an electric current to the material to induce heating. The heating, in turn, will increase the temperature of the material and induce strain recovery in the manner discussed Section 6.5.

A common model of the heat transfer associated with electrical heating (also known as *Joule heating*) of the wire is

$$(\rho A) c_p \frac{d\theta(t)}{dt} = i^2 R - h_c A_c \left[\theta(t) - \theta_\infty\right], \tag{6.61}$$

where ρ is the density of the shape memory material, A is the cross-sectional area, and c_p is the specific heat. The current is denoted i and the resistance per unit length of the material is R. The parameters h_c is the heat transfer coefficient and A_c is the circumferential area of the unit length of wire. The ambient temperature is denoted θ_∞.

Rewriting equation (6.61) in the form

$$\frac{d\theta(t)}{dt} + \frac{h_c A_c}{\rho A c_p}\theta(t) = \frac{R}{\rho A c_p} i^2 + \frac{h_c A_c}{\rho A c_p}\theta_\infty, \tag{6.62}$$

and assuming that the current and ambient temperature are constants, the solution to this differential equation is

$$\theta(t) - \theta_\infty = \frac{R}{h_c A_c}\left(1 - e^{-t/t_h}\right) i^2 + (\theta_o - \theta_\infty) e^{-t/t_h}. \tag{6.63}$$

The time constant associated with the heat transfer process is

$$t_h = \frac{\rho A c_p}{h_c A_c}. \tag{6.64}$$

Equation (6.63) is valid for any initial temperature. When the initial temperature is equal to the ambient temperature and the current is constant, which is often the case during heating for strain recovery, the temperature rise is modeled as

$$\theta(t) - \theta_\infty = \frac{R}{h_c A_c}\left(1 - e^{-t/t_h}\right) i^2. \tag{6.65}$$

The steady-state temperature, θ_{ss}, is obtained by letting the exponential term go to zero,

$$\theta_{ss} = \frac{R}{h_c A_c} i^2 + \theta_\infty. \tag{6.66}$$

The temperature will reach approximately 95% of the steady-state value when $t = 3t_h$. The time required to reach a desired temperature θ_d can be solved for explicitly from equation (6.65):

$$t_d = -t_h \ln\left(1 - \frac{\theta_d - \theta_\infty}{i^2}\frac{h_c A_c}{R}\right) \qquad \text{heating.} \tag{6.67}$$

Equation (6.67) can be combined with equation (6.66) to yield an expression for the time required in terms of the temperature differences:

$$t_d = -t_h \ln \frac{\theta_{ss} - \theta_d}{\theta_{ss} - \theta_\infty} \qquad \text{heating.} \tag{6.68}$$

A model for the cooling of a shape memory wire can be derived from equation (6.63) by setting $i = 0$ and solving for the temperature:

$$\theta(t) = \theta_\infty + (\theta_o - \theta_\infty) e^{-t/t_h}. \tag{6.69}$$

The steady-state value of the temperature when cooled is obviously the ambient temperature θ_∞. The time required to reach a desired temperature during cooling is

$$t_d = -t_h \ln \frac{\theta_d - \theta_\infty}{\theta_o - \theta_\infty} \qquad \text{cooling.} \tag{6.70}$$

The basic properties of the heating and cooling model are shown in Figure 6.14*a* and *b*. In general, the heating or cooling temperatures desired will be related to the start and finish temperatures of the material. In the example of heating, the steady-state temperature will be larger than the desired temperature to reduce the time required

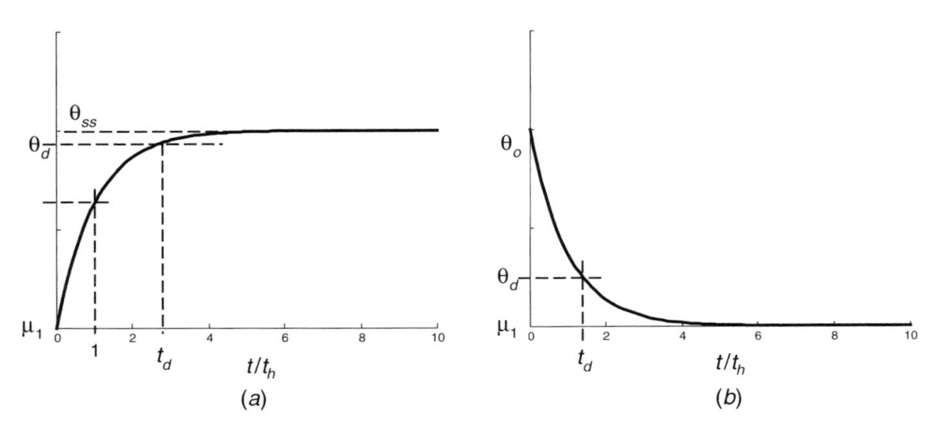

Figure 6.14 Representations of (a) heating and (b) cooling for a shape memory material

to reach the desired value. As shown in equation (6.66), the steady-state temperature is determined by the material properties but also by the current induced in the wire. Increasing the current will increase the steady-state value and reduce the amount of time required to reach the desired temperature. Unfortunately, no such 'knobs' are available in the cooling process since the cooling time is determined completely by the initial temperature and the time constant t_h.

Example 6.8 A circular shape memory wire with the properties listed in Table 6.5 is subjected to a current to induce heating. The diameter of the wire is 200 μm. (a) Determine the current required to heat the wire from an ambient temperature of 25°C to a steady-state value of 45°C. (b) Determine the time required to reach a desired temperature of 40°C with the result from part (a).

Solution (a) The relationship between current and steady-state temperature is shown in equation (6.66). Computing the steady-state temperature requres that we know the resistance per unit length of the wire. This is computed from the resistivity listed in Table 6.5 from the expression

$$R = \frac{76 \times 10^{-8}\ \Omega \cdot \text{m}}{\pi(200 \times 10^{-6}\ \text{m})^2/4} = 24.2\ \Omega/\text{m}.$$

Table 6.5 Representative shape memory properties for heat transfer analysis

Resistivity	$76\ \mu\Omega \cdot \text{cm}$
h_c	$150\ \text{J}/(\text{m}^2 \cdot {}^\circ\text{C} \cdot \text{sec})$
ρ	$6450\ \text{kg/m}^3$
c_p	$0.2\ \text{kcal/kg} \cdot {}^\circ\text{C}$

The current can be solved from equation (6.66):

$$
\begin{aligned}
i &= \sqrt{\frac{R}{h_c A_c}(\theta_{ss} - \theta_\infty)} \\
&= \sqrt{\frac{(150\ \mathrm{J/m^2 \cdot {}^\circ C \cdot s})(\pi \times 200 \times 10^{-6}\ \mathrm{m})}{24.2\ \Omega/\mathrm{m}}(45^\circ\mathrm{C} - 25^\circ\mathrm{C})} \\
&= 279.1\ \mathrm{mA}.
\end{aligned}
$$

(b) The time required to reach a desired temperature can be computed from equation (6.68). Dividing both sides of the equation by the time constant yields

$$
\begin{aligned}
\frac{t_d}{t_h} &= -\ln \frac{45^\circ\mathrm{C} - 40^\circ\mathrm{C}}{45^\circ\mathrm{C} - 25^\circ\mathrm{C}} \\
&= 1.39.
\end{aligned}
$$

The time constant for heating, t_h, is computed using equation (6.64) and the parameters listed in Table 6.5:

$$
\begin{aligned}
t_h &= \frac{(6450\ \mathrm{kg/m^3})(\pi/4)(200 \times 10^{-6}\ \mathrm{m})^2(0.2\ \mathrm{kcal/kg \cdot {}^\circ C})(4285.5\ \mathrm{J/kcal})}{(150\ \mathrm{J/m^2 \cdot {}^\circ C \cdot s})(\pi \times 200 \times 10^{-6}\ \mathrm{m})} \\
&= 1.84\ \mathrm{s}.
\end{aligned}
$$

Combining the two preceding expressions, we see that the time required to reach 40°C is approximately 2.56 s.

The previous analysis along with the example illustrate computations of the time constants associated with heating SMA material using a constant current. From the analysis we note that the heating of the material can be controlled by the induced current. Increasing the applied current i increases the steady-state temperature according to equation (6.63), which will increase the rate at which the temperature rises in the material. The rate at which the material cools will be a function of the desired temperature and the initial temperature at which the current is decreased to zero. The relationship between the temperature during heating and cooling is described in the following example.

Example 6.9 A shape memory alloy wire with a circular cross section is heated with a constant current of 350 mA for 1.5 s starting at $t = 1$ s. The initial temperature of the wire is equal to the ambient temperature of 25°C, and the diameter of the wire is 250 μm. Plot the temperature of the wire over the time interval 0 to 10 s. The properties of the shape memory alloy are listed in Table 6.5.

Solution The time constant for the shape memory alloy wire is computed from equation (6.64):

$$t_h = \frac{(6450\ \text{kg/m}^3)(\pi/4)(250 \times 10^{-6}\ \text{m})^2(0.2\ \text{kcal/kg}\cdot{}^\circ\text{C})(4285.5\ \text{J/kcal})}{(150\ \text{J/m}^2\cdot{}^\circ\text{C}\cdot\text{s})(\pi \times 250 \times 10^{-6}\ \text{m})}$$
$$= 2.30\ \text{s}.$$

The equation for the temperature in the wire is expressed by equation (6.65) due to the fact that the ambient temperature is equal to the initial temperature. The coefficient in the equation is computed from the material parameters:

$$R = \frac{76 \times 10^{-8}\ \Omega\cdot\text{m}}{\pi(250 \times 10^{-6}\ \text{m})^2/4} = 15.5\ \Omega/\text{m}$$

$$\frac{h_c A_c}{R} = \frac{15.5\ \Omega/\text{m}}{(150\ \text{J/m}^2\cdot{}^\circ\text{C}\cdot\text{s})(\pi \times 250 \times 10^{-6}\ \text{m})} = 131.6^\circ\text{C}/A^2$$

Solving equation (6.65) for the temperature and substituting in the parameter values yields

$$\theta(t) = [131.6^\circ\text{C}/A^2](1 - e^{-t/2.3})(0.35\ \text{A})^2 + 25$$
$$= 16.1(1 - e^{-(t-1)/2.3}) + 25 \qquad t \geq 1$$

At $t = 2.5$ s the applied current is set to zero. The temperature at this time is computed from the preceding expression:

$$\theta(t) = (16.1)(1 - e^{-1.5/2.3}) + 25 = 32.7^\circ\text{C}.$$

The temperature for $t > 2.5$ s is computed from equation (6.69) using an initial temperature of 32.7°C and an ambient temperature of 25°C,

$$\theta(t) = (32.7 - 25)\, e^{-(t-2.5)/2.3} + 25 \qquad t > 2.5$$

The total solution over the time interval 0 to 10 s can be expressed with the piecewise continuous function

$$\theta(t) = \begin{cases} 25 & 0 \leq t \leq 1 \\ -16.1e^{-(t-1)/2.3} + 41.1 & 1 \leq t \leq 2.5 \\ 7.7e^{-(t-2.5)/2.3} + 25 & t \geq 2.5. \end{cases}$$

The plot of temperature versus time is shown in Figure 6.15. The plot illustrates that the rate of temperature change is greater in heating than in cooling. Once again, this is due to the fact that the rate of heating can be controlled by the induced current, whereas the rate of cooling is only a function of the temperature difference between the current temperature and the ambient temperature.

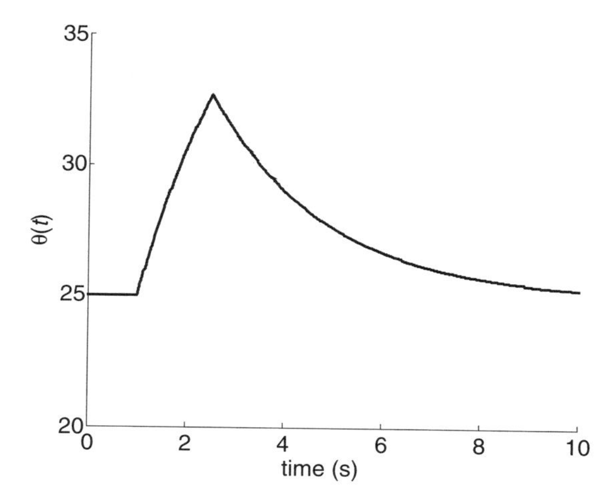

Figure 6.15 Temperature profile for Example 6.9.

6.7 DYNAMIC MODELING OF SHAPE MEMORY ALLOYS FOR ELECTRICAL ACTUATION

The response of shape memory alloys to thermal or electrical stimulus can be analyzed by combining the constitutive relationships described in Section 6.3 with the thermal heating and cooling model presented in Section 6.6. Combining the constitutive equations with the thermal model will allow us to analyze the time response of shape memory alloy materials to time-varying electrical inputs.

The case we consider is the one in which a constant preload is applied to an SMA material and an electrical stimulus is used to heat the wire to induce motion. The constant preload of the SMA wire can be visualized as a mass load on an SMA wire as shown in Figure 6.16. In this analysis we assume that the SMA material is initially in a state of zero stress and zero strain and the material has no stress-induced or temperature-induced martensite. Thus, the initial conditions of the analysis are

$$\mathrm{S}_0 = 0 \qquad \mathrm{T}_0 = 0 \qquad \xi_{S0} = 0 \qquad \xi_{T0} = 0 \tag{6.71}$$

Furthermore, we assume that the initial temperature is between the martensitic start temperature and the austenitic start temperature:

$$M_s < \theta_0 < A_s. \tag{6.72}$$

We denote the initial point as point a of the analysis. For this analysis we use the model illustrated in Figure 6.11 to define the relationship between the stress and critical temperatures of the material. From this figure it is clear that the stress required to

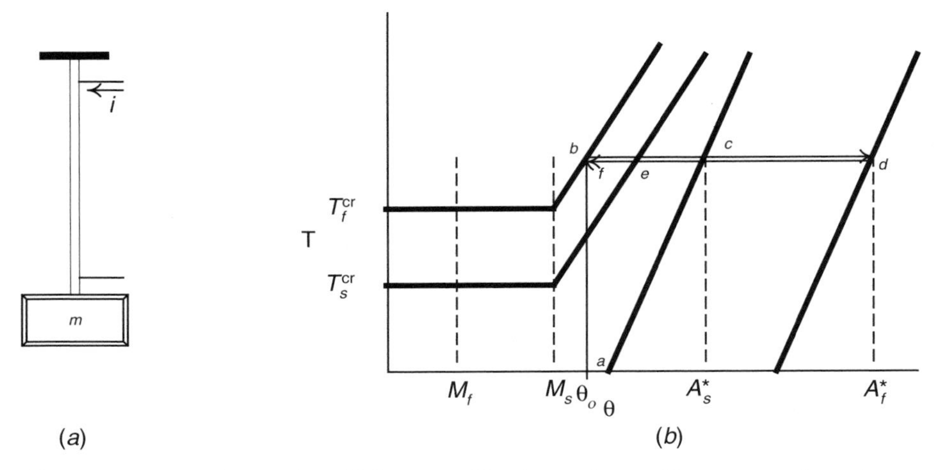

Figure 6.16 (*a*) Mass-loaded SMA wire with electric input; (*b*) corresponding stress–temperature diagram for the actuation cycle.

induce full transformation in the material is equal to

$$\mathrm{T} = \mathrm{T}_f^{\mathrm{cr}} + C_M\left(\theta_0 - M_s\right). \tag{6.73}$$

This is illustrated in Figure 6.16 by the vertical line. Smaller values of preload will induce only partial martensitic transformation. Assuming that the applied stress induces full transformation, the martenstic fraction parameters at point b of the analysis are

$$\xi_S^{\mathrm{b}} = 1 \qquad \xi_T^{\mathrm{b}} = 0. \tag{6.74}$$

From our discussion of the constitutive properties of SMA material, we know that increasing the temperature will induce phase transformation and strain recovery. Assuming that the temperature increase is being controlled by an induced current, we can use the analysis in Section 6.6 to relate the time-varying stimulus (the current) to the strain recovery of the SMA.

Assume that a constant current is applied to the preloaded material to induce a sufficient temperature rise for complete martensitic-to-austenitic transformation. From the solution of the heat transfer model when the initial temperature is equal to the ambient temperature, equation (6.65), the temperature rise in the material is known to be an exponential increase to the steady-state temperature θ_{ss}. The time required to begin the martensitic-to-austenitic phase transformation is denoted t^{c} and is computed from equation (6.68):

$$\frac{t^{\mathrm{c}}}{t_h} = -\ln\frac{\theta_{\mathrm{ss}} - A_s^*}{\theta_{\mathrm{ss}} - \theta_\infty}. \tag{6.75}$$

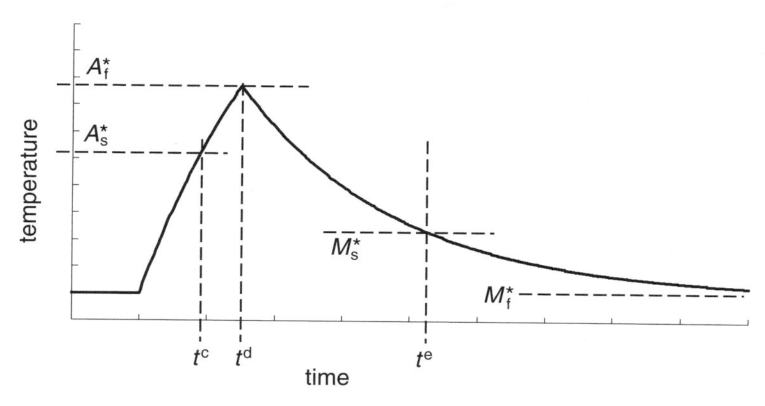

Figure 6.17 Representative heating–cooling cycle for SMA actuation illustrating critical temperatures and critical times.

Denoting the point at which full martensitic-to-austenitic phase transformation occurs, the time required to induce full transformation is

$$\frac{t^d}{t_h} = -\ln \frac{\theta_{ss} - A_f^*}{\theta_{ss} - \theta_\infty}. \tag{6.76}$$

Recall that the parameter in equations (6.75) and (6.76) that is controlled is the steady-state temperature θ_{ss}, which is proportional to the square of the applied current. Increasing the applied current will reduce the time required to start and finish the phase transformation. The parameters A_s^* and A_f^* are the critical temperatures for the applied preload, as shown in Figure 6.16. The heating cycle is represented in Figure 6.17 with the important temperature parameters labeled.

The strain response during the heating cycle can be modeled by applying the constitutive model expressed in equation (6.50):

$$\xi^{c\to d}(t) = \xi_S^{c\to d}(t) = \frac{1}{2}\left\{\cos\left[a_A\left(\theta(t) - A_s - \frac{T}{C_A}\right)\right] + 1\right\}, \tag{6.77}$$

where the time-dependent temperature is computed using equation (6.65). The strain during martensitic-to-austenitic transformation is obtained from equation (6.49):

$$S^{c\to d}(t) = S^c - S_L + S_L \xi_S^{c\to d}(t). \tag{6.78}$$

The cooling cycle will induce the austenitic-to-martensitic phase transformation due to the preload on the SMA material. This is illustrated in Figure 6.16 by the reversal in the stress-to-temperature diagram. Austenitic-to-martensitic phase transformation will start when the temperature has reached M_s^* and will continue until the material cools to M_f^*. This is illustrated as points c and d on Figure 6.16. Until the material cools to a temperature equal to M_s^*, the strain in the wire will remain constant. Continued cooling will induce phase transformation as the temperature of the wire approaches M_f^*.

The constitutive expressions in equation (6.51) can be modified to determine the transformation of the stress-induced martensitic fraction as a function of temperature when $M_f^* < \theta(t) < M_s^*$:

$$\xi^{e\to f}(t) = \xi_S^{e\to f}(t) = \frac{1}{2}\cos\left\{\frac{C_M \pi}{\mathrm{T_f^{cr}} - \mathrm{T}_s^{\mathrm{cr}}}[\theta(t) - \theta_0]\right\} + \frac{1}{2}. \tag{6.79}$$

The strain is computed by solving equation (6.49) assuming that the initial stress-induced martensitic fraction is zero:

$$\mathrm{S}^{e\to f}(t) = \mathrm{S}^e + \mathrm{S}_L \xi_S^{e\to f}(t). \tag{6.80}$$

The following examples illustrate the use of these expressions to compute an actuation cycle with an SMA wire.

Example 6.10 A 10-cm-long shape memory alloy wire with a circular cross section is preloaded with a weight to induce austenitic-to-martensitic phase transformation. The wire is initially in a state of zero stress- and temperature-induced martensite and is at an ambient temperature of 25°C. The diameter of the wire is 0.5 mm. The material properties of the wire are shown in Tables 6.4 and 6.5. (a) Determine the weight required to induce full austenitic-to-martensitic phase transformation such that M_f^* is equal to the ambient temperature. (b) Compute the current required to achieve a steady-state temperature of 75°C. (c) Compute the time required to induce martensitic-to-austenitic phase transformation and the time required to complete the phase transformation. (d) Assuming that the current is set to zero when the material completes the martensitic-to-austenitic phase transformation, plot the temperature, martensitic fraction, and strain of the wire as a function of time over the interval 0 to 30 s.

Solution (a) The stress required to induce full phase transformation is expressed in equation (6.73). Substituting the values from Table 6.4 into the expression yields

$$\begin{aligned} \mathrm{T} &= 170 \text{ MPa} + (8 \text{ MPa/°C})(25°\text{C} - 18°\text{C}) \\ &= 226 \text{ MPa}. \end{aligned}$$

The weight required to produce this stress is

$$\begin{aligned} W &= (226 \times 10^6 \text{ N/m}^2)\left[\frac{\pi(0.5 \times 10^{-3}\text{ m})^2}{4}\right] \\ &= 44.4 \text{ N}. \end{aligned}$$

(b) The relationship between current and steady-state temperature is shown in equation (6.66). This requires computing the resistance per unit length of the wire:

$$R = \frac{76 \times 10^{-8}\ \Omega \cdot \mathrm{m}}{\pi(500 \times 10^{-6}\ \mathrm{m})^2/4} = 3.87\ \Omega/\mathrm{m}.$$

For a steady-state temperature of 75°C, the current required is

$$\begin{aligned} i &= \sqrt{\frac{h_c A_c}{R}(\theta_{ss} - \theta_\infty)} \\ &= \sqrt{\frac{(150\ \mathrm{J/m^2 \cdot {}^\circ C \cdot s})(\pi \times 200 \times 10^{-6}\ \mathrm{m})}{3.87\ \Omega/\mathrm{m}}(75^\circ\mathrm{C} - 25^\circ\mathrm{C})} \\ &= 1.1\ \mathrm{A}. \end{aligned}$$

(c) The time required to induce and complete the martensitic-to-austenitic phase transformation is computed using the heat transfer model. To use equation (6.68) we need to compute the critical temperatures A_s^* and A_f^* for the specified load of 226 MPa. Using equation (6.11),

$$A_s^* = 35^\circ\mathrm{C} + \frac{226\ \mathrm{MPa}}{14\ \mathrm{MPa/^\circ C}} = 51.1^\circ\mathrm{C}$$

$$A_f^* = 49^\circ\mathrm{C} + \frac{226\ \mathrm{MPa}}{14\ \mathrm{MPa/^\circ C}} = 65.1^\circ\mathrm{C}.$$

The time constant associated with the material is

$$\begin{aligned} t_h &= \frac{(6450\ \mathrm{kg/m^3})(\pi/4)(500 \times 10^{-6}\ \mathrm{m})^2(0.2\ \mathrm{kcal/kg \cdot {}^\circ C})(4285.5\ \mathrm{J/kcal})}{(150\ \mathrm{J/m^2 \cdot {}^\circ C \cdot s})(\pi \times 500 \times 10^{-6}\ \mathrm{m})} \\ &= 4.61\ \mathrm{s}. \end{aligned}$$

The time required to induce the start of the phase transformation is computed using equation (6.68):

$$\begin{aligned} t^{\mathrm{c}} &= (4.61\ \mathrm{s})\ln\frac{75^\circ\mathrm{C} - 51.1^\circ\mathrm{C}}{75^\circ\mathrm{C} - 25^\circ\mathrm{C}} \\ &= 3.40\ \mathrm{s}. \end{aligned}$$

The time required to complete the phase transformation is

$$\begin{aligned} t^{\mathrm{d}} &= (4.61\ \mathrm{s})\ln\frac{75^\circ\mathrm{C} - 65.1^\circ\mathrm{C}}{75^\circ\mathrm{C} - 25^\circ\mathrm{C}} \\ &= 7.47\ \mathrm{s}. \end{aligned}$$

(d) Plotting the displacement of the wire requires computing the strain during the heating and cooling cycles. The strain during the heating cycle is computed from equation (6.78). This expression requires computation of the martensitic fraction using equation (6.77), which, in turn, is a function of the wire temperature. Thus, starting with the expression for the temperature during heating from equation (6.65),

$$\begin{aligned}\theta_h(t) &= \left[\frac{3.87\ \Omega/\text{m}}{(150\ \text{J/m}^2\cdot{}^\circ\text{C}\cdot\text{s})(\pi\times 200\times 10^{-6}\ \text{m})}\right](1.1\ \text{A})^2(1-e^{-t/4.61})+25\\ &= 50(1-e^{-t/4.61})+25,\end{aligned}$$

we can substitute the material parameters into equation (6.77):

$$\begin{aligned}\xi^{\text{c}\to\text{d}}(t) &= \frac{1}{2}\left\{\cos\left[\frac{\pi}{49^\circ\text{C}-35^\circ\text{C}}\left(\theta_h(t)-35-\frac{226\ \text{MPa}}{14\ \text{MPa}/^\circ\text{C}}\right)\right]+1\right\}\\ &= \frac{1}{2}\left\{\cos\left[\frac{\pi}{14^\circ\text{C}}(\theta_h(t)-51.1^\circ\text{C})\right]+1\right\}.\end{aligned}$$

The strain (relative to the initial strain in the deformed state) is computed from equation (6.78):

$$\text{S}^{\text{c}\to\text{d}}(t)-\text{S}_0 = 0.07\left[\xi^{\text{c}\to\text{d}}(t)-1\right].$$

Analogous expressions can be developed for the cooling cycle. The temperature during cooling is

$$\begin{aligned}\theta_c(t) &= 25+(65.1-25)\,e^{-t/4.61}\\ &= 25+40.1e^{-t/4.61}.\end{aligned}$$

Phase transformation will occur when the temperature becomes less than the martensitic start temperature. For a preload of 226 MPa, the martensitic start temperature is

$$M_s^* = 18^\circ\text{C}+\frac{226\ \text{MPa}}{14\ \text{MPa}/^\circ\text{C}} = 34.1^\circ\text{C}.$$

The martensitic fraction during phase transformation is

$$\xi^{\text{e}\to\text{f}}(t) = \frac{1}{2}\cos\left\{\frac{(8\ \text{MPa}/^\circ\text{C})\,\pi}{170\ \text{MPa}-100\ \text{MPa}}\left[\theta_c(t)-25\right]\right\}+\frac{1}{2},$$

and the strain during phase transformation is

$$\text{S}^{\text{e}\to\text{f}}-\text{S}_0 = (0.07)[\xi^{\text{e}\to\text{f}}(t)-1].$$

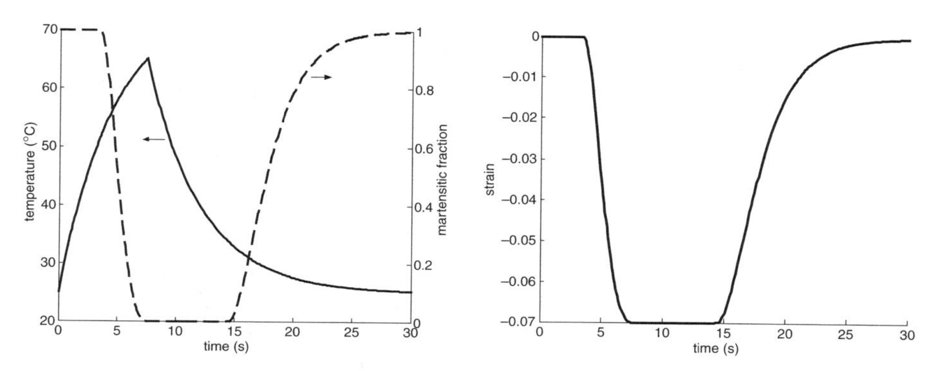

Figure 6.18 Temperature, martensitic fraction, and strain for Example 6.10.

The temperature, martensitic fraction, and strain are plotted in Figure 6.18. From the figure it is clear that the phase transformation does not occur until the critical temperatures are reached during the heating and cooling cycles. This introduces a time delay into the strain response due to the time required to heat the material to the austenitic start temperature. This time could be reduced at the expense of increasing the current input to the wire. The reverse phase transformation occurs when the material cools through the martensitic start temperature. From the figure we see that the reverse transformation requires longer, due to the fact that the rate of cooling is less than the rate of heating. The strain response correlates with the martensitic fraction and exhibits a different rate of heating and cooling. Note that negative strain implies *contraction* of the material because the material is already preloaded by the weight. Since we have assumed uniaxial strain throughout our discussion of shape memory materials, displacement of the wire can be computed by the product of the length and the maximum strain. For this example the peak contraction of the wire will be 7 mm for the 10-cm-long wire.

6.8 CHAPTER SUMMARY

In this chapter we focused on the fundamental properties and analysis techniques associated with shape memory alloys. Shape memory alloys are a class of material that exhibits large recoverable stress and strain due to the reversible conversion of martensitic and austenitic variants within the material. For thermomechanical shape memory alloys this conversion is triggered by heating and cooling of the material. The constititutive relationships for shape memory alloys require a definition of the martensitic fraction as a function of temperature and a definition of the state of stress of the material. For our discussion we focused solely on one-dimensional analysis for materials in tension. Under this assumption the pseudoelastic behavior and shape memory behavior of the material was analyzed with two basic constitutive models.

The first model used only the total martensitic fraction as the state variable, while the second used stress- and temperature-induced martensite as the state variables. These models coalesce under certain operating conditions, although the second model is able to predict a larger range of material behavior.

The analysis of shape memory alloys under electrical stimulus allowed us to model the basic actuation behavior of the materials. A heat transfer model was introduced that related the applied stimulus, in this case electrical current, to the temperature of the material. The analysis demonstrated that the rate of heating can often be faster than the rate of cooling. The difference in heating and cooling rates leads to differences in the strain response of the material during heating and cooling. In addition, the response time of thermomechanical shape memory materials is limited by the thermal time constant of the material. As shown through example, this time constant can often be on the order of 1 to 10 s, depending on the thermal properties and geometry of the shape memory material.

PROBLEMS

Unless noted otherwise, use the properties listed in Table 6.6 for these problems.

6.1. Compute the austenitic start and finish temperatures for an applied stress of 210 MPa.

6.2. A shape memory alloy wire is at an ambient temperature of 25°C and initially has a martensitic fraction of 1.

- **(a)** Compute the martensitic fraction when the material is heated to 38°C.
- **(b)** Compute the applied stress required to achieve the same martensitic fraction as in part (a) when the material is kept at an isothermal state at ambient temperature.

6.3. A shape memory alloy wire is at an ambient temperture of 25°C and initially has a martensitic fraction of 0.

- **(a)** Compute the stress required to begin the austenitic-to-martensitic phase transformation.
- **(b)** Compute the stress required to complete the austenitic-to-martensitic phase transformation.

6.4. A shape memory alloy wire is at an ambient temperature of 27°C and initially has a martensitic fraction of 0. Compute the martensitic fraction if a stress of 135 MPa is applied to the wire.

6.5.

- **(a)** Plot Figure 6.9 for the values listed in Table 6.6 over the stress range 0 to 300 MPa and the temperature range 0 to 60°C.
- **(b)** Sketch the relationship between stress and temperature on the plot in part (a) for a material that is undergoing the shape memory effect with full transformation.

Table 6.6 Shape memory alloy material properties

Elastic Properties	Transformation Temperatures	Transformation Constants	Maximum Recoverable Strain	Electrical/Material Properties
$Y_A = 62$ GPa	$M_f = 10°C$	$C_M = 7$ MPa/°C	$S_L = 0.06$	resistivity $= 66\ \mu\Omega \cdot$ cm
$Y_M = 20$ GPa	$M_s = 17°C$	$C_A = 11$ MPa/°C		$h_c = 140$ J/m$^2 \cdot$ °C $\cdot$ s
	$A_s = 31°C$	$T_s^{cr} = 105$ MPa		$\rho = 6450$ kg/m^3
	$A_f = 44°C$	$T_f^{cr} = 160$ MPa		$c_p = 0.25$ kcal/kg $\cdot$ °C

(c) Sketch the relationship on the plot in part (a) for a material that is undergoing the pseudoelastic effect with full transformation.

6.6. A shape memory material is at an ambient temperature of 23°C. Plot the stress–strain curve for the material as it is loaded and unloaded to produce maximum residual strain. Assume for this problem that the modulus of the material is equal to 62 GPa and does not vary with the martensitic fraction.

6.7. Problem 6.6 assuming that the elastic modulus varies as $Y(\xi) = Y_A + \xi\,(Y_M - Y_A)$.

6.8. A shape memory material is at an ambient temperature of 44°C. Plot the stress–strain curve for the material as it is loaded and unloaded to produce the pseudoelastic effect. Assume for this problem that the modulus of the material is equal to 62 GPa and does not vary with the martensitic fraction.

6.9. Repeat Problem 6.8 assuming that the elastic modulus varies as $Y(\xi) = Y_A + \xi\,(Y_M - Y_A)$.

6.10. A shape memory material is in the full martensite phase with 100% temperature martensite and no stress-induced martensite. The temperature of the material is 5°C. Compute the amount of stress-induced and temperature-induced martensite in the material if the temperature is increased to 35°C.

6.11. A shape memory material is in the full martensite phase with 100% temperature-induced martensite and no stress-induced martensite. The temperature of the material is 25°C. Compute the amount of stress and temperature-induced martensite if a stress of 145 MPa is applied to the material.

6.12. A shape memory material is initially in full martensite phase with 100% temperature-induced martensite and no stress-induced martensite. The ambient temperature is 5°Cand the material is at zero stress and zero strain. Plot the stress–strain curve of the material if it is loaded to 200 MPa and then the load is decreased to zero.

6.13. A shape memory wire is loaded and unloaded an room temperature of 23°C to produce a residual strain of 0.04%.

(a) Plot the recovery strain as a function of temperature as the material is heated to 44°C.

(b) Plot the recovery stress as the material is heated to 44°C.

(c) Combine the results of parts (a) and (b) produce a plot of blocked stress and free strain.

6.14. Plot the recovery stress and recovery strain as a function of temperature for a shape memory alloy wire that is attached to a spring for the following nondimensional spring constants: **(a)** $AY_A/kL = 0.1$; **(b)** $AY_A/kL = 1$; **(c)** $AY_A/kL = 10$.

6.15. Compute the time constant t_h for a wire with a 0.5-mm-diameter circular cross section.

6.16. Compute the steady-state temperature required to heat a shape memory material with properties listed in Table 6.6 to the austenitic finish temperature from 25°C in (**a**) 1 time constant; (**b**) 1/2 of a time constant; (**c**) 1/10 of a time constant.

6.17. Compute the constant current required to achieve the desired steady-state temperature for each part of Problem 6.16.

6.18. The ambient environment for a shape memory wire with properties listed in Table 6.6 has been cooled to 0°C. If the material is heated to its austenitic finish temperature with a constant current and then the current is reduced to zero, compute the number of time constants required to reach the martensitic finish temperature.

6.19. Repeat the computations in Example 6.10 using the properties listed in Table 6.6 for a shape memory wire of circular cross section and a diameter of 0.3 mm. Plot the temperature, martensitic fraction, and strain as a function of time.

NOTES

The bulk of the material in this chapter was drawn from the shape memory alloy modeling performed by Liang et al. [45–47]. Work by Brinson was also used extensively for the material in this chapter [48,49]. Additional material on the modeling of shape memory alloy materials may be found in the work of Boyd and Lagoudas [50,51].

7

ELECTROACTIVE POLYMER MATERIALS

The final material type that we consider, the class of materials known as *electroactive polymers*, comprises a wide range of material types that exhibit a variety of coupling mechanisms. We consider only those types that exhibit electromechanical coupling. Functionally, these materials are similar to the piezoelectric materials studied earlier in that they produce mechanical strain under the application of an electric potential and produce an electrical signal under the application of a mechanical stress. For certain materials this similarity will lead to modeling techniques that are very similar to those studied for piezoelectric materials.

One of the primary difference between electroactive polymers that exhibit electromechanical coupling and their piezoelectric counterparts is the stress and strain induced by the materials. As highlighted in earlier chapters, piezoelectric materials have a maximum free strain on the order of 0.1 to 0.2%, although single-crystal piezoelectrics have been shown to produce strains on the order of 1%. The blocked stress produced by ceramic piezoelectric materials can exceed 10 MPa, due to the fact that the modulus of the materials is on the order of 50 to 100 GPa.

The stress and strain induced by electroactive polymers is on the opposite end of the spectrum from to piezoelectric materials. As we discuss shortly, most electroactive polymers are soft materials. The elastic modulus of a majority of these materials is below 1 GPa, and certain materials that we will study (e.g., dielectric elastomers) have elastic modulus values on the order of 1 to 10 MPa. In contrast, the strain produced by these materials is generally greater than 1%, and certain materials can produce strain on the order of 50 to 100%. Thus, electroactive polymers generally fall in a category of high strain actuators.

In this chapter we discuss basic types of electroactive polymers and their relevant electromechanical coupling properties. As we shall see, the modeling of electroactive polymers is not necessarily as far advanced as models associated with piezoelectric and shape memory materials. For that reason in this chapter we focus primarily on phenomenological models that highlight the salient features of this class of smart materials.

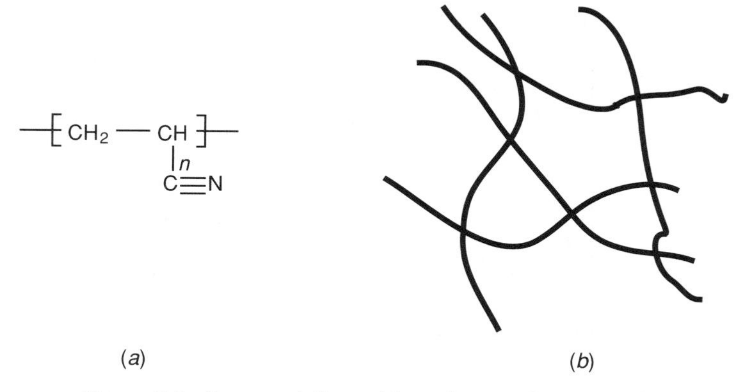

Figure 7.1 Representations of the polymer polyacrylonitrile.

7.1 FUNDAMENTAL PROPERTIES OF POLYMERS

One of the primary attributes that distinguishes electroactive polymer materials from piezoelectric ceramics and shape memory materials is their fundamental material structure. As the name implies, this class of materials is based on a class of materials known as *polymers*. The word *polymer* means "many parts," and this definition is a good way to visualize the fundamental structure of polymer materials. By definition, a polymer is a substance composed of molecules that have long sequences of one or more species of atoms or groups of atoms linked to one another by chemical bonds. The word *macromolecule* is a synonym for the word *polymer*. Polymers are formed by combining molecules known as *monomers* through chemical reactions. The process of linking monomers together to form a polymer is known as *polymerization*.

Polymers are represented by the chemical structure of the base monomer. For example, the monomer acrylonitrile is represented as $CH_2{=}CHCN$, and a polymer comprised of this monomer is represented by the structure shown in Figure 7.1*a*. The letters refer to the atoms that comprise the molecule. In the case of acrylonitrile, the constituent atoms are carbon (C), hydrogen (H), and nitrogen (N). As shown in Figure 7.1, a polymer consists of a large number of repeating monomer units. This chemical structure is known as a *polymer chain*, or simply, a *chain*. A visual representation of a chain is shown in Figure 7.1*b*. It typically consists of a wavy line that may or may not have any information about the chemical structure of the chain.

A critical feature of a polymer is the *topology* of the chemical links between monomer units. One of the simplest polymer topologies is a *linear polymer*, which may be represented by the wavy line shown in Figure 7.1*b*. A branched polymer has a *side chain*, or *branch*, that is of significant length and bonded to the main polymer chain at junctions known as *branch points*. *Network polymers* have a three-dimensional structure that are connected at junctions by *cross-links*. The amount of cross-linking in the polymer is characterized by the *crosslink density*.

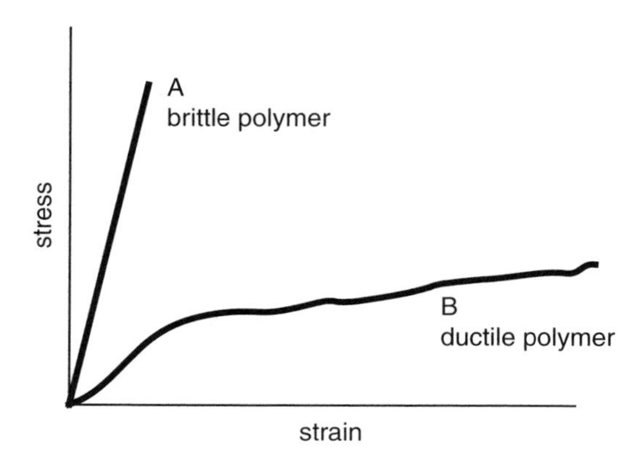

Figure 7.2 Comparison of the stress–strain behavior of a brittle and ductile polymer.

Topics such as methods for synthesis of polymers and the numerous methods for characterizing the structure of polymers are the realm of polymer science. In this book, what is important to understand is that the structure and topology of the polymer material can significantly affect the mechanical, electrical, and thermal properties. For example, a stress–strain test on polystrene would yield behavior that is representative of a brittle polymer; the stress–strain curve would exhibit predominantly linear behavior up until a point at which the polymer would fracture, as shown in Figure 7.2. Stress–strain tests on another polymer (e.g., a polyamide) would yield a more ductile behavior in which the linear stress–strain behavior at low values of strain would transform into a region of very soft behavior in which small increases in stress would yield very large increases in strain. This behavior is also represented in Figure 7.2. It is often the case that maximum strain in a ductile polymer is on the order of 100% or greater.

A number of polymer compositions also exhibit mechanical behavior known as *viscoelasticity*. A viscoelastic material is one in which the stress is a function not only of the strain but also of the strain rate. A viscoelastic material exhibits three distinct regions of stress–strain behavior. When cycled at low frequency, the material exhibits a stress–strain curve that is very similar to that of a purely elastic material; this is called the *rubbery regime*. Increasing the frequency of the cycling will produce a stress–strain response that exhibits an increasing amount of loss or hysteresis in addition to an increase in the elastic modulus of the material. This frequency regime is typically called the *transition regime*. Increasing the cyclic rate further will produce a stress–strain behavior that once again is predominantly elastic, but the elastic modulus of the material will be significantly higher than that of the material in the rubbery regime. This is called the *glassy regime* of the viscoelastic material. The strain rate dependence of the mechanical behavior also alters the stress–strain plots shown in Figure 7.2. At low strain rate a viscoelastic material may exhibit ductile behavior as shown in curve B of Figure 7.2. Increasing the strain rate, though, may produce very brittle behavior that results in fracture at much lower values of strain.

The chemical structure of a polymer also has a significant effect on the electronic properties of material. A majority of polymers are insulating materials in the same way that a piezoelectric ceramic or polymer is an insulator. Application of an electric field across the polymer will produce the rotation of bound charge and lead to a storage of electrical energy in the material; recall the discussion in Section 2.1.3. Chemical modification of certain materials can produce polymers that are *electronically conductive* in the same way that a copper metal or shape memory alloy is conductive. These materials, generally known as *conducting polymers*, are the basis for certain types of electroactive polymer actuators. Different chemical modifications can yield polymers that are *ionically conductive*. Instead of conducting electrons, as in the case of an electronic conductor, an ionic conductor will transport charged atoms and molecules within the polymer network. This type of polymer material is central to the development of energy conversion devices such as fuel cells and is also the basis for a class of electroactive polymers known as ionomeric transducers.

7.1.1 Classification of Electroactive Polymers

There are a number of different types of electroactive polymers that exhibit a variety of coupling mechanisms. In this book we study only those types of materials that exhibit coupling between an applied potential (or charge) and mechanical stress and strain. Even within this class of electromechanical materials there are significant differences in the properties and behavior of different types of electroactive polymers.

A common way to classify the basic types of electromechanical electroactive polymers is into electronic and ionic materials. This classification scheme defines *electronic electroactive polymer materials* as those that exhibit coupling due to polarization-based or electrostatic mechanisms. The electronic properties of these materials are very similar to those of piezoelectric materials since they are insulators that contain bound charge in the form of electronic dipoles. In contrast, *ionic electroactive polymers* exhibit electromechanical coupling due to the *diffusion*, or conduction, of charged species within the polymer network. This motion of charge species produces electromechanical coupling due to the accumulation of charge within the material. The electronic properties of ionic materials are substantially different from those of electronic materials. Ionic materials are more closely related to conductors, with the important exception that an ionic material is conducting charged atoms or molecules, whereas an electronic conductor is conducting electrons. These properties of electronic and ionic materials are discussed in more detail later in the chapter.

Before discussing detailed models of these materials, let us overview the basic operating characteristics of electronic and ionic electroactive polymers. One class of electroactive polymers that we have already introduced in this book is a class of high-strain piezoelectric polymers that exhibit electrostrictive behavior. As you recall from the discussion in Chapter 4, electrostrictive materials are those that exhibit quadratic coupling between polarization (or electric field) and strain. As is typical with piezoelectric ceramics, the maximum strain induced by the application of a field or electric polarization for a conventional material is on the order of 0.1 to 0.2%. This amount of maximum strain is also very typical of piezoelectric polymers that exhibit either linear piezoelectric coupling or quadratic electrostrictive coupling.

Recently, though, it has been demonstrated that chemical modification of a class of piezoelectric polymers produces a material that exhibits greatly increased maximum strain. Irradiation of piezoelectric polymers produces a material that exhibits electrostrictive behavior as opposed to the linear piezoelectric behavior of the original material. Furthermore, the electrostrictive coefficient of the material is $-14\ \mathrm{m^4/C^2}$, which is over three orders of magnitude larger than those listed for the representative electrostrictive ceramics in Table 4.4. Experimental results demonstrate that the irradiated piezoelectric polymers could produce 4% strain at electric field values on the order of 150 MV/m. More recent results have demonstrated that introducing a high-dielectric filler into the polymer will reduce the field required to achieve high strain. Results have been reported which indicate that 2% strain is achievable at electric field strengths on the order of 10 to 15 MV/m.

The electrostrictive behavior of irradiated piezoelectric polymers can be analyzed using the techniques introduced in Chapter 4. Recall from equation (4.183) that the relationship between polarization and electric field, which includes saturation at high fields, can be modeled as a hyperbolic tangent function. In a single dimension this relationship is

$$\mathrm{P} = \chi \mathrm{E_S} \tanh \frac{\mathrm{E}}{\mathrm{E_S}},$$

where χ is the *pseudosusceptibility* and $\mathrm{E_S}$ is the *saturation electric field*. Experimental data for the polarization as a function of electric field for this class of irradiated piezoelectric films is shown as circles in Figure 7.3*a*. The pseudosusceptibility of the polarization-to-field response is obtained by fitting a linear relationship to the data points near zero. The result from the linear fit indicates that $\chi = 5.98 \times 10^{-10}$ F/m. The linear relationship between polarization and field is shown as the dashed line in Figure 7.3*a*. The linear model fits the data well at field levels less than approximately 25 MV/m, but above this value the saturation of the polarization response results in substantial deviation between the experimental response and the linear model. A model that incorporates saturation is included as the solid line in Figure 7.3*a*. Using a saturation field of $\mathrm{E}_S = 90$ MV/m in equation (4.183) and the pseudosusceptibility from the linear fit results in a model that accurately represents the polarization response over the range -150 to $+150$ MV/m.

The mechanical response of an electrostrictive material is a quadratic relationship between polarization and strain. The strain response of the electrostrictive piezoelectric polymers is shown in Figure 7.3*b* using the model parameters. The electrostrictive coefficient of $-13.5\ \mathrm{m^4/C^2}$ is used to compute the strain response from the expression $\mathrm{S} = Q\mathrm{P}^2$. The result is shown as the solid line in Figure 7.3*b*, with circles representing experimental data of Zhang et al. [52]. In addition to predicting the strain response accurately, the model that incorporates saturation illustrates that the strain will saturate at approximately 4% when the field is increased to approximately 200 MV/m.

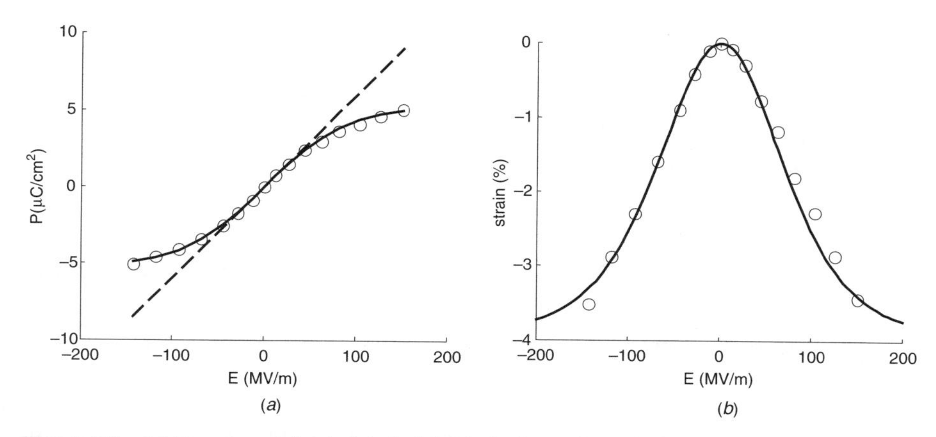

Figure 7.3 (*a*) Experimental data (circles) for the polarization-to-field response of an irradiated piezoelectric polymer [52] and the linear (dashed) and nonlinear (solid) model. (*b*) Corresponding strain-to-elective field relationship.

The properties of the irradiated piezoelectric polymers can be compared to conventional materials by computing the effective piezoelectric strain coefficient and the energy density. In Section 4.8.2 the relationship between the electrostrictive properites and the effective piezoelectric strain coefficient was derived. Using the model between strain and electric field that incorporated saturation, it was shown that the maximum effective piezoelectric strain coefficient, d, is obtained by oscillating the field about a bias field of $2\mathrm{E}_S/3$. At this operating condition the effective piezoelectric strain coefficient is $0.38Q\chi^2\mathrm{E}_S$. Using the parameters for irradiated PVDF, we obtain an effective piezoelectric strain coefficient of 165 pm/V. The maximum energy density of the material is obtained by computing the product of one-half the maximum stress and maximum strain. Using a modulus of 380 MPa [52] and a maximum strain of 4%, the volumetric energy density of an irradiated piezoelectric polymer is computed to be $\frac{1}{2}Y\mathrm{S}_{\mathrm{max}}^2 = 304\ \mathrm{kJ/m}^3$. Examining Table 4.3, we see that the volumetric energy density of a piezoelectric ceramic is generally on the order of 10 to 20 kJ/m^3 at a field of 1 MV/m. Assuming a maximum field of 2 to 3 MV/m, we see that the energy density of an irradiated piezoelectric polymer is on the order of 3 to 10 times higher than that of a piezoelectric ceramic.

Direct comparisons of relevant material properties of piezoelectric ceramics are shown in Table 7.1. As the discussion illustrates, the irradiated piezoelectric polymer

Table 7.1 Energy density of different types of piezoelectric materials in extensional and bending mode at an electric field of 1 MV/m

	d_{33} (pm/V)	d_{13} (pm/V)	Y^{E} (GPa)	Y_1^{E} (GPa)	E_v: Extensional (kJ/m^3)	E_v: Bending (kJ/m^3)
APC 856	620	260	58.8	66.7	11.7	1.3
PZT-5H	650	320	62.1	50	13.1	1.4

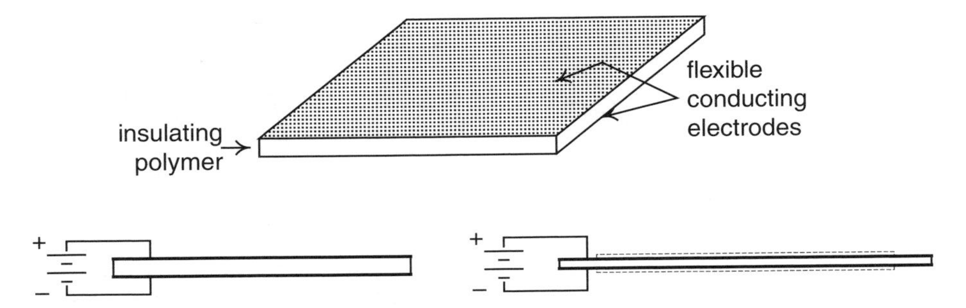

Figure 7.4 Basic operating properties of dielectric elastomer actuators.

has a modulus that is approximately two orders of magnitude smaller than that of a piezoelectric ceramic, but the maximum strain is approximately 40 times larger. The conventional piezoelectric polymer also has a modulus that is over an order of magnitude smaller than the ceramic, but the maximum strain is still on the order of 0.1%.

The discussion of irradiated piezoelectric polymers is a good introduction to the basic properties of many types of electroactive polymers. The techniques discussed earlier in the book can be used to analyze the properties of electrostrictive piezoelectric materials, but the primary difference between these materials and the piezoelectric and electrostrictive materials studied earlier in the book is the fact that the irradiated piezoelectric materials exhibit much larger strain than do their ceramic counterparts. The ability to produce strain in excess of $\approx 1\%$ is a typical characteristic of electroactive polymers, as are the soft elastic modulus properties as compared to ceramic materials or metal alloys.

Another type of soft electroactive polymer that produces large strains, a class of material known as *dielectric elastomers*, consist of a soft polymer material sandwiched between two conducting metal electrodes. We study dielectric elastomer materials in more detail in this chapter, but the basic characteristics of these materials are shown in Figure 7.4. As discussed in Section 2.1.3, Coulomb's law states that a force will develop between two charged plates when placed at a distance from one another. If the region between the charged plates is filled with an elastic material, the force induced by the electrostatic attraction (or repulsion) will produce stress on the interstitial material. This stress is called *Maxwell stress*, due to the fact that it arises from the electrostatic interactions of the charged plates. The Maxwell stress induced in the material will induce strain and hence a controllable mechanical deformation.

This concept has led to the development of a wide class of actuating materials that exhibit large strains under the application of large electric fields. Electric fields required to produce sufficient mechanical stress for actuation are often on the order of 50 to 250 MV/m. Electric field strengths of this order are similar to those required for irradiated piezoelectric materials but one to over two orders of magnitude larger than those required for a typical piezoelectric material. As we discuss in this chapter, this actuator configuration leads to strains that are typically much greater than 10% and in certain instances have been measured to be greater than 100%.

Electromechanical modeling of dielectric elastomer materials is challenging for a number of reasons. The large induced strain in the material requires more complex models of the constitutive relationships since they are generally operated in the nonlinear elastic regime. We will see shortly, as well, that the Maxwell stress induced in the material is a quadratic function of the applied field; therefore, the material behaves similar to an electrostrictive material in the sense that the induced stress is a nonlinear function of the applied voltage.

Irradiated piezoelectric materials and dielectric elastomers are both examples of electronically electroactive polymer materials. These materials are similar in the sense that they are soft materials that produce strains on the order of 1% to greater than 100% when subjected to electric fields greater than 10 MV/m (and sometimes greater than 250 MV/m). They are also similar in the fact that the electromechanical coupling arises from electrostatic interactions such as material polarization (as in the case of a piezoelectric) or Maxwell stress.

A second class of electroactive polymer materials that exhibit electromechanical coupling, called *ionic materials*, produce electromechanical coupling due to the transport of charged species within the material. Recall the discussion of electronic conduction in Section 2.1.3. Application of an electric field will produce forces on charged particles and in electronic conductors will result in the transport of negative charge (electrons) in the material. This is the basis for the generation of electric current in a conductor.

The same phenomenon will occur in a material, or more generally a medium, which contains postively or negatively charged atoms or molecules. From basic chemistry we might recall that when a solid compound is dissolved in a solution, it is broken down into its constituent atoms. A simple example of this is the dissolution of salt, or NaCl, in water. When dissolved in water, NaCl will dissolve into positively charged sodium ions, Na^+, and negatively charged chloride atoms, Cl^-, a process known as *disassociation*. Now consider placing electrodes in the salt solution and applying a specified potential between the electrodes as shown in Figure 7.5*a* and *b*. The application of a potential will produce electrostatic forces that will cause the ions to move in the solution. The positive ions (sodium in this example) will accumulate on the negative electrode and the negative ions (chloride) will accumulate on the positive electrode. The movement of charged ions due to the application of the electric field is called *migration*, and models for this process are introduced later in the chapter.

Migration of ions will also occur in polymer materials that contain charged species. One class of polymer materials that contains charged species are *hydrogels*, polymer networks that can contain side chains that terminate in a charged group. When placed in a salt solution these charged groups are neutralized by the disassociated ions. Application of an electric field across the hydrogel will produce transport of charge in a manner similar to the transport of charge in a salt solution.

Electromechanical coupling arises in polymer networks due to the charge imbalance that occurs as a result of charge transport. As in the case of a salt solution, charge transport produces charge accumulation within the hydrogel, which, in turn,

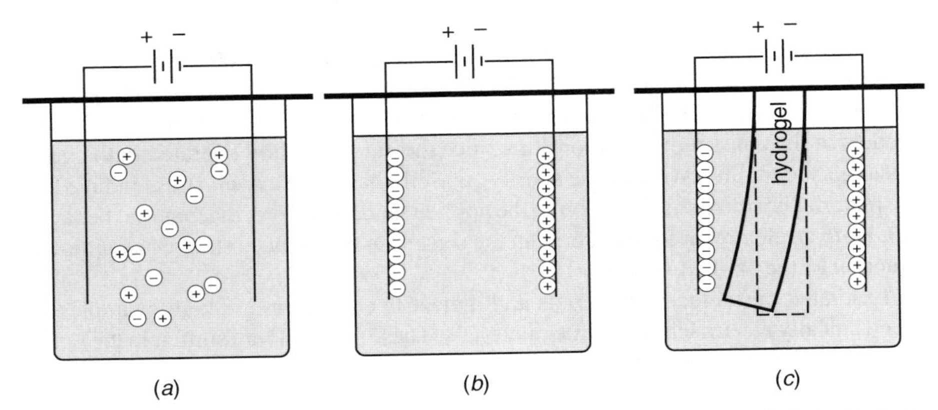

Figure 7.5 (*a, b*) Migration of ionic species in a salt solution due to an applied electric field; (*c*) bending of a hydrogel due to the osmotic forces induced by ionic migration.

produces an osmotic pressure that is proportional to the difference in concentration of the positive and negative ions. The osmotic pressure can induce bending in a hydrogel strip, as shown in Figure 7.5*c*.

Hydrogels are one type of polymer material that can produce electromechanical coupling due to the transport of charge. Two other types of polymer materials that produce electromechanical coupling due to ionic migration are conducting polymers and ionomeric polymers. As their name implies, *conducting polymers* are a class of polymer material that exhibits electronic conduction in a manner similar to that of conductive metals. *Ionomeric materials* also called *ion-exchange polymers*, are based on a class of polymer known as *ionomers*, polymers composed of macromolecules that have a small but significant portion of ionic groups. Ion-exchange polymers exchange positive or negative ions with the ionic components of a solution. An important attribute of ion-exchange polymers is that only one of the ionic groups within the polymer is capable of transport when the material is subjected to an electric field. The charged group that is bound to the polymer chain is an *immobile* ion, and will not transport in the presence of an electric field. The oppositely charged ion that is only weakly bound to the polymer chain is *mobile* in the presence of an electric field. For this reason these materials are sometimes known as *single-ion conductors*.

A transducer is fabricated from an ionomer by first plating the surfaces with conductive electrodes, typically composed of metal particulates that are incorporated into the ionomer film through a deposition process. Processes for deposition of the conductive electrode include chemical processes akin to electroless deposition and processes that utilize mechanical pressure to fuse the metal particulates to the ionomer film. The result of the electrode deposition process is a composite film in which the middle portion is an ionomeric film and the outer surfaces consist of a mixture of electronically conducting metal particles and the ionomer, as shown in Figure 7.6*a*.

The electromechanical coupling in ionomeric transducers is due to the migration of ionic species due to the application of an electric field. As with other ionic materials,

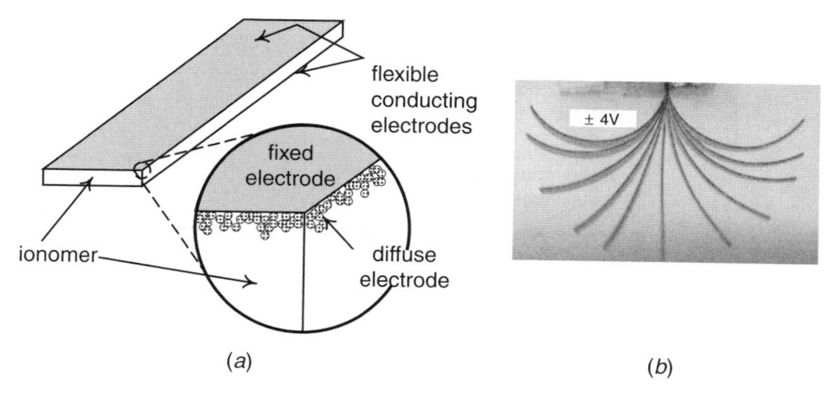

Figure 7.6 (*a*) Cross section of an ionomeric transducer, illustrating the electrode layer; (*b*) bending of an ionomeric transducer under the application of 4 V.

mechanical deformation can be induced by the application of a voltage that is generally less than 5 V. As we discuss later in this chapter, the application of a potential across the actuator thickness produces a spatially varying electric field in the material due to the existence of mobile charge in the polymer. This contrasts with materials that exhibit polarization-based electromechanical coupling, such as piezoelectrics or dielectric elastomers, in which the electric field is constant through the thickness when it is assumed that there is no mobile charge. As with other electroactive polymer actuators, bending induced by the application of a voltage is generally greater than 1%. A representative actuator response is shown in Figure 7.6*b* for an ionomeric material that is excited by a 4-V potential. The induced strain is greater than 2% in this example.

Throughout this discussion we have highlighted the important features of electroactive polymers as compared to other materials studied in this book. The primary feature of electroactive polymers is their ability to produce large strains, generally greater than 1% but in some instances (e.g., dielectric elastomers) much greater than 10%, with correspondingly lower forces, due to the low elastic modulus compared to electroactive ceramics. Within the class of electroactive polymers we see that there are electronic materials such as irradiated piezoelectric film and dielectric elastomers that require electric field strengths greater than 10 MV/m to operate. Electronic materials often require voltages on the order of 100 V to greater than 1 kV for typical film thicknesses. Ionic materials such as conductive polymers and ionomeric polymers exhibit electromechanical coupling due to charge migration. Ionic materials generally require voltages on the order of less than 5 V to operate. In the remainder of this chapter we study the fundamental properties and models of actuators based on dielectric elastomers, conducting polymers, and ionomeric polymers, and compare their properties to other electroactive materials introduced in the book.

7.2 DIELECTRIC ELASTOMERS

The mechanics model of a dielectric elastomer actuator will be developed using the energy approach introduced earlier. Unlike the analysis for piezoelectric materials,

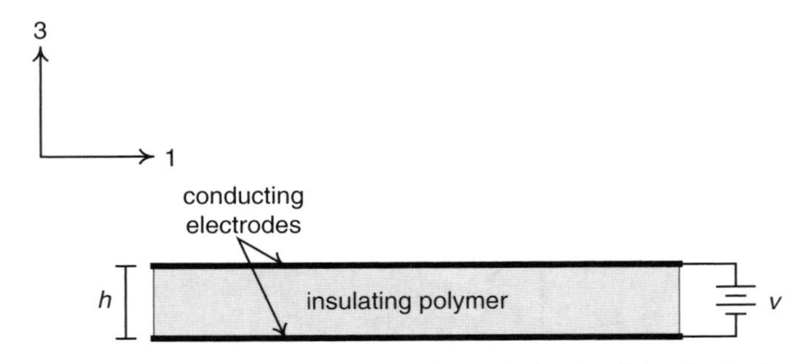

Figure 7.7 Definitions of the geometry for analysis of a dielectric elastomer.

where strain and electric displacement are defined as the independent variables, we base our analysis on an energy function that is a function of strain and electric field. The rationale for this decision is that typically in a dielectric actuator application the electric potential is the prescribed variable; therefore, we can prescribe the associated electric field. Referring to Table 5.2, we note that this requires a definition of the electric enthalpy function, H_2, with S and E as the independent variables.

Assuming that the material is linear elastic, the electric enthalpy function will be defined as

$$H_2 = \frac{1}{2}\mathbf{S}'c\mathbf{S} - \frac{1}{2}\mathbf{E}'\varepsilon\mathbf{E}. \tag{7.1}$$

We will make the following assumptions about the prescribed electric field. We assume that the electric field in the 1 and 2 directions are zero and that the only prescribed electric field is in the 3 direction. This geometry is shown in Figure 7.7.

Assuming that the dielectric elastomer material is a perfect dielectric, the electric field within the material is equal to the difference in the applied potential between the electrodes and the thickness of the material. We define the potential at the bottom face to be equal to zero and the potential at the top electrode is defined as v. The thickness of the actuator in the undeformed state is defined as h, and the thickness of the deformed actuator is defined as $h\,(1 + S_3)$. Combining these assumptions the electric field is written as

$$\begin{aligned} E_1 &= 0 \\ E_2 &= 0 \\ E_3 &= \frac{v}{h\,(1 + S_3)} \end{aligned} \tag{7.2}$$

We will denote v/h as E_o, which represents the electric field generated by the applied potential across the undeformed material. Substituting these assumptions into equation (7.1) produces the electric enthalpy function

$$H_2 = \frac{1}{2}\mathbf{S}'\mathrm{c}\mathbf{S} - \frac{1}{2}\varepsilon_3 \frac{\mathrm{E}_0^2}{(1+\mathrm{S}_3)^2}. \tag{7.3}$$

The constitutive equations are derived by taking the appropriate derivatives of the electric enthalpy function,

$$\begin{aligned}
\frac{\partial H_2}{\partial \mathrm{S}_1} &= \mathrm{T}_1 = c_{11}\mathrm{S}_1 + c_{12}\mathrm{S}_2 + c_{13}\mathrm{S}_3 \\
\frac{\partial H_2}{\partial \mathrm{S}_2} &= \mathrm{T}_2 = c_{12}\mathrm{S}_2 + c_{22}\mathrm{S}_2 + c_{23}\mathrm{S}_3 \\
\frac{\partial H_2}{\partial \mathrm{S}_3} &= \mathrm{T}_3 = c_{13}\mathrm{S}_1 + c_{23}\mathrm{S}_2 + c_{33}\mathrm{S}_3 + \varepsilon_3 \frac{\mathrm{E}_o^2}{(1+\mathrm{S}_3)^3} \\
\frac{\partial H_2}{\partial \mathrm{E}_3} &= -\mathrm{D}_3 = -\varepsilon_3 \frac{\mathrm{E}_o}{(1+\mathrm{S}_3)^2}.
\end{aligned} \tag{7.4}$$

As with other actuator materials, let us consider the two cases of *free strain* and *blocked stress*. The free strain response of the material is obtained by setting $\mathrm{T}_1 = \mathrm{T}_2 = 0$ in equation (7.4) and solving for the strain in the 1 and 2 directions,

$$\begin{aligned}
\mathrm{S}_1 &= -\nu \mathrm{S}_3 \\
\mathrm{S}_2 &= -\nu \mathrm{S}_3.
\end{aligned} \tag{7.5}$$

Substituting these two expressions into the third expression in equation (7.4) results in the expression

$$Y\mathrm{S}_3 + \varepsilon_3 \frac{\mathrm{E}_o^2}{(1+\mathrm{S}_3)^3} = \mathrm{T}_3. \tag{7.6}$$

The maximum applied stress is the stress induced when S_3 is constrained to be zero,

$$\mathrm{T}_a = \mathrm{T}_3|_{\mathrm{S}_3=0} = \varepsilon_3 \mathrm{E}_o^2. \tag{7.7}$$

One of the defining features of dielectric elastomers is that the stress induced by the applied field is a quadratic function of the prescribed potential. Thus, the material is similar to electrostrictive materials in its fundamental response characteristics. The free strain is solved for by setting $\mathrm{T}_3 = 0$ in equation (7.7):

$$\mathrm{S}_3(1+\mathrm{S}_3)^3 + \frac{\varepsilon_3 \mathrm{E}_o^2}{Y} = 0. \tag{7.8}$$

The final term on the left-hand side of equation (7.8) is the ratio of the maximum applied stress to the modulus of the material. We denote this the nondimensional

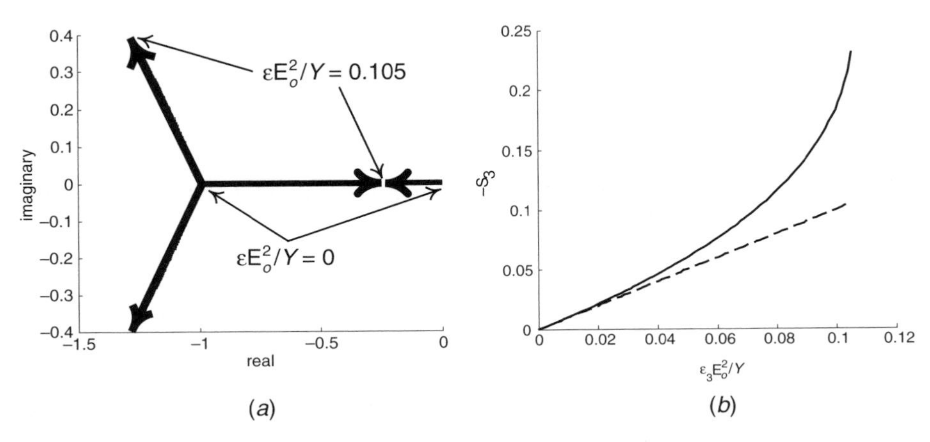

Figure 7.8 (*a*) Locus of solutions for the free strain of a dielectric elastomer actuator as a function of nondimensional applied stress; (*b*) free strain as a function of nondimensional applied stress.

applied stress and use this as the parameter in our analysis of the free strain response.

The nonlinear relationship between electric field and strain introduces a nonlinearity in the relationship between applied stress and free strain. Equation (7.8) has four solutions for each value of $\varepsilon_3 \mathrm{E}_o^2/Y$. The locus of solutions as a function of $\varepsilon_3 \mathrm{E}_o^2/Y$ is shown in Figure 7.8*a*. The trivial solution of $(0, -1, -1, -1)$ occurs when $\varepsilon_3 \mathrm{E}_o^2/Y = 0$. Increasing the nondimensional applied stress $\varepsilon_3 \mathrm{E}_o^2/Y$ results in two complex-conjugate solutions, which are not physical, and two solutions bounded by 0 and -1. One of the solutions tends to move from -1 toward -0.25 while the other moves from 0 to -0.25 as $\varepsilon_3 \mathrm{E}_o^2/Y \to 0.105$.

Using the solution that begins at 0 and moves towards -0.25 as the physical solution, the relationship between strain and induced stress (in nondimensional form) can be analyzed. The result illustrates that the dielectric elastomer actuator exhibits nonlinear behavior above strain values of approximately 0.02, or 2%, as shown in Figure 7.8*b*. The maximum strain is 25% when the nondimensional stress is approximately 0.105. Strain in the 1 and 2 directions is found from equation (7.5). For incompressible materials Poisson's ratio is $\nu = 0.5$ and the strain perpendicular to the electrodes is approximately 12.5% at maximum applied stress.

Example 7.1 A polyurethane material with a Young's modulus of 17 MPa and a relative dielectric constant of 5 is being used for the design of a dielectric elastomer. Determine the voltage required to produce 10% compressive free strain in an actuator with a thickness of 25 μm.

Solution The nondimensional stress required to achieve the free strain specified is computed from equation (7.8). Substituting $\mathrm{S}_3 = -0.1$ into the expression yields

$$(-0.1)(1 - 0.1)^3 + \varepsilon_3 \mathrm{E}_o^2/Y = 0.$$

The nondimensional stress is $\varepsilon_3 \mathrm{E}_o^2/Y = 0.0729$. Solving this expression for the electric field,

$$\mathrm{E}_0 = \sqrt{\frac{(0.0729)(17 \times 10^6 \ \mathrm{N/m^2})}{(5)(8.85 \times 10^{-12} \ \mathrm{F/m})}} = 167.4 \ \mathrm{MV/m}.$$

The electric potential that produces this electric field is computed from

$$\begin{aligned} v &= \mathrm{E}_0 h \\ &= (167 \times 10^6 \ \mathrm{V/m})(25 \times 10^{-6} \ \mathrm{m}) \\ &= 4185 \ \mathrm{V}. \end{aligned}$$

As discussed earlier in the chapter, the voltage required to achieve large strain in a dielectric elastomer actuator is on the order of kilovolts.

The electrostatic attraction of the electrodes in a dielectric elastomer produces strain in the two directions perpendicular to the thickness of the material, as shown in the preceding analysis. The quantity that is analogous to the blocked stress of other types of smart materials is the stress induced in the 1 and 2 directions by the applied field across the actuator thickness when the strain in these directions is constrained to be equal to zero. The blocked stress of the actuator is analyzed by setting $\mathrm{S}_1 = \mathrm{S}_2 = 0$ in equation (7.4). In addition, we assume that the applied stress $\mathrm{T}_3 = 0$ as well. These assumptions reduce the constitutive equations to

$$\begin{aligned} \mathrm{T}_{\mathrm{bl}} &= \begin{cases} c_{13}\mathrm{S}_3 \\ c_{23}\mathrm{S}_3 \end{cases} \\ 0 &= c_{33}\mathrm{S}_3 + \varepsilon_3 \frac{\mathrm{E}_o^2}{(1+\mathrm{S}_3)^3}. \end{aligned} \tag{7.9}$$

The first two equations are identical because $c_{13} = c_{23}$, indicating that the blocked stress perpendicular to the thickness direction is the same. If we assume that the material is isotropic, then the third expression in equation (7.9) can be rewritten as

$$\mathrm{S}_3 (1+\mathrm{S}_3)^3 + \frac{(1+\nu)(1-2\nu)}{1-\nu} \frac{\varepsilon_3 \mathrm{E}_o^2/Y}{(1+\mathrm{S}_3)^3} = 0. \tag{7.10}$$

Solving the third expression in equation (7.9) for S_3 and substituting into the second expression yields

$$\mathrm{T}_{\mathrm{bl}} = \frac{c_{23}}{c_{33}} \frac{\varepsilon_3 \mathrm{E}_o^2}{(1+\mathrm{S}_3)^3} = \frac{\nu}{1-\nu} \frac{\varepsilon_3 \mathrm{E}_o^2}{(1+\mathrm{S}_3)^3}. \tag{7.11}$$

The blocked stress is computed by first solving equation (7.10) for strain as a function of the nondimensional applied stress and then substituting the result into equation (7.11) to compute the blocked stress.

For an incompressible material, the strain due to the applied stress is equal to zero, due to the fact that the volume of an incompressible material is constant. Thus, if the strain in the 1 and 2 directions is equal to zero, S_3 must also be zero. In this case, the blocked stress expression reduces to

$$T_{bl} = \varepsilon_3 E_o^2, \tag{7.12}$$

which indicates that the blocked stress is equal to the applied stress when the material is incompressible.

Analyzing the induced strain as a function of applied stress requires solution of the constitutive equations. For this analysis we assume that the applied stress in the 2 and 3 directions is equal to zero and that the applied stress in the 1 directions is T_o. With these assumptions the constitutive equations reduce to

$$\begin{aligned} T_o &= c_{11}S_1 + c_{12}S_2 + c_{13}S_3 \\ 0 &= \begin{cases} c_{12}S_2 + c_{22}S_2 + c_{23}S_3 \\ c_{13}S_1 + c_{23}S_2 + c_{33}S_3 + \varepsilon_3 \dfrac{E_o^2}{(1+S_3)^3}. \end{cases} \end{aligned} \tag{7.13}$$

Solving the first two expressions for S_1 and S_2 yields

$$\begin{pmatrix} S_1 \\ S_2 \end{pmatrix} = -\begin{bmatrix} \nu \\ \nu \end{bmatrix} S_3 - \frac{1}{Y}\begin{bmatrix} (\nu-1)(\nu+1) \\ \nu(\nu+1) \end{bmatrix} T_o. \tag{7.14}$$

Substituting the results of equation (7.14) into equation (7.13) produces an equation that relates the stress induced by the electrostatic forces to the resisting stress in the 1 direction,

$$\nu\frac{T_o}{Y} + \frac{\varepsilon_3 E_o^2/Y}{(1+S_3)^3} + S_3 = 0. \tag{7.15}$$

Assuming that the materials are incompressible, $\nu = \frac{1}{2}$, and the pertinent expressions are written

$$\begin{aligned} S_1 &= -\frac{1}{2}S_3 + \frac{3}{4}\frac{T_o}{Y} \\ \frac{\varepsilon_3 E_o^2}{Y} &+ \left(S_3 + \frac{1}{2}\frac{T_o}{Y}\right)(1+S_3)^3 = 0. \end{aligned} \tag{7.16}$$

The second expression in equation (7.16) is solved first for a specified value of the nondimensional applied stress and nondimensional stress T_o/Y. Specifying these two

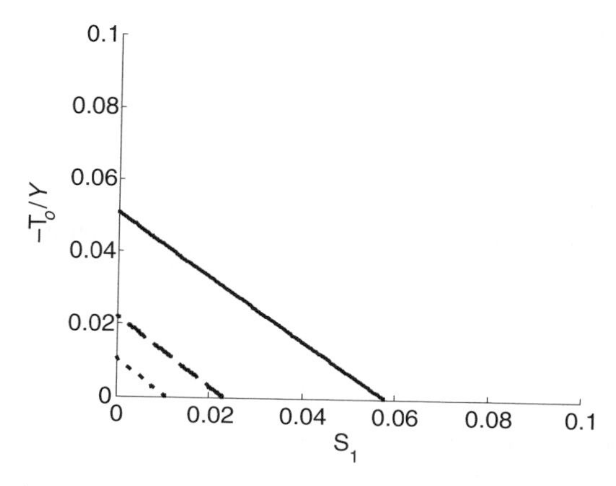

Figure 7.9 Induced stress and strain for an incompressible dielectric elastomer material for $\varepsilon_3 E_o^2/Y = 0.02$ (dotted), 0.04 (dashed), and 0.08 (solid).

parameters enables the computation of S_3 as a function of $\varepsilon_3 E_o^2/Y$ and T_o/Y. The result is then substitutied into the first expression in equation (7.16) to solve for the strain in the 1 direction, S_1.

Solving these two expressions allows us to compute the relationship between induced stress and strain due to the applied electrostatic forces of the dielectric elastomer material. Figure 7.9 illustrates this relationship for three values of the nondimensional applied stress. The free strain in the 1 direction is close to 6% for a nondimensional applied stress of 0.08, and the nondimensional blocked stress (the stress that reduces S_1 to zero) is approximately 0.05.

Example 7.2 A 0.025-mm-thick polyurethane material with modulus of 17 MPa and a dielectric constant of 5 is actuated with 3 kV across its thickness. Compute the free strain and blocked stress of the actuator in the 1 direction assuming that the stress in the 2 direction is zero. Assume that the material is incompressible.

Solution Equation (7.16) is used for the computation. To compute the free strain ($T_o = 0$) condition, first compute the nondimensional applied stress:

$$\frac{\varepsilon E_o^2}{Y} = \frac{(5 \times 8.85 \times 10^{-12}\ \text{F/m})(3000\ \text{V}/0.025 \times 10^{-3}\ \text{m})^2}{17 \times 10^6\ \text{Pa}} = 0.037.$$

Substituting this result and $T_o = 0$ into the second expression in equation (7.16) yields

$$0.037 + S_3 (1 + S_3)^3 = 0,$$

a fourth-order equation in the strain S_3. Computing the roots of the expression yields two real roots, -0.0421 and -0.6063. Since we are assuming that the material starts with zero strain, we take the root $S_3 = -0.0421$ as the result. The strain in the 1 direction is computed from

$$S_1 = -\frac{1}{2}S_3 = 0.021. \tag{7.17}$$

Thus, the free strain is equal to 2.1%.

The blocked stress is computed using the same set of equations. Solving the first expression in equation (7.16) for S_3 when $S_1 = 0$ yields

$$S_3 = \frac{3}{2}\frac{T_o}{Y}.$$

Substituting this result into the second expression of equation (7.16) yields

$$\frac{\varepsilon E_o^2}{Y} + 2\left(\frac{T_o}{Y}\right)\left(1 + \frac{3}{2}\frac{T_o}{Y}\right)^3 = 0.$$

Solving for the roots of this expression yields two real roots. The relevant root is $T_o/Y = -0.022$. Thus, the blocked stress is computed from

$$T_o = (0.022)(17 \times 10^6 \text{ Pa}) = 0.374 \text{ MPa}.$$

7.3 CONDUCTING POLYMER ACTUATORS

A separate class of electroactive polymer material that exhibits high-strain response is conducting polymers. Conducting polymers are materials that exhibit a reversible volume change due to electrochemical reactions caused by the introduction and removal of ions into the polymer matrix. The volume change due to electrochemical processes is controlled by the application of a low voltage, typically less than 5 V, to the polymer through electronically conducting electrodes. In this section we describe transducer models of conducting polymers and highlight the actuating properties of these materials.

Conducting polymer actuators can be synthesized in a number of forms, but typically they are arranged in thin films or as fibers. A conducting polymer actuator consists of the conducting polymer material, an electrolyte that serves as a source of ions, and two electrodes that control the ionic diffusion. A representative setup is shown in Figure 7.10. Application of a voltage potential between the working electrode and the counter electrode causes an electrochemical reaction, known as an *reduction–oxidation* or *redox reaction*, which in turn causes a volume change inside the conducting polymer. The volume change in the conducting polymer is transformed into mechanical deformation of the film or fiber, which can be utilized as actuation strain. The fibers are generally configured such that the predominant motion is along the length of the polymer and the primary actuation response is axial strain in the material. Films can be synthesized onto passive substrates to produce a bending

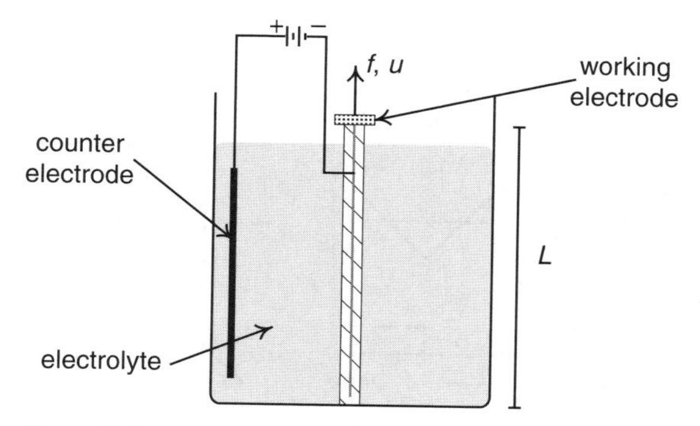

Figure 7.10 Elements of a conducting polymer actuator.

deformation when the conducting polymer undergoes a volume change. The physical mechanism for bending is identical to the mechanism associated with a piezoelectric bimorph or unimorph actuator.

7.3.1 Properties of Conducting Polymer Actuators

The fundamental mechanism of strain generation in conducting polymer actuators is the redox reaction that occurs upon application of an electric potential, or equivalently, the application of an induced current. Before discussing this phenomenon, let us discuss the relationship between applied potential and induced current for the conducting polymer material. The voltage–current relationship is measured with a measurement technique called *cyclic voltammetry*, in which the applied potential is controlled and the current induced on the material is measured. This technique is called *potentiostatic measurement* since the applied potential is the controlled variable. An alternative method is *galvanostatic measurement*, in which the induced current is controlled and the resulting potential is measured. Consider the case of potentiostatic cyclic voltammetry for linear electric circuit elements. For a resistor the current-to-voltage relationship is a proportionality:

$$i(t) = \frac{1}{R} v(t), \tag{7.18}$$

where R is the resistance. For a capacitor of capacitance C, the current-to-voltage relationship is

$$i(t) = C \frac{dv(t)}{dt}. \tag{7.19}$$

In most cases of potentiostatic cyclic voltammetry, the controlled voltage waveform is a triangular wave with a specified scan rate. The scan rate is the slope of the triangular waveform and is generally defined in terms of V/s or, more commonly, mV/s. For an

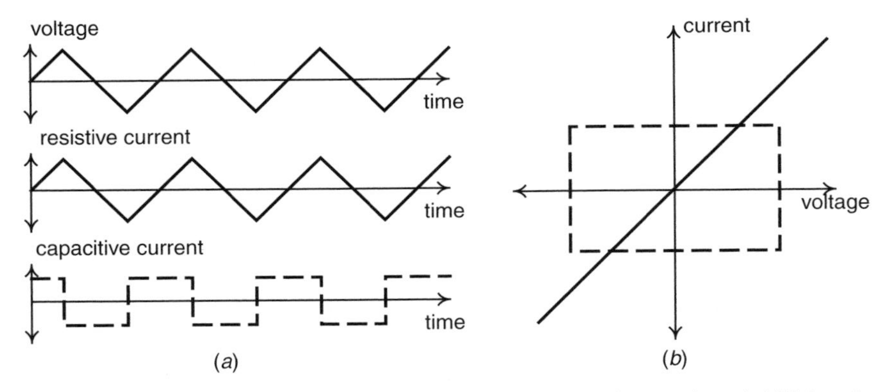

Figure 7.11 (*a*) Representative voltage and current waveforms for a resistor (middle) and capacitor (bottom); (*b*) ideal cyclic voltammogram for a resistor (solid) and a capacitor (dashed).

ideal resistor, the output current to the triangular voltage waveform will itself be a triangular waveform. The constant of proportionality between the applied potential and induced current is the resistance R.

Cyclic voltammetry measurements are typically displayed as a plot of current versus potential called a *voltammagram*. A representative voltammagram for an ideal resistor is shown in Figure 7.11. As expected, the relationship is simply a straight line whose slope is equal to the resistance R. For an ideal capacitor, a voltammagram with a triangular waveform will yield step changes in the current due to the fact that the slope of the applied potential is constant. When the slope of the waveform switches, the sign of the current will also switch. Thus, the voltammagram will be a rectangular box that switches between $+C dv/dt$ and $-C dv/dt$.

As with most materials, conductive polymers are neither purely resistive nor purely capacitive. A material that exhibits resistive and capacitive behavior may have a voltammagram similar to the one shown in Figure 7.12. This representative voltammagram exhibits behavior that is associated with both resistive elements and capacitive elements. Furthermore, this voltammagram may change as the scan rate is changed, due to the fact that some of the behavior of the material may be frequency dependent.

The ability of a conducting polymer actuator to induce stress and strain is measured in a manner that is similar to the methods discussed in Chapter 4 for piezoelectric materials. Two key figures of merit for piezoelectric materials are the blocked stress and free strain of the material due to the application of an electric potential. Considering the test setup shown in Figure 7.10, the free strain is defined as the linear strain in the material with negligible load applied. Denoting the displacement $u(t)$ and assuming uniaxial strain, the strain is defined as

$$S_1(t) = \frac{u(t)}{L}. \tag{7.20}$$

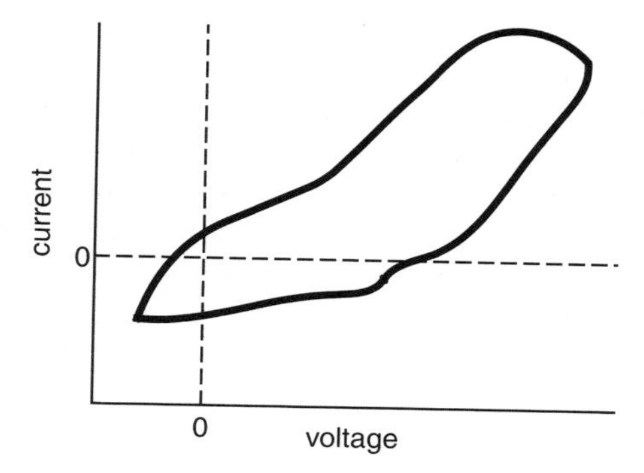

Figure 7.12 Representative voltammogram for a conducting polymer actuator.

It is typical for a small tensile load to be applied to the material so that the linear actuator does not go slack during the experiment. Denoting this prestrain as S_o, the dynamic strain of the material is defined as

$$\Delta S_1(t) = S_1(t) - S_o. \tag{7.21}$$

Representative results on a linear conducting polymer actuator are shown in Figure 7.13. In this measurement, the electric potential is varied as a triangular wave with a set voltage *scan rate*, similar to the wave shown at the top of Figure 7.11*a*. Typical scan rates are on the order of 1 to 100 mV/s, thus, the fundamental actuation frequency is quite low (< 0.1 Hz). In the result shown in Figure 7.13, the prestrain is on the order of 1%, and the peak-to-peak value of the dynamic strain is on the order of 0.5%. These values are reasonable for a linear conducting polymer actuator. Values in the range 2 to 5% have also been reported using actuators with reasonable

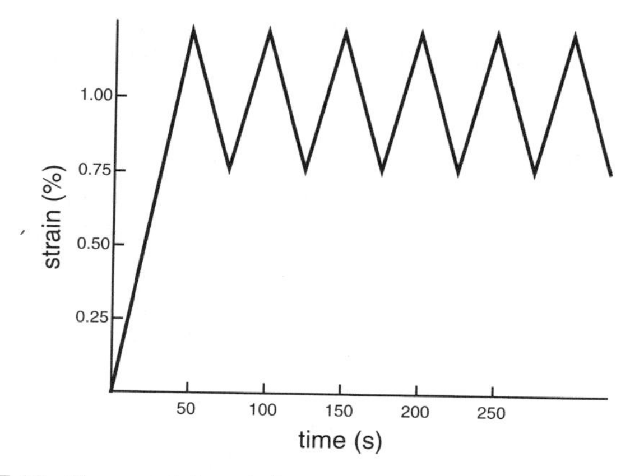

Figure 7.13 Representative strain output of a conducting polymer actuator.

durability, and values greater than 30% have been measured in actuators with very large ions for the redox reaction.

An important characteristic of conducting polymer actuators is that the strain induced in the material is approximately proportional to the charge induced during the electrochemical cycle. The charge induced is the time integral of the current, and measured results demonstrate that the strain induced in the linear actuator is approximately proportional to the charge induced per unit volume of the actuator. The amount of strain induced per unit volume is related to the size of the ion that is used during the redox cycling. Larger ions produce increased strain due to the larger volume change that occurs as the ion migrates in and out of the polymer volume.

Another important characteristic of conducting polymer actuators is the frequency dependence of the free strain response. The electromechanical coupling mechanism is based on the principle of ionic migration into and out of the polymer volume. The volume change caused by ionic migration produces the mechanical deformation that is correlated with applied electric potential. The migration of ions into and out of the polymer is controlled by the diffusion of ions through the electrolyte, and therefore the actuation properties of a conducting polymer actuator exhibit a classical diffusion-controlled frequency dependence.

It has been shown that a semi-infinite diffusion process has a frequency dependence that approaches $f^{1/2}$, where f is the actuation frequency. Correspondingly, experimental results demonstrate that the strain output of a conducting polymer actuator also approximates a power law of $f^{1/2}$. This is expressed as the relationship

$$|\mathrm{S}_1(f)| = |\mathrm{S}_1(f_o)| \left(\frac{f_o}{f} \right)^{1/2}, \tag{7.22}$$

where $|\mathrm{S}_1(f_o)|$ is the magnitude of the strain at f_o. This model, which is the only approximation that is valid at low frequencies, predicts that the strain output of a conducting polymer will drop by $\sqrt{10} \approx 3.16$ for every tenfold increase in the actuation frequency. Similarly, a factor of 10 decrease in the actuation frequency will cause the peak strain to rise by approximately 3.16. It is important to note that this model completely neglects any dynamic effects, such as resonance, in the response of the actuator.

Example 7.3 The linear strain in a conducting polymer actuator has been measured to be 0.65% at a frequency of 0.3 Hz. Assuming that a semi-infinite diffusion process controls the strain generation, compute the strain of the actuator at 10 Hz assuming that the voltage input is equal to that of the input at 0.3 Hz.

Solution Using the expression in equation (7.22), the strain at 10 Hz is

$$\begin{aligned} |\mathrm{S}_1(10)| &= (0.65\%) \left(\frac{0.3}{10} \right)^{1/2} \\ &= 0.11\%. \end{aligned}$$

Note that the strain at 10 Hz is approximately equal to the maximum strain generated by a polycrystalline piezoelectric material at maximum electric field.

7.3.2 Transducer Models of Conducting Polymers

In Section 7.3.1 the basic electrical and electromechanical properties of conducting polymers were introduced. The state of the art in modeling conducting polymers is not as far advanced as it is for other smart materials, but transducer models of linear actuators have been developed for the purposes of understanding the force–deflection behavior of conducting polymer actuators.

The force–deflection models introduced in this section are similar to those developed for piezoelectric transducers in Chapter 4. Considering the arrangement shown in Figure 7.10, the relationship between the applied force f and the elongation of the actuator u is derived assuming linear elasticity theory and one-dimensional mechanics. One of the important features of conducting polymer materials is that the elastic modulus is a function of the applied voltage. For this reason we denote Y_o as the elastic modulus when the potential is zero and Y_v as the elastic modulus when a voltage v is applied. The deflection can be written as a sum of three components,

$$u = u_o v + \frac{fL}{AY_o} + \frac{fL}{A}\left(\frac{1}{Y_v} - \frac{1}{Y_o}\right), \tag{7.23}$$

where the first component is the free strain due to the applied voltage, the second component is the static deflection due to the elasticity when the potential is zero, and the third term is the deflection that occurs due to the change in elastic modulus upon actuation. The term u_o is the free deflection per unit voltage. Simplifying equation (7.23) results in

$$u = u_o v + \frac{fL}{AY_v}. \tag{7.24}$$

The result is a transducer equation of a form that is identical to that for linear piezoelectric materials, except for the fact that we are accounting for the change in elastic modulus that occurs upon the application of a voltage. Using the nomenclature from Chapter 4, the free deflection and blocked force of the actuator is written as

$$\begin{aligned} \delta_o &= u_o v \\ f_{\mathrm{bl}} &= \frac{Y_v A}{L} u_o v. \end{aligned} \tag{7.25}$$

Equation (7.25) highlights the the fact that the blocked force of the conducting polymer actuator is a function of the elastic modulus when a potential is applied. The difference between the elastic moduli with and without the application of a potential depends on the type of conducting polymer used for the actuator. Typical values range between 20% of the passive modulus and approximately equal to the passive modulus.

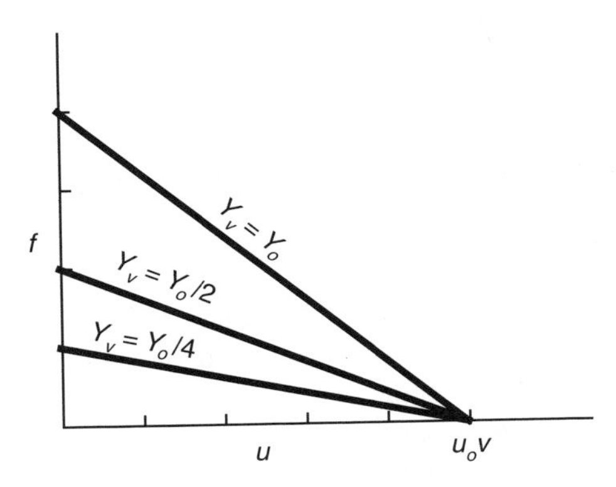

Figure 7.14 Variation in the force–deflection curve of a conducting polymer actuator as a function of the variation in elastic modulus upon application of an electric potential.

Under the assumptions of this analysis, the force–deflection curve for a conducting polymer actuator is analogous to those studied for piezoelectric materials, except for the fact that the slope of the line is related to the modulus when a potential is applied. Figure 7.14 illustrates this effect. If $Y_v = Y_o$, the blocked force of the actuator will be equal to the amount of preload required to elongate the actuator. Any reduction in the elastic modulus due to the applied potential will produce a reduction in the blocked force and a corresponding change in the slope of the force–deflection diagram.

Example 7.4 A conducting polymer configured as a linear actuator has a modulus of 85 MPa without a potential applied and an elastic modulus of 45 MPa with an applied potential of 1 V. The free strain in the actuator has been measured to be 3.5%/V. For an actuator 3 cm long with a circular cross section of radius of 150 μm, compute (a) the preload required to achieve a static deflection of 12 mm, and (b) the blocked force achieved for a 1-V actuation.

Solution (a) The preload required to achieve a static deflection of 12 mm is computed by multiplying the stiffness and the static deflection. The stiffness of the actuator is

$$k_a = \frac{Y_o A}{L} = \frac{(85 \times 10^6 \text{ N/m}^2)(\pi)(150 \times 10^{-6} \text{ m})^2}{30 \times 10^{-2} \text{ m}}$$
$$= 20.03 \text{ N/m}.$$

The required preload is

$$f_{\text{pl}} = k_a u_{\text{st}} = (20.03 \text{ N/m})\,(12 \times 10^3 \text{ m})$$
$$= 240 \text{ mN}.$$

(b) The blocked force is obtained using equation (7.25), which states that the blocked force is equal to the stiffness of the actuator when a voltage is applied multiplied by the free displacement. The free displacement is obtained from the free strain through the expression

$$\begin{aligned} u_o v &= (0.035 \text{ V}^{-1})(30 \times 10^{-2} \text{ m})(1 \text{ V}) \\ &= 10.5 \text{ mm}. \end{aligned}$$

The blocked force is

$$\begin{aligned} f_{\text{bl}} &= \frac{Y_v A}{L} u_o v \\ &= \left[\frac{(45 \times 10^6 \text{ N/m}^2)(\pi)(150 \times 10^{-6} \text{ m})^2}{30 \times 10^{-2} \text{ m}} \right] (10.5 \times 10^{-3} \text{ m}) \\ &= 111 \text{ mN}. \end{aligned}$$

7.4 IONOMERIC POLYMER TRANSDUCERS

Ionomeric polymer transducers are another class of material that exhibits electromechanical coupling. As discussed earlier in the chapter, ionomeric polymer transducers exhibit electromechanical coupling due to the motion of ionic species upon application of an electric field or mechanical deformation. Functionally, they are very similar to piezoelectric bimorphs in their sensing and actuation properties. In this book we concentrate on the development of models that enable design of systems that incorporate ionomeric polymer transducers as sensors or actuators. There have been a number of physics-based models of these materials, but our focus will be on the use of input–output transducer models that enable the prediction of relevant actuation and sensing properties such as free deflection, blocked force, and dynamic sensitivity.

7.4.1 Input–Output Transducer Models

Consider a cantilevered sample of an ionomeric polymer transducer with fixed electrodes at the clamped end. Application of a potential across the thickness of the transducer produces a bending response that is functionally similar to that of a piezoelectric bimorph. Similarly, inducing bending in the transducer produces an electrical response that can be measured as either a voltage or a charge output of the polymer. The similarity between the actuation and sensing response of an ionomeric polymer transducer and piezoelectric bimorphs is the motivation for a transducer-level model for analysis of ionomeric polymer transducers. Recall the discussion in Chapter 4, in which the transducer equations for a piezoelectric actuator were developed. In that chapter we developed a relationship between the force, displacement, charge, and voltage by making an assumption about the state of the material and then introducing the geometric parameters associated with the problem. A similar approach can be applied to the development of a transducer model for ionomeric polymers that bend

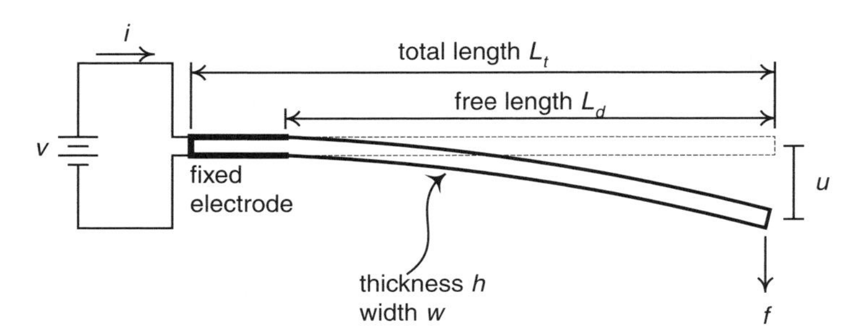

Figure 7.15 Geometry of the ionomeric polymer transducer analysis.

under the application of an electric field. Consider a cantilevered sample of material with the geometric parameters defined in Figure 7.15. One method of deriving the coupled equations for this transducer is to assume an *equivalent-circuit representation* of the ionomeric polymer. In this equivalent-circuit model, the important input variables and output variables are defined and the resulting analysis is to determine a matrix relationship that couples the input–output parameters. The result, as we will see, is a transducer model that is functionally similar to that derived for piezoelectric materials directly from first principles.

To begin the analysis, consider the equivalent-circuit representation shown in Figure 7.16. The electrical variables of interest are the applied potential, v, and the induced current i. The mechanical variables of interest are the resulting force, f, at the location of the measurement point and the velocity of the measurement point, $\dot{u}$. The input–output variables are related to one another through three impedance terms: the electrical impedance, $Z_p(j\omega)$, and two mechanical impedance terms, $Z_{m1}(j\omega)$ and $Z_{m2}(j\omega)$. The model is complete once the three impedance elements have been determined in terms of the material and geometric parameters of the polymer transducer.

Let us first analyze the electrical properties of the ionomeric polymer transducer. Experimental measurements of the transducers demonstrate that the electrical impedance exhibits resistive behavior at low frequencies (typically, in the megahertz range), capacitive behavior in the midfrequency range, and then resistive behavior at higher frequencies. The electrical impedance of an ionomeric polymer transducer is compared to the impedance of a purely capacitive device such as a piezoelectric

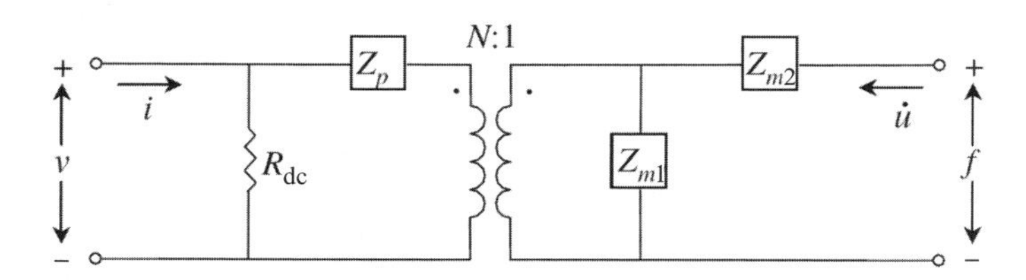

Figure 7.16 Equivalent-circuit model for the ionomeric polymer transducer analysis.

material. Examining the left-hand side of the equivalent-circuit model, we note that the relationship between applied voltage and induced current is written as

$$\frac{v}{i} = \frac{R_{\text{dc}} Z_p(\mathsf{s})}{R_{\text{dc}} + Z_p(\mathsf{s})}, \tag{7.26}$$

when $v_2 = 0$. The frequency-dependence of the electrical impedance is incorporated into equation (7.26) by expressing Z_p in the Laplace domain. The frequency-dependent component of the impedance can be represented as a parallel connection of resistor and capacitor networks whose Laplace-domain representation is

$$Z_p(\mathsf{s}) = \frac{1}{\sum_{i=1}^{n} [\mathsf{s}C_i/(1 + \mathsf{s}C_i R_i)]}, \tag{7.27}$$

where C_i and R_i are the individual capacitance and resistance terms of the network. The capacitive and resistive elements of the circuit are expressed in terms of the permittivity and resistivity terms through the expressions

$$\begin{aligned} R_{\text{dc}} &= \frac{\gamma_{\text{dc}} h}{L_t w} \\ R_i &= \frac{\gamma_i h}{L_t w} \\ C_i &= \frac{\eta_i L_t w}{h}. \end{aligned} \tag{7.28}$$

Substituting these expressions into equation (7.27) results in

$$Z_p(\mathsf{s}) = \frac{1}{\mathsf{s}} \frac{h/L_t w}{\sum_{i=1}^{n} [\eta_i/(1 + \mathsf{s}\eta_i \gamma_i)]}. \tag{7.29}$$

To simplify the expression we define a frequency-dependent permittivity as

$$\varepsilon(\mathsf{s}) = \sum_{i=1}^{n} \frac{\eta_i}{1 + \mathsf{s}\eta_i \gamma_i}. \tag{7.30}$$

This definition allows us to write the electrical impedance as

$$Z_p(\mathsf{s}) = \frac{1}{\mathsf{s}} \frac{h/(L_t w)}{\varepsilon(\mathsf{s})}. \tag{7.31}$$

The mechanical elements of the equivalent-circuit model are obtained by considering the right-hand side of the circuit with $f_2 = 0$. In this case the mechanical impedance terms are chosen to represent the static stiffness and inertial terms associated with the first transducer resonance. Assuming that the transducer is modeled as an Euler–Bernoulli beam, the static stiffness to a load applied at L_d is represented as

$$u = \frac{L_d^2}{3YI} f, \tag{7.32}$$

where $I = \frac{1}{12}wh^3$. Equation (7.32) is rewritten in the Laplace domain and Z_{m1} is expressed as

$$Z_{m1} = \frac{1}{\mathsf{s}} \frac{3YI}{L_d^3}. \tag{7.33}$$

The remaining mechanical impedance term is assumed to relate to the inertial properties of the cantilever beam. This term is chosen to model the first resonance of a cantilever beam,

$$Z_{m2} = \mathsf{s} \frac{3L_f^4 \rho w h}{L_d^3 \Gamma^4}, \tag{7.34}$$

where ρ is the material density and Γ is a parameter that relates to the boundary conditions of the beam. For a cantilever, $\Gamma = 1.875$.

The final component of the input–output transducer model is the turns ratio N of the equivalent-circuit model. This parameter is determined by assuming that the material exhibits a linear electromechanical coupling that is equivalent to a piezoelectric material. Although this model certainly does not model the physical mechanism associated with ion conduction due to an applied electric field, it does model the bending that occurs when a voltage is applied to the polymer. For this reason, the turns ratio is determined by first writing the relationship between charge, stress, and electric field for a material that exhibits linear coupling,

$$\mathrm{D} = d\mathrm{T} + \varepsilon \mathrm{E}. \tag{7.35}$$

Assuming that the sample can be modeled as an Euler–Bernoulli beam, the stress due to the applied force is written as

$$\mathrm{T} = \frac{f(L_d - x)h}{2I}. \tag{7.36}$$

The electric field is written as

$$\mathrm{E} = \frac{v}{h}, \tag{7.37}$$

and equations (7.36) and (7.37) can be substituted into equation (7.35) and integrated over the area of the transducer to obtain the total charge Q:

$$Q = \int_0^{L_t} \int_{-w/2}^{w/2} \left[d \frac{f(L_d - x)h}{2I} + \varepsilon \frac{v}{h} \right] dx_2\, dx_1. \tag{7.38}$$

The result after integration is

$$Q = 3\frac{dL_d^2}{h^2} f + \varepsilon \frac{L_t w}{h} v. \tag{7.39}$$

The turns ratio represents the amount of voltage that is produced for an applied force when the material is held in a short-circuit condition. Thus, we can set $Q = 0$ in equation (7.39) and solve for v/f, resulting in

$$N = \frac{3dL_d^2}{\varepsilon L_t w h}. \tag{7.40}$$

All four terms required for the input–output ionomeric transducer model have been defined: the electrical impedance, the two mechanical impedance terms, and the turns ratio associated with the transformer in the equivalent circuit. The input–output model is obtained by writing the equations that represent the equivalent circuit:

$$\begin{aligned} v &= R_{dc}(i - i_2) \\ Z_p i_2 + v_2 + R_{dc}(i_2 - i) &= 0 \\ f_2 + Z_{m1}(\dot{u}_2 - \dot{u}) &= 0 \\ f &= Z_{m2}\dot{u} + Z_{m1}(\dot{u} - \dot{u}_2) \\ v_2/N &= f \\ -i_2 N &= \dot{u}_2. \end{aligned} \tag{7.41}$$

Note that the Laplace variable has been omitted for clarity. A pair of coupled equations between voltage, force, current, and velocity is obtained by eliminating v_2, i_2, f_2, and $\dot{u}_2$ from equation (7.41). The result is

$$\begin{Bmatrix} v \\ f \end{Bmatrix} = \begin{bmatrix} \dfrac{R_{dc}(N^2 Z_{m1} + Z_p)}{R_{dc} + N^2 Z_{m1} + Z_p} & \dfrac{N R_{dc} Z_{m1}}{R_{dc} + N^2 Z_{m1} + Z_p} \\ \dfrac{N R_{dc} Z_{m1}}{R_{dc} + N^2 Z_{m1} + Z_p} & \dfrac{(Z_{m1} + Z_{m2})(R_{dc} + Z_p) + N^2 Z_{m1} Z_{m2}}{R_{dc} + N^2 Z_{m1} + Z_p} \end{bmatrix} \begin{Bmatrix} i \\ \dot{u} \end{Bmatrix}. \tag{7.42}$$

This set of equations can be viewed as a linear, coupled model of the ionomeric polymer transducer. The model has a number of similarities to the model for piezoelectric materials described earlier in the book. The coefficients in equation (7.42) can be simplified by introducing the assumption that the reflected mechanical impedance is negligible relative the the electrical impedance term Z_p. This assumption takes two forms, depending on whether the blocked or free boundary condition is considered. For the blocked boundary condition, the assumption can be expressed as

$$N^2 Z_{m1} << Z_p. \tag{7.43}$$

For the free boundary condition, the assumption is

$$N^2 \frac{Z_{m1} Z_{m2}}{Z_{m1} + Z_{m2}} << Z_p. \tag{7.44}$$

Employing the assumptions in equations (7.43) and (7.44), equation (7.42) becomes

$$\left\{ \begin{array}{c} v \\ f \end{array} \right\} = \left[\begin{array}{cc} \frac{Z_p}{1 + Z_p/R_{\mathrm{dc}}} & \frac{N Z_{m1}}{1 + Z_p/R_{\mathrm{dc}}} \\ \frac{N Z_{m1}}{1 + Z_p/R_{\mathrm{dc}}} & Z_{m1} + Z_{m2} \end{array} \right] \left\{ \begin{array}{c} i \\ \dot{u} \end{array} \right\}. \tag{7.45}$$

At certain frequencies the magnitude of Z_p/R_{dc} is much less than unity and the coupled equations can be reduced to

$$\left\{ \begin{array}{c} v \\ f \end{array} \right\} = \left[\begin{array}{cc} Z_p & N Z_{m1} \\ N Z_{m1} & Z_{m1} + Z_{m2} \end{array} \right] \left\{ \begin{array}{c} i \\ \dot{u} \end{array} \right\}. \tag{7.46}$$

Thus, we have three different coupled models for ionomeric polymer transducers. The most accurate under the assumptions of linearity and beam bending is equation (7.42). This form incorporates the most accurate representation of electromechanical coupling. Under the assumption that the coupling is small enough such that equations (7.43) and (7.44) are valid, equation (7.45) is used. At frequencies at which $|Z_p/R_{\mathrm{dc}}| << 1$, equation (7.46) can be used. In this case the equations are written as the matrix

$$\left\{ \begin{array}{c} v \\ f \end{array} \right\} = \frac{1}{\mathsf{s}} \left[\begin{array}{cc} \frac{h}{L_t w} \frac{1}{\varepsilon(\mathsf{s})} & \frac{3}{4} \frac{h^2}{L_d L_t} \frac{d(\mathsf{s}) Y(\mathsf{s})}{\varepsilon(\mathsf{s})} \\ \frac{3}{4} \frac{h^2}{L_d L_t} \frac{d(\mathsf{s}) Y(\mathsf{s})}{\varepsilon(\mathsf{s})} & \frac{Y(\mathsf{s}) w h^3}{4 L_d^3} + \mathsf{s}^2 \frac{3 L_f^4 \rho w h}{L_d^3 \Gamma^4} \end{array} \right] \left\{ \begin{array}{c} i \\ \dot{u} \end{array} \right\}. \tag{7.47}$$

This representation is similarity to the model developed for piezoelectric bimorphs. The primary difference in the models is that the material parameters for an ionomeric polymer transducer are frequency-dependent even at very low frequencies (<100 Hz).

This frequency dependence gives rise to a much different response than piezoelectric bimorphs.

7.4.2 Actuator and Sensor Equations

The input–output models developed for ionomeric polymer transducers are used to quantify the transducer performance. Typical performance parameters are the force output and deflection for a voltage or current input, or the sensor output for an imposed mechanical force or deflection.

As an example, consider computing the force output of the transducer when the velocity (or displacement) at the loading point is held equal to zero. In a manner similar to piezoelectric materials, we denote the variable held constant with a superscript:

$$\left(\frac{f}{v}\right)^{\dot{u}} \tag{7.48}$$

will represent the force-to-voltage relationship when the velocity is held equal to zero. This is equivalent to the blocked force condition for the transducer. The representation of the blocked force will depend on which input–output model is used in the analysis, or equivalently, which assumptions are made regarding the material coupling.

For the case in which no assumptions are made regarding the coupling, equation (7.42) with the assumption $\dot{u} = 0$, the equations are rewritten

$$\left\{\begin{array}{c} v \\ f \end{array}\right\} = \left[\begin{array}{c} \dfrac{R_{\text{dc}}(N^2 Z_{m1} + Z_p)}{R_{\text{dc}} + N^2 Z_{m1} + Z_p} \\ \\ \dfrac{N R_{\text{dc}} Z_{m1}}{R_{\text{dc}} + N^2 Z_{m1} + Z_p} \end{array}\right] i. \tag{7.49}$$

Solving the first expression for the current-to-voltage relationship and substituting into the second expression yields the blocked force relationship

$$\left(\frac{f}{v}\right)^{\dot{u}} = N \frac{Z_{m1}}{N^2 Z_{m1} + Z_p}. \tag{7.50}$$

If we assume that equation (7.43) is valid, we can reduce the expression for the blocked force to

$$\left(\frac{f}{v}\right)^{\dot{u}} = N \frac{Z_{m1}}{Z_p} = \frac{3}{4} \frac{wh}{L_d} d(\mathrm{s}) Y(\mathrm{s}). \tag{7.51}$$

A similar analysis can be performed for the free deflection. If we assume that $f = 0$ and solve for the displacement in terms of the applied voltage, we obtain a rather cumbersome expression unless we assume that the coupling is negligible.

Under this assumption the free deflection expression is

$$\left(\frac{u}{v}\right)^f = -N\frac{Z_{m1}}{sZ_p\left(Z_{m1}+Z_{m2}\right)}. \tag{7.52}$$

Substituting the expressions for the impedance functions into equation (7.52) results in the expression

$$\left(\frac{u}{v}\right)^f = \frac{-3L_d^2/h^2}{\Omega(\mathsf{s})\mathsf{s}^2+1}d(\mathsf{s}), \tag{7.53}$$

where

$$\Omega(\mathsf{s}) = \frac{12\rho L_f^4}{\Gamma^4 h^2 Y(\mathsf{s})}. \tag{7.54}$$

At frequencies much lower than the first resonance, the free deflection expression is

$$\left(\frac{u}{v}\right)^f = \frac{-3L_d^2}{h^2}d(\mathsf{s}). \tag{7.55}$$

Expressions for the sensing response of ionomeric polymer transducer are also derived from the input–output models. Two sensing expressions of interest are the relationship between output current to applied velocity, which is equivalent to the output charge to input deflection and the voltage output to an applied force. The relationship between current and velocity in the short-circuit ($v = 0$) condition is

$$\left(\frac{i}{\dot{u}}\right)^v = -N\frac{Z_{m1}}{Z_p} = -\frac{3}{4}\frac{wh}{L_d}d(\mathsf{s})Y(\mathsf{s}) \tag{7.56}$$

under the assumption that the reflected impedance is negligible. Similarly, the voltage output to a force input is

$$\left(\frac{v}{f}\right)^i = N\frac{R_{\mathrm{dc}}Z_{m1}}{\left(Z_{m1}+Z_{m2}\right)\left(R_{\mathrm{dc}}+Z_p\right)} \tag{7.57}$$

when the current is held equal to zero. If the inertial term is neglected compared to the static impedance, $|Z_{m2}/Z_{m1}| << 1$, and $\left|Z_p/R_{\mathrm{dc}}\right| << 1$, the expression reduces to

$$\left(\frac{v}{f}\right)^i = N = 3\frac{L_d^2}{L_t wh}\frac{d(\mathsf{s})}{\varepsilon(\mathsf{s})}. \tag{7.58}$$

7.4.3 Material Properties of Ionomeric Polymer Transducers

The input–output model depends on knowledge of three material parameters: the elastic modulus, the dielectric permittivity, and the strain coefficient. These parameters have identifical physical interpretation as in the case of piezoelectric materials, with a major difference being that these properties are assumed to be frequency dependent over the range of frequencies that is typically of interest to the analysis of ionomeric polymer transducers. These material parameters are generally determined through experiment. In this book we assume that the material parameters have been determined and are available for analysis.

Several studies of the material properties for ionomeric polymer transducers have shown that the representations have very similar forms for a wide range of transducer compositions. For analyses below approximately 50 Hz, the viscoelastic properties of the material are typically fairly small and the elastic modulus can be modeled with a constant value. The dielectric permittivity and the strain coefficient can be modeled with a Laplace domain representation of the form

$$\begin{aligned} d(\mathsf{s}) &= d_o \frac{\mathsf{s} + 1/\tau_{d1}}{(\mathsf{s} + 1/\tau_{d2})(\mathsf{s} + 1/\tau_{d3})} \\ \varepsilon(\mathsf{s}) &= \varepsilon_o \frac{\mathsf{s} + 1/\tau_{e1}}{(\mathsf{s} + 1/\tau_{e2})(\mathsf{s} + 1/\tau_{e3})}, \end{aligned} \tag{7.59}$$

where the frequency dependence is a represenation of the relaxation in the material behavior. It is this relaxation behavior that differentiates the material properties of ionomeric polymers with the material properties of piezoelectric materials.

The relaxation of the strain coefficient for ionomeric polymers generally occurs in the low- and high-frequency ranges. Thus, experiments have shown that

$$\tau_{d1} > \tau_{d2} > \tau_{d3}, \tag{7.60}$$

which results in a frequency response magnitude similar to the one shown in Figure 7.17*a*. The relaxation behavior of the dielectric properties of an ionomeric polymer transducer generally occurs at higher frequencies; therefore, the time constants associated with the dielectric permittivity are generally governed by

$$\tau_{e2} > \tau_{e1} > \tau_{e3} \tag{7.61}$$

Examining equations (7.51) and (7.55), we see that both the blocked force and free deflection are a function of the strain coefficient expression $d(\mathsf{s})$. Under the assumption that the elastic modulus is constant, we can express the response characteristics of the free deflection and blocked force in the time domain by performing partial fraction expansion on the Laplace domain expression in equation (7.59). The inverse

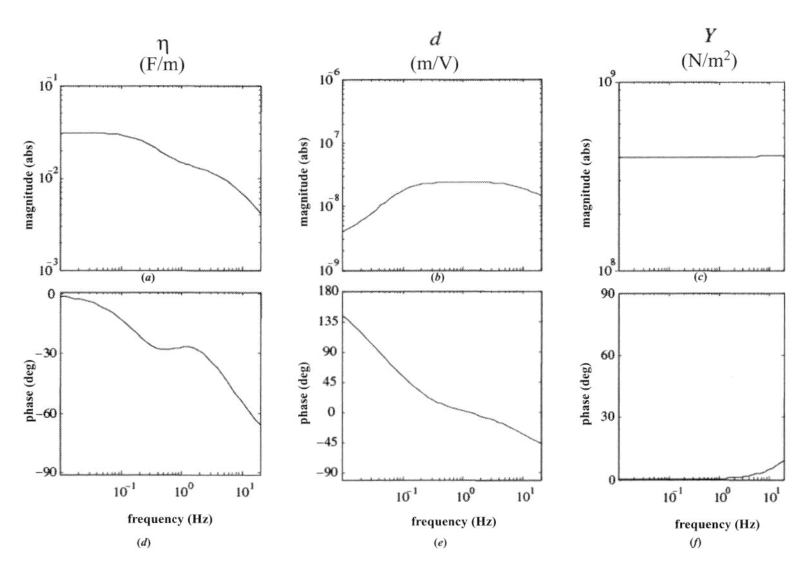

Figure 7.17 Representative frequency-dependent material properties for an ionomeric polymer transducer.

Laplace transform of the strain coefficient is

$$\mathcal{L}^{-1}\left[d(\mathrm{s})\right] = d_o \frac{\tau_{\mathrm{d1}}^{-1} - \tau_{\mathrm{d2}}^{-1}}{\tau_{\mathrm{d3}}^{-1} - \tau_{\mathrm{d2}}^{-1}} e^{-t/\tau_{\mathrm{d2}}} + d_o \frac{\tau_{\mathrm{d3}}^{-1} - \tau_{\mathrm{d1}}^{-1}}{\tau_{\mathrm{d3}}^{-1} - \tau_{\mathrm{d2}}^{-1}} e^{-t/\tau_{\mathrm{d3}}}. \tag{7.62}$$

The inverse Laplace transform illustrates the basic properties of the free deflection and blocked force response of the ionomeric polymer transducers. The fundamental response consists of a linear combination of exponential terms with time constants defined by τ_{d2} and τ_{d3}. The transmission zero defined by τ_{d1} determines the sign of the coefficients that multiply the exponential terms. Combining equation (7.62) with the expression for the quasistatic free deflection, equation (7.55), yields the inverse Laplace transform

$$\mathcal{L}^{-1}\left[\left(\frac{u}{v}\right)^f\right] = 3d_o \frac{L_d^2}{h^2}\left(A' e^{-t/\tau_{\mathrm{d2}}} + B' e^{-t/\tau_{\mathrm{d3}}}\right), \tag{7.63}$$

where

$$A' = \frac{\tau_{\mathrm{d1}}^{-1} - \tau_{\mathrm{d2}}^{-1}}{\tau_{\mathrm{d3}}^{-1} - \tau_{\mathrm{d2}}^{-1}}$$
$$B' = \frac{\tau_{\mathrm{d3}}^{-1} - \tau_{\mathrm{d1}}^{-1}}{\tau_{\mathrm{d3}}^{-1} - \tau_{\mathrm{d2}}^{-1}}. \tag{7.64}$$

Note that the negative sign in the free deflection expression has been omitted for clarity.

The free deflection to a step input at time zero with amplitude v_o is given by

$$\mathcal{L}^{-1}\left[\left(\frac{u}{v}\right)^f \frac{v_o}{\mathsf{s}}\right] = 3d_o v_o \frac{L_d^2}{h^2}\left[\tau_{d2}A'\left(1 - e^{-t/\tau_{d2}}\right) + \tau_{d3}B'\left(1 - e^{-t/\tau_{d3}}\right)\right]. \quad (7.65)$$

The free deflection can be written in a form that is amenable to nondimensional analysis by making the substitution $t = \tau_{d3}T$ into equation (7.65) and expressing the coefficients in front of the exponential terms as a function of τ_{d2}/τ_{d1} and τ_{d3}/τ_{d2}. The result is

$$\mathcal{L}^{-1}\left[\left(\frac{u}{v}\right)^f \frac{v_o}{\mathsf{s}}\right] = 3d_o v_o \frac{L_d^2}{h^2}\tau_{d3}\left[A(1 - e^{-T}) + B\left(1 - e^{-\tau_{d3}T/\tau_{d2}}\right)\right], \quad (7.66)$$

where

$$A = \frac{1 - (\tau_{d3}/\tau_{d2})(\tau_{d2}/\tau_{d1})}{1 - \tau_{d3}/\tau_{d2}}$$

$$B = \frac{\tau_{d2}/\tau_{d1} - 1}{1 - \tau_{d3}/\tau_{d2}}. \quad (7.67)$$

The general model expressed in equation (7.66) is useful for understanding the basic response characteristics of ionomeric polymer transducers. In this model there are two time constants of interest. The fundamental response time of the material is governed by the time constant τ_{d3}, while the ratio τ_{d3}/τ_{d2} governs the *relaxation* behavior of the material. Since $\tau_{d3}/\tau_{d2} < 1$, this relaxation occurs more slowly than the fundamental response time of the material.

If we assume that all time constants are positive and the inequality in equation (7.60) holds, then A is a positive value and B is negative. This result demonstrates that the relaxation will be in the direction opposite the initial motion. In this case the free deflection of the material will exhibit a fast rise governed by the time constant τ_{d3} and a slow relaxation in the opposite direction. As discussed above, the time response of the relaxation behavior is related to the ratio of time consants τ_{d3}/τ_{d2}.

Analysis of the time constants illustrates that the amount of relaxation is related directly to the ratio of the time constants. As $\tau_{d2}/\tau_{d1} \to 1$, the amount of relaxation decreases. If the ratio of τ_{d2}/τ_{d1} increases, the amount of relaxation increases as well, and the peak deflection of the transducer is decreased as shown in Figure 7.18*a*. The magnitude of the corresponding frequency responses is shown in Figure 7.18*b*.

Example 7.5 An ionomeric polymer transducer has been determined to have a frequency-dependent strain coefficient modeled by the function

$$d(\mathsf{s}) = 240{,}000 \frac{\mathsf{s} + 0.1}{(\mathsf{s} + 0.4)(\mathsf{s} + 10)} \text{ pm/V}.$$

The dimensions of the transducer are a free length of 30 mm, a thickness of 0.2 mm, and a width of 5 mm. The transducer is in a cantilevered configuration. (a) Determine

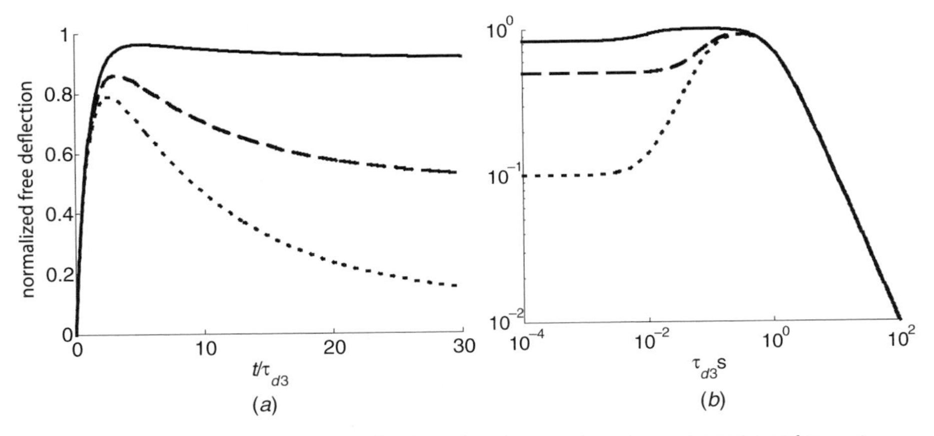

Figure 7.18 (*a*) Normalized free deflection of an ionomeric polymer transducer for $\tau_{d3}/\tau_{d2} = 1/10$ and three values of τ_{d2}/τ_{d1}: 1/1.1 (solid), 1/2 (dashed), and 1/10 (dotted); (*b*) corresponding frequency response magnitude.

the expression for the free deflection of the transducer for a 1-volt input. (b) Plot the deflection as a function of time.

Solution (a) The solution is obtained by first placing the expression in the form analyzed in equation (7.65). The time constants are determined to be

$$\begin{aligned} \tau_{d1} &= 10 \\ \tau_{d2} &= 2.5 \\ \tau_{d3} &= 0.1. \end{aligned}$$

Substituting this result in equation (7.67), we have

$$\begin{aligned} A &= \frac{1 - (0.1/2.5)(2.5/10)}{1 - 0.1/2.5} = 1.0313 \\ B &= \frac{2.5/10 - 1}{1 - 0.1/2.5} = -0.7813. \end{aligned}$$

The coefficient that multiplies the free deflection expression is

$$3 d_o v_o \frac{L_d^2}{h^2} \tau_{d3} = 3(240{,}000 \times 10^{-12}\ \mathrm{m \cdot s/V})(1\ \mathrm{V}) \left(\frac{30^2\ \mathrm{mm}^2}{0.2^2\ \mathrm{mm}^2} \right) (0.1\ \mathrm{s}^{-1}) = 1.6\ \mathrm{mm}.$$

The complete expression for the free deflection is

$$\delta(t) = (1.6)(1.3013)[(1 - e^{-10t}) - (0.7813)(1 - e^{-0.4t})]\ \mathrm{mm}.$$

(b) The free deflection plot for this problem is shown in Figure 7.19. The peak displacement of 1.44 mm occurs at approximately 0.4 s and the relaxation reduces the steady-state displacement to approximately 0.4 mm.

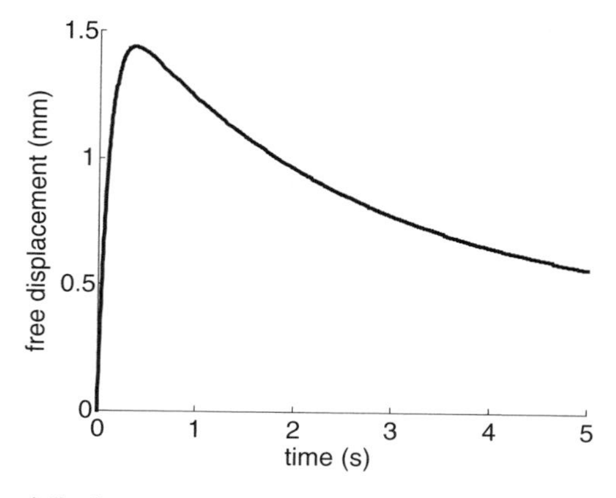

Figure 7.19 Free deflection response for the ionomeric polymer studied in Example 7.5.

One of the advantages of analyzing the strain coefficient in this manner is that the same attributes that apply to the free deflection also apply to the blocked force. Examining equation (7.51), we see that the blocked force expression is also directly related to the strain coefficient. Modeling the strain coefficient with the two-pole model shown in equation (7.59) allows us to write the step response of the blocked force as

$$\mathcal{L}^{-1}\left[\left(\frac{f}{v}\right)^{u}\frac{v_o}{s}\right] = \frac{3}{4}Y d_o v_o \frac{wh}{L_d}\tau_{d3}\left(Ae^{-T} + Be^{-\tau_{d3}T/\tau_{d2}}\right). \tag{7.68}$$

This result assumes that the elastic modulus of constant (i.e., viscoelastic) properties have been ignored. The relaxation behavior that is exhibited by the free deflection will also occur in the blocked force response. The primary difference in the computation of the blocked force and the free deflection is the geometric relationships. As in the case of a piezoelectric bimorph, increasing the ratio of the free length to the thickness will increase the free deflection but reduce the blocked force. This is illustrated in the following example.

Example 7.6 Compute the expression for the blocked force of the transducer analyzed in Example 7.5. Assume that the modulus of the material is 220 MPa. Also compute the peak blocked force.

Solution The coefficient in front of equation (7.68) is computed first:

$$\begin{aligned}\frac{3}{4}d_o v_o \frac{wh}{L_d}\tau_{d3} &= \frac{3}{4}(220 \times 10^6\ \mathrm{N/m^2})(240{,}000 \times 10^{-12}\ \mathrm{m \cdot s/V})(1\ \mathrm{V}) \\ &\quad \times \frac{(5 \times 10^{-3}\ \mathrm{m})(0.2 \times 10^{-3}\ \mathrm{m})}{30 \times 10^{-3}\ \mathrm{m}}(0.1\ \mathrm{s^{-1}}) \\ &= 0.132\ \mathrm{mN}.\end{aligned}$$

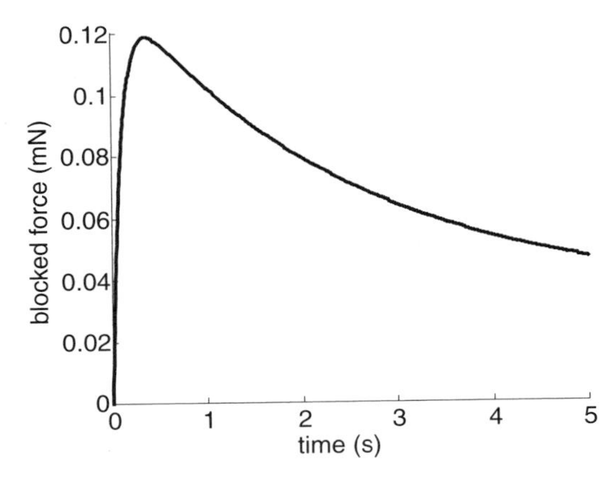

Figure 7.20 Blocked force response for the ionomeric polymer studied in Example 7.6.

The result of the analysis of the free deflection can now be incorporated with the previous computation. The result is

$$f(t) = (0.132)[(1.3013)(1 - e^{-10t}) - (0.7813)(1 - e^{-0.4t})] \text{ mN}.$$

The blocked force response is shown in Figure 7.20. The peak value is approximately 0.12 mN, and as is the case with the free deflection, the response exhibits relaxation to a smaller steady-state value.

7.5 CHAPTER SUMMARY

Electroactive polymers are an emerging class of smart materials that exhibit a wide range of coupling mechanisms. This chapter focused on the fundamental behavior of polarization-based or electrostatic electroactive polymers and those that exhibit coupling due to the migration of ionic species. In both instances the key facet of electroactive polymers is that they exhibit a large strain at the cost of (generally) lower induced stress than piezoelectric ceramics or shape memory alloys. Polarization-based materials and those that utilize electrostatic forces require a large electric field to operate. Typical values are on the order of 10 MV/m to greater than 100 MV/m. For this reason these materials need to be operated at high voltages to induce maximum stress and strain. These materials are characterized by free strain values on the order of 4% to greater than 100%, making them ideal actuators for high-displacement applications.

Conducting polymers and ionomeric transducers are the two types of ionic electroactive polymers studied in this book. In contrast to other electroactive polymers, these materials require only low voltage ($<$10 V) to operate but correspondingly higher current. Induced strain output of these materials are generally on the order of 1 to 10%. Basic phenomenological models of these actuator materials were presented in this book to illustrate design principles for these materials. Characteristics of the materials, such as the back relaxation of ionomeric polymers, was studied in relation to transfer function models of the transducers.

PROBLEMS

7.1. Find references that list the chemical composition of three types of polymers that are used for actuators: polyurethane, polypyrrole, and Nafion.

7.2. A dielectric elastomer actuator with an elastic modulus of 13 MPa and a relative dielectric constant of 4.5 is actuated by an applied electric field of 125 MV/m. Compute the strain in the material in the 3 direction. Assume that the material is incompressible.

7.3. A dielectric elastomer actuator is fabricated from an incompressible material that has an elastic modulus of 22 MPa and a relative dielectric constant of 8. Compute the blocked stress of an actuator that is 30 μm thick with an applied potential of 3 kV.

7.4. Plot the relationship between actuation stress and actuation strain in a dielectric elastomer actuator fabricated from an incompressible material that has an elastic modulus of 15 MPa and a relative dielectric constant of 5. The applied potential field is 150 MV/m.

7.5. Plot the cyclic voltammagram of a material that is modeled by the electric impedance function

$$Z(j\omega) = R_1 + \frac{R_2}{R_2 C \mathrm{s} + 1},$$

where $R_1 = 10$ kΩ, $R_2 = 20$ kΩ, and $C = 150$ μF, at frequencies of 0.1, 1, and 10 Hz.

7.6. The strain of a conducting polymer actuator has been measured to be 4.5% at a frequency of 0.05 Hz. Estimate the strain in the material at 30 Hz assuming that the assumption of semi-infinite diffusion is valid for the actuation model of the material.

7.7. Compute the volumetric energy density of the actuator studied in Example 7.4.

7.8. A linear conducting polymer actuator of length 40 cm and a diameter of 200 μm has been measured to have a free displacement of 15 mm. The elastic modulus of the material is 95 MPa when zero potential is applied and when a potential is applied.

(a) Compute the blocked force of the actuator.

(b) Compute the volumetric energy density of the actuator.

7.9. Repeat Problem 7.8 under the assumption that the elastic modulus changes to 55 MPa when a potential is applied.

7.10. Compute the blocked force and free deflection of a piezoelectric transducer of the same dimensions as the ionomeric transducer studied in Example 7.5. Use PZT-5H for the piezoelectric material properties.

7.11. Repeat Example 7.5 assuming that the strain coefficient of the material is modeled by the function

$$d(s) = 220{,}000 \frac{\mathrm{s} + 0.2}{(\mathrm{s} + 0.3)\,(\mathrm{s} + 20)}.$$

7.12. Repeat Example 7.6 assuming that the strain coefficient of the material is modeled by the function

$$d(s) = 220{,}000 \frac{\mathrm{s} + 0.2}{(\mathrm{s} + 0.3)\,(\mathrm{s} + 20)}.$$

NOTES

An excellent overview of electroactive polymer materials is the book edited by Bar-Cohen [53]. It contains a large number of articles on topics ranging from the various types of materials, to test methods, to applications. The seminal reference on irradiated PVDF materials is the work by Zhang et al. [52]. An article by Pelrine et al. [54] provides the most important reference to the development of dielectric elastomer materials, although these materials had been developed several years prior to the publication of this work [55]. An excellent reference on mechanical modeling of dielectric elastomer materials is that of Goulbourne et al. [56]. Conducting polymer actuators have been studied for a number of years. Some of the early work on these materials may be found in Baughman [57], Santa et al. [58], and Madden et al. [59]. More recent articles on the modeling and fabrication of conducting polymer actuators were the basis for the discussion in this chapter [60,61]. Probably the most seminal reference in the field of ionomeric polymer transducers is the work of Oguro et al. [62]. This group subsequently developed transfer function–based models of ionomeric transducers [63,64]. One of the most cited references in the field is the work of Shahinpor et al. [65]; this paper is an excellent overview of the early developments of this field. In recent years there have been numerous advances in this field, most notably the development of dry ionomeric actuators [66–68]. Physics-based models of ionomeric transducers have also been proposed. The work of Nemat-Nasser and his group has provided some of the most complete physics-based models of ionomeric polymer transducers to date [69–71]. The modeling sections of this chapter were based on the work of the author's group in the development of transducer models of these materials. Most of the work is based on publications by Newbury and Leo [72–74].

8

MOTION CONTROL APPLICATIONS

One of the most common uses of smart materials is in the field of motion control. Motion control applications are ubiquitious in modern society. Control surfaces on aircraft, printer heads in inkjet printers, and nanoprecision positioning of semiconductor wafers for microelectronics fabrication are all examples of important engineering applications that require controlled positioning.

A typical motion control application requires an actuator to provide the motive force and a sensor to measure the position, velocity, or acceleration of the object being moved. Traditional motion control applications might incorporate an electric, hydraulic, or pneumatic motor as the actuator, due to the fact that they are readily available from a number of vendors and the force and motion can be scaled to accommodate a wide range of applications. Sensing elements might include an electric or magnetic sensor that measures the relevant physical quantity, for example, an LVDT is an electromagnetic device that can output a voltage proportional to displacement or velocity. Positioning is achieved through either open-loop control or closed-loop feedback utilizing a control system to increase positioning accuracy or to reduce the effects of undesirable attributes such as deadband and stiction in the system.

Smart materials such as piezoelectric devices, shape memory alloys, or electroactive polymers can often provide advantages for motion control applications as compared to more traditional technologies. Many motion control applications require *high bandwidth*, meaning that the system must respond very quickly to changes in the input command. As discussed earlier, smart materials such as piezoelectric or electrostrictive devices can have response times in the millisecod or even microsecond range, making them advantageous in certain applications. Many motion control applications also require micrometer or nanometer positioning accuracy, which is often difficult to achieve with traditional hydraulic or pneumatic technology. Solid-state ceramic materials such as piezoelectrics and electrostrictives can be designed with submicrometer positioning accuracy. Finally, certain applications require large motion in a compact space. In this case shape memory alloys or electroactive polymers might be advantageous, due to their ability to produce large strain upon electrical or thermal stimulus.

In this chapter we utilize the theory developed in the second section of this book to analyze the use of smart materials for motion control applications. The analysis will focus on basic concepts of shaping the time response of the actuator to commands such as step changes in the input, ramp inputs for velocity control, and following a harmonic input.

8.1 MECHANICALLY LEVERAGED PIEZOELECTRIC ACTUATORS

In Chapter 4 the constitutive equations were used to derive the transducer expressions for piezoelectric stack and piezoelectric bimorph actuators. The analysis highlighted the fact that piezoelectric stacks are generally limited to outputs on the order of submicrometers to tens of micrometers, while piezoelectric bimorph actuators can generate output displacements on the order of hundreds of micrometers to approximately a millimeter. The primary trade-off in increasing the output displacement using a bimorph actuator is a substantial reduction in the output force compared to a piezoelectric stack. In Chapter 4 we presented detailed derivations of the transducer equations for a piezoelectric stack and piezoelectric bimorphs. From the prospective of device design using piezoelectric materials, much of the derivation can be summarized by the information listed in Table 4.2. For a piezoelectric stack, the free deflection is proportional to the number of layers and the blocked force is proportional to the cross-sectional area; for a piezoelectric bimorph, the free deflection is proportional to the square of the length-to-thickness ratio. The blocked force of a bimorph is proportional to the width of the actuator but inversely proportional to the length-to-thickness ratio. These relationships summarize the fundamental design principles for stacks and cantilevered bimorphs.

Piezoelectric devices can be purchased from a number of vendors in a variety of standard and customized forms. In many instances a vendor will have a standard set of devices based on the compositions of piezoelectric material they manufacture. In the case of stack actuators, these standard products generally consist of a few cross-section shapes (e.g., square, circular, or annular) that are manufactured with varying numbers of layers. Piezoelectric stacks are also generally low-voltage (100 to 200 V)

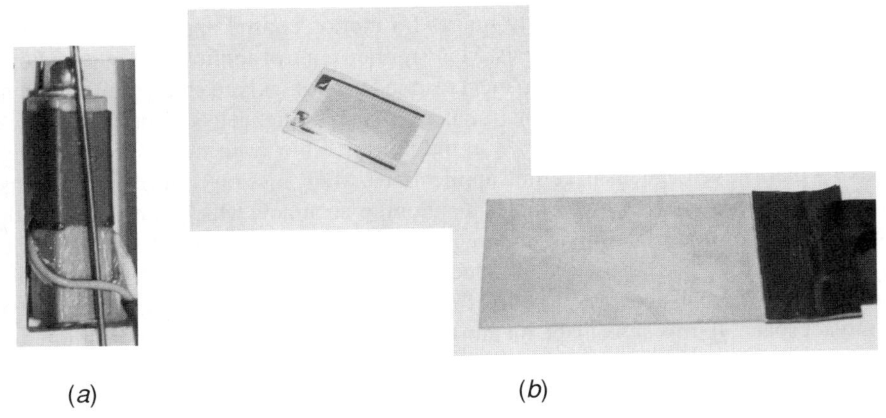

(*a*) (*b*)

Figure 8.1 (*a*) Commercially available piezoelectric stacks and (*b*) bimorph bender.

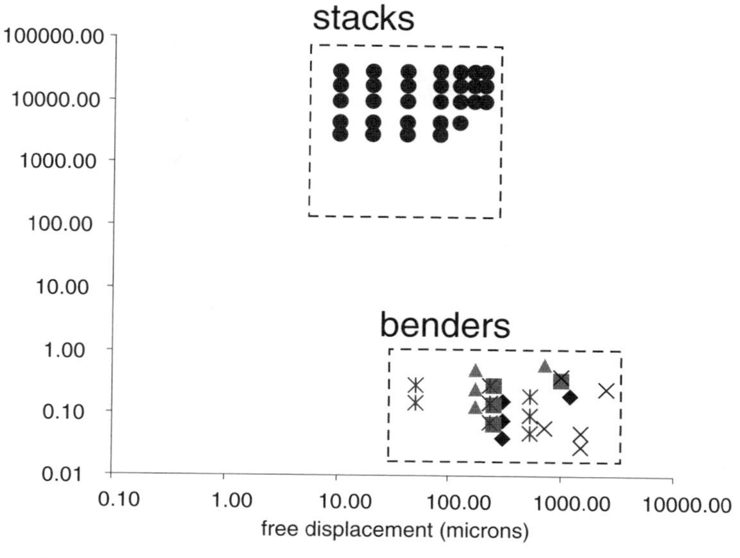

Figure 8.2 Representative force–deflection specifications for piezoelectric stacks and benders.

or high-voltage ($\approx$1000 V), depending on the thickness of the individual stack layers. An example of a commercially available stack is shown in Figure 8.1*a*.

Piezoelectric ceramic bimorphs are also available in a number of standard configurations. A majority of the devices are designed to be cantilever bimorphs, although these devices can also be used in a simply supported arrangement with proper design of the boundary conditions. A vendor will often have a series of bimorph devices whose lengths and widths vary so that the devices span a range of force and deflection requirements (see Figure 8.1*b*). Additionally, piezoelectric bimorph actuators are generally sold as either a *parallel* or *series* arrangement. In Chapter 4 the parallel connection of bimorph actuators was studied in detail and transducer equations were derived. In a series arrangement, the two piezoelectric layers are attached to the substrate such that their polarization directions are opposite one another. The layers are connected electrically so that one face of the piezoelectric layer is ground and the potential is applied to the opposing face on the second layer. Changing the wiring from series to parallel does not alter the basic performance of a piezoelectric bender; it only alters the amount of voltage that is required to produce a specified free deflection or output force. In a series configuration, the voltage requirements are twice that of the requirements for parallel operation, and the current requirements are one-half that of the parallel arrangement. Additionally, the capacitance of the series transducer will be one-fourth of the capacitance of the parallel transducer of equal geometry.

Typical blocked force and free deflection specifications for piezoelectric stacks and benders are also plotted in Figure 8.2. Plotting representative values for these two specifications illustrates that stacks and benders generally group into two regions in the design space of force and deflection. Bender actuators generally fall in the lower-right portion of the design space in which force is between 0.01 and 1 N and free

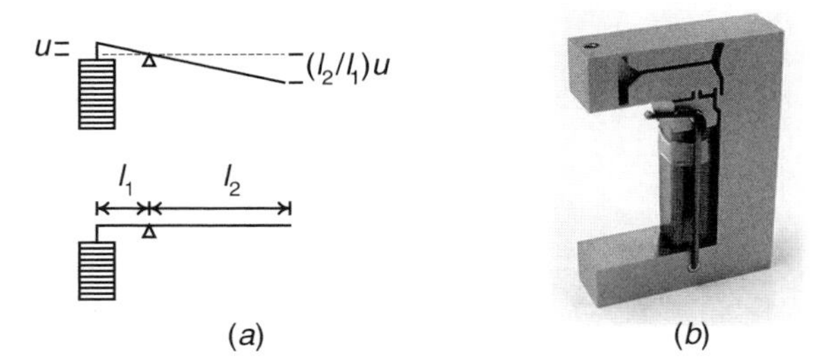

Figure 8.3 (*a*) Ideal concept of a mechanically leveraged piezoelectric stack; (*b*) commercially available stack [(*b*) Courtesy of Dynamic Structures and Materials].

displacement is between 50 μm and about 3 mm. Stacks generally produce blocked forces on the order of 100 N to well over 10,000 N but are limited in their free deflection to less than a few hundred micrometers.

During the past 10 to 15 years there has been substantial research and development of *leveraged* piezoelectric actuators. As the name implies, a leveraged piezoelectric actuator utilizes a mechanical interface between the piezoelectric material and the load to transform the force–deflection properties of the bare material. For piezoelectric materials, this generally results in an increase in the displacement output at the expense of a reduction in the force output. In this manner in entire range of devices can be fabricated that spans a broader spectrum of force–deflection characteristics than those shown in Figure 8.2.

One concept for increasing the displacement output of a piezoelectric device is to use a mechanical lever between the actuator and the load. Consider a rigid mechanical lever as shown in Figure 8.3*a* with lever arm ratio l_2/l_1. The motion u of the stack will be amplified to $(l_2/l_1)u$ on the opposite side of the lever. Since the work performed on both sides of the lever must be equal, this increase in deflection will be accompanied by a decrease in the output force by the inverse of the amplification ratio.

An example of a commercially available piezoelectric-lever actuator is shown in Figure 8.3*b*. The mechanical lever is not nearly as simple as the concept shown in Figure 8.3*a*. The device shown in Figure 8.3*b* produces approximately 100 μm of displacement, which indicates that the amplification ratio is on the order of 10:1. The lever must be designed to produce the desired amplification ratio in addition to having the necessary strength for the internal stresses that are generated by the flexure of the lever caused by the actuator displacement.

Another class of leveraged piezoelectric stack actuators are called *flextensional actuators*. Similar to the mechanical lever concept illustrated in Figure 8.3*a*, a flextensional actuator uses a specially designed housing to increase the output deflection caused by elongation of a piezoelectric stack. The primary attribute of a flextensional actuator is that the elongation of the stack produces flexure in the housing, which, in turn, produces an amplified linear displacement of the device. Proper design of the housing can lead to a controllable amplification ratio similar to the amplification ratio of a mechanical lever.

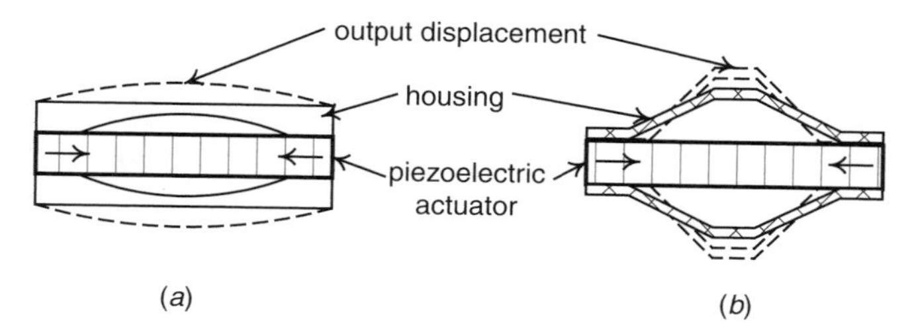

Figure 8.4 (*a*) Moonie and (*b*) Cymbal actuator concepts [76,77].

The concept of a flextensional device was originally proposed in the early 1970s by Royster. There was a renewed interest in this field in the late 1980s and 1990s, and two recent embodiments of the concept are the Moonie actuator and the Cymbal actuator. The basic concepts of Moonie and Cymbal actuators are shown in Figure 8.4. Both utilize the extension of a piezoelectric actuator (usually, a stack but could be a disk) to flex a housing. The primary difference between the designs is the shape of the housing. The housing can be designed to optimize the deflection and generative force of the actuator. Typical values of free displacement and blocked force for the Cymbal design are 160 μm and 15 N, respectively.

More sophisticated housing designs have been used to create a class of flextensional actuators that have a wide range of blocked force and free deflection specifications. Series of actuators available from vendors such as Dynamic Structures and Materials utilize a housing that incorporates flexure points to tailor the output deflection and force of the device (Figure 8.5). This concept enables a wide range of actuators with specified free deflection and blocked force characteristics. Actuators in this class range from devices that have free deflections between approximately 150 μm and over 2 mm and blocked force specifications between tens and hundreds of newtons. Examining Figure 8.2, we see that this range of output specifications bridges the gap between conventional multilayer designs and bimorph benders.

Most of the leveraged actuators discussed to this point have amplified the 33 operating mode of the transducer. As discussed in Chapter 4, operation in 33 mode advantageous compared to using the 31 mode, due to the higher volumetric energy density of a stack actuator compared to a bender actuator. Representative

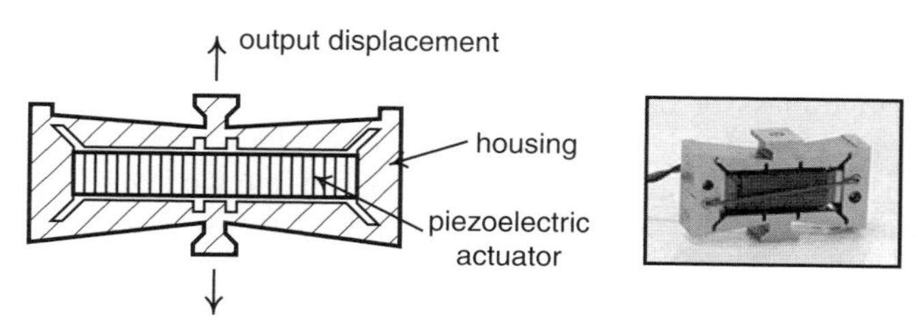

Figure 8.5 Flextensional design that utilizes flexure points in the housing to tailor the force–deflection characteristics. (Courtesy of Dynamic Structures and Materials.)

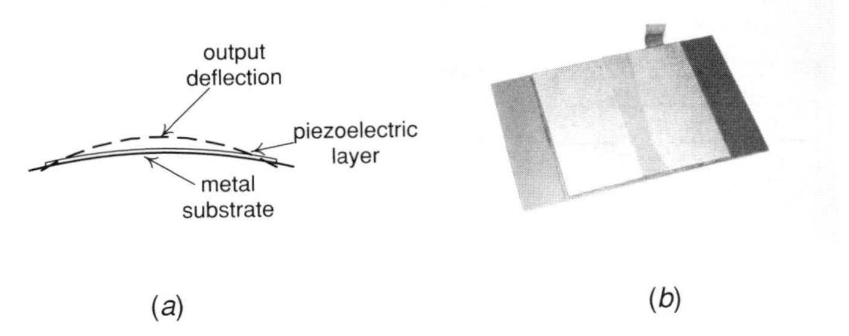

(*a*) (*b*)

Figure 8.6 (*a*) Prestressed unimorph actuator concept; (*b*) actuator.

computations in that chapter demonstrated that the volumetric energy density of a multilayer stack is on the order of 8 to 10 times higher than that of a bimorph bender.

Although the energy density is smaller for a bimorph bender, it is still useful to apply mechanical levering techniques to bending actuators for the purpose of increasing their displacement output to values that are even greater than that of a typical cantilevered bimorph. A large class of *unimorph* actuators have been developed that utilize the extension of a piezoelectric material in the 31 mode to actuate a substrate whose geometry has been designed to amplify the output deflection in the direction normal to the piezoelectric layer. As the name *unimorph* implies, these actuators generally use only a single piezoelectric layer offset from the neutral axis of the actuator–substrate composite to produce the response.

One class of unimorph actuators uses prestressed curved sections to amplify the displacement response. These actuators are fabricated by bonding a piezoelectric actuator to a metal substrate at elevated temperature. Upon cooling, the differing thermal expansion coefficients of the active and nonactive layers produces a curved shape in which both the substrate and the piezoelectric material are prestressed (Figure 8.6). Application of an electric field to the piezoelectric layer induces a bending moment in the piezoelectric–metal composite and induces a normal deflection in the actuator. Actuator materials in this class include the Rainbow, Cerambow, Crescent, and Thunder actuators. Thunder actuators have been studied extensively and are commercially available from Face International Corporation. They are sold in a variety of dimensions with free displacement specifications that range from 100 μm to 7 mm and blocked force specifications that range from 3 to over 100 N. Figure 8.6*b* is a picture of a commercially available actuator.

An innovative class of bimorph devices that enable scalable performance are building-block actuators. One type of actuator is the C-block, which consists of curved sections of piezoelectric ceramic (or polymer) shaped into the letter C. Another type of building-block actuator is the Recurve actuator. Both types are scalable in the sense that they can be connected mechanically in series or in parallel to tailor the force–deflection characteristics of the combined device. Dynamic experiments on single Recurve elements constructed with piezoceramic material measured displacements on the order of 70 μm for drive voltages in the linear range.

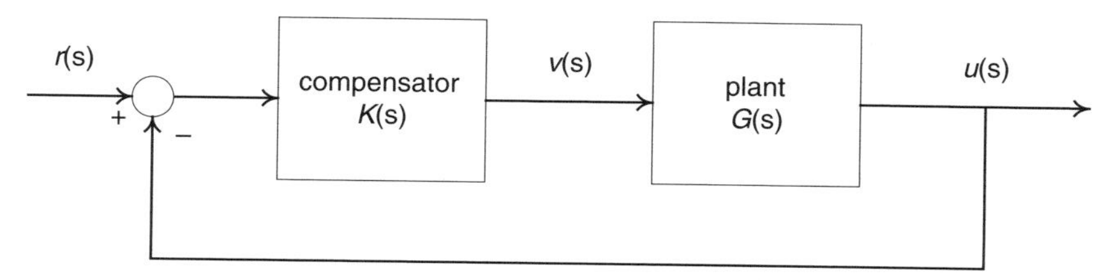

Figure 8.7 Block diagram for servo control of a piezoelectric device.

8.2 POSITION CONTROL OF PIEZOELECTRIC MATERIALS

One of the primary results of Chapter 4 was determining that the response of a linear piezoelectric device could be modeled as a second-order harmonic oscillator with voltage as the input and displacement as the state variable. The equation of motion for a piezoelectric device with an inertial load (i.e., simply a mass) is shown in equation (4.155) and repeated below for convenience:

$$m\ddot{u}(t) + k_a^{\mathrm{E}} u(t) = k_a^{\mathrm{E}} x_v v(t),$$

where $u(t)$ is the displacement, $v(t)$ is the applied voltage, m is the inertial mass, and k_a is the short-circuit stiffness of the actuator. A viscous damping term $c\dot{u}(t)$ could also be added to model the effects of energy dissipation in the system. As shown in Figure 4.31, the response of a piezoelectric actuator to a step change in the applied voltage generally exhibits substantial oscillation and overshoot before reaching the steady-state displacement, particularly when the amount of energy dissipation in the system is small.

Feedback control is one method of shaping the time response of a piezoelectric actuator so that the output more closely tracks the input command. The block diagram of a standard servo control system for an actuator is shown in Figure 8.7, where *plant* represents the piezoelectric actuator and *compensator* represents the control system that is designed to tune the output response. Both the plant and the compensator are represented in the Laplace domain, so that standard block diagram manipulations can be used to represent the response of the closed-loop system to the reference commands $r(t)$. Standard notation for feedback control systems specifies the plant as $G(\mathsf{s})$ and the compensator as $K(\mathsf{s})$.

The plant is represented by the transfer function of the equations of motion:

$$\frac{u(\mathsf{s})}{v(\mathsf{s})} = \frac{k_a^{\mathrm{E}}/m}{\mathsf{s}^2 + k_a^{\mathrm{E}}/m} x_v. \tag{8.1}$$

Recall that x_v is the free deflection per unit voltage of the piezoelectric actuator. The ratio of the stiffness to the mass is replaced by $k_a^{\mathrm{E}}/m = \omega_n^{\mathrm{E}^2}$ and the expression for the plant is nondimensionalized by substituting $\mathsf{s} = \omega_n^{\mathrm{E}} \sigma$, where σ is nondimensional frequency in which 1 represents the undamped short-circuit natural frequency of the plant. Making these substitutions into equation (8.1) and introducing a linear damping term, $2\zeta\sigma$, into the expression for the plant yields

$$\frac{u(\sigma)}{v(\sigma)} = \frac{1}{\sigma^2 + 2\zeta\sigma + 1} x_v. \tag{8.2}$$

The expression for the plant transfer function is now in a form that allows analysis using standard servo control, or classical control techniques.

8.2.1 Proportional–Derivative Control

The simplest type of compensator that yields substantial change in the response of the piezoelectric actuator system is *proportional–derivative control*. As the name implies, proportional–derivative (PD) control is a combination of proportional control, which is used to vary the speed of response of the closed-loop system, and derivative control, which is used to minimize the overshoot and settling time of the system.

The compensator of a PD controller is of the form

$$K(\sigma) = k_p + k_d\sigma, \tag{8.3}$$

where k_p is the proportional gain and k_d is the derivative gain. The *loop transfer function* of the system with PD control is

$$K(\sigma)G(\sigma) = k_p \frac{1 + \sigma/\tau_{\text{pd}}}{\sigma^2 + 2\zeta\sigma + 1} x_v. \tag{8.4}$$

The numerator of the loop transfer function is shown in this form to highlight the fact that the ratio of the proportional gain to the derivative gain, $\tau_{\text{pd}} = k_p/k_d$, determines the frequency of the zero of the loop transfer function. The zero, in turn, determines the frequency at which the loop transfer function begins to exhibit phase lead.

It is well known from classical control theory that a proportional–derivative controller will exhibit 45° of phase lead at $\sigma = \tau_{\text{pd}}$. For this reason it is most desirable from the standpoint of reducing overshoot and reducing settling time to place the zero at a frequency near to, or below, the natural frequency of the system. Decreasing the value of τ_{pd} will increase the amount of phase lead near the resonance of the piezoelectric device and will provide sufficient phase lead to produce the necessary phase margin at the gain crossover frequency (Figure 8.8*a*). The overall gain of the loop transfer function is then set with the proportional gain k_p. Increasing the gain of $K(\sigma)G(\sigma)$ increases the gain crossover frequency, as shown in Figure 8.8*b*.

The closed-loop transfer function between the reference input and the output is obtained using standard block diagram manipulation,

$$\frac{u(\sigma)}{r(\sigma)} = \frac{KG}{1 + KG} = \frac{k_p[(1 + \sigma/\tau_{\text{pd}})/(\sigma^2 + 2\zeta\sigma + 1)]x_v}{1 + k_p[(1 + \sigma/\tau_{\text{pd}})/(\sigma^2 + 2\zeta\sigma + 1)]x_v}. \tag{8.5}$$

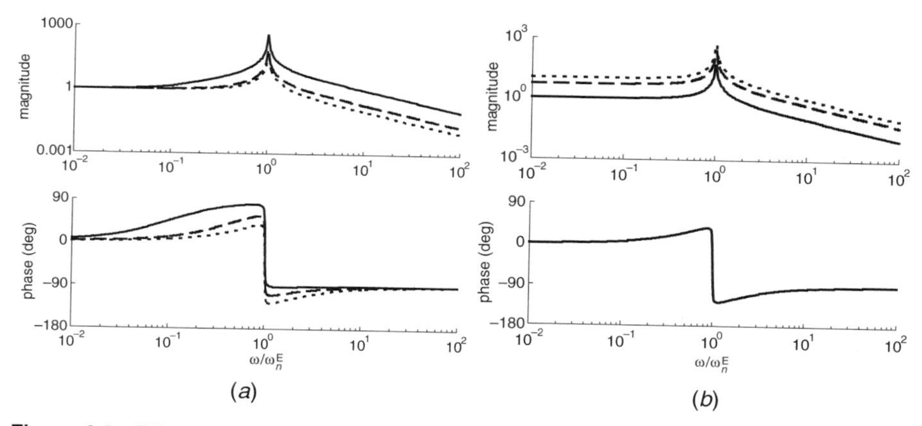

Figure 8.8 Effect of PD control parameters on the loop gain and phase of a piezoelectric device: (*a*) $k_p x_v = 1$ and $\tau_{pd} = 0.1$ (solid), $\tau_{pd} = 0.5$ (dashed), and $\tau_{pd} = 1$ (dotted); (*b*) $\tau_{pd} = 1$ and $k_p x_v = 1$ (solid), $k_p x_v = 5$ (dashed), and $k_p x_v = 10$ (dotted).

Equation (8.5) is manipulated to yield

$$\frac{u(\sigma)}{r(\sigma)} = \frac{k_p x_v \left(\sigma/\tau_{pd} + 1\right)}{\sigma^2 + (2\zeta + k_p x_v/\tau_{pd})\,\sigma + 1 + k_p x_v}. \tag{8.6}$$

Equation (8.6) illustrates that the closed-loop transfer function for a piezoelectric device with PD control is itself a second-order system with additional numerator dynamics. The poles of the system can be chosen arbitrarily by noting that

$$\begin{aligned} \zeta_{cl} &= \frac{\zeta + k_p x_v/2\tau_{pd}}{\sqrt{1 + k_p x_v}} \\ \frac{\omega_{cl}^2}{\omega_n^{E2}} &= 1 + k_p x_v. \end{aligned} \tag{8.7}$$

Applying the final value theorem to equation (8.6) for a step input of value R to the system produces the steady-state response

$$\frac{u(\infty)}{R} = \frac{k_p x_v}{1 + k_p x_v} = \frac{1}{(1/k_p x_v) + 1}. \tag{8.8}$$

The steady-state error is

$$e(\infty) = u(\infty) - R = \frac{-R}{1 + k_p x_v}, \tag{8.9}$$

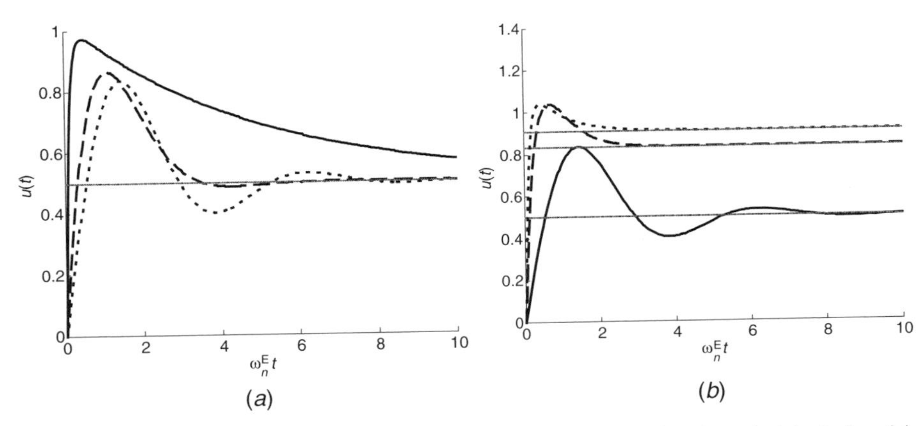

Figure 8.9 Effect of PD control parameters on the step response of a piezoelectric device: (*a*) $k_p x_v = 1$ and $\tau_{pd} = 0.1$ (solid), $\tau_{pd} = 0.5$ (dashed), and $\tau_{pd} = 1$ (dotted); (*b*) $\tau_{pd} = 1$ and $k_p x_v = 1$ (solid), $k_p x_v = 5$ (dashed), and $k_p x_v = 10$ (dotted).

which shows that increasing the proportional gain reduces the steady-state error of the system. The negative sign indicates that the closed-loop system settles to a final value that is less than the value desired.

Step response plots for two values of τ_{pd} and multiple gain values are shown in Figure 8.9. Decreasing the value of the PD time constant from 1 to 0.1 has a detrimental effect on the settling time of the closed-loop system. Comparing Figure 8.9*a* and *b* it is clear that decreasing the value of τ_{pd} tends to increase the speed of the initial rise but to slow down the response as it approaches the steady-state value. This effect would become even more pronounced if τ_{pd} was decreased below 0.1.

Example 8.1 A piezoelectric stack actuator with short-circuit stiffness of 22 N/μm and a free deflection of 0.03 μm/V is driving a 10-g load. (a) Compute the proportional derivative gains such that the closed-loop system has a damping ratio of 0.707 and a natural frequency of 8.5 kHz. (b) Plot the step response to a step command of 1 μm. Estimate the time to peak and settling time of the response and compute the steady-state error.

Solution (a) The proportional gains required to achieve a specified closed-loop damping ratio and natural frequency are listed in equation (8.8). The short-circuit natural frequency of the device is

$$\omega_n^E = \sqrt{\frac{22 \times 10^6 \text{ N/m}}{10 \times 10^{-3} \text{ kg}}} = 46{,}904 \text{ rad/s}.$$

The desired natural frequency is $\omega_{cl} = 8500 \times 2\pi = 53,407$ rad/s. Substituting the definition $\tau_{pd} = k_p/k_d$ into equation (8.8) yields a set of two equations and two unknowns for the compensator gains,

$$\frac{(53{,}407)^2}{(46{,}904)^2} = 1 + k_p \left(0.03 \times 10^{-6}\ \text{m/V}\right)$$
$$0.707 = \frac{k_d \left(0.03 \times 10^{-6}\ \text{m/V}\right)/2}{53{,}407/46{,}904}.$$

Solving the two equations yields the gains

$$k_p = 9.884 \times 10^6$$
$$k_d = 5.367 \times 10^7.$$

(b) The step response is computed by first determining the closed-loop transfer function between the reference input and the output from equation (8.6). Computing the time constant of the PD compensator from the gains yields

$$\tau_{\text{pd}} = \frac{9.884 \times 10^6}{5.367 \times 10^7} = 0.18.$$

Substituting the gains computed in part (a) and τ_{pd} into equation (8.6) produces

$$\frac{u(\sigma)}{r(\sigma)} = \frac{\left(9.884 \times 10^6\right)\left(0.03 \times 10^{-6}\ \text{m/V}\right)(\sigma/0.18 + 1)}{\sigma^2 + 1.41\sqrt{1 + \left(9.884 \times 10^6\right)\left(0.03 \times 10^{-6}\ \text{m/V}\right)}\sigma + 1 + \left(9.884 \times 10^6\right)\left(0.03 \times 10^{-6}\ \text{m/V}\right)}$$
$$= \frac{1.65\sigma + 0.30}{\sigma^2 + 1.61\sigma + 1.30}.$$

The expression for the transfer function is in nondimensional frequency. To transform back into a dimensional frequency for time-domain analysis, first substitute $\sigma = \mathsf{s}/\omega_n^{\text{E}}$ and then multiply the numerator and denominator by $\omega_n^{\text{E}^2}$:

$$\frac{u(\mathsf{s})}{r(\mathsf{s})} = \frac{1.65\omega_n^{\text{E}}\mathsf{s} + 0.30\omega_n^{\text{E}^2}}{\mathsf{s}^2 + 1.61\omega_n^{\text{E}}\mathsf{s} + 1.30\omega_n^{\text{E}^2}}.$$

The step response is obtained by substituting $r(\mathsf{s}) = 1\ \mu\text{m}/\mathsf{s}$ into the transfer function and solving for the Laplace representation of the output:

$$u(\mathsf{s}) = \frac{1.65\omega_n^{\text{E}}\mathsf{s} + 0.30\omega_n^{\text{E}^2}}{\mathsf{s}\left(\mathsf{s}^2 + 1.61\omega_n^{\text{E}}\mathsf{s} + 1.30\omega_n^{\text{E}^2}\right)}(1 \times 10^{-6}).$$

The step response is plotted in Figure 8.10. The time to peak is estimated to be approximately 25 μs and the settling time is approximately 150 μs.

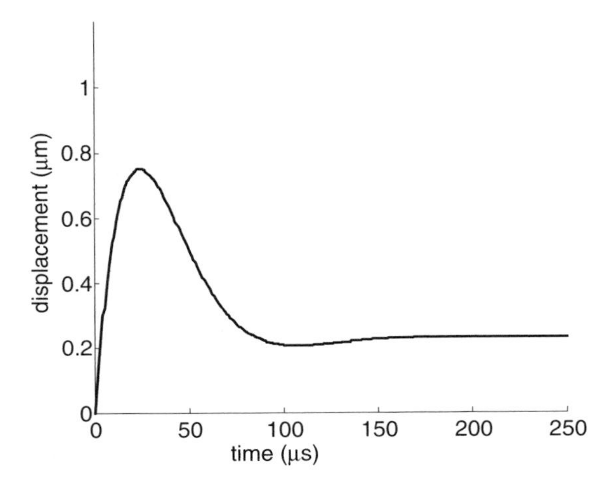

Figure 8.10 Step response of the closed-loop system analyzed in Example 8.1.

The steady-state error to the 1-μm step input command is computed from equation (8.9):

$$e(\infty) = \frac{-1\ \mu\text{m}}{1 + \left(9.884 \times 10^6\right)\left(0.03 \times 10^{-6}\ \text{m/V}\right)}$$
$$= -0.77\ \mu\text{m},$$

which is consistent with the response shown in Figure 8.10.

8.2.2 Proportional–Integral–Derivative Control

A proportional–derivative controller maintains the response speed of a piezoelectric device. This is illustrated in Figure 8.8 by the fact that the gain crossover frequency is near the resonance frequency of the actuator; therefore, the closed-loop response speed will be similar to that of the device in the open loop. The primary advantage of proportional–derivative control compared to simply using the open-loop response to position the device is that the overshoot and settling time can be tuned by proper choice of the compensator parameters.

The primary drawback of using PD control for positioning a piezoelectric device is the nonzero steady-state error. Steady-state error can be reduced by increasing the proportional gain of the compensator [see equation (8.9)], but there will be a practical limit to the amount that k_p can be increased due to noise limitations and stability limits caused by higher-frequency dynamics of the device.

A standard classical control method for eliminating steady-state error in the commanded displacement is to introduce *integral control* into the compensator. As the name implies, integral control produces a control output that is proportional to the error between the reference input and the displacement. Design rules for proportional–integral–derivative (PID) compensation have a long history and are well documented in numerous textbooks on feedback control theory. In this section we concentrate the discussion on the basic properties of PID compensation as they apply to the feedback control of a piezoelectric device modeled as a second-order oscillator.

The transfer function for a PID compensator is (in nondimensional frequency)

$$K(\sigma) = \frac{k_i}{\sigma} + k_p + k_d\sigma. \tag{8.10}$$

The terms in the transfer function can be placed into a single ratio of polynomials by finding a common denominator,

$$K(\sigma) = k_p \frac{(k_d/k_p)\sigma^2 + \sigma + k_i/k_p}{\sigma}. \tag{8.11}$$

Comparing equation (8.11) with equation (8.3) illlustrates the differences between the two types of compensation. A PID compensator includes not a single zero, as with PD control, but two zeros whose frequencies are determined by the ratio of the compensator gains k_i/k_p and k_d/k_p. Moreover, the PID compensator includes an integral term that is expressed as the single pole at $\sigma = 0$ in the denominator. It is this attribute of PID compensation that will produce zero steady-state error in the step response of the actuator.

The loop transfer function is obtained by forming the product of equation (8.11) with equation (8.2):

$$K(\sigma)G(\sigma) = k_p \frac{(k_d/k_p)\sigma^2 + \sigma + k_i/k_p}{\sigma(\sigma^2 + 2\zeta\sigma + 1)} x_v. \tag{8.12}$$

The closed-loop transfer function between the reference input and the output is

$$\frac{u(\sigma)}{r(\sigma)} = \frac{KG}{1+KG} = \frac{(k_d/k_p)\sigma^2 + \sigma + (k_i/k_p)}{\sigma^3 + (2\zeta + k_d x_v)\sigma^2 + (1 + k_p x_v)\sigma + k_i x_v} k_p x_v. \tag{8.13}$$

The integrator in the compensator raises the order of the denominator by 1 so that the closed-loop transfer function is third order and thus has three poles. Applying the final value theorem to equation (8.13) for a step input of R produces the steady-state value of the closed-loop system,

$$\frac{u(\infty)}{R} = \frac{k_i/k_p}{k_i x_v} k_p x_v = 1. \tag{8.14}$$

Equation (8.14) illustrates that the steady-state error to a step input is equal to zero for all values of the control parameters. This is a clear distinction with PD control studied in Section 8.2.1, in which the steady-state error was finite and could

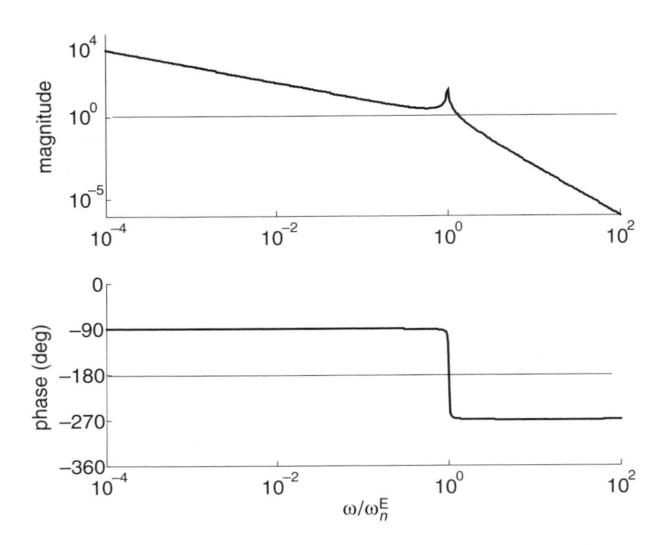

Figure 8.11 Magnitude and phase response of a piezoelectric actuator with unity gain integral control.

only be reduced by increasing the proportional gain of the controller. For PID control, the steady-state error is always zero no matter how we choose the compensator parameters.

PID compensation allows us to tune the transient response of the closed-loop system independent of the requirement to maintain steady-state error. This is a clear advantage over PD control, so a question might arise as to what the drawback is to implementing a PID compensator for controlling a piezoelectric device compared to a PD compensator. The primary drawback is that integral control introduces a 90° phase lag into the forward loop KG. The additional phase lag due to the integral compensation makes the stability of the closed-loop system a function of the phase lag caused by the resonant dynamics of the piezoelectric actuator. In Figure 8.11, the magnitude and phase of the loop transfer function KG is shown for pure integral compensation. The 90° phase lag in conjuction with the 180° phase lag that occurs at resonance introduces a total of 270° of phase loss into the control system. Pure integral compensation (as shown in the figure) would result in an unstable closed-loop system due to the fact that the gain margin and phase margin of the forward loop are negative.

A simple method of compensating for the phase lag introduced by integral control would be to reduce the integral gain to recover postive gain and phase margins. For pure integral control with gain $k_i \approx 1/75$, the closed-loop response of the device exhibits a stable response with a settling time on the order of $t = 300/\omega_n^{\mathrm{E}}$. Recall that the settling times for a PD compensator were on the order of $2/\omega_n^{\mathrm{E}}$ to $10/\omega_n^{\mathrm{E}}$, and it is clear that pure integral control produces a severe reduction in the closed-loop speed of response of the device for the added benefit of eliminating steady-state error.

The next step is to apply tuning rules for PID compensators to shape the closed-loop step response. There are numerous methods for PID design, and there is not one accepted method of tuning the compensator for a particular type of plant. With the advent of automated design tools and fast microprocessors in the last 10 to 15 years, much of the emphasis has shifted to automated and adaptive design of PID controllers. Computer-aided control design enables fast iteration of control design for a particular application.

The present discussion focuses on the basic properties of design for a piezoelectric device modeled as a second-order oscillator. Figure 8.11 clearly illustrates that the fundamental problem in PID design for a piezoelectric actuator is that the phase lag associated with integral control can introduce negative stability margins near the resonance of the device. Stability margins can be recovered by introducing phase lead using the zeros of the PID compensator. The zeros are obtained by setting the numerator of equation (8.11) equal to zero and solving for the roots of the polynomial. Doing so results in the expressions

$$\sigma_{z1}, \sigma_{z2} = \frac{k_p}{2k_d}\left(-1 \pm \sqrt{1 - 4\frac{k_d}{k_p}\left(\frac{k_i}{k_p}\right)}\right). \tag{8.15}$$

Equation (8.15) demonstrates that the center frequency of the zeros is shifted by varying k_p/k_d and that the zeros are a pair of purely root zeros if $1 > 4(k_d/k_p)(k_i/k_p)$ and are two complex-conjugate pairs if $1 < 4(k_d/k_p)(k_i/k_p)$. One design method is to choose k_p/k_d to set the center frequency of the compensator zeros and then choose k_i/k_p to vary the sharpness of the phase lead. Choosing two purely real roots will produce a more gradual phase lead in the frequency domain.

As discussed above, the parameters chosen for a particular design are heavily dependent on the application. Certain applications will stress, for example, the rise time of the actuator but will not be as concerned with the settling time. Other designs will have the opposite requirements, where the settling time of the closed-loop system is the most important design constraint. In any case, a computer-aided design package will greatly assist in the design iterations.

Representative step responses using PID compensation on a piezoelectric actuator are obtained by applying the design concepts discussed above. By choosing the compensator to have two repeated real roots, it is specified that the compensator will provide 90° of phase lead at the center frequency of the zeros. This phase lead will eventually rise to 180° at one decade above the center frequency. Representative step responses are obtained by applying this concept and varying the center frequency in relation to the resonant response of the actuator.

In design 1 the center frequency of the zeros is chosen to be at the resonance of the actuator by setting $k_i/k_p = k_d/k_p = 1/2$. This choice produces 90° of phase lead at $\sigma = 1$. Varying the gain from $k_p x_v = 1/5$ to $k_p x_v = 5$ changes the gain crossover (Figure 8.12*a*) and increases the speed of response of the system (Figure 8.12*b*). For all choices of parameters the steady-state response settles to a value of 1 (it is difficult to see those for the case of $k_p x_v = 1/5$, but simulations to longer times indicate that

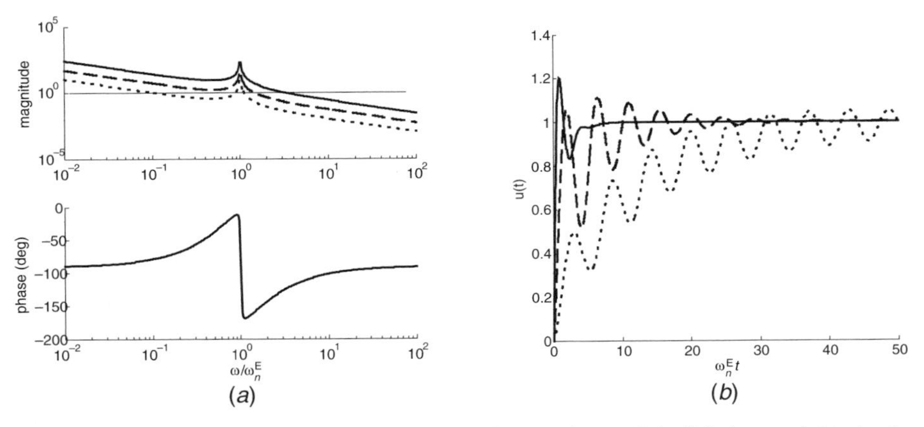

Figure 8.12 (*a*) Loop gain and phase for PID design 1: $k_p x_v = 5$ (solid), $k_p x_v = 1$ (dashed), $k_p x_v = 1/5$ (dotted); (*b*) step response for PID design 1: $k_p x_v = 5$ (solid), $k_p x_v = 1$ (dashed), $k_p x_v = 1/5$ (dotted).

this is the case) as expected for PID compensation. The approximate settling time of the system decreases from greater than $50/\omega_n^E$ to approximately $10/\omega_n^E$ as the gain is increased. The settling time for the PID design is generally greater than that for the PD designs studied earlier in the section (compare Figure 8.12 with Figure 8.9).

The second representative PID design analyzed the effects of lowering the center frequency of the PID zeros. In design 2, values of $k_d/k_p = 2$ and $k_i/k_p = 1/8$ are chosen. This choice of parameters sets the center frequency of the zeros to be $\sigma = 0.25$ and makes the two zeros purely real repeated roots. Choosing the center frequency of the zeros to be well below resonance increases the phase lead near the resonance (Figure 8.13*a*). As a consequence, the step response of the closed-loop system does exhibit the oscillations that were visible in the response when the center frequency was at the resonance of the actuator. Comparing Figure 8.12*b* with Figure 8.13*b* it is clear that the trade-off in reducing the oscillations is an increase in the settling time of the system. Just as in the case of PD control, when the compensator is placed well below resonance, placing the compensator zero center frequency in a PID compensator increases the settling time of the step response. For the PID compensators studied in this design, the settling time was between $50\ /\omega_n^E$ and $\approx 20/\omega_n^E$ for the two highest-gain cases.

Example 8.2 The actuator studied in Example 8.1 had a short-circuit stiffness of $22\ \text{N}/\mu\text{m}$ and a free displacement per unit voltage of $0.03\ \mu\text{m}/\text{V}$. The parameters from PID design 1, $k_p x_v = 1$, are used to design a PID compensator for this actuator. (a) Compute the transfer function for the compensator in dimensional form as a function of the Laplace variable s. (*b*) Use Figure 8.12*b* to estimate the time required to stay within the bounds of $\pm 1\%$ of the steady-state response of the actuator. Compare the result to the settling time obtained with the PD control design studied in Example 8.1.

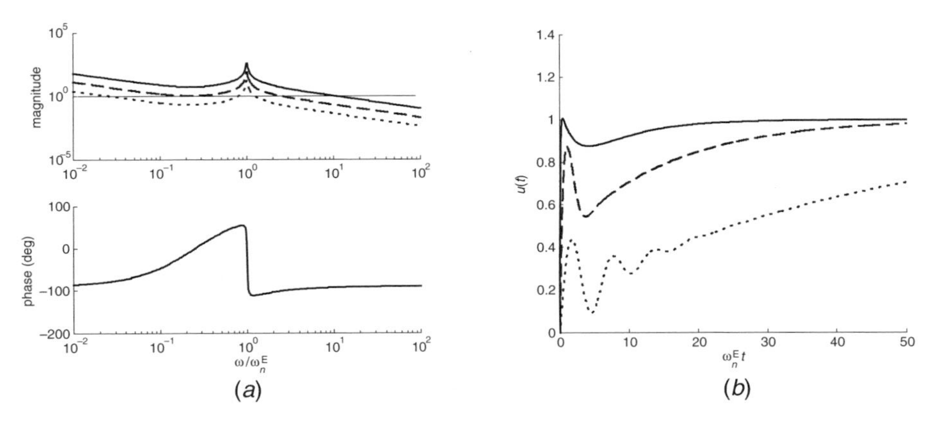

Figure 8.13 (*a*) Loop gain and phase for PID design 2: $k_p x_v = 5$ (solid), $k_p x_v = 1$ (dashed), $k_p x_v = 1/5$ (dotted); (*b*) step response for PID design 2: $k_p x_v = 5$ (solid), $k_p x_v = 1$ (dashed), $k_p x_v = 1/5$ (dotted).

Solution (a) The short-circuit natural frequency was computed in Example 8.1 to be 46,904 rad/s. The nondimensional form of the compensator is shown in equation (8.11). Substituting the definition of σ into equation (8.11) yields

$$K(\mathsf{s}) = k_p \frac{(k_d/k_p)\left(\mathsf{s}/\omega_n^{\mathrm{E}}\right)^2 + \left(\mathsf{s}/\omega_n^{\mathrm{E}}\right) + k_i/k_p}{\mathsf{s}/\omega_n}.$$

The proportional gain is computed from $k_p x_v = 1$:

$$k_p = \frac{1}{0.03 \times 10^{-6}\ \mathrm{m/V}} = 3.33 \times 10^7\ \mathrm{V/m}.$$

Substituting $k_p = 3.33 \times 10^7$, $k_d/k_p = 1/2$, and $k_i/k_p = 1/2$ into the expression along with the value for the natural frequency yields

$$\begin{aligned} K(\mathsf{s}) &= (3.33 \times 10^7)\frac{\mathsf{s}^2(1/2/46,904)^2 + \mathsf{s}(1/46,904) + 1/2}{\mathsf{s}/46,904} \\ &= \frac{355.33\mathsf{s}^2 + 3.33 \times 10^7\mathsf{s} + 7.81 \times 10^{11}}{\mathsf{s}}, \end{aligned}$$

which is the transfer function of the compensator. This simple example shows the value of working with parameters in nondimensional form first rather than accounting for the natural frequency of the actuator throughout the analysis.

(b) Judging from Figure 8.12*b*, the settling time to reach $\pm 1\%$ of the steady-state value of 1 is approximately $30/\omega_n^{\mathrm{E}}$. Thus,

$$t_s \approx \frac{30}{46,904} \approx 640\ \mu\mathrm{s}.$$

This value is about four times larger than the value obtained with PD control. Of course, the trade-off is that the closed-loop system will track the step input with zero steady-state error, whereas the PD controller had large steady-state error to a step input (77%).

8.3 FREQUENCY-LEVERAGED PIEZOELECTRIC ACTUATORS

One of the primary weaknesses of piezoelectric materials is small strain. Most polycrystalline piezoelectric materials are limited to free strain values on the order of 0.1 to 0.2%. A notable exception is the single-crystal piezoelectric materials, which have been measured to produce approximately 1% strain at the expense of a modulus that is approximately four times lower than that of polycrystalline ceramics. In any application that requires large displacements, the small strain of piezoelectric materials will be a limiting factor.

Earlier in this chapter the concept of mechanically leveraged piezoelectric materials was introduced. The motivation for developing mechanically leveraged piezoelectric materials is to increase and amplify the motion of the material to increase the output displacement of the device. The range of devices available that use mechnical leverage concepts produces a class of actuators that can produce displacements close to 10 mm. Blocked force values for these types of actuators are on the order of tens of newtons.

An alternative method for overcoming the limited stroke of piezoelectric materials is to use *frequency leveraging* to increase the output displacement of a device. As has been mentioned many times in the book, piezoceramic materials can respond quickly to changes in the applied potential. The response time of small piezoelectric elements can be on the order of microseconds. Even for larger devices such as piezoelectric stacks or flextensional actuators, response times on the order of 0.1 to 1 ms are not uncommon.

Frequency leveraging is a concept that utilizes the fast response of a piezoelectric actuator to overcome the stroke limitations of mechanically leveraged devices. Envision operating a piezoelectric stack actuator at a high cycle rate: for example, 10,000 Hz. Even if the displacement each cycle is only 1 μm, the actuator *could* produce 10,000 μm or 10 mm of displacement if the small per-cycle motion could be *accumulated*. If the accumulation could continue, this device would produce 50 mm of displacement in 5 s of operation.

One way of accumulating the oscillatory displacement of the piezoelectric material is to attach the piezoelectric to a *mechanical rectifier* or *ratchet*. Using the same concept that is used in a ratcheting screwdriver, the piezoelectric material can oscillate a lever that is attached to a mechanical device that moves in only one direction. When

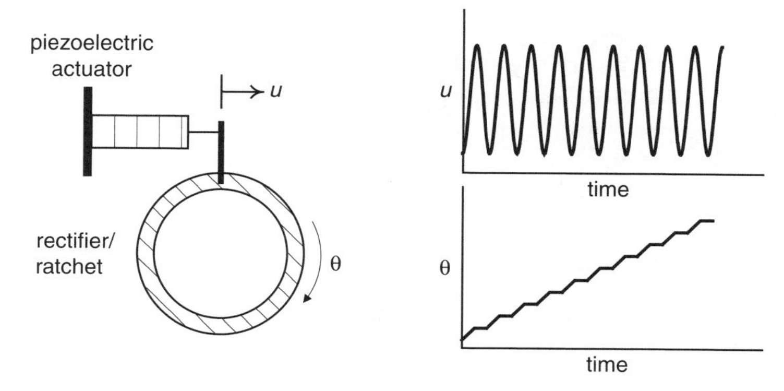

Figure 8.14 Concept of mechanical rectification or ratcheting using an oscillatory piezoelectric actuator.

the piezoelectric material moves forward, the ratchet is pushed ahead one "click." When the piezoelectric material retracts, the ratchet does not move backward but is stationary. If the motion of the lever and the output of the ratchet are plotted with respect to time (Figure 8.14), it shows that the ratchet output is accumulating the oscillatory displacement of the piezoelectric device. The average speed of the device can be estimated by taking the slope of the staircase waveform output of the ratchet. The speed of the ratchet output can be varied by changing the oscillation frequency of the piezoelectric device. Another common method of rectifying the oscillatory motion of a piezoelectric material into unidirectional motion is to use the inchworm concept. The inchworm concept was developed and patented by Burleigh Instruments in the 1970s. One embodiment of the inchworm concept is shown in Figure 8.15. The concept consists of two piezoelectric brakes and a single piezoelectric extender. At rest both brakes are engaged. In the first step, brake A disengages, and in the second step the piezoelectric extender actuates to move forward. Brake A then engages to lock the mechanism, and in step four, brake B disengages to release the opposite side of the piezoelectric actuator. In the fifth step the piezoelectric

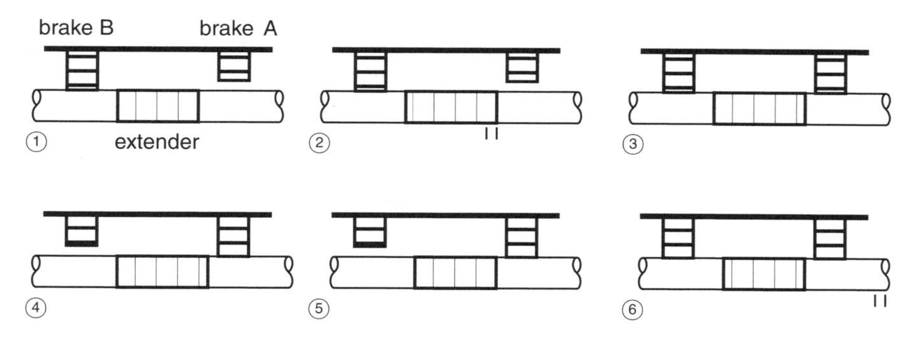

Figure 8.15 Inchworm concept using piezoelectric actuators.

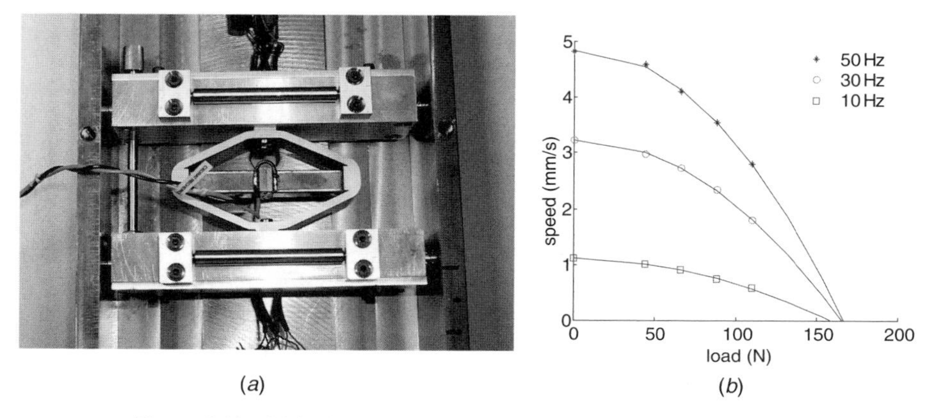

Figure 8.16 (*a*) Inchworm actuator; (*b*) measured speed–load curves.

extender contracts and in the final step, brake B engages to lock the mechanism once again. Thus, after one complete cycle of motion the device has moved forward one increment. This motion can be repeated to produce unidirectional motion. Also, the direction of the device can be reversed simply by changing the order of the brake action.

The incremental motion of the inchworm device is related to the displacement of the piezoelectric actuator. At no load the displacement of the actuator is equal to the free displacement of the extender. Increasing the resistance load decreases the incremental motion of the extender and decreases the average speed of the device. For this reason a piezoelectric inchworm is generally characterized by a speed-versus-load curve in much the same way that an electric motor is characterized. The speed–load curve is analogous to the force–displacement curve for a piezoelectric actuator, except that any point on the curve does not represent the work performed by the actuator, but it represents the output *power* of the device.

One design of an inchworm-type actuator is shown in Figure 8.16 along with the force-to-load curve at three operating frequencies. At low loads the speed is approximately proportional to the operating frequency, as expected. At higher load values the velocity output of the device converges to approximately zero at a load of approximately 160 N. This is attributed to compliance in the extender mechanism which causes increased backstepping at higher load values.

Several vendors sell piezoelectric inchworm-type devices. Generally, the devices are sold as positioning stages or piezoelectric motors. High-resolution devices such as the Burleigh TSE-820 are sold that have less than 1-nm motor resolution and a speed range between nm/s and 1.5 mm/s. The maximum force associated with this device is specified at 15 N.

Another type of frequency-leveraged device is a *piezohydraulic actuator*. As the name implies, piezohydraulics is the combination of a piezoelectric actuator with a hydraulic circuit. The fundamental concept is very similar to an inchworm-type device in the sense that the oscillation of the piezoelectric actuator is transformed into

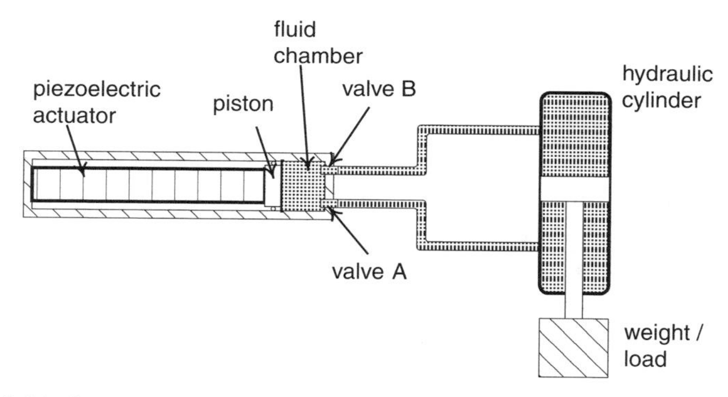

Figure 8.17 Components of a piezohydraulic actuation system with hydraulic cylinder.

unidirectional displacement with a step-and-repeat motion. The primary difference is that rectification of the oscillatory motion is performed using a hydraulic circuit. The hydraulic circuit consists of a set of valves that only allow fluid flow in one direction. The opening and closing of the valves is controlled by the pressure differential across the valve. When the pressure differential is sufficiently large, the valve opens and allows fluid flow in one direction. In this manner the valves take the place of the brakes that rectify the motion of an inchworm-type device.

A hydraulic transmission has some advantages compared to the mechanical drive in an inchworm device. Hydraulic power can be transmitted effectively through hydraulic lines. In addition, many control systems are driven hydraulic actuators; therefore, the development of a piezohyrdraulic device would enable a seamless interface with other hydraulic motion control systems. One of the motivations for replacing a conventional hydraulic system with one driven by piezoelectrics is that the centralized pump of a conventional system could be replaced with localized pumping devices, thus eliminating the losses associated with hydraulic lines and eliminating the weight of hydraulic piping.

The fundamental component of a piezohydraulic system is a piezoelectric actuator and a closed, fluid-filled chamber. Extension of the piezoelectric actuator increases the pressure in the chamber until the pressure differential across valve A is large enough to open the valve (Figure 8.17). Once the valve is opened, continued extension of the stack will produce fluid flow in the hydraulic circuit. The pressurized fluid is used to move a hydraulic cylinder, which itself could be raising a weight. Movement of the hydraulic cylinder pressurizes the opposite side of the hydraulic circuit. As the piezoelectric actuator contracts, the pressure differential on valve B is high enough to open the valve and allow fluid to return to the chamber. Thus, one cycle of motion produces incremental motion of the hydraulic cylinder.

It is instructive to analyze the motion of a piezohydraulic system in relation to the force–deflection diagram for the piezoelectric actuator. Hydraulic systems are generally pressurized to reduce the risk of air cavitation in the fluid and to increase the bulk modulus of the fluid. This creates a minimum low-pressure force on the

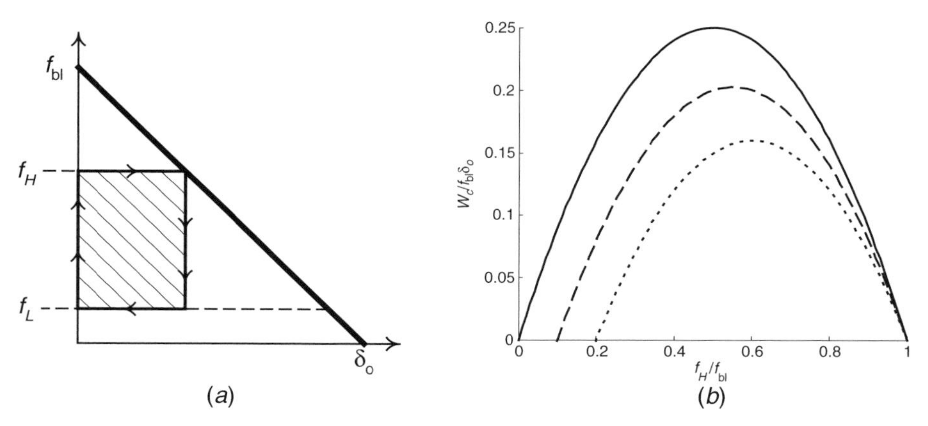

Figure 8.18 (*a*) Work cycle for an ideal piezohydraulic actuator; (*b*) nondimensional work per cycle for an ideal piezohydraulic actuator for $\tilde{f}_L = 0$ (solid), $\tilde{f}_L = 0.1$ (dashed), and $\tilde{f}_L = 0.2$ (dotted).

piezoelectric actuator, which we denote f_L. Application of a field across the actuator increases the force on the fluid. In the ideal case in which the fluid is incompressible, there will be no extension of the stack due to the application of the field. The force in the stack will increase until it is great enough to overcome the pressure differential of valve A, denoted f_H. At this point the valve will open and the stack will extend to the deflection value associated with the resistance force, x_H. Once the stack has reached maximum extension, the field is reduced until the force in the stack–fluid chamber is small enough to allow inflow of the fluid.

The area associated with the shaded region in Figure 8.18*a* is the work performed by the piezohydraulic actuator during each cycle. The work per cycle, W_c, is quantified by writing

$$W_c = (f_H - f_L)x_H. \tag{8.16}$$

From the transducer equations for a piezoelectric actuator, the displacement is

$$x_H = -\frac{1}{k_a} f_H + x_v v. \tag{8.17}$$

Recalling that the free displacement, δ_o, is equal to $x_v v$ and that the blocked force is equal to $k_a x_v v$, the work per cycle normalized to the product of the blocked force and the free deflection is obtained by combining equations (8.16) and (8.17):

$$\tilde{W}_c = \frac{W_c}{f_{bl}\delta_o} = \left(\tilde{f}_H - \tilde{f}_L\right)\left(1 - \tilde{f}_H\right) = -\tilde{f}_H^2 + \left(1 + \tilde{f}_L\right)\tilde{f}_H - \tilde{f}_L, \tag{8.18}$$

where $\tilde{f}_H = f_H/f_{\text{bl}}$ and $\tilde{f}_L = f_L/f_{\text{bl}}$. The value of $\tilde{f}_H$ that maximizes the work per cycle is

$$\tilde{f}_H^* = \frac{1 + \tilde{f}_L}{2}, \tag{8.19}$$

and the maximum value of the nondimensional work is

$$\tilde{W}_c^* = \frac{1}{4} - \frac{1}{2}\tilde{f}_L + \frac{1}{4}\tilde{f}_L^2. \tag{8.20}$$

Plots of the nondimensional work per cycle for a piezohydraulic actuator are shown in Figure 8.18*b*. As might be expected, the maximum work output of the piezohydraulic actuator is 1/4 $f_{\text{bl}}\delta_o$ when the low-pressure force, $\tilde{f}_L$, is equal to zero. The maximum occurs when the force at high pressure is equal to half the blocked force. Increasing the low-pressure force decreases the maximum work per cycle of the piezohydraulic system.

Example 8.3 A piezoelectric stack has a blocked force of 10,000 N and a free displacement of 30 μm at maximum voltage. It is being considered for a piezohydraulic actuation system in which the low-pressure force is 1250 N. Compute (a) the high-pressure force that maximizes the output work per cycle. (b) the output power of the ideal system if it is operating at 300 Hz.

Solution (a) The nondimensional high-pressure force that maximizes the output work per cycle is computed from equation (8.19):

$$\tilde{f}_H^* = \frac{1 + 1250/10,000}{2} = 0.5625.$$

The high-pressure force that maximizes the output work is

$$f_H = (0.5625)(10,000 \text{ N}) = 5625 \text{ N}. \tag{8.21}$$

(b) The output power must be computed from the output work. The nondimensional output work per cycle at the optimum load is

$$\tilde{W}_c^* = \frac{1}{4} - \frac{1}{2}\left(\frac{1250}{10,000}\right) + \frac{1}{4}\left(\frac{1250}{10,000}\right)^2 = 0.1914.$$

The work per cycle at optimum load is

$$W_c^* = (0.1914)(10,000 \text{ N})(30 \times 10^{-6} \text{ m}) = 0.0574 \text{ J/cycle}.$$

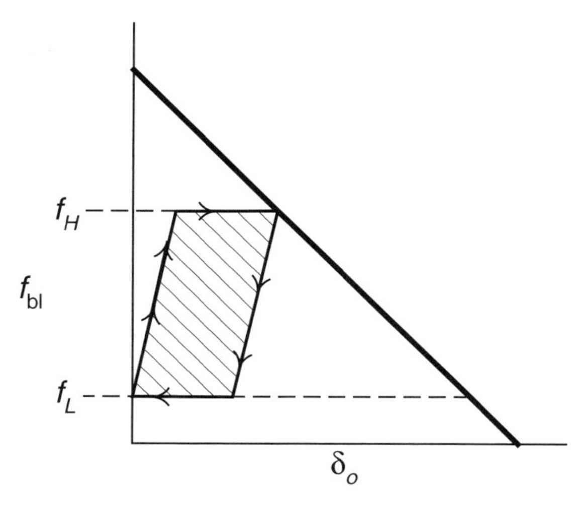

Figure 8.19 Work cycle for a piezohydraulic system that exhibits fluid compressibility.

The power output is the product of the operating frequency and the output power per cycle,

$$P = (0.0574 \text{ J/cycle})(300 \text{ Hz}) = 17.23 \text{ W}.$$

The analysis thus far of piezohydraulic actuators has made a number of assumptions about the ideal behavior of the system. One of the most critical is the assumption of incompressibility in the fluid. Detailed analyses of piezoelectric–hydraulic systems have been published. These analyses combine an electromechanical model of the piezoelectric with either an incompressible or compressible model of the fluid. One of the primary results of these analyses has been that compressibility of the fluid has a significant effect on the output power of a piezohydraulic system particular in high-frequency operation.

Although a coupled fluid model is beyond the scope of this book, some simplifying assumptions can yield a qualitative understanding of the effects of fluid compressibility on the output work and power of a piezohydraulic system. Reviewing Figure 8.18*a*, it is noted that the increase in the force with zero extension of the actuator is a consequence of the assumption of fluid incompressibility. If we assume that the fluid is acting as a linear spring load on the actuator (which is an acceptable discussion for the sake of illustration, but not fully accurate; see Nasser and Leo [86] for a more thorough analysis), the work diagram of Figure 8.18*a* changes to that of Figure 8.19. The slope of the line during pressurization of the fluid is a consequence of the fluid compressibility. As the compressibility gets smaller and the stiffness of the fluid increases, the work cycle diagram approaches that of the ideal system shown in Figure 8.18*a*. Comparing Figure 8.18*a* and Figure 8.19, it is clear that the work per cycle is smaller when compressibility is included in the analysis for the same low- and high-pressure forces. The reduction is work is due to the storage of mechanical energy in the fluid during operation.

Piezohydraulic actuation is not to the same level of commercial availability as mechanically leveraged actuators or inchworm-type devices. Two companies that are developing piezohydraulic actuation systems are Kinetic Ceramics, Inc. and CSA Engineering, Inc.

8.4 ELECTROACTIVE POLYMERS

Electroactive polymers are also useful for motion control when large displacements may be needed and substantial force is not required. Recall from Chapter 7 that EAP materials are generally characterized by large strains, and hence large displacement, but limited stress, resulting in limited force. As discussed in that chapter as well, ionic EAP materials are characterized by a relatively slow response due to the relationship between charge diffusion and electromechanical coupling. Feedback control allows the design of a compensator that will sacrifice displacement in the interest of increasing the response time.

8.4.1 Motion Control Using Ionomers

Motion control using electroactive polymers will be studied with the actuator model of ionomeric transducers developed in Chapter 7. Combining equations (7.55) and (7.59) results in a transfer function between displacement and voltage:

$$\left(\frac{u}{v}\right)^f = G(\mathsf{s}) = \frac{-3L_d^2 d_o}{h^2} \frac{\mathsf{s} + 1/\tau_{d1}}{(\mathsf{s} + 1/\tau_{d2})(\mathsf{s} + 1/\tau_{d3})}. \tag{8.22}$$

Note that we have dropped the negative sign for convenience. A proportional–integral–derivative controller of the form in equation (8.11) results in the loop gain transfer function:

$$K(\mathsf{s})G(\mathsf{s}) = \frac{3L_d^2 d_o}{h^2} \frac{(k_d \mathsf{s}^2 + k_p \mathsf{s} + k_i)(\mathsf{s} + 1/\tau_{d1})}{\mathsf{s}(\mathsf{s} + 1/\tau_{d2})(\mathsf{s} + 1/\tau_{d3})}. \tag{8.23}$$

The relaxation properties of a typical ionomer material specify that $\tau_{d2}/\tau_{d1} < 1$ and $\tau_{d2}/\tau_{d3} > 1$.

One approach to the compensation would be to use the compensator zeros to cancel the poles of the ionomer. This is obtained by setting

$$\begin{aligned} k_d &= 1 \\ k_p &= \frac{1}{\tau_{d2}} + \frac{1}{\tau_{d3}} \\ k_i &= \frac{1}{(\tau_{d2}\tau_{d3})}. \end{aligned} \tag{8.24}$$

Using this compensator design the loop transfer function is

$$K(\mathsf{s})G(\mathsf{s}) = \frac{3L_d^2 d_o}{h^2}\frac{(\mathsf{s} + 1/\tau_{d1})}{\mathsf{s}}. \tag{8.25}$$

The closed-loop transfer function for this compensator is

$$H(\mathsf{s}) = \frac{K(\mathsf{s})G(\mathsf{s})}{1 + K(\mathsf{s})G(\mathsf{s})} = \frac{u(\mathsf{s})}{r(\mathsf{s})} = \frac{A'(\mathsf{s} + 1/\tau_{d1})}{(1 + A')\mathsf{s} + A'/\tau_{d3}}, \tag{8.26}$$

where

$$A' = \frac{3L_d^2 d_o}{h^2}. \tag{8.27}$$

The inverse Laplace transform of equation (8.26) to a step voltage $r(\mathsf{s}) = R/\mathsf{s}$ is

$$u(t) = \frac{R}{1 + A'}\left(1 + A' - e^{-A't/\tau_{d1}(1+A')}\right). \tag{8.28}$$

The closed-loop response of the system for this compensator is first-order with additional zero dynamics. The steady-state error of the system is zero; therefore, this compensator will enable tracking of a step input. Note that the closed-loop pole of the system is not a function of the compensator gains. Thus, the closed-loop time response is governed only by the transducer dynamics. In general, the closed-loop time response using a compensator with pole–zero cancellation is *slower* than the open-loop time response, although it will track the input with zero steady-state error and will not exhibit any relaxation behavior. Thus, it is often more beneficial to use a compenator that does not cancel the dynamics of the transducer. This is studied in the following example.

Example 8.4 An ionomeric transducer of length 40 mm and thickness 0.2 mm has the following properties: $d_o = 65,000 \times 10^{-12}$ m · s/V, $\tau_{d1} = 100$ rad/s, $\tau_{d2} = 10$ rad/s, $\tau_{d3} = 1$ rad/s. (a) Compute the open-loop step response to a 1-V input. (b) Compute the closed-loop response using a PID compensator that cancels the poles of the open-loop transducer to a step reference voltage of 5 mm. (c) Compute and plot the closed-loop response for a PI controller with the values $k_d = 0$, $k_p = 200$, and $k_i = 100$.

Solution (a) The open-loop response of the transducer is obtained from equation (7.65). The result is

$$u(t) = 3(65{,}000 \times 10^{-12}\ \mathrm{m \cdot s/V})\left(\frac{40\ \mathrm{mm}}{0.2\ \mathrm{mm}}\right)^2 [(10)A'(1 - e^{-t/10}) + (1)B'(1 - e^{-t})],$$

where

$$A' = \frac{1/100 - 1/10}{1 - 1/10} = -0.1$$
$$B' = \frac{1 - 1/100}{1 - 1/10} = 1.1.$$

Combining the results yields

$$\begin{aligned} u(t) &= 7.8 \times 10^{-3} \left[(1.1)(1 - e^{-t}) - \left(1 - e^{-t/10}\right)\right] \text{ m} \\ &= 7.8 \times 10^{-3} \left[0.1 - 1.1e^{-t} + e^{-t/10}\right] \text{ m.} \end{aligned}$$

(b) The closed-loop response for a compensator that cancels the open-loop dynamics is shown in equation (8.28). Using the values given in the problem, we have

$$\begin{aligned} u(t) &= \frac{5 \times 10^{-3}}{1 + 0.0078} \left(1 + 0.0078 - e^{-0.0078t/(100)(1+0.0078)}\right) \\ &= 4.9613 \times 10^{-3} \left(1.0078 - e^{7.7396 \times 10^{-5} t}\right). \end{aligned}$$

Using a compensator that cancels the open-loop dynamics produces a closed-loop time constant that is greater than 10,000 s. Obviously, this type of control is not useful for positioning the ionomeric actuator.

(c) The loop transfer function for the PI compensator is obtained by substituting the values into equation (8.23):

$$K(\mathsf{s})G(\mathsf{s}) = 7.8 \times 10^{-3} \left[\frac{(200\mathsf{s} + 100)(\mathsf{s} + 0.01)}{\mathsf{s}(\mathsf{s} + 0.1)(\mathsf{s} + 1)}\right].$$

The closed-loop transfer function for the compensated system is

$$H(\mathsf{s}) = \frac{u(\mathsf{s})}{r(\mathsf{s})} = \frac{K(\mathsf{s})G(\mathsf{s})}{1 + K(\mathsf{s})G(\mathsf{s})} = \frac{1.56\mathsf{s}^2 + 0.7956\mathsf{s} + 0.0078}{\mathsf{s}^3 + 2.66\mathsf{s}^2 + 0.8956\mathsf{s} + 0.0078}.$$

For the step input function $r(\mathsf{s}) = 0.005/\mathsf{s}$, the inverse Laplace transform of the closed-loop response yields

$$u(t) = \left(5 - 3.23e^{-2.27t} - 1.24e^{-0.385t} - 0.53e^{-8.95 \times 10^{-3} t}\right) \times 10^{-3} \text{ m.}$$

The time constant of the closed-loop response with the PI controller is much faster than those of the compensator that utilized pole–zero cancellation. For the PI compensator that fastest time constant is on the order of 0.5 s.

The closed-loop response using PI control is plotted in Figure 8.20 and compared to the open-loop response. The results demonstrate that feedback control is able to eliminate the relaxation behavior associated with the ionomer response. The

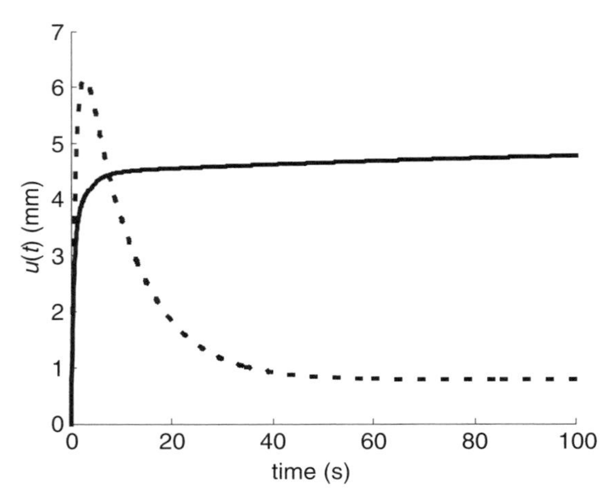

Figure 8.20 Open-loop (dashed) and closed-loop (solid) response of the ionomeric transducer studied in Example 8.4.

plot also demonstrates that the steady-state error is not reached until time greater than 100 s. This is consistent with the fact that there is a exponential term in the time response solution with a time constant greater than 100 s. Thus, although PI control is able to speed up the response (compared to control using pole–zero cancellation), the steady-state error of the system is not near zero for several hundred seconds.

8.5 CHAPTER SUMMARY

Open- and closed-loop performance of motion control using piezoelectric and ionomeric actuators were studied in this chapter. A variety of methods for amplifying the output displacement of piezoelectric materials were introduced. These methods consisted of mechanical levers and frequency-leveraged actuators. A mechanical lever increased the output displacement at the expense of output force due to the conservation of mechanical work. Frequency-leveraged piezoelectric actuators such as inchworm actuators and piezoelectric pumps maintained constant power; therefore, they were able to maintain the force output of the actuator with increased displacement.

The closed-loop behavior of piezoelectric and ionomeric actuators was also studied. A second-order model of a piezoelectric positioner was developed from the transducer equations introduced earlier. Proportional–derivative and proportional–integral–derivative compensators were analyzed. Proportional–derivative compensators were shown to enable the control of the speed and damping in the closed-loop piezoelectric actuator; proportional–integral–derivative compensation enabled zero steady-state error. Analysis of the closed-loop response of an ionomeric actuator to proportional–integral–derivative control showed that the relaxation behavior of ionomeric materials could be overcome through the use of feedback.

PROBLEMS

8.1. Prepare a list of 10 commercially available piezoelectric stack actuators. For each actuator, specify the blocked force, free displacement, volume, and mass. Compute the energy density of all the actuators.

8.2. Prepare a list of 10 commercially available piezoelectric bender actuators. For each actuator, specify the blocked force, free displacement, volume, and mass. Compute the energy density of all the actuators.

8.3. A piezoelectric stack actuator has a short-circuit stiffness of 58 N/μm and a free deflection of 13 μm at a peak voltage of 100 V. The load on the actuator has a mass of 40 g.

(a) Compute the second-order model of the stack actuator assuming zero internal damping.

(b) Compute the PD compensator gains that produce a closed-loop damping ratio of 0.2 and a closed-loop resonance frequency that is 50% larger than the open-loop resonance.

(c) Plot the closed-loop response to a 1-μm reference input.

8.4. A piezoelectric bender actuator has a short-circuit stiffness of 130 N/m and a free deflection of 1.5 mm at a peak voltage of 120 V. The load on the actuator has a mass of 5 g.

(a) Compute the second-order model of the stack actuator assuming zero internal damping.

(b) Compute the PD compensator gains that produce a closed-loop damping ratio of 0.2 and a closed-loop resonance frequency that is 50% larger than the open-loop resonance.

(c) Plot the closed-loop response to a 100-μm reference input.

8.5. A piezoelectric stack actuator has a short-circuit stiffness of 58 N/μm and a free deflection of 13 μm at a peak voltage of 100 V. The load on the actuator has a mass of 40 g.

(a) Compute the second-order model of the stack actuator assuming zero internal damping.

(b) Plot the closed-loop response for the PID compensator gains $k_p = 10$, $k_d = 100$, and $k_i = 10$ to a step reference input of 1 μm.

8.6. A piezoelectric stack actuator has a short-circuit stiffness of 45 N/μm and a free deflection of 80 μm at a peak voltage of 100 V. The load on the actuator has a mass of 100 g.

(a) Compute the second-order model of the stack actuator assuming zero internal damping.

(b) Choose the PID compensator gains such that the closed-loop step response has a rise time of less than 1 ms, an overshoot of less than 20%, and zero steady-state error.

8.7. A piezoelectric cantilever for atomic force microscopy has a short-circuit stiffness of 5 N/m and a free deflection of 40 nm at a voltage of 50 V. The effective mass of the cantilever is 8 pg.

(a) Compute the short-circuit resonant frequency.

(b) Compute the peak displacement when excited with 50 V at resonance assuming an internal damping ratio of 0.01.

8.8. Find the specifications for a commercially available piezoelectric pump and a piezoelectric inchworm actuator.

8.9. Compute the control voltage for the compensator designed in Example 8.4 for an ionomeric transducer actuator.

8.10. An ionomeric transducer of length 30 mm and thickness 0.2 mm has the following properties: $d_o = 45,000 \times 10^{-12}$ m · s/V, $\tau_{d1} = 100$ rad/s, $\tau_{d2} = 10$ rad/s, and $\tau_{d3} = 1$ rad/sec.

(a) Compute the open-loop step response to a 1-V input.

(b) Design a PID compensator that achieves a rise time of less than 1 s, no overshoot, and zero steady-state error.

(c) Plot the closed-loop response for a step reference input of 2 mm.

NOTES

There are a number of design references for motion control applications using piezoelectric materials. Vendors of piezoelectric materials such as Piezo Systems, Tokin, and American PiezoCeramic have technical information and design guides that are useful for understanding the basic performance characteristics of piezoelectric actuators. Physik Instrumente and Dynamic Structures and Materials are two companies that sell piezoelectric positioners, and Physik Instrumente has a set of technical notes related to the design and implementation of precise positioning devices.

Research in this field over the past several years has been concentrated on the development of piezoelectric devices that trade force for stroke to increase the motion output of an actuator. An excellent overview from the mid-1990s is the work of Near [88]. A more recent overview of the state of the art in piezoelectric actuation is the article by Niezrecki et al. [89], which contains a large number of good references on the topic from work in the 1980s and 1990s. Characteristics of C-block actuators discussed in this chapter can be found in the work of Brei's group [82,83,90], and their work on recurve architectures is described in a paper of Ervin and Brei [85]. Research in the combination of piezoelectric materials with hydraulic systems has received renewed interest recently for the development of high-power devices. Although the concept has been discussed for a number of years, recent work that initiated many of the current developments was published in papers by Mauck and Lynch [87, 91]. Recently, a group at the University of Maryland has published a number of articles on the topic, analyzing the various design parameters associated with piezohydraulic systems [92]. The author's work in this field is described in detail by Nasser and Leo [86] andTan et al. [93].

The sections on control of piezoelectric actuators are based on general control methodologies for second-order systems (see, e.g., Franklin et al. [22]). The control of ionomeric materials is studied in work of Mallavarapu and Leo [94] and Kothera and Leo [95].

9

PASSIVE AND SEMIACTIVE DAMPING

This chapter focuses on the use of smart materials as energy-dissipating elements in structural systems. We first analyze the use of piezoelectric materials as energy-dissipating elements using passive and semiactive damping methods. This analysis will require that we combine the results discussed in earlier chapters into a system-level model of a structure. Once the system-level model is developed, we can analyze the use of piezoelectric elements for energy dissipation.

9.1 PASSIVE DAMPING

Structures that have very little inherent damping often suffer from excessive or unwanted vibrations. As detailed in Chapter 3, the steady-state response of a structure to harmonic excitations can become amplified when the excitation is near the structural natural frequencies, and the amount of amplification near resonance is strongly a function of the structural damping. For the simple model of a single-degree-of-freedom oscillator, the amplification that occurs at resonance is inversely proportional to the damping ratio. In the time domain, the lack of structural damping increases the number of vibration cycles that occur before the amplitude of the response decays to a small value. Increasing the damping will produce a faster decay of the oscillatory vibration.

Energy-dissipating materials are often incorporated into structures to reduce the effects of unwanted or excessive vibration. A common type of damping element are viscoelastic materials that dissipate mechanical energy when subjected to cyclic motion. Viscoelastic materials can be incorporated into structural elements and, when properly designed, substantially reduce the free and forced vibration of a structure.

As the name implies, a viscoelastic material is characterized by its combination of viscous and elastic properties. For a purely elastic material, the state of stress is a function of the state of strain alone. In a viscoleastic material, the state of stress is a function of both the strain and the rate of strain in the material. This property produces a frequency dependence in the stress–strain relationship (Figure 9.1). Application of

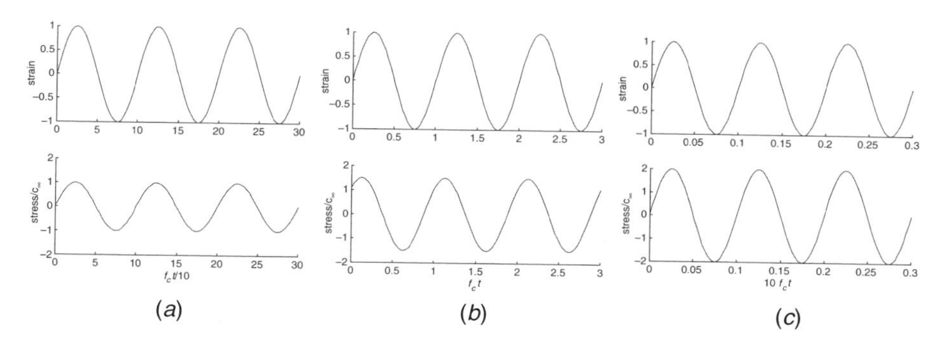

(*a*) (*b*) (*c*)

Figure 9.1 Frequency dependence of stress and strain in a viscoelastic material.

harmonic strain to linear viscoelastic material well below this critical frequency, denoted f_c, produces a stress response that is scaled by the elastic modulus of the material and is approximately in phase with the strain input. Increasing the frequency of the input causes the stress response to increase in magnitude and a phase shift develops between the stress and strain time histories. Further increases in frequency and we measure a stress response whose magnitude continues to increase but with decreasing phase shift.

Plotting the stress–strain relationships explicitly produces the figures shown in Figure 9.2. For frequencies much lower than the critical frequency the stress–strain curve is approximately linear, which is consistent with the fact that the material is exhibiting linear elastic properties. When the frequency is increased to the critical frequency, the stress–strain plot exhibits hysteresis that indicates the existence of energy dissipation in the mechanical response. The material returns to a linear elastic material when the frequency is increased to well above f_c.

A mechanical model of linear viscoelastic material can be obtained by combining the properties of an elastic spring element and a viscous damping element. If we combine a linear elastic spring in series with a linear viscous damper, we produce the mechanical model shown in Figure 9.3*a*. The model parameters are the elastic

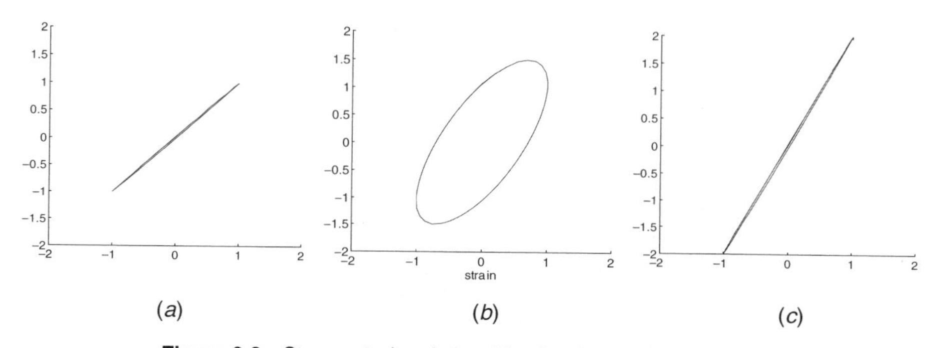

(*a*) (*b*) (*c*)

Figure 9.2 Stress–strain relationships for viscoelastic materials.

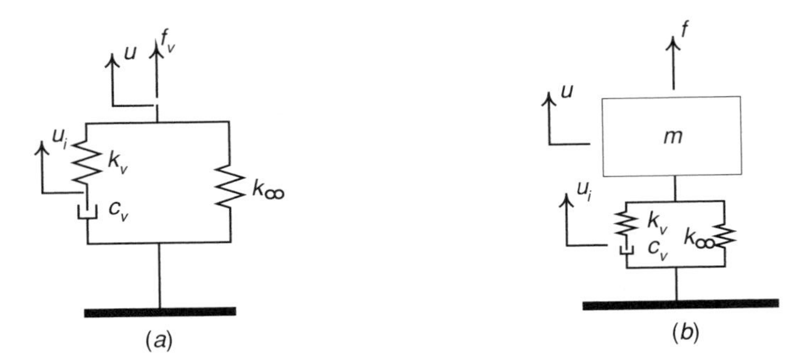

Figure 9.3 (*a*) Mechanical model of a viscoelastic structural element; (*b*) single-degree-of-freedom mass–spring system with a viscoelastic damping element.

stiffness of the spring, k_v, and the viscous damping coefficient of the damper, c_v. The motion of the damping element is parameterized as a function of the displacement, u, and the *internal* displacement, u_i.

Summing forces at the location between the spring and damper produces the relationships

$$k_v(u_i - u) = -c_v \dot{u}_i. \tag{9.1}$$

Summing the forces at the top node yields the expressions

$$k_v(u - u_i) + k_\infty u = f_v. \tag{9.2}$$

A single-degree-of-freedom vibration model of a system with viscoelastic damping is shown in Figure 9.3*b*. The mass and stiffness of the vibrating system are modeled with m and k, respectively. Using summation of forces to determine the model yields the equations

$$m\ddot{u} = f - f_v = f - k_v(u - u_i) - k_\infty u. \tag{9.3}$$

Rearranging equation (9.3) expression and combining it with equation (9.1) produces

$$\begin{aligned} m\ddot{u} + (k_v + k_\infty)u - k_v u_i &= f \\ c_v \dot{u}_i - k_v u + k_v u_i &= 0. \end{aligned} \tag{9.4}$$

The expressions in equation (9.4) illustrate the means by which the viscoelastic element introduces damping into the vibrating system. The spring k_∞ acts as elastic stiffness of the system in the presence of a negligibly small viscoelastic stiffness k_v. The viscoelastic stiffness serves to couple the viscous damping of the member with the vibration of the mass. This is illustrated by the fact that the viscoelastic stiffness k_v couples the degrees of the freedom u and u_i.

9.2 PIEZOELECTRIC SHUNTS

In Section 9.1 we described a common mechanical model of a vibrating system with viscoelastic damping. To understand how active materials can be utilized for damping applications, consider the equations derived for a piezoelectric stack transducer operating in the 33 mode,

$$\begin{aligned} u_3 &= \frac{s_{33}^{\mathrm{E}} L}{A} f + \frac{d_{33} L}{t} v \\ q &= \frac{d_{33} L}{t} f + \frac{n \varepsilon_{33}^{\mathrm{T}} A}{t} v. \end{aligned} \tag{9.5}$$

To maintain consistency between this derivation and Section 9.1, denote $u_3 = u$ and interchange the dependent and independent variables:

$$\begin{pmatrix} f \\ v \end{pmatrix} = \frac{1}{s_{33}^{\mathrm{E}} \varepsilon_{33}^{\mathrm{T}} - d_{33}^2} \begin{bmatrix} \dfrac{\varepsilon_{33}^{\mathrm{T}} A}{L} & -d_{33} \dfrac{t}{L} \\ -d_{33} \dfrac{t}{L} & \dfrac{s_{33}^{\mathrm{E}} t}{n A} \end{bmatrix} \begin{pmatrix} u \\ q \end{pmatrix}. \tag{9.6}$$

All of the elements can be divided by $s_{33}^{\mathrm{E}} \varepsilon_{33}^{\mathrm{T}}$ and the coefficient in front of the matrix can be rewritten $1 - k_{33}^2$; therefore,

$$\begin{pmatrix} f \\ v \end{pmatrix} = \frac{1}{1 - k_{33}^2} \begin{bmatrix} \dfrac{A}{s_{33}^{\mathrm{E}} L} & -\dfrac{d_{33}}{s_{33}^{\mathrm{E}} \varepsilon_{33}^{\mathrm{T}}} \dfrac{t}{L} \\ -\dfrac{d_{33}}{s_{33}^{\mathrm{E}} \varepsilon_{33}^{\mathrm{T}}} \dfrac{t}{L} & \dfrac{t}{n \varepsilon_{33}^{\mathrm{T}} A} \end{bmatrix} \begin{pmatrix} u \\ q \end{pmatrix}. \tag{9.7}$$

Applying the definitions

$$\begin{aligned} s_{33}^{\mathrm{D}} &= s_{33}^{\mathrm{E}} \left(1 - k_{33}^2\right) \\ \varepsilon_{33}^{\mathrm{S}} &= \varepsilon_{33}^{\mathrm{T}} \left(1 - k_{33}^2\right) \end{aligned} \tag{9.8}$$

and denoting

$$g_{33} = \frac{d_{33}}{s_{33}^{\mathrm{D}} \varepsilon_{33}^{\mathrm{T}}} \frac{t}{L}, \tag{9.9}$$

we have the relationships

$$\begin{pmatrix} f \\ v \end{pmatrix} = \begin{bmatrix} \dfrac{A}{s_{33}^{\mathrm{D}} L} & -g_{33} \\ -g_{33} & \dfrac{t}{n \varepsilon_{33}^{\mathrm{S}} A} \end{bmatrix} \begin{pmatrix} u \\ q \end{pmatrix}. \tag{9.10}$$

The equations can be simplified further if we denote the top left element as the open-circuit stiffness of the actuator, k_a^{D}, and the bottom right element as the inverse of the strain-free capacitance, $1/C^{\mathrm{S}}$. With these definitions we have

$$\begin{pmatrix} f \\ v \end{pmatrix} = \begin{bmatrix} k_a^{\mathrm{D}} & -g_{33} \\ -g_{33} & \dfrac{1}{C^{\mathrm{S}}} \end{bmatrix} \begin{pmatrix} u \\ q \end{pmatrix}. \tag{9.11}$$

Consider the piezoelectric stack connected to a mass–spring system with mass m and passive stiffness k. Summing forces on the mass and coupling them to the previous set of expressions yields

$$\begin{pmatrix} f - m\ddot{u} - ku \\ v \end{pmatrix} = \begin{bmatrix} k_a^{\mathrm{D}} & -g_{33} \\ -g_{33} & \dfrac{1}{C^{\mathrm{S}}} \end{bmatrix} \begin{pmatrix} u \\ q \end{pmatrix}. \tag{9.12}$$

Combining these expressions with the fact that

$$v = -R\dot{q} \tag{9.13}$$

produces the coupled expressions

$$m\ddot{u} + ku + k_a^{\mathrm{D}} u - g_{33} q = f \tag{9.14}$$

$$R\dot{q} - g_{33} u + \frac{1}{C^{\mathrm{S}}} q = 0. \tag{9.15}$$

Comparing equations (9.4) and (9.15) illustrates that the two expressions are identical in form. The open-circuit stiffness of the actuator is equivalent to the elastic stiffness elements of the viscoelastic damper. The coupling term of the piezoelectric device, g_{33}, acts as the coupling stiffness in the same manner as k_v for a viscoelastic element. Finally, the resistance of the external circuit is analogous to the viscous damping properties of the viscoelastic material.

The equations that represent a piezoelectric shunt can be analyzed using the concepts introduced in Chapter 3. Applying the Lapace transform to equation (9.15) (assuming zero initial conditions) yields

$$\begin{aligned} \left(m\mathsf{s}^2 + k + k_a^{\mathrm{D}}\right) u(\mathsf{s}) - g_{33} q(\mathsf{s}) &= f(\mathsf{s}) \\ (R\mathsf{s} + 1/C^{\mathrm{S}}) q(\mathsf{s}) - g_{33} u(\mathsf{s}) &= 0. \end{aligned} \tag{9.16}$$

Solving the second expression in equation (9.16) and substituting it into the first expression results in a relationship between the force and displacement,

$$\left[m\mathsf{s}^2 + k + k_a^{\mathrm{D}} - \frac{g_{33}^2}{R\mathsf{s} + 1/C^{\mathrm{S}}}\right] u(\mathsf{s}) = f(\mathsf{s}). \tag{9.17}$$

It can be shown that $g_{33}^2 C_p^S = k_{33}^2 k_a^D$, therefore we can rewrite the previous expression as

$$\left[m\mathsf{s}^2 + k + k_a^D\left(1 - \frac{k_{33}^2}{RC^S\mathsf{s}+1}\right)\right]u(\mathsf{s}) = f(\mathsf{s}). \tag{9.18}$$

Equation (9.18) provides insight into the effect of the piezoelectric material on the mechanical system. The piezoelectric shunt is acting as a frequency-dependent stiffness in the same manner as the viscoelastic material discussed at the outset of the chapter. The frequency dependence of the stiffness will produce energy dissipation in the mechanical system and result in structural damping. Solving for the ratio of $u(\mathsf{s})/f(\mathsf{s})$ yields

$$\frac{u(\mathsf{s})}{f(\mathsf{s})} = \frac{1/m(\mathsf{s}+1/RC^S)}{\mathsf{s}^3 + (1/RC^S)\mathsf{s}^2 + (k_a^D/m)\mathsf{s} + (1/RC^S k_a^D/m)\left(1-k_{33}^2\right)}. \tag{9.19}$$

The poles of the system with a piezoelectric shunt are obtained by solving for the roots of the third-order denominator of equation (9.19). We can gain some insight into the problem by letting the shunt resistance approach zero and infinity to determine the transfer function in the limiting cases of short-circuit ($R \to 0$) and open-circuit ($R \to \infty$) electrical boundary conditions. In these two cases we have

$$\text{short circuit:} \qquad R \to 0, \quad \frac{u(\mathsf{s})}{f(\mathsf{s})} = \frac{1/m}{\mathsf{s}^2 + (k_a^D/m)\left(1-k_{33}^2\right)} \tag{9.20}$$

$$\text{open circuit:} \qquad R \to \infty, \quad \frac{u(\mathsf{s})}{f(\mathsf{s})} = \frac{1/m}{\mathsf{s}^2 + k_a^D/m}. \tag{9.21}$$

The limiting analysis demonstrates that the vibrating system is undamped with both short- and open-circuit boundary conditions. This is consistent with our analysis of piezoelectric materials, due to the fact that the compliance of the material will change depending on the electrical boundary condition.

An alternative way to express the dynamic response of the system is to normalize it with respect to the static displacement. The static displacement u_{st} is solved for by substituting $\mathsf{s} = 0$ into equation (9.18),

$$u_{\text{st}} = \frac{f}{k + k_a^D\left(1-k_{33}^2\right)} = \frac{f}{k + k_a^E}. \tag{9.22}$$

With this result in mind, first rewrite equation (9.18) as

$$\left[m\mathsf{s}^2 + k + k_a^E - k_a^E + \frac{k_a^E}{1-k_{33}^2}\left(1 - \frac{k_{33}^2}{RC^S\mathsf{s}+1}\right)\right]u(\mathsf{s}) = f(\mathsf{s}). \tag{9.23}$$

Equation (9.23) can be simplified as

$$\left(m\mathrm{s}^2 + k + k_a^{\mathrm{E}} + \frac{k_{33}^2 k_a^{\mathrm{E}}}{1 - k_{33}^2} \frac{RC^{\mathrm{S}}\mathrm{s}}{RC^{\mathrm{S}}\mathrm{s} + 1}\right) u(\mathrm{s}) = f(\mathrm{s}). \tag{9.24}$$

Denoting the short-circuit mechanical resonance as

$$\omega_m^{\mathrm{E}^2} = \frac{k + k_a^{\mathrm{E}}}{m}, \tag{9.25}$$

we can remove $k + k_a^{\mathrm{E}}$ from the expression in parentheses and write

$$\left[\left(\frac{\mathrm{s}}{\omega_m^{\mathrm{E}}}\right)^2 + 1 + \frac{k_a^{\mathrm{E}}}{k + k_a^{\mathrm{E}}} \frac{k_{33}^2}{1 - k_{33}^2} \frac{RC^{\mathrm{S}}\mathrm{s}}{RC^{\mathrm{S}}\mathrm{s} + 1}\right] u(\mathrm{s}) = \frac{f(\mathrm{s})}{k + k_a^{\mathrm{E}}} = u_{\mathrm{st}}. \tag{9.26}$$

The dynamic response normalized with respect to the static displacement can be written as a function of the nondimensional frequency $\sigma = \mathrm{s}/\omega_m^{\mathrm{E}}$,

$$\frac{u(\sigma)}{u_{\mathrm{st}}} = \frac{RC_p^{\mathrm{S}}\omega_m\sigma + 1}{(\sigma^2 + 1)\left(RC_p^{\mathrm{S}}\omega_m\sigma + 1\right) + K_{33}^2 RC_p^{\mathrm{S}}\omega_m\sigma}. \tag{9.27}$$

A nondimensional parameter denoted the *generalized coupling coefficient* has been introduced in equation (9.27):

$$K_{33}^2 = \frac{k_a^{\mathrm{E}}}{k + k_a^{\mathrm{E}}} \frac{k_{33}^2}{1 - k_{33}^2}. \tag{9.28}$$

The generalized coupling coefficient is analogous to the coupling coefficient for the material, but note that it is a function of the material properties *and* the relative stiffness of the piezoelectric actuator and the original system. Physically, it is related to the amount of strain energy that is stored in the piezoelectric material compared to the strain energy stored in the system stiffness. The form of the equation indicates that even if the material has a high coupling coefficient, the overall coupling of the actuator to the system can be low if the piezoelectric is soft compared to the system stiffness. Only a combination of high material coupling and high relative stiffness will produce large coupling between the actuator and the original system.

The rationale for this strange transformation is that equation (9.27) illustrates that the design of a piezoelectric shunt for a single-mode vibrating system is only a function of two nondimensional parameters: the generalized piezoelectric coupling coefficient K_{33} and the design parameter $RC_p^{\mathrm{S}}\omega_m^{\mathrm{E}}$. Of these two parameters, the first is fixed by the material properties of the piezoelectric and the structure, while the second can be changed by varying the shunt resistance R.

The frequency response normalized to the static deflection is plotted to illustrate the effect of changing the nondimensional design paramaters. Figure 9.4 is a plot of

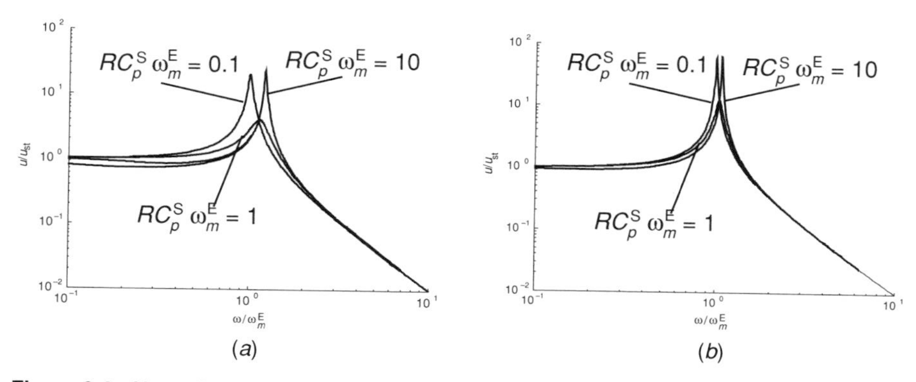

Figure 9.4 Normalized frequency response functions for a piezoelectric shunt: (*a*) $K_{33} = 0.7$; (*b*) $K_{33} = 0.4$.

the normalized frequency responses as a function of $RC_p^S\omega_m^E$ for two values of the generalized coupling coefficient. Comparing the two figures, we see that a decrease in the generalized piezoelectric coupling coefficient reduces the difference in the open- and short-circuit natural frequencies. This is expected from our discussion in Chapter 4, since a decrease in the coupling coefficient produces a decrease in the compliance variation from open-circuit to short-circuit boundary conditions.

Comparing the two figures we also see that a decrease in the piezoelectric coupling coefficient produces a decrease in the maximum achievable damping. A generalized coupling coefficient of 0.7 (Figure 9.4*a*) produces a broad, flat frequency response near resonance when $RC_p^S\omega_m^E = 1$. In contrast, the frequency response is much sharper in this region when the coupling coefficient is reduced to 0.4, as shown in Figure 9.4*b*. The broad, flat frequency response in this region is indicative of a larger mechanical damping coefficient in the vibrating system.

The damping introduced by the shunt can be quantified by solving for the roots of the system characteristic equation. Under the assumption that the roots include a single real root and a pair of complex conjugate roots, we can investigate the damping by computing the damping ratio associated with the complex-conjugate pair as a function of $RC^S\omega_m^E$ for a specified value of K_{33}. This result is shown in Figure 9.5*a*. From this plot we see that increasing the coupling coefficient produces an increase in the achievable damping. For $K_{33} = 0.7$, we see that the maximum achievable damping is approximately 0.11 or 11% critical and the maximum damping occurs at a value of $RC^S\omega_m^E = 0.75$. Reducing the piezoelectric coupling coefficient to 0.3 reduces the maximum achievable damping to 2.2% critical and changes the optimum value of $RC^S\omega_m^E$ to approximately 0.96.

The design of a piezoelectric shunt can be studied succinctly by plotting the maximum achievable damping and optimal value of $RC^S\omega_m^E$ as a function of the piezoelectric coupling coefficient. Figure 9.5*b* is a plot of the maximum achievable damping and the corresponding value of $RC^S\omega_m^E$ as a function of the piezoelectric coupling coefficient. The result clearly illustrates that the maximum achievable damping

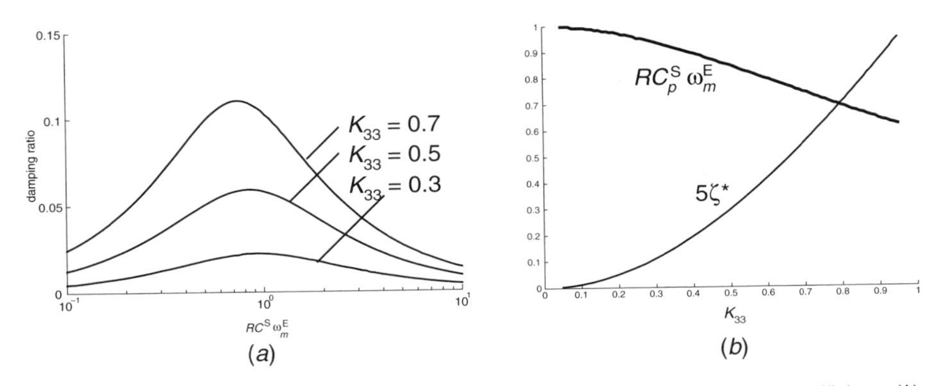

Figure 9.5 (*a*) Damping as a function of $RC^{S}\omega_{m}^{E}$ for three values of coupling coefficient; (*b*) maximum achievable damping and optimal value of the damping ratio.

increases with increasing coupling coefficient, while the value of resistance that achieves maximum damping drops from approximately 1 to approximately 0.6.

The curves plotted in Figure 9.5*b* can be curve-fit as a function of K_{33} to obtain design curves for single-mode piezoelectric shunts. The curve for maximum achievable damping, denoted ζ^*, can be fit accurately with the second-order polynomial

$$\zeta^* = 0.1823K_{33}^2 + 0.0320K_{33}, \tag{9.29}$$

and the optimal resistance value can be obtained from the fit:

$$R^* = \frac{0.4372K_{33}^3 - 0.8485K_{33}^2 + 0.0117K_{33} + 1}{C^S\omega_m^E}. \tag{9.30}$$

Both of these equations are valid approximations for generalized coupling coefficients lower than $K_{33} = 0.95$.

Example 9.1 A piezoelectric shunt is being designed using a material with a generalized coupling coefficient of 0.55. The short-circuit natural frequency of the shunt is 85 Hz and the strain-free capacitance is 0.7 μF. Estimate the maximum achievable damping and the corresponding resistance value.

Solution The solution can be obtained from Figure 9.5*b* or from equations (9.29) and (9.30). Using the polynomial fits, we obtain a maximum damping value of

$$\begin{aligned}\zeta^* &= (0.1823)(0.55)^2 + (0.0320)(0.55)\\ &= 0.0727.\end{aligned}$$

The maximum achievable damping is approximately 7%. The resistance value that achieves maximum damping is obtained from equation (9.30):

$$R^* = \frac{(0.4372)(0.55)^3 - (0.8485)(0.55)^2 + (0.0117)(0.55) + 1}{(85)(2)(\pi)(0.7 \times 10^{-6})\text{F}}$$
$$= 2200\ \Omega.$$

9.2.1 Inductive–Resistive Shunts

A resistive shunt provides some measure of damping through heating of the electrical element due to mechanical vibration. As Figure 9.5 illustrates, the damping is directly related to the coupling coefficient of the piezoelectric material. Studying Figure 9.5*b*, we note that the damping is less than 6% when the material coupling coefficient is less than 0.5. As we will see in discussing other modes of resistive shunting, the generalized coupling coefficient of the material system can often be in the range 0.1 to 0.3, therefore limiting the amount of damping that can be introduced into the vibrating system.

One method of overcoming the limitations of a purely resistive shunt is to incorporate an inductive element into the electrical circuit. The resistive–inductive shunt in series with the piezoelectric material produces a resistive–inductive–capacitive (RLC) circuit, which can be tuned to introduce damping into the mechanical vibration of the system.

A piezoelectric stack with an inductive–resistive network produces the following set of equations:

$$m\ddot{u} + k + k_a^{\text{D}} u - g_{33} q = f \tag{9.31}$$

$$L\ddot{q} + R\dot{q} - g_{33} u + \frac{1}{C^{\text{S}}} q = 0, \tag{9.32}$$

where L is the inductance of the electrical circuit. Transforming equation (9.32) into the Laplace domain and writing the relationship between $u(\mathsf{s})$ and $f(\mathsf{s})$ produces

$$\left[m\mathsf{s}^2 + k + k_a^{\text{D}}\left(1 - \frac{k_{33}^2}{LC^{\text{S}}\mathsf{s}^2 + RC^{\text{S}}\mathsf{s} + 1}\right)\right] u(\mathsf{s}) = f(\mathsf{s}). \tag{9.33}$$

The analysis is slightly more complicated than for the case of a purely resistive shunt, due to the existence of the inductive element in the electrical circuit. The analysis can be simplified somewhat by realizing that the inductive–resistive shunt essentially creates an electronic tuned-mass damper with the piezoelectric coupled to the inductor and resistor acting as the secondary mass–spring–damper. For these reasons, introduce the following variables into the analysis:

$$\omega_m^{E^2} = \frac{k + k_a^{\text{E}}}{m} \tag{9.34}$$

$$\omega_e^2 = \frac{1}{LC^{\text{S}}}, \tag{9.35}$$

where ω_m^{E} is the *short-circuit mechanical resonance* of the system and ω_e is the *electrical resonance*. Defining the ratio of the electrical-to-mechanical resonance as

$$\alpha = \frac{\omega_e}{\omega_m}, \tag{9.36}$$

we can write the frequency response between dynamic deflection and static response as

$$\frac{u(\mathrm{s})}{u_{\mathrm{st}}} = \frac{\sigma^2 + \alpha^2 RC^{\mathrm{S}}\omega_m^{\mathrm{E}}\sigma + \alpha^2}{(\sigma^2+1)\left(\sigma^2 + \alpha^2 RC^{\mathrm{S}}\omega_m^{\mathrm{E}}\sigma + \alpha^2\right) + K_{33}^2\left(\sigma^2 + \alpha^2 RC^{\mathrm{S}}\omega_m^{\mathrm{E}}\sigma\right)}. \tag{9.37}$$

The nondimensional frequency is once again denoted σ and the static deflection u_{st} is defined in equation (9.26). Although equation (9.37) is not necessarily a simpler form of the transfer function, this expression highlights two important facts about the inductive–resistive shunts:

1. The term $\omega_m^{\mathrm{E}} RC^{\mathrm{S}}$ is still an important parameter in the analysis.
2. The additional critical term in the analysis is the ratio of the electrical to mechanical natural frequency, α.

Thus, we can parameterize the effectiveness of the inductive–resistive shunt in terms of the parameter $\omega_m^{\mathrm{E}} RC^{\mathrm{S}}$ and the ratio of natural frequencies.

The choice of these parameters can be assisted by remembering that the inductive–resistive shunt is essentially a tuned-mass damper. The optimization of tuned-mass damper properties is well studied in vibration theory, and results from that discipline indicate that a tuning ratio α of approximately 1 increases the damping achievable in the mechanical system. Setting the tuning ratio equal to 1 in equation (9.37) results in the expression

$$\frac{u(\mathrm{s})}{u_{\mathrm{st}}} = \frac{\sigma^2 + RC^{\mathrm{S}}\omega_m^{\mathrm{E}}\sigma + 1}{(\sigma^2+1)\left(\sigma^2 + RC^{\mathrm{S}}\omega_m^{\mathrm{E}}\sigma + 1\right) + K_{33}^2\left(\sigma^2 + RC^{\mathrm{S}}\omega_m^{\mathrm{E}}\sigma\right)}. \tag{9.38}$$

Figure 9.6*a* is a plot of the magnitude of the transfer function for various values of $\omega_m^{\mathrm{E}} RC^{\mathrm{S}}$ when the generalized coupling coefficient is set to 0.7 and the tuning ratio is equal to 1. The figure illustrates that a resistance value that is well below 1 produces a system with two lightly damped resonances. These resonances correspond to the resonances associated with the mechanical and electrical properties of the coupled system. Increasing the resistance such that $\omega_m RC^{\mathrm{S}}$ is 0.15 produces a system with two damped resonances. Increasing the resistance further, though, increases the peak response of the system and produces a result that resembles a damped mass–spring oscillator. Figure 9.6*b* is a magnitude plot for $K_{33} = 0.4$. The trends for a generalized coupling coefficient of 0.4 are similar to that of $K_{33} = 0.7$, but the variation in the resonance peaks as a function of the shunt resistance is smaller.

The trends displayed in Figure 9.6 illustrate that there is a clear trade-off in the choice of the resistance value for the inductive–resistive shunt. A value that is too

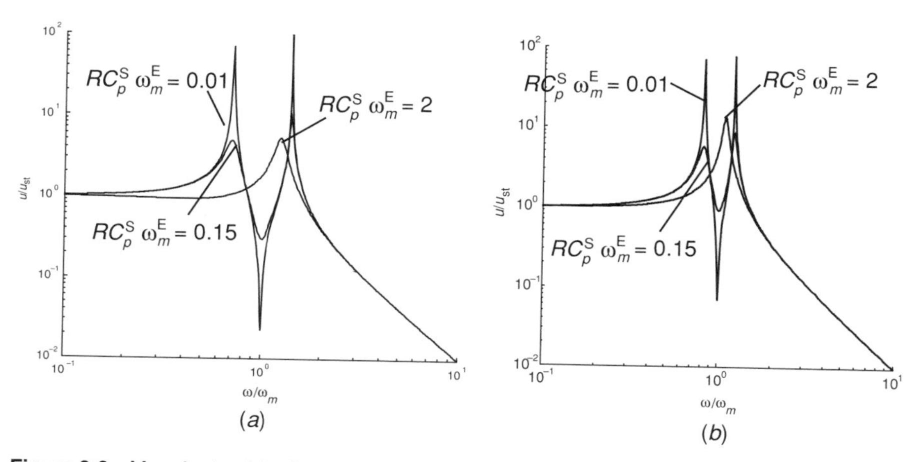

Figure 9.6 Magnitude of the frequency response of an RL shunt for (*a*) $K_{33} = 0.7$ and (*b*) $K_{33} = 0.4$ and $\alpha = 1$ for various values of $\omega_m RC^S$.

small produces two lightly damped resonances, while a value that is too large produces an amplification peak that is larger than the peak obtained with a smaller resistance value. These trends are similar to those exhibited by a purely resistive shunt, although the addition of the inductor produces an additional resonance peak in the response due to the electrical inertia.

We can analyze the performance of the inductive–resistive shunt in a manner similar to our method of analyzing a purely resistive network. From equation (9.37) we see that the characteristic polynomial of the system is fourth order, which yields two separate second-order systems, each with an associated damping ratio. The minimal damping ratio will determine the oscillatory decay of the system; therefore, maximizing the minimal damping ratio will produce the smallest decay time in the vibrating system. Denoting the minimal damping ratio as

$$\zeta^* = \min\{\zeta_1, \zeta_2\}, \tag{9.39}$$

we can analyze an RL shunt numerically to determine the minimal damping ratio as a function of the generalized coupling.

Numerical analysis of the characteristic equation can be performed to determine the damping ratio at the optimal choice of tuning ratio and $\omega_m^E RC^S$. Figure 9.7 is a plot of ζ^* at the optimum conditions as a function of the piezoelectric coupling coefficient. We can obtain an accurate estimate of the maximum damping using an inductive–resistive shunt from the expression

$$\zeta^* \approx \frac{K_{33}}{2}, \tag{9.40}$$

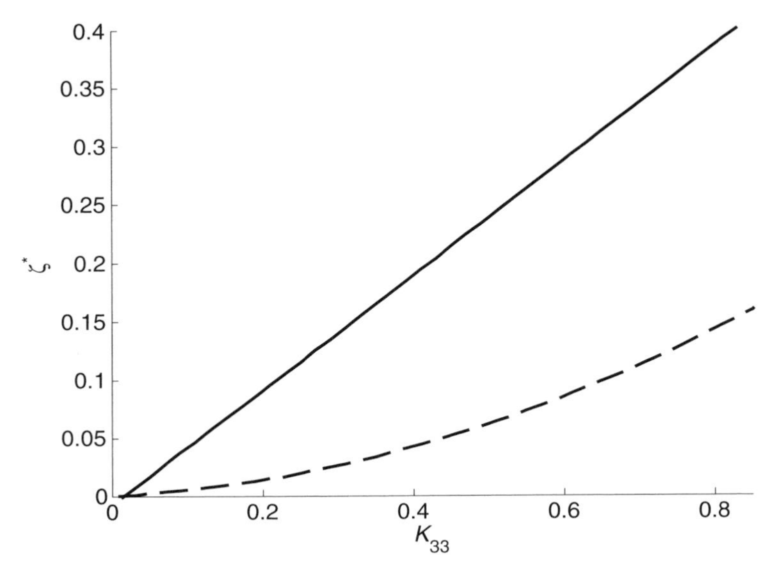

Figure 9.7 Maximum damping obtained with a resistive–inductive shunt (solid) and a resistive shunt (dashed) as a function of the generalized coupling coefficient.

which is accurate over the range $0.05 \leq K_{33} \leq 0.8$. The shunt parameters associated with the optimal damping are shown in Figures 9.8 as a function of the generalized coupling coefficient. The analysis demonstrates that the tuning ratio for the inductive–resistive shunt is approximately 1 for low values of the coupling coefficient and increases to approximately 2 as K_{33} approaches 1. The resistive value of the shunt is small for low values of the coupling coefficient but approaches 0.8 and then decreases

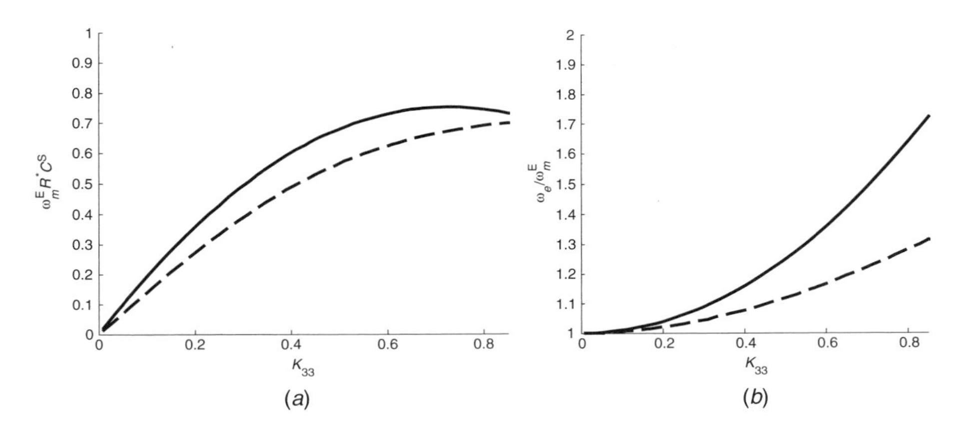

Figure 9.8 Optimal parameters for an inductive-resistive shunt: maximum damping (solid) and frequency response approach (dashed).

slightly. The shunt parameters that achieve maximum damping can be approximated by the polynomials

$$R^*_{\text{DAMP}} = \frac{-1.4298K_{33}^2 + 2.0748K_{33}}{\omega_m^{\text{E}} C^{\text{S}}} \tag{9.41}$$

$$\left(\frac{\omega_e}{\omega_m^{\text{E}}}\right)^*_{\text{DAMP}} \approx K_{33}^2 + 1. \tag{9.42}$$

Example 9.2 An RL shunt is being designed for the mechanical system studied in Example 9.1. Estimate the maximum achievable damping in the system and the values of R and L that produce the maximum damping.

Solution The maximum damping achievable can be computed from equation (9.40) for the stated value of $K_{33} = 0.55$:

$$\begin{aligned} \zeta^* &\approx \frac{0.55}{2} \\ &= 0.275. \end{aligned}$$

The maximum damping achievable is estimated to be 27.5% for an RL shunt. The optimal resistance value is computed from equation (9.41):

$$\begin{aligned} R^*_{\text{DAMP}} &= \frac{(-1.4298)(0.55)^2 + (2.0748)(0.55)}{(85)(2)(\pi)(0.7 \times 10^{-6})} \\ &= 1895\ \Omega, \end{aligned}$$

and the tuning ratio that achieves maximum damping is computed from

$$\begin{aligned} \left(\frac{\omega_e}{\omega_m}\right)^*_{\text{DAMP}} &\approx (0.55)^2 + 1 \\ &= 1.302. \end{aligned}$$

The electrical resonance can then be computed from

$$\omega_e = (1.302)(85)(2)\pi = 695 \text{ rad/s}, \tag{9.43}$$

and the inductance is computed by rewriting equation (9.35):

$$\begin{aligned} L &= \frac{1}{(695)^2\,(0.7 \times 10^{-6})} \\ &= 2.95 \text{ H}. \end{aligned}$$

Thus, the choice of optimal parameters leads to $R = 1895\ \Omega$ and $L = 2.95$ H.

An alternative method for choosing the optimal values of the shunt parameters has been derived using the frequency response u/u_{st}. In a manner analogous to that used by Den Hartog for tuned-mass dampers, values that optimize the frequency response can be derived by noting that there are two frequencies at which all of the magnitudes of the frequency response functions cross. Defining the optimal values of the frequency ratio α and nondimensional resistance $RC^S\omega_m^E$ as the values at which the magnitude of the transfer function is equal at these two points, the nondimensional tuning parameters are found to be

$$\begin{aligned} \alpha_{FRF}^* &= \sqrt{1 + K_{33}^2} \\ R_{FRF}^* C^S \omega_m^E &= \frac{\sqrt{2} K_{33}}{1 + K_{33}^2}. \end{aligned} \tag{9.44}$$

The tuning parameters for maximizing damping and for optimizing the frequency response are compared in Figure 9.8. Maximizing the damping generally requires a higher value of resistance and a higher natural frequency than simply equalizing the peaks of the frequency response magnitude. Also, the result demonstrates that equalizing the peaks of the frequency response will not lead to maximum damping.

The difference in the frequency response is examined by computing the magnitude of u/u_{st} for various values of the optimal parameters. In Figure 9.9, lines (*a*) and (*d*)

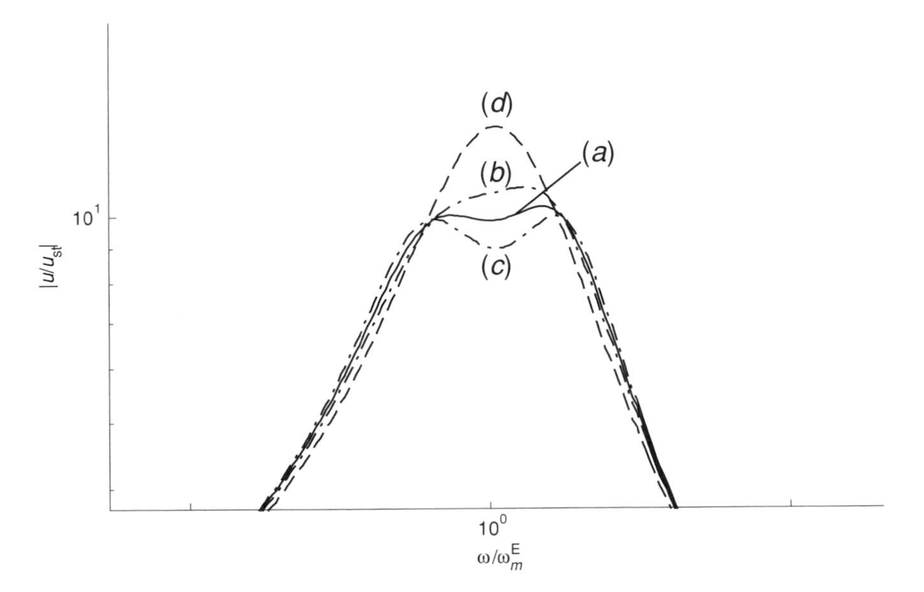

Figure 9.9 Comparison of frequency response magnitudes for inductive-resistive shunts ($K_{33} = 0.14$): (*a*) optimal frequency response parameters $\alpha = 1.01$ and $RC^S\omega_m^E = 0.196$; (*b*) $\alpha = 1.01$ and $RC^S\omega_m^E = 0.178$; (*c*) $\alpha = 1.01$ and $RC^S\omega_m = 0.216$; (*d*) optimal parameters for maximum damping.

represent the frequency response magnitudes when the tuning parameters are chosen for optimal frequency response and maximum damping, respectively. Comparing the two results, we see that optimizing the frequency response will produce a minimum peak value of $|u/u_{st}|$, although the damping in the resonant modes may be higher than that for a system optimized by maximizing the damping in both modes. Also shown in Figure 9.9 are two plots, (*b*) and (*c*), that illustrate the effect of mistuning on the frequency response. In these responses the frequency ratio α has been tuned properly but the nondimensional resistance value has been mistuned. This creates variations in the frequency response that produce changes in the breadth of the magnitude response in this region as well as the peak values.

9.2.2 Comparison of Shunt Techniques

Two different methods of adding damping using piezoelectric devices have been described. The first requires that we connect the piezoelectric element to a resistor, while the second requires that we connect the piezoelectric element to an inductor–resistor series network. In both techniques we can analyze the system to obtain values of the electrical parameters that lead to the maximum damping in the mechanical systems.

The trade-offs between the two techniques can be illustrated by comparing the shunt performance and the shunt implementation. Comparing Figure 9.5*b* with Figure 9.7 we see that an RL shunt that is tuned for maximum damping will produce a larger value of ζ^* than a purely resistive shunt for exactly the same generalized coupling coefficient. Dividing equation (9.40) by equation (9.29) demonstrates that an RL shunt will produce approximately six times the damping for values of the coupling coefficient below 0.4 and between two and three times the damping for coupling coefficients above 0.5. Thus, we can conclude that an RL shunt is most beneficial in systems that have a low piezoelectric coupling coefficient.

The increase in achievable damping is offset by the need to incorporate an inductor into the electrical shunt circuit. Inductors are electromagnetic devices whose size increases with increasing inductance. If we rewrite equation (9.35) to solve for the inductance,

$$L = \frac{1}{\omega_e^2 C^S}, \tag{9.45}$$

we find that the inductance is proportional to the inverse of the natural frequency squared and the material strain-free capacitance. As shown in Figure 9.8*b*, the electrical resonance is generally within a factor of 2 of the mechanical resonance; therefore, we can conclude that the optimal circuit inductance increases as the natural frequency of the mechanical system decreases.

The implementation of an RL shunt circuit can be impractical for energy dissipation in a system that exhibits mechanical resonance below a few hundred hertz. As we saw in Example 9.2, the inductance required to achieve maximum damping can be on the order of henrys when the mechanical resonance is on the order of 100 Hz. Inductors of this size can be prohibitively large and bulky and can make the implementation of an RL shunt circuit impractical due to size or volume constraints. This problem can be overcome by using what is termed a *synthetic inductor*, a small electronic circuit that mimics a passive inductor over a circuit frequency range. Although this solution overcomes the size and volume problems associated with an actual inductor, it introduces the need to power an electronic circuit (albeit with small power) and produces a solution that is exact only over a finite range of frequencies.

9.3 MULTIMODE SHUNT TECHNIQUES

The analysis of resistive and resistive–inductive shunts has concentrated on the effect of shunts on a single vibration mode of a structure. In many applications a structure will exhibit multiple resonant frequencies within some bandwidth of interest for a particular application. In this case it is desirable to use passive shunting methods to control multiple structure modes of vibration.

We know from the discussion in Chapter 5 that the coupling of a piezoelectric element to various structural modes of vibration is a function of a number of parameters. The geometry and material properties of the piezoelectric element relative to the host structure will affect the relative coupling of the piezoelectric element to the multiple vibration modes. In addition, placement of the piezoelectric element relative to the modal response of the structure greatly changes the relative coupling between the piezoelectric element and the structure. In certain instances, as discussed in Chapter 5, the piezoelectric element is designed to couple to only a single mode of structure vibration using the concept of modal filters.

There are two approaches to extending the analysis of Section 9.2 to systems with multiple structural resonances. First, we can envision that a single piezoelectric element is used as a shunt for multiple structural resonances. The disadvantage of this approach is that a single piezoelectric element might not couple into all modes equally well, thus limiting the amount of achievable vibration suppression. Another approach might be to implement single-mode shunts with multiple piezoelectric elements. The use of multiple piezoelectric elements allows greater freedom in design since each element can be optimized for a particular mode, although, as we will see, the design is complicated by the coupling that exists between the structural modes of vibration.

The design of a multimode shunt with a single piezoelectic element or a small number of elements is very similar to the design of feedback control elements. For this reason, discussion of this approach is delayed until Chapter 10, where vibration control is discussed in detail. Furthermore, this approach is discussed more efficiently in terms of state-space analysis, which is also emphasized in Chapter 10.

In this section we provide an example of a multimode shunt that utilizes multiple piezoelectric elements, each with a single shunt damper. Multimode shunt theory is based on the undamped second-order matrix equations derived in Chapter 5 and stated in equations (5.133) and (5.134). Structural damping could be added, but for the purpose of illustration we utilize the undamped equations of motion. In multi-degree-of-freedom equations there is the potential to have an arbitrary number of piezoelectric elements; therefore, the number of generalized coordinates for the charge variables is arbitrary. As shown in equation (5.134), the voltage inputs to piezoelectrics act on the system through the input matrix B_v. In general form we can write the applied voltage as a function of the total charge collected for each piezoelectric element:

$$\mathbf{v} = -L\ddot{\mathbf{q}}_t - R\dot{\mathbf{q}}_t, \tag{9.46}$$

where L and R are now matrices of inductive and resistive components. Assuming that the charge is collected in the same manner as the voltage is applied, we can write $\mathbf{q}_t = B_v'\mathbf{q}$ and substitute the expression into equation (9.46) to produce

$$\mathbf{v} = -LB_v'\ddot{\mathbf{q}} - RB_v'\dot{\mathbf{q}}. \tag{9.47}$$

Substituting equation (9.47) into equation (5.134) produces second-order linear matrix expressions for a multimode shunt:

$$(M_s + M_p)\ddot{\mathbf{r}} + (K_s + K_p)\mathbf{r} - \Theta\mathbf{q} = B_f\mathbf{f} \tag{9.48}$$

$$B_vLB_v'\ddot{\mathbf{q}} + B_vRB_v'\dot{\mathbf{q}} - \Theta'\mathbf{r} + {C_p}^{-1}\mathbf{q} = 0. \tag{9.49}$$

Writing the equations in this manner clearly identifies the physics associated with a multimode shunt circuit. The inductive terms are adding *electronic mass*, and the resistive terms are adding *electronic damping*. Coupling between the mechanical and electrical degrees of freedom is introduced through the stiffness elements quantified by the matrix θ. The design parameters for the multimode shunt are the inductance and resistance matrices, L and R, respectively.

The system under consideration is a slender cantilever beam with multiple piezoelectric elements bonded to the surface of the beam as shown in Figure 9.10. Each piezoelectric element is defined by its location and material and geometric parameters. The electrical connections are defined such that the voltage applied to each pair of piezoelectric elements is positive for the upper layer and negative for the lower layer. Thus, the input matrix for the piezoelectric elements is written as

$$B_v\mathbf{v} = \begin{bmatrix} \begin{pmatrix} 1 \\ -1 \end{pmatrix} & \cdots & 0 \\ \vdots & \ddots & \vdots \\ 0 & \cdots & \begin{pmatrix} 1 \\ -1 \end{pmatrix} \end{bmatrix} \begin{pmatrix} v_1 \\ \vdots \\ v_{N_p} \end{pmatrix}, \tag{9.50}$$

where N_p is the number of piezoelectric elements.

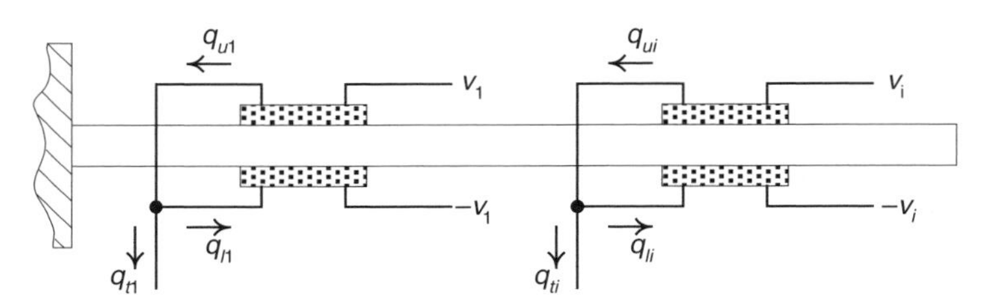

Figure 9.10 Beam with multiple piezoelectric elements for multimode shunting.

The charge output of the piezoelectric elements can be written

$$\begin{pmatrix} q_{t1} \\ \vdots \\ q_{tN_p} \end{pmatrix} = \begin{bmatrix} (1 & -1) & \cdots & 0 \\ \vdots & \ddots & \vdots \\ 0 & \cdots & (1 & -1) \end{bmatrix} \begin{pmatrix} q_{u1} \\ q_{l1} \\ \vdots \\ q_{uN_p} \\ q_{lN_p} \end{pmatrix} = \mathrm{B}_\mathrm{v}'\mathbf{q}, \tag{9.51}$$

which demonstrates that the charge output is being measured in the same manner as voltage is applied.

For illustrative purposes we assume that the network of piezoelectric shunts are *decoupled*, which means that the voltage applied to each piezoelectric element is solely a function of the charge measured from the same element. In the case of a decoupled network, the matrix of inductance values and resistance values are diagonal matrices of the form

$$\mathrm{L} = \begin{bmatrix} L_1 & \cdots & 0 \\ \vdots & \ddots & \vdots \\ 0 & \cdots & L_{N_p} \end{bmatrix} \qquad \mathrm{R} = \begin{bmatrix} R_1 & \cdots & 0 \\ \vdots & \ddots & \vdots \\ 0 & \cdots & R_{N_p} \end{bmatrix}. \tag{9.52}$$

Combining equation (9.52) with the definitions of the input matrix B_v enables the definition of equations of motion for the multimode shunt. The matrices required for the equations of motion are

$$\mathrm{B}_\mathrm{v}\mathrm{L}\mathrm{B}_\mathrm{v}' = \begin{bmatrix} \begin{bmatrix} L_1 & -L_1 \\ -L_1 & L_1 \end{bmatrix} & \cdots & 0 \\ \vdots & \ddots & \vdots \\ 0 & \cdots & \begin{bmatrix} L_{N_p} & -L_{N_p} \\ -L_{N_p} & L_{N_p} \end{bmatrix} \end{bmatrix}. \tag{9.53}$$

Table 9.1 Geometric and material parameters for the multimode shunt example

Geometric	Substrate	Piezoelectric
$L = 38.1$ cm	$t_s = 1.6$ mm	$t_p = 0.5$ mm
$w = 3.79$ cm	$\rho_s = 2700$ kg/m^3	$\rho_p = 7800$ kg/m^3
	$c_{s11} = 69$ GPa	$c_{11}^{\mathrm{D}} = 130$ GPa
		$h_{13} = -1.35 \times 10^9$ N/C
		$\beta_{33}^{\mathrm{S}} = 1.5 \times 10^8$ m/F

The matrix for the resistance can be written in the same manner:

$$\mathrm{B_v R B_v'} = \begin{bmatrix} \begin{bmatrix} R_1 & -R_1 \\ -R_1 & R_1 \end{bmatrix} & \cdots & 0 \\ \vdots & \ddots & \vdots \\ 0 & \cdots & \begin{bmatrix} R_{N_p} & -R_{N_p} \\ -R_{N_p} & R_{N_p} \end{bmatrix} \end{bmatrix}. \tag{9.54}$$

The block diagonal structure of the matrices is due to the assumption of decoupling within the piezoelectric network.

The matrices in equations (9.53) and (9.54) can be incorporated into the equations of motion to analyze the effect of a multimode shunt on the piezoelectric material system. The geometric and material parameters for the example are listed in Table 9.1.

One of the primary results from the discussion of shunts for single-mode systems is that the piezoelectric coupling coefficient plays a key role in determining the amount of damping that can be introduced by the shunt circuit. One method of determining suitable locations for the piezoelectric elements is to analyze the generalized coupling coefficient for each of the vibration modes of interest. As discussed in Chapter 5, the generalized coupling coefficient is an extension of the concept of a single vibration mode to systems of multiple modes, and it quantifies the coupling of the piezoelectric element to the separate modes of vibration.

The approach to multimode shunt design is first to analyze the generalized coupling coefficients for the modes of interest to determine suitable locations for the piezoelectric elements. Once the location of the elements has been determined, the results for single-mode vibration presented in Section 9.2 will be applied to determine the values of the inductance and resistance for each element. The frequency response of the complete system is then analyzed with the equations of motion, equations (9.48) and (9.49), to analyze the dynamic response. The placement of the piezoelectric elements are studied by computing the generalized coupling coefficients for the first three modes of the beam. For illustration we assume that the length of the piezoelectric element is $0.0833L$, which is meant to satisfy a constraint that the total length of the piezoelectric material added to the beam is no greater than 25% of the beam length. The geometric and material parameters listed in Table 9.1 are used to define the mass, stiffness, and coupling matrices for the piezoelectric material system using

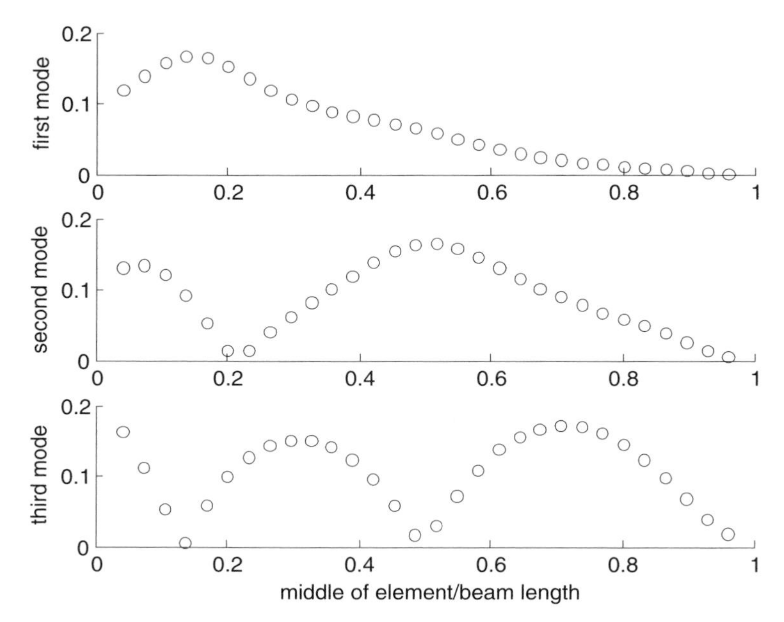

Figure 9.11 Generalized coupling coefficients for the first three modes of the beam used in the multimode shunt example.

the methods derived in Chapter 5. At each location of the piezoelectric, the short- and open-circuit natural frequencies are computed, and then the generalized coupling coefficient is computed using equation (5.211). The analysis is repeated for multiple locations of the piezoelectric element to yield a plot of the generalized coupling coefficients for the first three modes.

The analysis demonstrates that the location of the piezoelectric strongly affects the generalized coupling coefficient. Figure 9.11 is a plot of the generalized coupling coefficient for the first three modes of the beam. The peak coupling coefficient for the first mode occurs near the clamped end of the beam, while the peak coupling coefficient for the second and third modes occurs farther along the span of the beam. The locations of the peak coupling coefficients are determined to be $0.14L$, $0.52L$, and $0.70L$ for the first three modes, respectively, using Figure 9.11. The peak coupling coefficient for all three of these locations is approximately 0.17.

Before analyzing the design of the individual shunt circuits, it is instructive to look at the general trends that are exhibited by Figure 9.11. First, analysis of the generalized coupling coefficients illustrates the trade-offs associated with choosing a single location for the shunt element. For example, if a location that maximizes coupling to the first mode is chosen, the coupling coefficients for the remaining two modes are 0.09 and 0.01, respectively. Choosing this location would produce good coupling into the first vibration mode but would yield poor coupling into the second and third modes. This trend would hold if locations were chosen to maximize coupling into the second and third modes as well. Second, the plot of the generalized coupling coefficients exhibits a periodicity that is related to the mode shape of the structure. The plot of the generalized coupling coefficient for the second mode has a single node, or location that yields approximately zero coupling, while the plot of k_{ij} for the third mode has two nodal plots. This is not a coincidence since it is related directly to integration of the strain over the mode shape of the structure. Recalling from the discussion in Chapter 5 that the coupling coefficients in the matrix θ are directly related to the difference in slopes at the ends of the piezoelectric element. Positioning the piezoelectric at a location in which the slopes at the end are equal will produce a zero in the coupling matrix for the particular vibration mode. Physically, this indicates that integration of the strain over the volume of the piezoelectric element is equal to zero. This will result in a minimum in the generalized coupling coefficient at that location. The periodicity in Figure 9.11 is related directly to the periodicity of the structural mode shape for the beam.

Returning to the design of the shunt circuits, the next step in the procedure is to compute the values of the shunt components. In this example we utilize an inductive–resistive shunt to maximize the amount of structural damping introduced in the modes. The analysis presented earlier in the chapter will be applied to determine the values of the resistance and inductance required for maximum damping. The values are computed using the polynomial approximations in equations (9.41) and (9.42) for a coupling coefficient of 0.17:

$$\begin{aligned} R^* \omega_m^{\mathrm{E}} C_p^{\mathrm{S}} &= (-1.4298)(0.17)^2 + (2.0748)(0.17) \\ &= 0.3114 \end{aligned} \tag{9.55}$$

$$\begin{aligned} \left(\frac{\omega_e}{\omega_m}\right)^* &= (0.17)^2 + 1 \\ &= 1.0289 \end{aligned} \tag{9.56}$$

The equations of motion are derived using the methods introduced in Chapter 5. Analysis of the multiple piezoelectric elements results in the matrices

$$\mathrm{M_s} + \mathrm{M_p} = \begin{bmatrix} 0.07107 & 0.008295 & -0.007541 \\ 0.008295 & 0.07366 & -0.003592 \\ -0.007541 & -0.003592 & 0.07314 \end{bmatrix} \text{kg} \tag{9.57}$$

$$\mathrm{K_s} + \mathrm{K_p} = \begin{bmatrix} 315.8 & -21.97 & 298.1 \\ -21.97 & 13{,}270.0 & -8{,}261.0 \\ 298.1 & -8261.0 & 89770.0 \end{bmatrix} \text{N/m} \tag{9.58}$$

$$\theta = \begin{bmatrix} -12{,}130.0 & 12{,}130.0 & -4{,}760.0 & 4{,}760.0 & -2{,}059.0 & 2{,}059.0 \\ -31{,}860.0 & 31{,}860.0 & 67{,}330.0 & -67{,}330.0 & 49{,}360.0 & -49{,}360.0 \\ 11{,}690.0 & -11{,}690.0 & -34{,}100.0 & 34{,}100.0 & -19{,}6700.0 & 19{,}6700.0 \end{bmatrix} \text{N/C} \tag{9.59}$$

Table 9.2 Natural frequency and shunt parameters for the multimode analysis

Mode	f_n^{D} (Hz)	R^*(kΩ)	L^*(H)
First	10.59	76.618	3433
Second	65.97	12.299	88
Third	177.67	4.567	12

$$\mathbf{C}_{\mathrm{p}}^{\mathrm{S}} = \begin{bmatrix} 3.212 & 0 & 0 & 0 & 0 & 0 \\ 0 & 3.212 & 0 & 0 & 0 & 0 \\ 0 & 0 & 3.212 & 0 & 0 & 0 \\ 0 & 0 & 0 & 3.212 & 0 & 0 \\ 0 & 0 & 0 & 0 & 3.212 & 0 \\ 0 & 0 & 0 & 0 & 0 & 3.212 \end{bmatrix} \times 10^{-8} \text{ F.} \tag{9.60}$$

The undamped natural frequencies are found using equation (5.144) and are listed in Table 9.2. A structural damping matrix is added to the equations of motion to model the small amount of inherent damping in the beam. For this analysis a damping factor of 0.005 is assumed for each mode. The result is

$$\mathbf{D}_{\mathrm{s}} = \begin{bmatrix} 0.03477 & 0 & 0 \\ 0 & 0.2179 & 0 \\ 0 & 0 & 0.6103 \end{bmatrix} \mathrm{N \cdot s/m}. \tag{9.61}$$

The resistance value for each of the piezoelectric shunts is computed from equation (9.55). For the first mode,

$$R^* = \frac{0.3275}{[(10.59)(2)(\pi \text{ rad/s})](6.424 \times 10^{-8} \text{ F})} = 76.618 \text{ k}\Omega. \tag{9.62}$$

Note that the capacitance used in the computation is the total capacitance of the piezoelectric shunt, which is equal to twice the capacitance of each element in $\mathbf{C}_{\mathrm{p}}^{\mathrm{S}}$. The value for the shunt inductance is computed from equation (9.56) combined with equation (9.45):

$$L^* = \frac{1}{[(1.012)(10.59)(2\pi)]^2(6.424 \times 10^{-8} \text{ F})} = 3433 \text{ H}. \tag{9.63}$$

The results for the remaining two modes are listed in Table 9.2. These values are combined with equations (9.53) and (9.54) to compute the inductance and resistance matrices for the shunt network. The results are

$$L = \begin{bmatrix} 3,433.0 & -3,433.0 & 0 & 0 & 0 & 0 \\ -3,433.0 & 3,433.0 & 0 & 0 & 0 & 0 \\ 0 & 0 & 88.0 & -88.0 & 0 & 0 \\ 0 & 0 & -88.0 & 88.0 & 0 & 0 \\ 0 & 0 & 0 & 0 & 12.0 & -12.0 \\ 0 & 0 & 0 & 0 & -12.0 & 12.0 \end{bmatrix} \text{H} \tag{9.64}$$

$$R = \begin{bmatrix} 76,620.0 & -76,620.0 & 0 & 0 & 0 & 0 \\ -76,620.0 & 76,620.0 & 0 & 0 & 0 & 0 \\ 0 & 0 & 12,300.0 & -12,300.0 & 0 & 0 \\ 0 & 0 & -12,300.0 & 12,300.0 & 0 & 0 \\ 0 & 0 & 0 & 0 & 4,567.0 & -4,567.0 \\ 0 & 0 & 0 & 0 & -4,567.0 & 4,567.0 \end{bmatrix} \Omega. \tag{9.65}$$

The equations of motion for the multimode shunt are defined completely once the inductance and resistance matrices are defined. The equations of motion can then be solved in the time or frequency domain to examine the performance of the shunt network. To compare the results of the multimode shunt directly with the single-mode shunt, the equations of motion are solved in the frequency domain to examine the reduction in the amplitude of the steady-state response with introduction of the shunt network. Equations (9.48) and (9.49) are transformed into the frequency domain and the deflection at the tip of the beam (which is also the point of force application) is computed as a function of frequency for the system without a shunt and with the shunt parameters defined in Table 9.2.

The frequency-domain analysis demonstrates that the multimode shunt is able to introduce energy dissipation into all three structural modes. Figure 9.12 is the magnitude of the frequency response between the point force at the tip and deflection at the tip. The dashed line is the frequency response with only structural damping ($\zeta_i = 0.005$) and the solid line is the frequency response with the multimode shunt. The insets in Figure 9.12 demonstrate that the peak response at frequencies near the structural resonances are reduced significantly and the peak is rounded due to the additional damping. The analysis also illustrates that similar levels of damping are introduced into all three structural modes. This is a result of using three separate piezoelectric elements, each tuned to a particular structural mode.

Practical implementation of the resistive–inductive shunt is hampered by the large inductance values that are required for each shunt element. As Table 9.2 illustrates, the inductance values computed for this application range from 10 to over 1000 henrys, due to the low natural frequencies of the structure and the low capacitance of the piezoelectric elements. Inductance values on the order of henrys are difficult

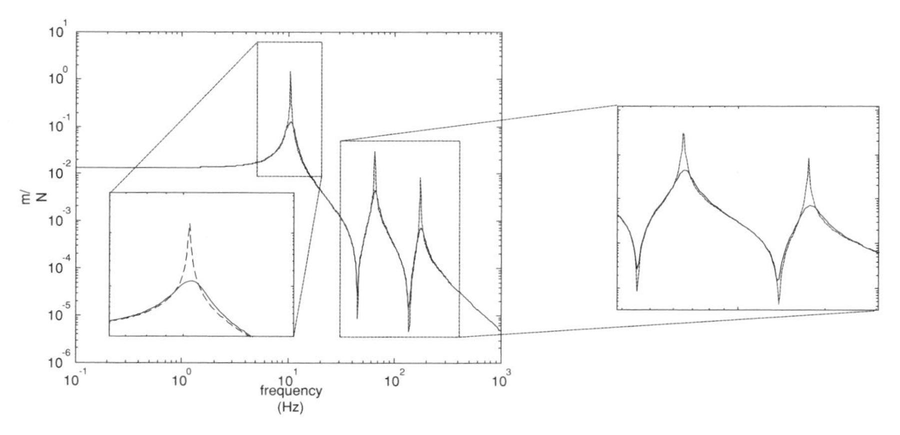

Figure 9.12 Frequency response for the system with structural damping (dashed) and the resistive–inductive shunt network (solid).

to achieve with passive elements and it is often required to implement a synthetic inductor in the shunt network. A synthetic inductor is an active electronic circuit that mimics the response of an inductor over a specified frequency range.

9.4 SEMIACTIVE DAMPING METHODS

An important parameter that arises in the analysis of passive shunts using piezoelectric materials is the tuning of the resistance and inductance of the shunt circuit. The design methodology is based on the proper choice of resistance and inductance values. As illustrated in Section 9.3, the proper choice of these parameters depends on the amount of coupling between the piezoelectric element and the structural mode. For multimode structures the coupling is affected by the material properties, geometry, and (probably most important) the location of the element relative to the strain distribution in the structural mode.

In cases in which the structural resonances are known a priori and relatively constant with time, the method introduced earlier is a good way to choose the parameters of the piezoelectric shunt. In many cases, though, the resonances of a structure are not known a priori or the resonances actually change over the operating life of the structure. Unknown or uncertain resonant properties might be due to the difficultly or cost of building an accurate structural model, and variations in the resonant properties might occur due to changes in the operating conditions or variations in mechanical properties due to such factors as aging. In these cases it is of interest to develop a method that automatically tunes the parameters of a passive shunt network to reduce structural vibration.

Vibration suppression in the presence of uncertain or time-varying structural resonances leads naturally to a discussion of *adaptive* shunt networks. Adaptive shunt networks are those whose parameters change as a function of time for the purpose of achieving maximum performance. The parameter variation can be introduced through implementation of the shunt network with programmable circuit elements or through implementation of the shunt circuit in a digital microprocessor.

9.4.1 System Norms for Performance Definition

A critical aspect of using adaptation to tune the parameters of a shunt circuit automatically lies in defining the meaning of *maximum performance*. In the discussion of single-mode shunts and then in the development of multimode networks, performance was defined in terms of the amount of damping introduced into the structural modes. Although this might be an appropriate method of defining *performance* in a general case, there are a number of other definitions of performance that yield equally valid means of choosing shunt circuit parameters.

The performance of a dynamic system is often characterized using *norms* to quantify the definition of "good" and "poor" performance. The mathematical basis of system norms is a large field of study that is beyond the scope of this book, but we draw on the basic concepts to understand how they can be applied to the automated design of shunt networks. One interpretation of norms is that they quantify the size of input–output relationships in a dynamic system. Consider a linear dynamic system with input y and output x defined as the Laplace representation

$$x(\mathsf{s}) = T_{xy}(\mathsf{s})y(\mathsf{s}). \tag{9.66}$$

There are a number of norms that can be associated with the input–output relationship $T_{xy}(\mathsf{s})$. For example, one could imagine defining a norm that represents the peak output of the system to a particular type of input, or one could envision defining a norm that quantified the average output of the system to a type of input.

Our discussion will utilize a norm that is commonly applied to the analysis of dynamic systems with uncertain or broadband excitations. One definition of the *average* size of a time-dependent signal $y(t)$ is the mean-square value

$$< y >^2 = \frac{1}{T}\int_0^T y^2(t)\,dt. \tag{9.67}$$

Since the integrand of equation (9.67) is squared, it is clear that the mean-square value of a signal can only be zero or positive. An analogous definition exists for multivariate signals $\mathbf{y}(t)$,

$$< y >^2 = \frac{1}{T}\int_0^T \mathbf{y}'(t)\mathbf{y}(t)\,dt. \tag{9.68}$$

The *one-sided power spectral density* function of $y(t)$, $S_{yy}(f)$, is related to the mean-square value of the signal through the relationship

$$< y >^2 = \int_0^\infty S_{yy}^2(f)\,df. \tag{9.69}$$

The power spectral density (PSD) of a function is computed using Fourier transforms of the time-dependent signal. Physically, the PSD of a signal quantifies the energy content of the signal as a function of frequency. A signal with high energy content in a particular frequency range will have a high PSD.

Once the PSD of the input signal is defined, the PSD of the output of the dynamic system can be written

$$S_{xx}(f) = |T_{xy}(f)|S_{yy}(f). \tag{9.70}$$

In an expression analogous to equation (9.69), the mean-square value of the output signal is written as the function

$$< x >^2 = \int_0^\infty |T_{xy}(f)|^2 S_{yy}^2(f)\,df. \tag{9.71}$$

Equation (9.71) quantifies how the size of the input–output relationship T_{xy} is related to the size of the signal in the time domain. Assuming that the power spectral density function of the input is known, the mean-square value of the output is the integration of the square of the magnitude of T_{xy} multiplied by the square of the input PSD. Essentially, the frequency response of the dynamic system between the input and the output is acting as a filter to the energy content of the input signal. Frequency ranges in which the magnitude of T_{xy} is high, such as the resonant frequencies of structures, will amplify the energy content of the input signal and contribute to the average size of the output in the time domain.

A particular type of input that is often used to characterize dynamic systems is a *white noise input*, a signal that has an equal energy content at all frequencies and is represented by a constant spectral density function. Denoting the amplitude of the PSD of the white noise input as the *root mean-square* value of the input, $< y >$, equation (9.71) is rewritten as

$$\frac{< x >^2}{< y >^2} = \int_0^\infty |T_{xy}(f)|^2 df, \tag{9.72}$$

which indicates that integration of the square of the frequency response magnitude is equivalent to the ratio of the mean-square values of the output and the input when the input signal is white noise.

In practice, it is difficult to achieve a perfect white noise signal (i.e., a signal with equal energy content over a large frequency range). Thus, in design, the concept of white noise and its relationship to the frequency response of a dynamic system is often approximated by assuming that the input signal has equal energy content over a limited range and negligible energy content outside this range. In this case the integration stated in equation (9.72) can be written as an integration between a set of frequency limits that define the energy content of the input signal. In this case the mean-square value of the output is simply equal to integration of the square of the frequency response between the minimum and maximum frequency range in the analysis.

9.4.2 Adaptive Shunt Networks

The discussion of system norms and their relationship to quantifying performance provides the foundation for developing automated methods for adapting shunt parameters. Examining equation (9.72) it becomes clear that one benefit of utilizing a shunt circuit for vibration suppression is that it reduces the mean-square output of the system to white noise excitations. For example, Figure 9.12 clearly shows that introduction of the shunt network produces a reduction in the magnitude of the input–output frequency response. If we were to integrate the square of the frequency response for the system with and without shunt damping, we would find that the ratio defined by equation (9.72) has been reduced from 0.377 to 0.034, indicating that the average size of the dynamic response to white noise excitations has been reduced by approximately a factor of 10.

How can these concepts be utilized for the development of automated methodologies? Often it is difficult to compute the frequency response function of interest because the input excitations are not measurable. If we assume that it is possible to measure the output time histories at locations of interest on the structure, *and* we assume that the input excitations are white noise with a constant mean-square amplitude, we know from equation (9.72) that a reduction in the input–output frequency response will produce a reduction in the mean-square value of the ouput.

Before discussing adaptive shunt techniques for a multiple-mode system, it is instructive to investigate the variation in the mean-square response for the single-mode shunt analyzed to illustrate the adaptation of the shunt parameters. The Laplace transform of the input–output response for a single-mode shunt is specifed by equation (9.37) for an inductive–resistive circuit. The variation in the mean-square response can be computed from this expression to illustrate how the mean-square response changes as a function of the shunt parameters. The mean-square response is obtained by computing the frequency response and then integrating the magnitude according to equation (9.72) over a range of nondimensional shunt parameters. Plotting the contours of the mean-square response illustrates that the mean-square response varies significantly as the shunt parameters change. Figure 9.13*a* is a plot of the mean-square response contours for a coupling coefficient of 0.7. The circular region near (1,1) signifies the minimal value of the mean-square response and therefore represents the optimal value of the shunt parameters for $k = 0.7$. The sharp increase in shunt parameters when the parameters are much less than 1 indicates that the mean-square response will rise substantially if the nondimensional parameters are less than unity. At values of α much greater than 1, the contours become approximately straight, indicating that the mean-square response does not vary substantially as the inductance value is changed. The contours for a coupling coefficient of $k = 0.3$ are similar in shape (Figure 9.13*b*), although the optimal value of $\omega_m^{\mathrm{E}} R C_p^{\mathrm{S}}$ lies at much less than 1 (it is approximately 0.3). This result is consistent with the results plotted in Figure 9.8, which illustrate that the optimal value of $\omega_m^{\mathrm{E}} R C_p^{\mathrm{S}}$ decreases as the coupling coefficient becomes smaller.

The concept of mean-square response is the foundation for the development of adaptive shunt networks. In optimization theory, the first step is to define a *cost function* that quantifies the meaning of "performance." Generally, the cost function

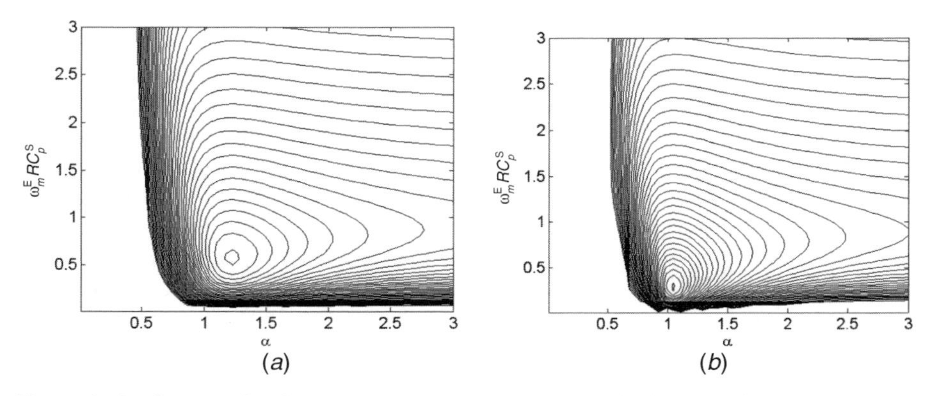

Figure 9.13 Contour plots for the mean-square response of a single-mode shunt network as a function of the shunt parameters: (*a*) $k_{33} = 0.7$; (*b*) $k_{33} = 0.3$.

is defined such that a large number indicates poor performance, and minimization of the cost function results in the optimal solution. Defining the cost function for the adaptive shunt as the mean-square displacement of the system, $< x >^2$, the cost function can be computed in real time with a sampled version of the displacement using the approximation

$$< x >^2 \approx \frac{1}{N} \sum_{k=1}^{N} x(k)^2. \tag{9.73}$$

The size of the sample, N, is chosen suitably large such that this approximation will approach the actual mean-square response. Defining the cost function as

$$f(r, \alpha) =< x >^2, \tag{9.74}$$

where the notation indicates that the mean-square response is a function of the nondimensional resistance r and the ratio of electrical to mechanical frequencies. The gradient of the cost function defines the direction of maximum increase in $f(r, \alpha)$,

$$\nabla f(r, \alpha) = \frac{\partial f}{\partial r} d\hat{r} + \frac{\partial f}{\partial \alpha} d\hat{\alpha}. \tag{9.75}$$

Decreasing the cost function requires that we step in the negative gradient direction; thus,

$$\text{step direction} = -\nabla f(r, \alpha) = -\frac{\partial f}{\partial r} d\hat{r} - \frac{\partial f}{\partial \alpha} d\hat{\alpha}. \tag{9.76}$$

For a general optimization problem in which the cost function is written explicitly in terms of the optimization variables, the gradient is computed by taking the derivatives of the cost function. For an adaptive control problem the gradients must be computed in real time to determine the step directions. This necessitates the computation of gradient approximations using measured signals. For adaptive shunt networks the gradient approximations are computed from

$$\begin{aligned}\frac{\partial f}{\partial r} &\approx \frac{f(r+\Delta r,\alpha)-f(r,\alpha)}{\Delta r}\\ \frac{\partial f}{\partial \alpha} &\approx \frac{f(r,\alpha+\Delta\alpha)-f(r,\alpha)}{\Delta\alpha},\end{aligned} \tag{9.77}$$

where each of the terms in the gradient approximation is computed from equation (9.73) using measured signals. Thus, determining the proper step direction requires two computations of the mean-square response per shunt. Once the gradient approximation is measured and computed, the step length is chosen in the negative gradient direction, although, in general, the size of the step is varied from the magnitude of the gradient approximation. If this variable is denoted β, the step length is

$$\text{step length} = \beta\left[-\nabla f(r,\alpha)\right]. \tag{9.78}$$

All of the steps required for the shunt adaptation are determined. The algorithm is:

Initialize $\alpha = \alpha_o$ and $r = r_o$
Measure $x(k)$
Compute $f_o(r_o,\alpha_o) = <x>^2$ using equation (9.73)

while tol > prescribed value (e.g. 0.01)
  Increment $\alpha_i = \alpha_i + \Delta\alpha$
  Measure $x(k)$
  Compute $f(r_i,\alpha_i+\Delta\alpha) = <x>^2$

  Increment $r_i = r_i + \Delta r$
  Measure $x(k)$
  Compute $f(r_i+\Delta r_i,\alpha_i) = <x>^2$

  Compute gradient approximations using equation (9.78)
  Compute gradient magnitude $|\nabla f| = \sqrt{(\partial f_i/\partial r)^2 + (\partial f_i/\partial\alpha)^2}$

  Increment $i = i + 1$

  Compute $r_i = r_{i-1} - \beta\dfrac{\partial f_i/\partial r}{|\nabla f|}$

  Compute $\alpha_i = \alpha_{i-1} - \beta\dfrac{\partial f_i/\partial\alpha}{|\nabla f|}$

  Compute $\text{tol} = \dfrac{f(r_i,\alpha_i) - f(r_{i-1},\alpha_{i-1})}{f(r_{i-1},\alpha_{i-1})}$

end

Table 9.3 Simulation parameters for the shunt adaptation

N	10000
α_o	0.5
r_o	0.5
β	1/50
Tolerance	0.001

The adaptive shunt algorithm was applied to a model system consisting of the one-degree-of-freedom shunt network derived earlier in the chapter. The system is modified to include the effects of structural damping on the dynamic response. Including a structural damping term in the original equations of motion and solving for the response normalized with respect to the static response yields the system

$$\frac{u(\sigma)}{u_{\text{st}}} = \frac{\sigma^2 + \alpha^2 r\sigma + \alpha^2}{(\sigma^2 + 2\zeta\sigma + 1)(\sigma^2 + \alpha^2 r\sigma + \alpha^2) + K_{33}^2(\sigma^2 + \alpha^2 r\sigma)}, \tag{9.79}$$

where ζ represents the damping ratio.

The algorithm is simulated by applying a random input to the dynamic system defined in equation (9.79) and using this result for the measured $u(k)$. The simulation parameters are listed in Table 9.3. The simulation demonstrates the ability of the adaptation algorithm to vary the resistance and natural frequency to obtain near-optimal performance in the presence of initial mistuning. Figure 9.14*a* illustrates that the adaptation is able to change the shunt parameters such that the final values are nearly optimal in terms of the mean-square response. The contour plot demonstrates that the adaptation generally follows the gradient of the optimization space. It also illustrates that the adaptation is not perfect since the value of the optimized parameters

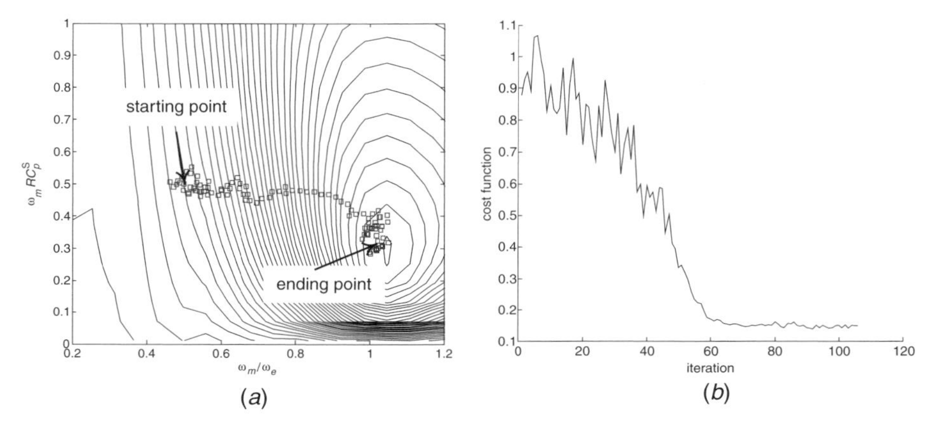

Figure 9.14 (*a*) Contour plot of the adaptation space for the shunt optimization (squares indicate values obtained during parameter adaptation); (*b*) cost function for the shunt adaptation.

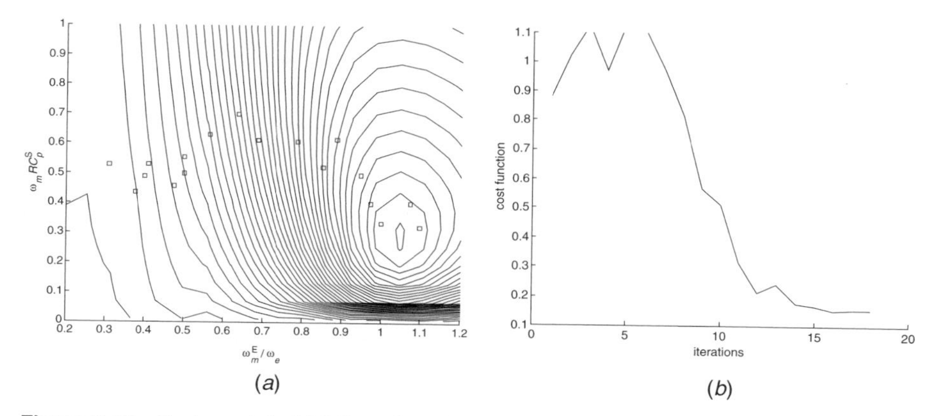

Figure 9.15 Contour plot of (*a*) the adaptation and (*b*) the cost function convergence for the case of tolerance = 0.01 and $\beta = 1/10$.

parameters seems to oscillate in regions where the gradient is not as steep. For example, near the starting point the optimization "hunts" near the initialized values before finally following the gradient toward the optimal solution. This is illustrated in Figure 9.14*b* as "noise" in the cost function early in the adaptation. Also important is the fact that the adaptation takes a number of iterations to reach convergence with the chosen tolerance of 0.001.

The rate of convergence in the adaptation is varied by changing the step length and tolerance in the optimization. Reducing the tolerance to 0.01 and increasing the step length by a factor of 5 ($\beta = 1/10$) results in the adaptation shown in Figure 9.15. The convergence rate is reduced to less than 20 iterations, due to the fact that the adaptation takes larger steps in each iteration and the requirement for convergence has been reduced by increasing the toleration to 0.01 instead of 0.001. As in the previous case, the optimization does converge to a nearly optimal solution, as shown in Figure 9.15*a*, resulting in approximately the same cost function (Figure 9.15*b*).

9.4.3 Practical Considerations for Adaptive Shunt Networks

The adaptation algorithm developed in Section 9.4.2 is based on classic optimization theory. Simulations demonstrate that it is able to adapt the shunt parameters and find a nearly optimal solution and that the convergence rate is controlled by proper choice of the adaptation parameters.

One of the most important practical considerations in the adaptation algorithm is proper estimation of the mean-square response and the cost function gradients. Correct mean-square estimates are obtained only with sufficiently long sample records due to transients that occur due to variation in the shunt parameters. In the simulations presented in Section 9.4.2, the number of time samples was set to 10,000 to obtain proper mean-square estimates. Since nondimensional parameters are used in the simulation, this corresponds to 10,000 periods of the system of the undamped structure.

For example, if the undamped natural frequency is 100 Hz, this corresponds to 100 s of time data to obtain estimates of the mean-square response.

A second important consideration is the cost function chosen for the adaptation. The simulation in Section 9.4.2 minimized the mean-square response of the system to determine the performance of the shunt network. One of the main assumptions in using this cost function is that the mean-square value of the input was constant with respect to time. In situations which the mean-square value of the forcing function was not constant, it would not be possible to use the mean square of the output as the cost function for the optimization.

This problem was overcome by using a cost function consisting of the ratio of the mean-square output to the mean-square voltage of the shunt network. Using the ratio of the output to the shunt voltage eliminated the problems associated with variations in the input signal because both signals would be affected equally by a change in forcing input. In this case the cost function was *maximized* when the shunt was tuned optimally. They used this concept to adapt the parameters of a piezoelectric vibration absorber on a representative structure. The system adapted the tuning parameter between the mechanical and electrical resonance frequency by changing the inductance of a synthetic inductor. Experimental results verified the effectiveness of the algorithm and demonstrated vibration suppression even in the presence of sudden changes in the natural frequency of the system.

9.5 SWITCHED-STATE ABSORBERS AND DAMPERS

Piezoelectric shunts are an excellent means of providing vibration reduction in a structure. Using either a resistive network or an inductive–resistive network, it is possible to achieve an appreciable amount of energy dissipation with a simple electrical circuit. The energy dissipation can be targeted toward a single structural mode or distributed to numerous structural modes with multiple shunt networks. If operating conditions are uncertain or time varying, an adaptive shunt network can be synthesized that will tune the parameters to maintain nearly optimal performance.

Although shunts are an effective means of vibration suppression, there are drawbacks that limit their performance. One of the most substantial drawbacks is that shunts provide vibration reduction only near the resonance frequencies of the structure. Figure 9.12 illustrates this phenomenon clearly. Examining the figure, we see that the peak response of the structure is decreased considerably near resonance, but the response at all other frequencies is unaffected by the shunt. In certain cases there is even an increase in the structural response near the zeros, or antiresonances, of the structural response, due to the addition of a shunt network. Thus, the energy dissipation provided a shunt network is considered *narrowband* in the sense that the reduction in the vibrational response occurs only in limited bands in the structural response.

Another limitation of a shunt network is that some knowledge of structural dynamics is required for effective design. For a fixed-parameter shunt knowledge of the structural resonances is required to tune the shunt parameters accurately. Performance degrades considerably if these performance parameters are mistuned even by only 20 or 30%, particularly for an inductive–resistive shunt. In the case of an adaptive shunt, some knowledge of the structural resonances is required to provide a reasonable initialization of the adaptation algorithm. The further the initialization is from the optimal parameters, the less likely it is that the shunt will tune itself for optimal performance.

State-switched absorbers and dampers have been utilized as a means of overcoming these limitations in piezoelectric shunts. A state-switched device is defined as one in which the constitutive properties are changed as a function of external stimuli. The concept of a state-switched device is similar to that of an adaptive device, except for the fact that the parameters of the system only switch between discrete states, which are typically prescribed a priori. Switched-state absorbers and dampers that utilize piezoelectric material are based on the fundamental principles of electromechanical coupling. One of the basic properties of a material that exhibits electromechanical coupling is that the mechanical constants are a function of the electrical boundary conditions, and the electrical properties are a function of the mechanical boundary conditions. The amount by which these properties change when the boundary conditions change is quantified by coupling coefficient k.

State-switched piezoelectric devices are based on the concept of switching between short- and open-circuit electrical boundary conditions for the purpose of reducing the vibrational response of a structural system. Switching the electrical boundary conditions produces a variable stiffness spring which can be stiffened or softened, depending on the electrical state of the piezoelectric material.

To analyze the utility of a state-switched piezoelectric device, consider once again the equations of motion for a piezoelectric mass–spring system, equation (9.12). The equations for the open- and short-circuit system are

$$
\begin{aligned}
m\ddot{u} + \left(k + \frac{k_a^{\mathrm{E}}}{1-k^2}\right) u &= f \\
m\ddot{u} + \left(k + k_a^{\mathrm{E}}\right) u &= f.
\end{aligned}
\tag{9.80}
$$

Using the procedure introduced earlier in the chapter, these equations can be written

$$
\begin{aligned}
\frac{1}{\omega^{\mathrm{E}^2}}\ddot{u} + (1+K^2)u &= \frac{f}{k+k_a^{\mathrm{E}}} \\
\frac{1}{\omega^{\mathrm{E}^2}}\ddot{u} + u &= \frac{f}{k+k_a^{\mathrm{E}}}.
\end{aligned}
\tag{9.81}
$$

The expressions are nondimensionalized by introducing the variable

$$
\tau = \omega^{\mathrm{E}} t,
\tag{9.82}
$$

and rewriting the time derivative as

$$\frac{d^2u}{dt^2} = \omega^{E^2} \frac{d^2u}{d\tau^2} = \omega^{E^2} u''. \tag{9.83}$$

This results in the nondimensional expressions

$$\begin{aligned} u'' + (1 + K^2)u &= \frac{f}{k + k_a^E} \\ u'' + u &= \frac{f}{k + k_a^E}. \end{aligned} \tag{9.84}$$

Equations (9.84) are the nondimensional equations of motion for the open- and short-circuit piezoelectric mass–spring systems. The equations are written in terms of the generalized coupling coefficient, K^2, in the same manner as the equations for the piezoelectric shunts studied earlier in the chapter, to facilitate comparison with the analysis of resistive and inductive–resistive shunts.

In the case of piezoelectric shunts, it was required to choose the shunt type and the algorithm associated with the parameter adaptation. Similarly, for switched-state devices we must specify the switching algorithm. A heuristic algorithm based on the concept of maintaining equilibrium works very well for switching between open- and short-circuit conditions. Knowing that the equilibrium state of the system is zero deflection, a simple switching algorithm that will tend to drive the system toward equilibrium is based on the following rules:

- If the system is moving away from equilibrium, switch to an open-circuit condition to maximize the stiffness.
- If the system is moving toward equilibrium, switch to a short-circuit condition and shunt the electrical energy to ground.

The rules can be stated succintly as a switching algorithm:

$$\text{open circuit:} \qquad uu' > 0 \tag{9.85}$$

$$\text{short circuit:} \qquad uu' < 0. \tag{9.86}$$

The time response of a state-switched system is simulated by transforming equation (9.85) into state space and using a standard numerical integration routine to compute u and u'. The state-space representation of the open-circuit equations of motion is

$$\begin{pmatrix} u' \\ u'' \end{pmatrix} = \begin{bmatrix} 0 & 1 \\ -(1 + K^2) & 0 \end{bmatrix} \begin{pmatrix} u \\ u' \end{pmatrix} + \begin{pmatrix} 0 \\ \dfrac{f}{k + k_a^E} \end{pmatrix} \tag{9.87}$$

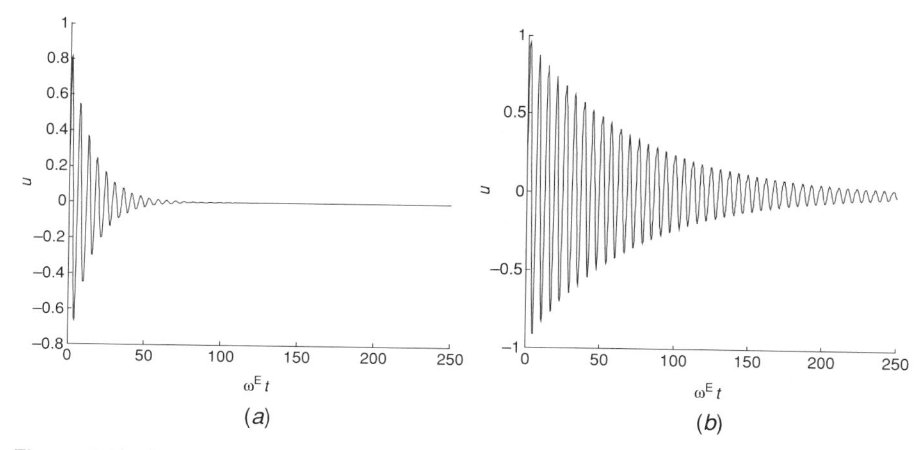

Figure 9.16 Initial-condition response of a piezoelectric switched-state system: (*a*) $K = 0.7$, (*b*) $K = 0.3$.

and the state-space representation of the short-circuit equations of motion is

$$\begin{pmatrix} u' \\ u'' \end{pmatrix} = \begin{bmatrix} 0 & 1 \\ -1 & 0 \end{bmatrix} \begin{pmatrix} u \\ u' \end{pmatrix} + \begin{pmatrix} 0 \\ \dfrac{f}{k + k_a^{\mathrm{E}}} \end{pmatrix}. \qquad (9.88)$$

Equations (9.87) and (9.88) are incorporated into a numerical integration algorithm to compute the time response to initial conditions. An impulse excitation is modeled by the initial conditions

$$u(0) = 0 \quad u' = 1. \qquad (9.89)$$

Simulation of the initial-condition response demonstrates that the switching rule effectively introduces damping into the system. Figure 9.16*a* and *b* are simulation results of the switched-state system for two values of the generalized coupling coefficient. As might be expected, a larger generalized coupling coefficient produces a larger effective damping constant, as exhibited by the faster rate of decay of the deflection u. The number of cycles required for u to fall below ≈ 0.05 drops from 40 for $K = 0.3$ to 10 for $K = 0.7$.

One of the major limitations of a resistive or inductive–resistive shunt is the sensitivity to the tuning parameters (recall that this was the motivation for the development of adaptive shunts). The problem is particularly acute when the generalized coupling coefficient is small. To compare the frequency response of a piezoelectric shunt with a state-switched device, assume that the forcing function is equal to

$$\frac{f}{k + k_a} = u_{\mathrm{st}} \sin\left(\frac{\omega}{\omega_e} t\right), \qquad (9.90)$$

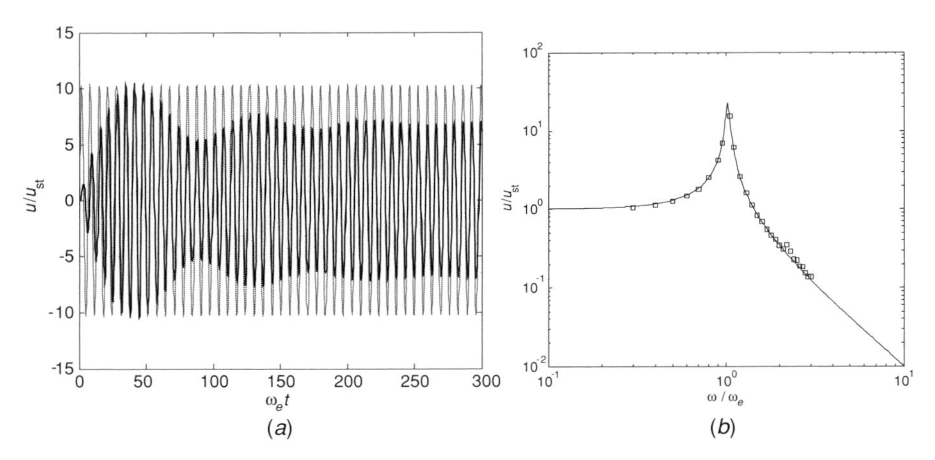

Figure 9.17 (*a*) Response of a piezoelectric mass–spring system with a state-switched damper (solid) and without a state-switched damper (dotted); (*b*) comparison of the frequency response of a an optimally tuned resistive shunt and the steady-state response of a state-switched damper (squares).

where u_{st} is the static deflection to the forcing function and ω/ω_e is the frequency of the forcing function normalized with respect to the short-circuit natural frequency. Simulations of the system with and without a state-switched damper for a normalized excitation frequency of 0.95 and $K = 0.3$ illustrate that the steady-state response of the system is reduced using the state-switched device (Figure 9.17*a*).

Simulating the forced response over a range of normalized frequencies and computing the steady-state amplitude of the state-switched system demonstrates that a state-switched device performs approximately the same as an optimally tuned resistive piezoelectric shunt. Figure 9.17*b* compares the frequency response of an optimally tuned resistive shunt and the steady-state response of a state-switched system. Over the frequency range, both systems have approximately the same steady-state response. The analysis demonstrates that a state-switched device performs similarly to a tuned resistive shunt. In both cases, also, the amount of reduction is related directly to the generalized coupling coefficient of the piezoelectric system. This coefficient is, in turn, related to the material coupling coefficient and the relative stiffness of the piezoelectric material and the structure.

Since the performance is similar, it is fair to ask why it is beneficial to use a state-switched device instead of a resistive shunt. The primary difference is that to be effective, a state-switched device does not require information about the structural dynamics. There is no need to *tune* a state-switched device because the electrical boundary condition is related directly to the measured response of the displacement and velocity. The primary drawback of a state-switched system is the need to measure the response to switch the state of the piezoelectric device. In contrast to a shunt, which simply requires a resistor, a state-switched device will require measurements of the structural response and some type of logic device to switch between states.

Detailed analyses of more sophisticated state-switched systems have been performed. Clark has shown that the performance can be improved by switching between an open-circuit piezoelectric and a resistively shunted boundary condition. Both the impulse response and the frequency response improve when this configuration is used. Cunefare has shown that state-switched systems suffer from additional mechanical transients when switching occurs at a state with nonzero strain energy. This is particularly important for multimode systems in which a higher-frequency resonance might be excited by the switching of the piezoelectric material. Cunefare et al. proposed new switching algorithms to reduce the transient mechanical vibrations due to the state switching.

An interesting question arises when one considers the physics of a state-switched device. It might seem impossible for a system that switches between two separate stiffnesses, neither of which dissipates energy, producing an effect that is approximately the same as that of an energy-dissipative device (i.e., a resistor). The question is investigated by considering the relationship between force and displacement in the state-switched device. As discussed at the outset of this chapter, a dissipative element such as a viscoelastic material will produce a force–displacement relationship (or stress–strain relationship) that exhibits cyclic energy dissipation (see Figure 9.1).

The force applied by the piezoelectric material can be expressed in nondimensional coordinates as

$$\begin{aligned} &\text{open circuit:} \qquad f_p = \frac{k_a^{\mathrm{E}}}{1-k^2}u \\ &\text{short circuit:} \qquad f_p = k_a^{\mathrm{E}}u, \end{aligned} \tag{9.91}$$

where the switching occurs according to the algorithm expressed in equation (9.86). If the force–deflection curve is plotted, we note that switching causes the piezoelectric material *to load and unload along a different path*. This change in the load path depending on the state of the piezoelectric material produces hysteresis in the force–deflection curve and results in energy dissipation. Although each state of the piezoelectric material is characterized by an elastic stiffness, the state-switching produces energy dissipation through hysteresis in the force–deflection curve. This explains how the state-switched device can act like a damper even though it is switching between two states that when examined separately, do not exhibit energy dissipation.

9.6 PASSIVE DAMPING USING SHAPE MEMORY ALLOY WIRES

Earlier in the chapter we demonstrated that piezoelectric materials are useful for energy dissipation in structural systems. Energy dissipation is achieved using passive shunts, adaptive networks, and switched-state absorbers.

In Chapter 6 the constitutive properties of shape memory alloy materials were stated and used to explain the stress–strain behavior of SMA materials. The pseudoelastic effect for shape memory alloys was shown to produce hysteresis behavior in an

SMA wire held at a constant temperature. Recall the result shown in Figure 6.1, which illustrates the hysteresis loop for an SMA wire undergoing full austenitic-to-martensitic phase transformation due to the stress applied.

The hysteresis induced in shape memory materials can be used as an energy-dissipation mechanism for structural control. Hysteresis in the shape memory material will produce energy loss during cyclic loading. The energy loss produces a nonlinear structural damping mechanism that can be used to reduce the vibration level.

One of the challenges associated with using shape memory materials as structural control elements is the nonlinearity of the constitutive properties. As introduced in Chapter 6, the constitutive behavior of an SMA wire is a function of the martensitic fraction in the material. The martensitic fraction is itself a function of the temperature of the wire and the states of stress of the wire. The martensitic fraction is a function of the loading history (both of stress and temperature) and the initial conditions of the material. All of these interrelationships produce a nonlinear stress–strain behavior that needs to be modeled to design structural energy dissipation elements using shape memory material.

In this section we focus on using the pseudoelastic effect in shape memory alloy wires as a means of introducing energy dissipation into a structure. Restricting ourselves to the pseudoelastic effect will allow us to develop a general model of a single-degree-of-freedom vibrational system that incorporates SMA wires for structural control.

9.6.1 Passive Damping via the Pseudoelastic Effect

The pseudoelastic effect for shape memory alloys is obtained in a shape memory alloy when the temperature is held constant at a value higher than the austenitic finish temperature of the material. The application of stress to the material induces nonlinear stress–strain behavior upon loading due to the austenitic-to-martensitic phase transformation. Upon unloading, the material undergoes the martensitic-to-austenitic phase transformation, the net result being a hysteresis loop in the stress–strain behavior of the material.

Consider a representative system that consists of a mass m with an external dynamic force $f_e(t)$ attached to ground through a shape memory alloy wire as shown in Figure 9.18. The force induced in the shape memory wire is denoted $f_{\text{sma}}(t)$. A constant bias force f_b is applied to the system to induce a prestrain on the shape memory alloy wires. The total displacement of the system and the wires is denoted

$$u_t(t) = u(t) + u_{\text{st}}, \tag{9.92}$$

where u_{st} is the constant displacement caused by the prestrain. Summing forces on the mass yields the equation of motion

$$m\frac{d^2u_t(t)}{dt^2} = f_e(t) - f_{\text{sma}}(t) + f_b. \tag{9.93}$$

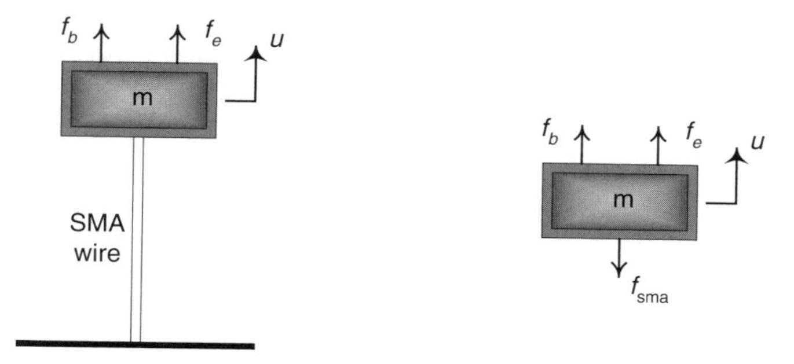

Figure 9.18 Representative system for studying passive structural control using shape memory alloy wires.

Substituting equation (9.92) into equation (9.93) results in the differential equation

$$m\frac{d^2u(t)}{dt^2} = f_e(t) - f_{\text{sma}}(t) + f_b, \tag{9.94}$$

due to the fact that the static displacement due to the prestrain is constant and its time derivative is zero. Equation (9.94) is the differential equation that is to be solved to determine the dynamic displacement of the system, while equation (9.92) is used to compute the total displacement.

The force induced in the shape memory alloy is governed by the constitutive relationships and kinetic law for the pseudoelastic effect. The constitutive equations and the associated kinetic law are summarized in Table 6.3. The challenging aspect of incorporating the pseudoelastic effect is that the stress in the shape memory wire is a function of the induced strain and the martensitic fraction, which is itself a function of the induced stress. Thus, the stress at each time step must be solved iteratively to determine the values of T and ξ that satisfy the equations.

The force in the SMA wire can be rewritten in terms of the cross-sectional area of the wire, A, and the stress in the wire, T_{sma}:

$$f_{\text{sma}}(t) = A\text{T}_{\text{sma}}(t), \tag{9.95}$$

where the stress in the wire is a function of the strain (or displacement) and the martensitic fraction. The stress in the shape memory alloy wires is assumed to be composed of a dynamic stress, $\text{T}(t)$, and a bias stress, T_b:

$$\text{T}_{\text{sma}}(t) = \text{T}(t) + \text{T}_b. \tag{9.96}$$

Substituting equations (9.96) and (9.95) into equation (9.94) results in the expression

$$m\frac{d^2u(t)}{dt^2} = f_e(t) - A\text{T}(t) + f_b - A\text{T}_b. \tag{9.97}$$

Assuming that the damper is designed such that the prestress in the wires equilibrates the bias force, $\mathrm{T}_b = f_b/A$, equation (9.97) is reduced to an equation for the dynamic response of the damper to the external force,

$$m\frac{d^2u(t)}{dt^2} = f_e(t) - A\mathrm{T}(t). \tag{9.98}$$

Several methods are available to solve equation (9.98). A numerical simulation package with a graphical user interface can be used, or the equations can be incorporated into a numerical integration routine such as Runge–Kutta to solve for the displacement as a function of time. To emphasize the methods of incorporating the pseudoelastic behavior into the equations of motion, in this section we utilize one of the most basic numerical approximations to solve the equations of motion. Assume that we define a fixed sampling time Δt, then recall that the derivative of a function can be approximated as

$$\frac{du(t)}{dt} \approx \frac{u(n\Delta t) - u((n-1)\Delta t)}{\Delta t}, \tag{9.99}$$

where n is an integer value greater than or equal to zero. Since the sampling time is assumed to be fixed, the notation for the approximation is typically stated as

$$\frac{du(t)}{dt} \approx \frac{u(n) - u(n-1)}{\Delta t}. \tag{9.100}$$

Applying the derivative approximation twice for the second derivative yields

$$\frac{d^2u(t)}{dt^2} \approx \frac{u(n) - 2u(n-1) + u(n-2)}{\Delta t^2}. \tag{9.101}$$

Substituting the approximation of the second derivative into equation (9.98) yields an *algebraic* equation of the form

$$m\frac{u(n) - 2u(n-1) + u(n-2)}{\Delta t^2} = f_e(n) - A\mathrm{T}(n). \tag{9.102}$$

The expression can be rearranged to yield an equation for $u(n)$,

$$\frac{m}{\Delta t^2}u(n) - 2\frac{m}{\Delta t^2}u(n-1) + \frac{m}{\Delta t^2}u(n-2) = f_e(n) - A\mathrm{T}(n). \tag{9.103}$$

In certain models it is useful to add a linear damping term (typically, small) to the equations of motion to represent additional energy dissipation mechanisms in the structure. In this case the equations of motion are

$$m\frac{d^2u(t)}{dt^2} + c\frac{du(t)}{dt} = f_e(t) - A\mathrm{T}(t), \tag{9.104}$$

where c is the linear viscous damping coefficient. Incorporating the approximations for the first and second derivatives yields

$$m\frac{u(n) - 2u(n-1) + u(n-2)}{\Delta t^2} + c\frac{u(n) - u(n-1)}{\Delta t} = f_e(n) - A\mathrm{T}(n). \quad (9.105)$$

Equation (9.105) can be grouped according to the terms of $u(n)$ as

$$\left(\frac{m}{\Delta t^2} + \frac{c}{\Delta t}\right)u(n) - \left(2\frac{m}{\Delta t^2} + \frac{c}{\Delta t}\right)u(n-1) + \frac{m}{\Delta t^2}u(n-2) = f_e(n) - A\mathrm{T}(n). \quad (9.106)$$

A general expression for the discretized equation for the dynamic response of the oscillator is

$$\tilde{A}u(n) + \tilde{B}u(n-1) + \tilde{C}u(n-2) = f_e(n) - A\mathrm{T}(n), \quad (9.107)$$

where the coefficients $\tilde{A}$, $\tilde{B}$, and $\tilde{C}$ are defined by the assumptions regarding damping in the oscillator.

The dynamic response of the oscillator with the shape memory alloy wire is obtained by solving equation (9.107) as a function of n. The complication in solving the discrete expressions arises due to the dependence of stress on the shape memory effect of the wire. Recalling the kinetic law for shape memory transformation (see Table 6.3), we note that there are two regimes that define the stress–strain behavior of the shape memory wire. When the martensitic fraction ξ is constant, the stress–strain behavior is linear, whereas when ξ is changing due to the material transformation, the stress–strain behavior is nonlinear.

Using the results listed in Table 6.3, we note that when the shape memory alloy is in the linear regime,

$$\mathrm{T}(n) = Y\mathrm{S}(n) + Y\mathrm{S}_L\xi. \quad (9.108)$$

Assuming one-dimensional strain, equation (9.108) can be rewritten in terms of the displacement as

$$\mathrm{T}(n) = \frac{Y}{L}u(n) + Y\mathrm{S}_L\xi. \quad (9.109)$$

When the stress–strain behavior in the shape memory wire is linear, equation (9.109) is substituted into equation (9.107) and the eqution can be solved explicitly for $u(n)$. The expressions are

$$\left(\tilde{A} + \frac{YA}{L}\right)u(n) = -\tilde{B}u(n-1) - \tilde{C}u(n-2) + f_e(n) + YA\mathrm{S}_L\xi. \quad (9.110)$$

Recall that the total stress and strain must be computed by adding the bias stress and strain to the result from the solution of equation (9.110).

In the case in which the stress–strain behavior is nonlinear, the displacement must be solved in a different manner. When the shape memory material is undergoing a phase transformation, the three equations that define the response at each time step are

$$\begin{gathered}\tilde{A}u(n) + \tilde{B}u(n-1) + \tilde{C}u(n-2) = f_e(n) - A\mathrm{T}(n) \\ \mathrm{T}(n) = \frac{Y}{L}u(n) + YS_L\xi \\ \xi = \frac{1}{2}\left\{\cos\left[a_M(\theta_0 - M_f) - \frac{a_M}{C_M}(\mathrm{T}(n) + \mathrm{T}_b)\right] + 1\right\}.\end{gathered} \tag{9.111}$$

These three equations must be solved for the three unknowns $u(n)$, $\mathrm{T}(n)$, and ξ to yield a solution at each time step. One method of solving them is to solve the first expression for $u(n)$:

$$u(n) = \frac{-\tilde{B}u(n-1) - \tilde{C}u(n-2) + f_e(n)}{\tilde{A}} - \frac{A}{\tilde{A}}\mathrm{T}(n). \tag{9.112}$$

Recognizing that the first term on the right-hand side of equation (9.112) is a constant, we can denote

$$\Gamma(n) = \frac{-\tilde{B}u(n-1) - \tilde{C}u(n-2) + f_e(n)}{\tilde{A}} \tag{9.113}$$

and rewrite the expression as

$$u(n) = \Gamma(n) - \frac{A}{\tilde{A}}\mathrm{T}(n). \tag{9.114}$$

Substituting equation (9.114) into the second expression in equation (9.111) results in

$$\mathrm{T}(n) = \frac{Y}{L}\Gamma(n) - \frac{YA}{L\tilde{A}}\mathrm{T}(n) + YS_L\xi. \tag{9.115}$$

Solving the expression for stress yields

$$\mathrm{T}(n) = \frac{Y/L}{1 + YA/L\tilde{A}}\Gamma(n) + \frac{YS_L}{1 + YA/L\tilde{A}}\xi. \tag{9.116}$$

This expression is substituted into the kinetic law for austenitic-to-martensitic phase transformation to yield an expression that can be solved for the martensitic fraction:

$$\xi - \frac{1}{2}\left\{\cos\left[a_m(\theta_0 - M_f) - \frac{a_M}{C_M}\left(\frac{Y/L}{1 + YA/L\tilde{A}}\Gamma(n) + \frac{Y S_L}{1 + YA/L\tilde{A}}\xi + \mathrm{T}_b\right)\right] + 1\right\} = 0. \tag{9.117}$$

Equation (9.117) is a transcendental expression; thus, it cannot be solved explicitly. The expression contains multiple solutions due to the periodic expression; therefore, care must be taken to determine the solution that is nearest to the present solution for ξ.

Solving for the response during the martensitic-to-austenitic phase transformation proceeds in a similar manner. The equations that must be solved are

$$\tilde{A}u(n) + \tilde{B}u(n-1) + \tilde{C}u(n-2) = f_e(n) - A\mathrm{T}(n)$$
$$\mathrm{T}(n) = \frac{Y}{L}u(n) + Y S_L \xi$$
$$\xi = \frac{\xi_0}{2}\left\{\cos\left[a_A(\theta_0 - A_s) - \frac{a_A}{C_A}(\mathrm{T}(n) + \mathrm{T}_b)\right] + 1\right\}. \tag{9.118}$$

One of the differences in the analysis is that we assume that the material does not necessarily go through a full phase transformation. The variable ξ_0 is the martensitic fraction when the phase transformation back to austenite begins.

Using the same procedure as described above results in a transcendental equation:

$$\xi - \frac{\xi_0}{2}\left\{\cos\left[a_A(\theta_0 - A_s) - \frac{a_A}{C_A}\left(\frac{Y/L}{1 + YA/L\tilde{A}}\Gamma(n) + \frac{Y S_L}{1 + YA/L\tilde{A}}\xi + \mathrm{T}_b\right)\right] + 1\right\} = 0. \tag{9.119}$$

that must be solved for ξ at each time step.

This analysis provides the equations required to solve for the displacement, stress, and martensitic fraction (if it is changing) at each time step of the analysis. The last component of the damping analysis states the triggers that switch the material among states. For this analysis we assume that the material is initially in a prestressed state defined by the bias stress and strain and that the material is initially in its austenitic phase ($\xi = 0$). Under this assumption the material will initially respond in the linear stress–strain regime until the stress becomes greater than the stress required to induce martensitic phase transformation:

$$\mathrm{T}(n) + \mathrm{T}_b > C_M(\theta_0 - M_s). \tag{9.120}$$

After phase transformation is induced, equation (9.117) must be solved for the martensitic fraction at each time step. As the material is undergoing austenitic-to-martensitic phase transformation ξ will be increasing as the phase transformation occurs. The analysis reverts to a linear analysis when the martensitic phase transformation begins to decrease. Recall that when xi begins to decrease, the material will unload with a linear stress–strain relationship until the martensitic-to-austenitic phase transformation begins. Switching to a linear stress–strain relationship occurs when

$$\xi(n) - \xi(n-1) < 0, \tag{9.121}$$

and a linear unloading of the material will continue until

$$\mathrm{T}(n) + \mathrm{T}_b < C_A(\theta_0 - A_s), \tag{9.122}$$

at which time the material will begin to revert to its austenitic phase. During this transformation, equation (9.119) must be solved at each time step to determine the martensitic fraction. This transformation will continue until $\xi = 0$, at which time the material will be in its full austenite phase and will respond in the linear stress–strain regime.

9.6.2 Parametric Study of Shape Memory Alloy Passive Damping

The analysis in Section 9.6.1 will be used to perform a parametric study of the use of shape memory alloy wires as passive damping elements. The representative system shown in Figure 9.18 will be used for the study. The parameters used in the study are listed in Table 9.4.

For this analysis we assume that the input is a harmonic function of the form

$$f_e(n) = F_e \sin(\omega_e n \,\Delta t), \tag{9.123}$$

where F_e is the amplitude of the input and ω_e is the forcing frequency. The variable Δt is the time step chosen for the analysis. The stress values at the phase transitions

Table 9.4 Values used for the SMA passive damping study

m	25 kg	Y	13 GPa
L	50 cm	A	3.14 mm^2
c	40.8 N·s/m	θ_0	27°C
$C_M = C_A$	11 MPa/°C	S_L	0.07
M_f	8°C	M_s	13°C
A_s	15°C	A_f	17°C

are computed from

$$C_M(\theta_0 - M_s) = 154 \text{ MPa}$$
$$C_M(\theta_0 - M_f) = 209 \text{ MPa}$$
$$C_A(\theta_0 - A_s) = 132 \text{ MPa}$$
$$C_A(\theta_0 - A_f) = 110 \text{ MPa}.$$

The first design decision to make is what value to choose for the bias stress of the shape memory alloy wire. For this study we choose the bias stress to be equal to

$$\mathrm{T}_b = 110 \text{ MPa}, \tag{9.124}$$

which yields a bias strain of $\mathrm{S}_b = \mathrm{T}_b/Y = 0.85\%$. This value is chosen to produce a regime of linear stress–strain response before inducing the shape memory effect. The coefficients of the kinetic law for the shape memory transformation are computed:

$$a_A = \frac{\pi}{A_f - A_s} = 1.571/{}^\circ\mathrm{C}^{-1}$$
$$a_M = \frac{\pi}{M_s - M_f} = 0.628/{}^\circ\mathrm{C}^{-1}.$$

The first parametric study we perform is to analyze the passive damping properties of the shape memory alloy as a function of the forcing frequency. The stiffness of the shape memory wire in the linear regime is equal to

$$k_{\text{sma}} = \frac{YA}{L} = 81.7 \text{ kN/m}, \tag{9.125}$$

and the linear natural frequency of the mass–spring system is

$$\omega_n = \sqrt{\frac{k_{\text{sma}}}{m}} = 57.2 \text{ rad/s}, \tag{9.126}$$

yielding a natural frequency of 9.1 Hz.

Solving for the response assuming that $F_e = 20$ N and $\omega_e = 10\pi$ (5 Hz) yields the response shown in Figure 9.19*a* and *b*. The mean value of the strain response is the bias value induced by the prestress. The peak-to-peak value of the strain is only approximately 0.2%. From Figure 9.19*b* we see that the stress induced by the external force is not large enough to induce the shape memory transformation. Also plotted is the strain and stress response for a linear system with a spring stiffness that is equivalent to the static stiffness of the wire. Overlaying the two results demonstrates that the response of the system is approximately equal to the response of the linear system due to the fact that no phase transformation is induced.

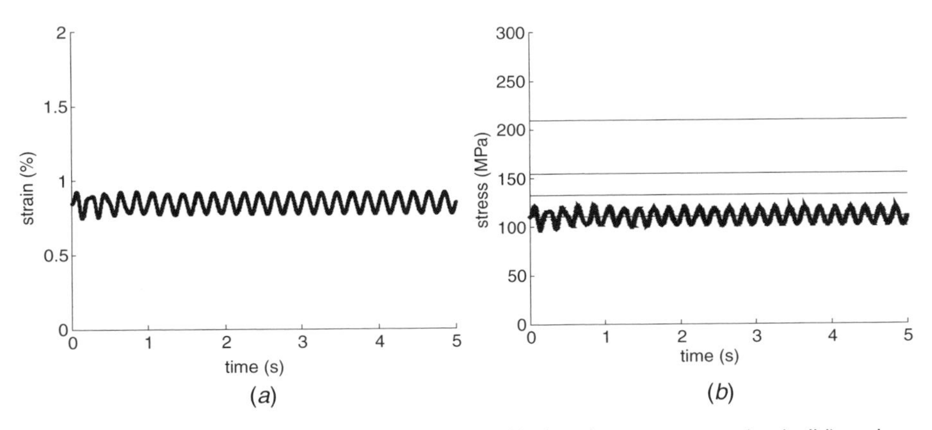

Figure 9.19 (*a*) Strain response of the system with the shape memory wire (solid) and an equivalent linear system (dotted) for $F_e = 20$ N and $\omega_e = 10\pi$; (*b*) stress response in the wire.

Increasing the excitation frequency to $\omega_e = 18\pi$ (9 Hz) produces a response that is close to the natural frequency of the system. In Figure 9.20*a* the linear response illustrates the characteristic amplification due to the resonance of the structure. The shape memory alloy wire significantly reduces the peak-to-peak response of the system even when the excitation frequency is near the system resonance. The reduction in the strain response is due to the hysteresis that occurs in the shape memory wire due to phase transformation. The hysteresis loops for this set of conditions is shown in Figure 9.20*b*. The area contained within the stress–strain response is equal to the energy dissipated during the cyclic excitation. This energy dissipation leads to the damped response of the system even though it is excited near resonance.

The hysteresis loops of the SMA wire indicate that the full phase transformation is not required to achieve significant levels of damping near resonance. Full

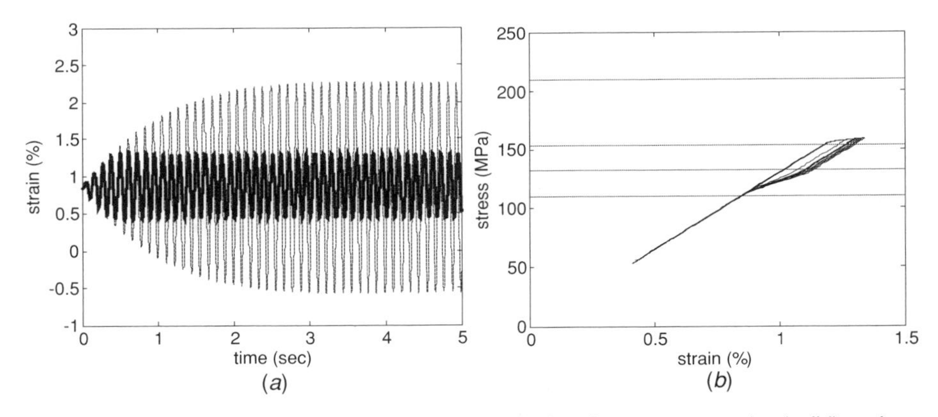

Figure 9.20 (*a*) Strain response of the system with the shape memory wire (solid) and an equivalent linear system (dotted) for $F_e = 20$ N and $\omega_e = 18\pi$; (*b*) stress–strain response in the wire, illustrating the hysteresis induced by the shape memory transformation.

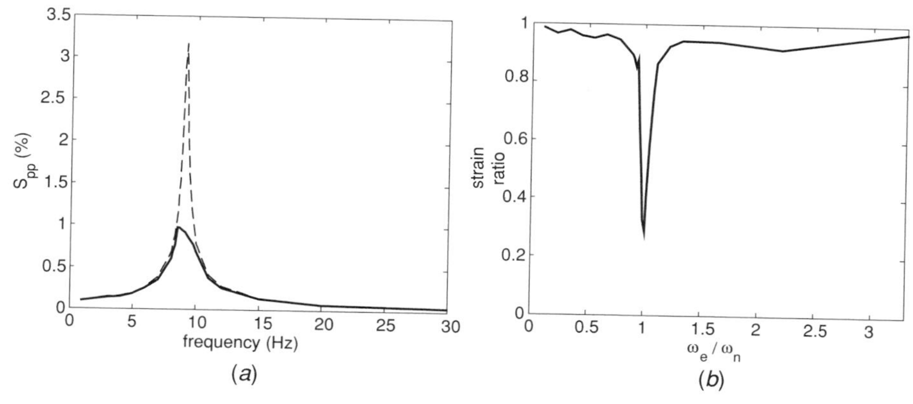

Figure 9.21 (*a*) Strain response for the linear system (dotted) and the system with the SMA wire (solid) as a function of frequency for $F_e = 20$ N; (*b*) ratio of the strain in the system with the SMA wire to the strain in the linear system, illustrating the frequency range at which the SMA wire is most effective.

austentic-to-martensitic phase transformation will yield approximately 7% strain recovery, whereas we see from Figure 9.20*b* that the peak strain in the wire is only approximately 1.3%. A computation of the martensitic fraction indicates that the peak values of ξ are only in the range 1 to 2%. This result demonstrates that significant levels of damping are achieved even with only a small portion of the phase transformation occurring during the excitation.

A plot analogous to a frequency response is obtained by solving for the time response over a range of frequencies and computing the peak-to-peak strain for the linear system and the system with the shape memory wire. Figure 9.21*a* illustrates that the linear system exhibits a characteristic amplification in the strain response near the resonance of the system, whereas the system with the SMA wire exhibits a much smaller amplification near resonance. Taking the ratio of the strain in the system with the SMA wire and the strain in the linear system is a measure of the effectiveness of the SMA wire in reducing the strain output due to the external force. Figure 9.21*b* illustrates that for this parametric analysis the SMA wire is able to reduce the strain response in the wire by approximately a factor of 3 due to the damping induced by SMA hysteresis.

Plotting the effectiveness of the shape memory alloy wire in reducing strain response illustrates that the wire naturally introduces damping near the resonance of the system. The damping is introduced in the resonant frequency range because the strain amplification at resonance automatically induces hysteresis in the shape memory wires. This effect is similar to the effect of other types of passive and semiactive dampers, such as the piezoelectric shunts described earlier in the chapter. One difference, though, is that the frequency response of the system with the shape memory wire does not exhibit the characteristic response of a damped linear system near resonance, as shown in Figure 9.21*a*.

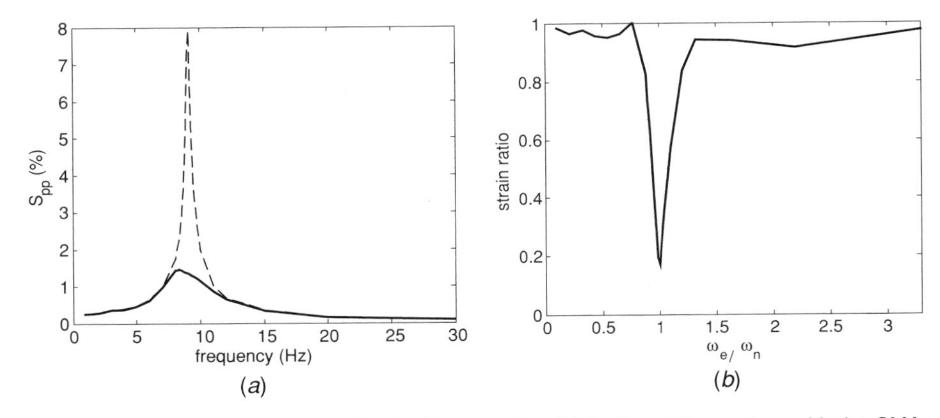

Figure 9.22 (*a*) Strain response for the linear system (dotted) and the system with the SMA wire (solid) as a function of frequency for $F_e = 50$ N; (*b*) ratio of the strain in the system with the SMA wire to the strain in the linear system, illustrating the frequency range at which the SMA wire is most effective.

Another important feature of passive damping with shape memory wires is the nonlinear properties of the damping effectiveness. Since the damping is induced by the hysteresis of the wire, it would be intuitive that larger input excitations will yield increased shape memory transformation and increased damping effectiveness. Recomputing the frequency response for an excitation input of $F_e = 50$ N yields the response shown in Figure 9.22*a*. Indeed, the response of the linear system now increases to approximately 8% peak-to-peak strain, while the response of the system with the shape memory wire is maintained at approximately 1.5% peak-to-peak strain. Plotting the damping effectiveness (Figure 9.22*b*) illustrates that increasing the input amplitude of the excitation now results in a strain reduction that is greater than a factor of 5. The increase in the damping effectiveness is due to the fact that the shape memory wire is undergoing increased phase transformation and is dissipating more energy per vibration cycle.

9.7 CHAPTER SUMMARY

The use of smart materials as passive and semiactive dampers was analyzed in this chapter. Equations for resistive and inductive–resistive piezoelectric shunts were analyzed to determine the frequency response as a function of the shunt parameters. Several methods for determining an optimal shunt were discussed. A comparison of a resistive and inductive–resistive shunt demonstrated that inductive–resistive shunts produce higher damping for equivalent generalized coupling coefficients, at the expense of requiring an additional passive element (the inductor). The need for an inductor is problematic at low frequencies, due to the inverse relationship between inductor size and resonant frequency of the structure. Multimode shunt techniques were analyzed to understand how to introduce vibration suppression in structures using multiple piezoelectric shunt elements.

The concept of passive damping was extended to include adaptive and state-switched shunt techniques. Adaptive shunt techniques rely on the slow variation of the shunt parameters to ensure optimal tuning. The tuning parameters of the shunt are varied in response to the output of the structure. Gradient-based algorithms were analyzed to determine how to vary the shunt parameters in real time to minimize structure vibration. State-switched shunt techniques rely on the electromechanical coupling inherent in piezoelectric materials. The stiffness of the shunt is varied to produce damping-type behavior in the system. The benefit of state-switched techniques compared to passive shunts or adaptive shunts is that the state-switched absorber can reduce the vibration of the structure off resonance.

The use of shape memory alloy wires was also studied using the model developed earlier in the book for shape memory transformation. Analyzing a system that utilizes shape memory wires for damping is complicated by the nonlinear stress–strain behavior of the shape memory materials. The analysis presented in this chapter was used in a parametric study to demonstrate important features of passive damping with shape memory materials. The study demonstrated that the damping effectiveness is related to the hysteresis induced in the shape memory material. This characteristic leads to a frequency-dependent behavior in which the shape memory wire is most effective near the resonance of the system. Similarly, the effectiveness of the shape memory wire increases with increasing excitation due to the increased hysteresis in the material.

PROBLEMS

9.1. Solve for the transfer function $u(\mathsf{s})/f(\mathsf{s})$ in equation (9.4) and compare the result to the frequency response of a resistive piezoelectric shunt, equation (9.19).

9.2. A piezoelectric shunt with short-circuit stiffness 130 N/m is incorporated into a structure that has a passive stiffness of 300 N/m. The structure is excited with a harmonic excitation $f(t) = 15 \sin 10t$.

(a) Compute the steady-state displacement.

(b) The shunt has a coupling coefficient of 0.32. Compute the generalized coupling coefficient of the shunt.

9.3. Write a computer code to replicate the results in Figure 9.4.

9.4. A piezoelectric shunt of capacitance 540 nF and a generalized coupling coefficient of 0.23 is incorporated into a structure with a coupled resonance of 135 Hz. Compute the shunt resistance value that maximizes the damping in the shunted structure.

9.5. Write a computer code to replicate Figure 9.6.

9.6. A resistive–inductive shunt with a generalized coupling coefficient of 0.30 is incorporated into a structure with a coupled resonance frequency of 450 Hz. The shunt capacitance is 320 nF.

(a) Estimate the maximum damping achievable with the resistive–inductive shunt.

(b) Compute the shunt resistance that maximizes the damping.

(c) Compute the shunt inductance that maximizes the damping.

9.7. Repeat Problem 9.6 for a structure that has a coupled resonance of 4.5 Hz. Compare the values of the resistance and inductance that maximize damping. State any problems that you foresee associated with the implementation of this RL shunt for a structure with a low-frequency resonance.

9.8. Repeat the calculation to produce Figure 9.17 for a generalized coupling coefficient of 0.7 and a normalized frequency of 0.5.

9.9. Plot the load path associated with equation (9.92).

NOTES

Additional references on the design of passive damping systems are those of Beranek and Ver [97] and Harris [98]. The seminal reference on the use of finite element techniques for the design of passive damping systems is the article by Johnson and Kienholz [99], but more recent historical reviews are referenced by Johnson [100]. The discussion of passive electrical networks using piezoelectric elements is based on work by Hagood and von Flotow [96]. More recent references on the design of piezoelectric shunts are those of Lesieutre [101], Tsai and Wang [102], Park and Inman [103], and Park et al. [105]. The theory of multimode shunting is based on the work by Hollkamp et al. [105]. The subject of networks of piezoelectric elements for vibration suppression has been studied by Morhan and Wang [106, 107]. Adaptive shunting networks were studied by Hollkamp [108]. References on switched-state absorbers include Cunefare et al. [109], Cunefare [110], and Holdhusen and Cunefare [111]. Additional work has been performed by Clark [112] and Corr and Clark [113].

10

ACTIVE VIBRATION CONTROL

One of the most sizable research areas for smart materials is the field of active vibration control. The coupling properties of smart materials makes them a natural candidate for problems in which the vibration of structural components must be controlled in real time and potentially with a high level of accuracy. Furthermore, the ability to integrate materials such as piezoelectric ceramics and shape memory alloys into structural materials enables the development of systems that seamlessly combine sensing, actuation, and control.

The focus of this chapter is to build on the models introduced in earlier chapters to analyze the problem of actively controlling structural vibration. In Chapter 9 we introduced the concept of active–passive vibration control. In this chapter we extend these concepts to include systems that utilize feedback as a means of changing the dynamic properties of structural systems. We begin with the structural models introduced in Chapter 5 for piezoelectric material systems to analyze the problem of *low authority control* of structures. These concepts lead us to an analysis of *self-sensing actuation* as a means of controlling structural vibration. This analysis will be a foundation for a control-theoretic analysis of structural control using pole placement methods and linear control and observation.

10.1 SECOND-ORDER MODELS FOR VIBRATION CONTROL

We found in previous chapters that linear models of smart material systems are often reduced to a set of coupled second-order differential equations of the form

$$\begin{aligned} \mathrm{M}_s\ddot{\mathbf{u}}(t) + \mathrm{K}_s^{\mathrm{D}}\mathbf{u}(t) - \Theta\mathbf{q}(t) &= \mathrm{B}_f\mathbf{f}(t) \\ -\Theta'\mathbf{u}(t) + \mathrm{C}_p^{\mathrm{S}^{-1}}\mathbf{q}(t) &= \mathrm{B}_v\mathbf{v}(t). \end{aligned} \tag{10.1}$$

Equation (10.1) assumes that there is no viscous damping in the system. This is a reasonable assumption for problems in vibration control since we are often trying to add damping to lightly damped systems and the damping introduced by the feedback control system is often much greater than the inherent damping in the structure.

The fact that there are no inertia terms associated with the charge coordinates allows us to simplify equation (10.1) into a single set of coupled second-order equations. Solving the second expression for the charge coordinates yields

$$\mathbf{q}(t) = \mathrm{C}_p^{\mathrm{S}}\mathrm{B}_v\mathbf{v}(t) + \mathrm{C}_p^{\mathrm{S}}\Theta'\mathbf{u}(t) \tag{10.2}$$

and substituting into the first expression yields a set of equations

$$\mathrm{M}\ddot{\mathbf{u}}(t) + \mathrm{K}^{\mathrm{E}}\mathbf{u}(t) = \mathrm{B}_f\mathbf{f}(t) + \Theta\mathrm{C}_p^{\mathrm{S}}\mathrm{B}_v\mathbf{v}(t). \tag{10.3}$$

The matrix K^{E} is the short-circuit stiffness matrix. It has the form

$$\mathrm{K}^{\mathrm{E}} = \mathrm{K}_s^{\mathrm{D}} - \Theta\mathrm{C}_p^{\mathrm{S}}\Theta'. \tag{10.4}$$

For simplicity, let us define

$$\mathrm{B}_c = \Theta\mathrm{C}_p^{\mathrm{S}}\mathrm{B}_v \tag{10.5}$$

as the control input vector and rewrite equation (10.3) as

$$\mathrm{M}\ddot{\mathbf{u}}(t) + \mathrm{K}^{\mathrm{E}}\mathbf{u}(t) = \mathrm{B}_f\mathbf{f}(t) + \mathrm{B}_c\mathbf{v}(t). \tag{10.6}$$

Now the equation is written as a set of undamped coupled second-order differential equations.

10.1.1 Output Feedback

The first control law that we study is a form of feedback in which the voltage is a linear combination of output measurements. Assume that the output measurements are of the form

$$\begin{aligned} \mathbf{y}_d(t) &= \mathrm{H}_d\mathbf{u}(t) \\ \mathbf{y}_v(t) &= \mathrm{H}_v\dot{\mathbf{u}}(t) \\ \mathbf{y}_a(t) &= \mathrm{H}_a\ddot{\mathbf{u}}(t), \end{aligned} \tag{10.7}$$

where the output matrices are defined by the location of the measurement points. Output feedback is a control law that takes the form

$$\mathbf{v}(t) = -\mathrm{G}_d\mathbf{y}_d(t) - \mathrm{G}_v\mathbf{y}_v(t) - \mathrm{G}_a\mathbf{y}_a(t), \tag{10.8}$$

where the matrices G_d, G_v, and G_a are *gain matrices* that define the feedback control law. Combining equations (10.7) and (10.8), we have

$$\mathbf{v}(t) = -G_d H_d \mathbf{u}(t) - G_v H_v \dot{\mathbf{u}}(t) - G_a H_a \ddot{\mathbf{u}}(t). \tag{10.9}$$

Substituting equation (10.9) into equation (10.6) and rearranging the terms yields

$$(M + B_c G_a H_a)\ddot{\mathbf{u}}(t) + B_c G_v H_v \dot{\mathbf{u}}(t) + (K^E + B_c G_d H_d)\mathbf{u}(t) = B_f \mathbf{f}(t). \tag{10.10}$$

Equation (10.10) illustrates the relationship between the terms in the output feedback and the physical effect on the system. The displacement feedback adds terms to the stiffness matrix of the system; therefore, displacement feedback can be thought of as a way to change the stiffness using feedback. Similarly, acceleration feedback effectively adds mass to the system through variations in the mass matrix. Velocity feedback adds a linear viscous damping term to the model.

A second type of output feedback relies on the use of charge as the feedback variable. This feedback is common to systems that are using the piezoelectric material as the sensing element as well as the elements for actuation. In the general case the sensing and actuation elements do not need to be collocated with one another, and the form of the feedback law is

$$\mathbf{v}(t) = -G_d H_d \mathbf{q}(t) - G_v H_v \dot{\mathbf{q}}(t) - G_a H_a \ddot{\mathbf{q}}(t). \tag{10.11}$$

Substituting this feedback law into equation (10.1) yields

$$\begin{aligned} M_s \ddot{\mathbf{u}}(t) + K_s^D \mathbf{u}(t) - \Theta \mathbf{q}(t) &= B_f \mathbf{f}(t) \\ B_v G_a H_a \ddot{\mathbf{q}}(t) + B_v G_v H_v \dot{\mathbf{q}}(t) - \Theta' \mathbf{u}(t) + \left(C_p^{S^{-1}} + B_v G_d H_d\right)\mathbf{q}(t) &= 0. \end{aligned} \tag{10.12}$$

Now we see that charge feedback adds terms to the closed-loop system that mimic mass, damping, and stiffness terms in the equations of motion for the piezoelectric element. This contrasts with displacement, velocity, and acceleration feedback, which adds feedback terms directly to the second-order equations for the structure. Adding mass, stiffness, and damping terms to the equations for the piezoelectric elements influences the equations for the structure through the coupling terms in the stiffness matrix. This effect is seen more clearly by rewriting equation (10.12) as a set of matrix second-order equations of the form

$$\begin{aligned} &\begin{bmatrix} M_s & 0 \\ 0 & B_v G_a H_a \end{bmatrix} \begin{Bmatrix} \ddot{\mathbf{u}}(t) \\ \ddot{\mathbf{q}}(t) \end{Bmatrix} + \begin{bmatrix} 0 & 0 \\ 0 & B_v G_v H_v \end{bmatrix} \begin{Bmatrix} \dot{\mathbf{u}}(t) \\ \dot{\mathbf{q}}(t) \end{Bmatrix} \\ &+ \begin{bmatrix} K_s^D & -\Theta \\ -\Theta' & C_p^{S^{-1}} + B_v G_d H_d \end{bmatrix} \begin{Bmatrix} \mathbf{u}(t) \\ \mathbf{q}(t) \end{Bmatrix} = \begin{bmatrix} B_f \\ 0 \end{bmatrix} \mathbf{f}(t). \end{aligned} \tag{10.13}$$

From equation (10.13) it is clear that the matrices for the closed-loop system contain mass and damping terms in the equations for the piezoelectric elements even though the original equations did not contain these terms. Furthermore, the closed-loop equations illustrate the fact that these terms in the lower partition of the matrices couple to the upper partition through the coupling terms Θ in the stiffness matrix.

A third type of feedback control is *collocated charge feedback*, which is based on the assumption that the sensors and actuators are located in identical places on the structure. Mathematically, this is represented by assuming that the influence matrix for the feedback control law is simply the transpose of the input matrix B_v. Under this assumption the feedback control law becomes

$$\mathbf{v}(t) = -\mathrm{G}_d B_v' \mathbf{q}(t) - \mathrm{G}_v B_v' \dot{\mathbf{q}}(t) - \mathrm{G}_a B_v' \ddot{\mathbf{q}}(t). \tag{10.14}$$

Substituting equation (10.14) into equation (10.1) and rewriting as a matrix equation yields

$$\begin{bmatrix} \mathrm{M}_s & 0 \\ 0 & \mathrm{B}_v \mathrm{G}_a B_v' \end{bmatrix} \begin{Bmatrix} \ddot{\mathbf{u}}(t) \\ \ddot{\mathbf{q}}(t) \end{Bmatrix} + \begin{bmatrix} 0 & 0 \\ 0 & \mathrm{B}_v \mathrm{G}_v B_v' \end{bmatrix} \begin{Bmatrix} \dot{\mathbf{u}}(t) \\ \dot{\mathbf{q}}(t) \end{Bmatrix}$$
$$+ \begin{bmatrix} \mathrm{K}_s^{\mathrm{D}} & -\Theta \\ -\Theta' & \mathrm{C}_p^{\mathrm{S}^{-1}} + \mathrm{B}_v \mathrm{G}_d B_v' \end{bmatrix} \begin{Bmatrix} \mathbf{u}(t) \\ \mathbf{q}(t) \end{Bmatrix} = \begin{bmatrix} \mathrm{B}_f \\ 0 \end{bmatrix} \mathbf{f}(t). \tag{10.15}$$

The analyses associated with output feedback for second-order systems illustrate that the closed-loop system is represented as a set of coupled second-order differential equations that are characterized by a mass, damping, and stiffness matrix. The control design consists of choosing the type of sensor output and the sensor location. Choosing the sensor type and location specifies the influence matrices for the feedback control law. The second component of the design is to choose the feedback gain matrices such that the stability and performance requirements for the control system is met.

The preceding analyses demonstrate that all three types of output feedback control laws result in a set of second-order matrix equations whose terms are a function of the sensor influence matrices and the feedback gain matrices. Examining equations (10.10), (10.13), and (10.15), we see that the closed-loop equations can be written as

$$\mathrm{M}\ddot{\mathbf{x}}(t) + \mathrm{D}\dot{\mathbf{x}}(t) + \mathrm{K}\mathbf{x}(t) = \mathbf{B}\mathbf{f}(t), \tag{10.16}$$

where the exact form of the matrices is a function of the type of output feedback. The vector $\mathbf{x}(t)$ is a generic displacement vector whose form also depends on the type of output feedback.

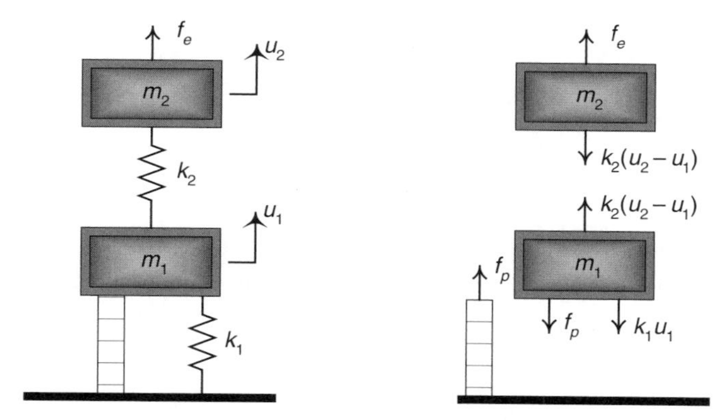

Figure 10.1 Representative piezoelectric system utilized throughout this chapter.

10.2 ACTIVE VIBRATION CONTROL EXAMPLE

Throughout this chapter we utilize a representative model of a piezoelectric system as a means of illustrating the control concepts introduced in this chapter. The representative system consists of two mass–spring combinations, with a piezoelectric stack attached between ground and the first mass (Figure 10.1).

The equations of motion for this system are derived using force balance equations on each mass. For the first mass the force balance is

$$\begin{aligned} m_1\ddot{u}_1 &= -f_p - k_1 u_1 - k_2(u_1 - u_2) \\ m_2\ddot{u}_2 &= -k_2(u_2 - u_1). \end{aligned} \tag{10.17}$$

The equations for the piezoelectric stack actuator are obtained by considering the transducer equations for a 33 device from Chapter 4:

$$\begin{aligned} u_1 &= \frac{1}{k_p^{\mathrm{E}}} f_p + \frac{d_{33}L_p}{t_p} v \\ q &= \frac{d_{33}L_p}{t_p} f_p + C_p^{\mathrm{T}} v. \end{aligned} \tag{10.18}$$

The equations of motion for the combined system are obtained most easily by interchanging the dependent and independent variables in equation (10.18):

$$\begin{aligned} f_p &= k_p^{\mathrm{D}} u_1 - \frac{d_{33}L_p}{t_p}\frac{k_p^{\mathrm{E}}}{C_p^{\mathrm{S}}} q \\ v &= -\frac{d_{33}L_p}{t_p}\frac{k_p^{\mathrm{E}}}{C_p^{\mathrm{S}}} u_1 + \frac{1}{C_p^{\mathrm{S}}} q, \end{aligned} \tag{10.19}$$

where

$$\begin{aligned} k_p^{\mathrm{D}} &= \frac{k_p^{\mathrm{E}}}{1 - k_{33}^2} \\ C_p^{\mathrm{S}} &= \left(1 - k_{33}^2\right) C_p^{\mathrm{T}}. \end{aligned} \tag{10.20}$$

Combining equations (10.17) and (10.20) results in the equations of motion for the system:

$$\begin{aligned} m_1 \ddot{u}_1 + \left(k_1 + k_2 + k_p^{\mathrm{D}}\right) u_1 - k_2 u_2 - \frac{d_{33} L_p}{t_p} \frac{k_p^{\mathrm{E}}}{C_p^{\mathrm{S}}} q &= 0 \\ m_2 \ddot{u}_2 + k_2 u_2 - k_2 u_1 &= f_e \\ -\frac{d_{33} L_p}{t_p} \frac{k_p^{\mathrm{E}}}{C_p^{\mathrm{S}}} u_1 + \frac{1}{C_p^{\mathrm{S}}} q &= v. \end{aligned} \tag{10.21}$$

The equations of motion can be placed in the form of equation (10.1) by defining the matrices

$$\mathbf{M}_s = \begin{bmatrix} m_1 & 0 \\ 0 & m_2 \end{bmatrix} \tag{10.22}$$

$$\mathbf{K}_s^{\mathrm{D}} = \begin{bmatrix} k_1 + k_2 + k_p^{\mathrm{D}} & -k_2 \\ -k_2 & k_2 \end{bmatrix} \tag{10.23}$$

$$\Theta = \begin{bmatrix} \left(\dfrac{d_{33} L_p}{t_p}\right) \dfrac{k_p^{\mathrm{E}}}{C_p^{\mathrm{S}}} \\ 0 \end{bmatrix} \tag{10.24}$$

$$C_p^{\mathrm{S}^{-1}} = \left[\frac{1}{C_p^{\mathrm{S}}}\right] \tag{10.25}$$

$$\mathbf{B}_f = \begin{bmatrix} 0 \\ 1 \end{bmatrix} \tag{10.26}$$

$$\mathbf{B}_v = [1]. \tag{10.27}$$

It is interesting to note that the equations of motion derived from a force balance take the same form as those derived from energy principles in Chapter 5.

The equations of motion are put into the form expressed in equation (10.6) by writing

$$\mathrm{B}_c = \Theta \mathrm{C}_p^{\mathrm{S}} \mathrm{B}_v = \begin{bmatrix} \dfrac{d_{33} L_p}{t_p} k_p^{\mathrm{E}} \\ 0 \end{bmatrix}. \tag{10.28}$$

Example 10.1 Compute the equations of motion for the piezoelectric system in Figure 10.1 using the values listed in Table 10.1. Express the equations of motion in a form consistent with equation (10.1).

Solution The first step is to compute the parameters associated with the piezoelectric stack. The short-circuit stiffness is

$$k_p^{\mathrm{E}} = \frac{Y_{33}^{\mathrm{E}} A}{L_p} = \frac{(60 \times 10^9 \ \mathrm{N/m^2})(4 \times 10^{-6} \ \mathrm{m^2})}{15 \times 10^{-3} \ \mathrm{m}} = 16 \ \mathrm{N}/\mu\mathrm{m}.$$

The stress-free capacitance is

$$C_p^{\mathrm{T}} = n \frac{\epsilon_{33}^{\mathrm{T}} A}{t_p} = \left(\frac{15 \ \mathrm{mm}}{0.25 \ \mathrm{mm}} \right) \frac{(39.8 \times 10^{-9} \ \mathrm{F/m})(4 \times 10^{-6} \ \mathrm{m^2})}{0.25 \times 10^{-3} \ \mathrm{m}} = 38.2 \ \mathrm{nF}.$$

The coupling coefficient of the material is

$$k_{33} = \frac{d_{33}}{\sqrt{\epsilon_{33}^{\mathrm{T}} s_{33}^{\mathrm{E}}}} = \frac{(650 \times 10^{-12} \ \mathrm{m/V}) \sqrt{60 \times 10^9 \ \mathrm{N/m^2}}}{\sqrt{39.8 \times 10^{-9} \ \mathrm{F/m}}} = 0.798.$$

The open-circuit stiffness of the piezoelectric stack is

$$k_p^{\mathrm{D}} = \frac{k_p^{\mathrm{E}}}{1 - k_{33}^2} = \frac{16 \ \mathrm{N}/\mu\mathrm{m}}{1 - 0.798^2} = 44.1 \ \mathrm{N}/\mu\mathrm{m}$$

Table 10.1 Parameters for Example 10.1

Structure	Piezoelectric Stack
$m_1 = 0.5$ kg	$L_p = 15$ mm
$m_2 = 0.5$ kg	$t_p = 0.25$ mm
$k_1 = 15\ \mathrm{N}/\mu\mathrm{m}$	$d_{33} = 650$ pm/V
$k_2 = 30\ \mathrm{N}/\mu\mathrm{m}$	$A = 4\ \mathrm{mm}^2$
	$\epsilon_{33}^{\mathrm{T}} = 39.8$ nF/m
	$Y_{33}^{\mathrm{E}} = 60$ GPa

and the zero-strain capacitance of the stack is

$$C_p^{\mathrm{S}} = \left(1 - k_{33}^2\right) C_p^{\mathrm{T}} = (1 - 0.798^2)(38.2\ \mathrm{nF}) = 13.9\ \mathrm{nF}.$$

Computing all of the stack parameters allows us to compute the matrices of the equations of motion using equations (10.22) to (10.25):

$$\mathrm{M}_s = \begin{bmatrix} 0.5 & 0 \\ 0 & 0.5 \end{bmatrix} \mathrm{kg}$$

$$\mathrm{K}_s^{\mathrm{D}} = \begin{bmatrix} 15 + 30 + 44.1 & -30 \\ -30 & 30 \end{bmatrix} \mathrm{N}/\mu\mathrm{m} = \begin{bmatrix} 89.1 & -30 \\ -30 & 30 \end{bmatrix} \times 10^6\ \mathrm{N/m}$$

$$\Theta = \begin{bmatrix} \dfrac{d_{33} L_p}{t_p} \dfrac{k_p^{\mathrm{E}}}{C_p^{\mathrm{S}}} \\ 0 \end{bmatrix} = \begin{bmatrix} 650 \times 10^{-12}\ \mathrm{m/V} \left(\dfrac{15\ \mathrm{mm}}{0.25\ \mathrm{mm}} \right) \dfrac{16 \times 10^6\ \mathrm{N/m}}{13.9 \times 10^{-9}\ \mathrm{F}} \\ 0 \end{bmatrix}$$

$$= \begin{bmatrix} 44.9 \\ 0 \end{bmatrix} \times 10^6\ \mathrm{N/C}$$

$$C_p^{\mathrm{S}^{-1}} = \begin{bmatrix} \dfrac{1}{C_p^{\mathrm{S}}} \end{bmatrix} = \frac{1}{13.9 \times 10^{-9}\ \mathrm{F}} = 71.9 \times 10^6\ \mathrm{F}^{-1}.$$

The equations of motion can now be written in a form consistent with equation (10.1):

$$\begin{bmatrix} 0.5 & 0 \\ 0 & 0.5 \end{bmatrix} \begin{pmatrix} \ddot{u}_1 \\ \ddot{u}_2 \end{pmatrix} + \begin{bmatrix} 89.1 \times 10^6 & -30 \times 10^6 \\ -30 \times 10^6 & 30 \times 10^6 \end{bmatrix} \begin{pmatrix} u_1 \\ u_2 \end{pmatrix} - \begin{bmatrix} 44.9 \times 10^6 \\ 0 \end{bmatrix} q = \begin{bmatrix} 0 \\ 1 \end{bmatrix} f_e$$

$$- \begin{bmatrix} 44.9 \times 10^6 & 0 \end{bmatrix} \begin{pmatrix} u_1 \\ u_2 \end{pmatrix} + 71.9 \times 10^6 q = v.$$

Example 10.2 Compute the magnitude of the frequency response between the output displacement of each mass and the external force input and the voltage input for the parameters computed in Example 10.1. Plot the results.

Solution The frequency response is computed by first transforming the equations of motion into the form expressed in equation (10.6). First compute

$$\mathrm{K}^{\mathrm{E}} = \mathrm{K}_s^{\mathrm{D}} - \Theta C_p^{\mathrm{S}} \Theta' = \begin{bmatrix} 61 & -30 \\ -30 & 30 \end{bmatrix} \times 10^6\ \mathrm{N/m}$$

$$\mathrm{B}_c = \Theta C_p^{\mathrm{S}} \mathrm{B}_v \begin{bmatrix} -0.6240 \\ 0 \end{bmatrix} \mathrm{N/V}.$$

The dynamic stiffness matrix is defined as

$$\begin{aligned}\Delta(\omega) &= \mathrm{K}^{\mathrm{E}} - \omega^2 \mathrm{M}_s \\ &= \begin{bmatrix} 61\times 10^6 - 0.5\omega^2 & -30\times 10^6 \\ -30\times 10^6 & 30\times 10^6 - 0.5\omega^2 \end{bmatrix}.\end{aligned}$$

The frequency response between the external force and the displacement is computed from

$$\begin{aligned}\mathbf{u}_f &= \Delta^{-1}(\omega)\mathrm{B}_f \\ &= \begin{bmatrix} 61\times 10^6 - 0.5\omega^2 & -30\times 10^6 \\ -30\times 10^6 & 30\times 10^6 - 0.5\omega^2 \end{bmatrix}^{-1} \begin{bmatrix} 0 \\ 1 \end{bmatrix} \\ &= \frac{1}{0.25\omega^4 - (45.5\times 10^6)\omega^2 + 930\times 10^{12}} \begin{bmatrix} 30\times 10^6 \\ 61\times 10^6 - 0.5\omega^2 \end{bmatrix}.\end{aligned}$$

The frequency response between the input voltage and the displacement is computed from

$$\begin{aligned}\mathbf{u}_v &= \Delta^{-1}(\omega)\mathrm{B}_c \\ &= \begin{bmatrix} 61\times 10^6 - 0.5\omega^2 & -30\times 10^6 \\ -30\times 10^6 & 30\times 10^6 - 0.5\omega^2 \end{bmatrix}^{-1} \begin{bmatrix} -0.6240 \\ 0 \end{bmatrix} \\ &= \frac{1}{0.25\omega^4 - (45.5\times 10^6)\omega^2 + 930\times 10^{12}} \begin{bmatrix} -0.6240(30\times 10^6 - 0.5\omega^2) \\ -18.72\times 10^6 \end{bmatrix}.\end{aligned}$$

The frequency response magnitude for each result is plotted in Figure 10.2. The sharp peaks in the frequency response are the short-circuit resonance frequencies of the piezoelectric system. Note that the amplitude near the peak is not representative of the magnitude since the model does not incorporate damping in the system.

10.3 DYNAMIC OUTPUT FEEDBACK

In Section 10.2 we studied feedback control laws which did not add any additional states to the dynamic system. This type of feedback is called *static feedback* since the feedback signal is simply a linear combination of sensor signals. Another type of feedback control, *dynamic output feedback*, is useful for the design of active vibration control systems. Dynamic output feedback utilizes control systems that add additional states to the equations of motion, thus, the feedback signal is a function of the output of a dynamic *compensator* designed to meet certain stability and performance criteria.

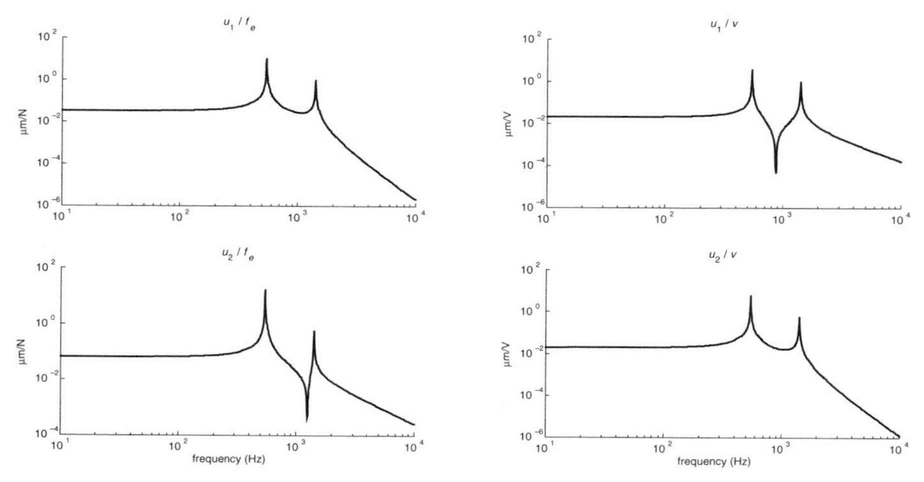

Figure 10.2 Frequency response functions for the representative piezoelectric system studied in Examples 10.1 and 10.2.

Before discussing the use of piezoelectric material systems with dynamic output feedback, let us focus on a representative second-order model to understand the basic control concepts. Consider a second-order structural model of the form

$$\mathbf{M}\ddot{\mathbf{u}}(t) + \mathbf{K}\mathbf{u} = \mathrm{B}_f\mathbf{f}(t) + \mathrm{B}_c\mathbf{f}_c, \tag{10.29}$$

where B_c is the influence matrix for the control forces and $\mathbf{f}_c$ are the control forces. The feedback control law is of the form

$$\mathbf{f}_c = \mathrm{G}_c\boldsymbol{\eta}(t), \tag{10.30}$$

where $\boldsymbol{\eta}(t)$ is the coordinate associated with the dynamic system

$$\mathrm{M}_c\ddot{\boldsymbol{\eta}}(t) + \mathrm{D}_c\dot{\boldsymbol{\eta}}(t) + \mathrm{K}_c\boldsymbol{\eta}(t) = G_c'\mathbf{y}(t). \tag{10.31}$$

The dynamic system represented in equation (10.31) is a *virtual system*, implemented in electronics, to shape the response of the closed-loop system. The input to this system is the output of the real dynamic system represented by equation (10.29). Assuming that the control forces are collocated with the system output, the output is expressed as

$$\mathbf{y}(t) = B_c'\mathbf{u}(t). \tag{10.32}$$

Combining equations (10.29) to (10.32), we obtain a coupled set of second-order equations of the form

$$\begin{bmatrix} \mathrm{M} & 0 \\ 0 & \mathrm{M}_c \end{bmatrix} \begin{Bmatrix} \ddot{\mathbf{u}}(t) \\ \ddot{\eta}(t) \end{Bmatrix} + \begin{bmatrix} 0 & 0 \\ 0 & \mathrm{D}_c \end{bmatrix} \begin{Bmatrix} \dot{\mathbf{u}}(t) \\ \dot{\eta}(t) \end{Bmatrix} + \begin{bmatrix} \mathrm{K} & -\mathrm{B}_c\mathrm{G}_c \\ -G_c'B_c' & \mathrm{K}_c \end{bmatrix} \begin{Bmatrix} \mathbf{u}(t) \\ \eta(t) \end{Bmatrix} = \begin{bmatrix} \mathrm{B}_f \\ 0 \end{bmatrix} \mathbf{f}(t). \tag{10.33}$$

One of the defining features of dynamic output feedback is that the control law adds additional states to the open-loop system. This is represented in equation (10.33) by states associated with the coordinate $\eta(t)$. Another important feature of dynamic output feedback is the damping introduced by the control system. We have assumed, for simplicity, that the open-loop system does not contain viscous damping terms, yet the closed-loop system does contain damping through the introduction of viscous damping into the control system. Damping in the control system couples to the real dynamic system through the coupling term $\mathrm{B}_c\mathrm{G}_c$ in the stiffness matrix.

It is important to recognize that the control law discussed above is not the most general dynamic output feedback control law. In general, a feedback control law can be constructed that also incorporates coupling terms in the mass and damping matrices. In addition, the control expressed in equation (10.33) assumes collocated feedback.

Insight into the physics of dynamic output feedback is gained by examining the control law in combination with a modal model of the system. Making the substitution

$$\mathbf{u}(t) = \mathrm{P}\mathbf{r}(t), \tag{10.34}$$

where P is the matrix of mass-normalized eigenvectors, we can rewrite equation (10.29) as

$$\mathrm{I}\ddot{\mathbf{r}}(t) + \Lambda\mathbf{r}(t) = \mathrm{P}'\mathrm{B}_f\mathbf{f}(t) + \mathrm{P}'\mathrm{B}_c\mathbf{f}_c(t), \tag{10.35}$$

where Λ is a diagonal matrix of eigenvalues. The terms $\mathrm{P}'\mathrm{B}_f$ and $\mathrm{P}'\mathrm{B}_c$ can be interpreted as a set of *modal influence coefficients*, defined as

$$\begin{aligned} \mathrm{P}'\mathrm{B}_f &= \Phi_f \\ \mathrm{P}'\mathrm{B}_c &= \Phi_c, \end{aligned} \tag{10.36}$$

where each term represents the input term to the ith mode from the jth input. Substituting the definitions of the modal influence coefficients into the equations of motion and rewriting the expressions for the closed-loop system, equation (10.33), we have

$$\begin{bmatrix} \mathrm{I} & 0 \\ 0 & \mathrm{M}_c \end{bmatrix} \begin{Bmatrix} \ddot{\mathbf{r}}(t) \\ \ddot{\eta}(t) \end{Bmatrix} + \begin{bmatrix} 0 & 0 \\ 0 & \mathrm{D}_c \end{bmatrix} \begin{Bmatrix} \dot{\mathbf{r}}(t) \\ \dot{\eta}(t) \end{Bmatrix} + \begin{bmatrix} \Lambda & -\Phi_c\mathrm{G}_c \\ -\mathrm{G}'_c\Phi'_c & \mathrm{K}_c \end{bmatrix} \begin{Bmatrix} \mathbf{r}(t) \\ \eta(t) \end{Bmatrix} = \begin{bmatrix} \Phi_f \\ 0 \end{bmatrix} \mathbf{f}(t). \tag{10.37}$$

The goal of the control design is to choose the coefficients of the control law to achieve the desired stability and performance objectives. The control design simplified by also assuming that the compensator consists of a diagonal set of control coefficients. Thus, if we assume that

$$\begin{aligned} \mathrm{M}_c &= \mathrm{I} \\ \mathrm{D}_c &= \Delta_c = \mathrm{diag}\{2\zeta_{c1}\omega_{c1}, \ldots, 2\zeta_{cn_c}\omega_{cn_c}\} \\ \mathrm{K}_c &= \Lambda_c = \mathrm{diag}\{\omega_{c1}^2, \ldots, \omega_{cn_c}^2\} \end{aligned} \tag{10.38}$$

substituting these definitions into equation (10.37) yields

$$\begin{bmatrix} \mathrm{I} & 0 \\ 0 & \mathrm{I} \end{bmatrix} \begin{Bmatrix} \ddot{\mathbf{r}}(t) \\ \ddot{\eta}(t) \end{Bmatrix} + \begin{bmatrix} 0 & 0 \\ 0 & \Delta_c \end{bmatrix} \begin{Bmatrix} \dot{\mathbf{r}}(t) \\ \dot{\eta}(t) \end{Bmatrix} + \begin{bmatrix} \Lambda & -\Phi_c \mathrm{G}_c \\ -G_c' \Phi_c' & \Lambda_c \end{bmatrix} \begin{Bmatrix} \mathbf{r}(t) \\ \eta(t) \end{Bmatrix} = \begin{bmatrix} \Phi_f \\ 0 \end{bmatrix} \mathbf{f}(t). \tag{10.39}$$

The rationale for introducing the assumption of diagonal compensator matrices into the feedback control law is that it reduces the number of terms to choose in the control design. In addition, it adds physical insight into the control design problem.

To understand the basic properties of this type of dynamic output feedback control, let us examine a single-mode model of a structure coupled to a single-mode model of the controller. Under this assumption the closed-loop equations of motion are

$$\begin{bmatrix} 1 & 0 \\ 0 & 1 \end{bmatrix} \begin{Bmatrix} \ddot{r}(t) \\ \ddot{\eta}(t) \end{Bmatrix} + \begin{bmatrix} 0 & 0 \\ 0 & 2\zeta_c\omega_c \end{bmatrix} \begin{Bmatrix} \dot{r}(t) \\ \dot{\eta}(t) \end{Bmatrix} + \begin{bmatrix} \omega_n^2 & -\omega_n\omega_c g_c \\ -\omega_n\omega_c g_c & \omega_c^2 \end{bmatrix} \begin{Bmatrix} r(t) \\ \eta(t) \end{Bmatrix} = \begin{bmatrix} \omega_n^2 \\ 0 \end{bmatrix} f(t). \tag{10.40}$$

The control system is stable if the stiffness matrix is positive definite. This occurs when the expression

$$\omega_n^2\omega_c^2 - \omega_n^2\omega_c^2 g_c^2 = \omega_n^2\omega_c^2\left(1 - g_c^2\right) > 0 \tag{10.41}$$

is satisfied. This occurs when

$$g_c^2 < 1. \tag{10.42}$$

The frequency response between the output displacement and the input force is obtained by transforming equation (10.40) into the Laplace domain and solving for $R(\mathsf{s})/F(\mathsf{s})$. The result is

$$\frac{R(\mathsf{s})}{F(\mathsf{s})} = \frac{\omega_n^2\left(\mathsf{s}^2 + 2\zeta_c\omega_c\mathsf{s} + \omega_c^2\right)}{\left(\mathsf{s}^2 + \omega_n^2\right)\left(\mathsf{s}^2 + 2\zeta_c\omega_c\mathsf{s} + \omega_c^2\right) - \omega_n^2\omega_c^2 g_c^2}. \tag{10.43}$$

The analysis is facilitated by introducing the nondimensional parameters

$$\tilde{\mathsf{s}} = \frac{\mathsf{s}}{\omega_n} \qquad \alpha_c = \frac{\omega_c}{\omega_n}. \tag{10.44}$$

Substituting these expressions into equation (10.45) yields

$$\frac{R(\tilde{\mathsf{s}})}{F(\tilde{\mathsf{s}})} = \frac{\tilde{\mathsf{s}}^2 + 2\alpha_c\zeta_c\tilde{\mathsf{s}} + \alpha_c^2}{(\tilde{\mathsf{s}}^2 + 1)\left(\tilde{\mathsf{s}}^2 + 2\alpha_c\zeta_c\tilde{\mathsf{s}} + \alpha_c^2\right) - \alpha_c^2 g_c^2}. \tag{10.45}$$

The denominator of the frequency response is expanded to produce a fourth-order polynomial:

$$\frac{R(\tilde{\mathsf{s}})}{F(\tilde{\mathsf{s}})} = \frac{\tilde{\mathsf{s}}^2 + 2\alpha_c\zeta_c\tilde{\mathsf{s}} + \alpha_c^2}{\tilde{\mathsf{s}}^4 + 2\alpha_c\zeta_c\tilde{\mathsf{s}}^3 + \left(\alpha_c^2 + 1\right)\tilde{\mathsf{s}}^2 + 2\alpha_c\zeta_c\mathsf{s} + \alpha_c^2\left(1 - g_c^2\right)}. \tag{10.46}$$

Nondimensional analysis highlights the fact that there are three design parameters for a positive-position feedback controller. The design requires the choice of the ratio of the natural frequencies, α_c, the damping ratio, ζ_c, and the control gain g_c. In addition, we note the similarity between the positive-position feedback controller and other types of semiactive or passive control systems. The frequency response of a single-mode system with positive-position feedback is similar to the frequency response of a system with a tuned-mass damper or an inductive–resistive shunt since the characteristics polynomial is fourth order in all cases. One important difference, though, is that a positive-position controller has a finite stability margin, $g_c < 1$, whereas a tuned-mass damper or inductive–resistive shunt is guaranteed to be stable for all values of the design parameters.

Continuing with this analogy, we recall that the design of inductive–resistive shunts or tuned-mass dampers require that the natural frequency of the control system be tuned to the natural frequency of the structure. For this reason, let us analyze the positive-position controller for the case of $\alpha_c = 1.1$. One of the most instructive ways to examine the performance of the controller is to plot the roots of the characteristic polynomial for specific values of ζ_c over the range of gains 0 to 1. One plot is shown in Figure 10.3*a*. We note that there are two sets of roots. One set of roots corresponds to the pole of the control filter and the second set of roots corresponds to the pole of the structure. As the gain of the controller is increased, the filter pole moves toward the $j\omega$ axis (i.e., to the right) while the structural pole moves farther into the left-half plane. The movement of the structural pole into the left-half plane is desirable since it indicates that the damping ratio of the structural pole is increasing with an increase in the control gain. Increasing the gain closer to 1 (Figure 10.3*b*) produces a situation in which the filter pole moves closer to the $j\omega$ axis and the structural pole becomes overdamped. Increasing the gain further causes instability because the structural pole moves into the right-half plane for values of the control gain greater than 1. From this plot we note that there is a value of the gain in which both the filter pole and the structural pole have the same damping ratio. This situation is desirable since the

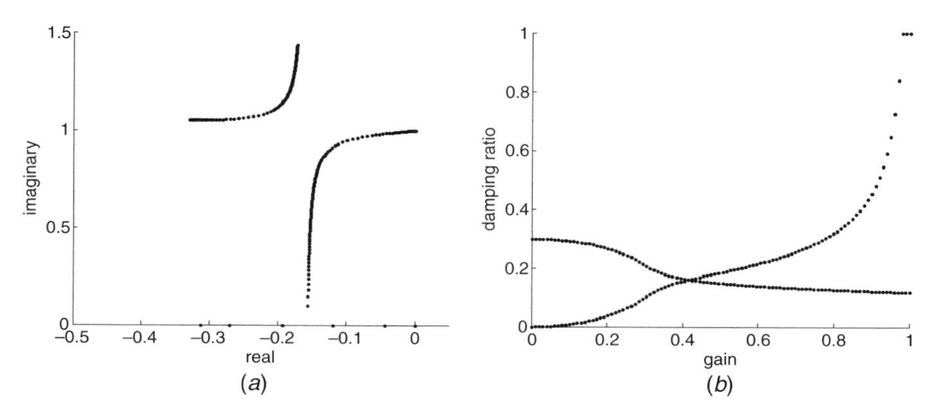

Figure 10.3 (*a*) Roots of the characteristic polynomial for a positive-position feedback controller for $\zeta_c = 0.3$ $\alpha_c = 1.1$; (*b*) corresponding damping ratios of the filter and structure illustrating the convergence of the damping ratios at a particular gain.

filter pole will produce a response in the closed loop between the input force and output displacement. Thus, a filter pole with negligible damping will be the dominant response in the closed loop if the gain is chosen to be too large. Thus, one definition of an *optimal* controller would be that the filter and structural poles have identical damping ratios. For the single-mode system represented by equation (10.40), the damping ratio of both the filter and structure poles is maximized when the following equations are solved,

$$
\begin{aligned}
2\zeta_c\omega_c &= 4\zeta_{cl}\omega_{cl} \\
\omega_n^2 + \omega_c^2 &= 2\omega_{cl}^2 + 4\zeta_{cl}^2\omega_{cl}^2 \\
2\zeta_c\omega_c\omega_n^2 &= 4\zeta_{cl}\omega_{cl}^3 \\
\omega_n^2\omega_c^2(1 - q_c^2) &= \omega_{cl}^4
\end{aligned}
\tag{10.47}
$$

is chosen as the controller design parameters.

The analysis of a single-mode system is representative of a multimode design, due to the fact that the equations of motion can be decoupled. A common way to achieve multimode vibration control is to design each filter sequentially by choosing the design parameters for each structural mode. A reasonable starting point for each controller parameter is the optimal values shown in equation (10.47). Typically, though, for a multimode design, the controller natural frequency must be moved closer to the natural frequency of the structural mode being controlled, and the damping ratio must be decreased to achieve the maximum damping in the structural mode.

10.3.1 Piezoelectric Material Systems with Dynamic Output Feedback

The analysis in Section 10.2 was a general treatment of one form of dynamic output feedback, positive position feedback, for a structural system. To apply this concept to a piezoelectric material system, begin with equation (10.6) and apply the feedback control law:

$$\mathbf{v}(t) = \mathrm{G}_c \boldsymbol{\eta}(t). \tag{10.48}$$

Combining equations (10.6) and (10.48) results in the expression

$$\mathrm{M}\ddot{\mathbf{u}}(t) + \mathrm{K}^{\mathrm{E}}\mathbf{u}(t) = \mathrm{B}_f \mathbf{f}(t) + \Theta \mathrm{C}_p^{\mathrm{S}} \mathrm{B}_v \mathrm{G}_c \boldsymbol{\eta}(t). \tag{10.49}$$

Assuming that we use collocated charge feedback as the input to the control law, the expression for the control system is

$$\mathrm{M}_c \ddot{\boldsymbol{\eta}}(t) + \mathrm{D}_c \dot{\boldsymbol{\eta}}(t) + \mathrm{K}_c \boldsymbol{\eta}(t) = \mathrm{G}_c' \mathrm{B}_v' \mathbf{q}(t). \tag{10.50}$$

The expression for the charge is given in equation (10.2) and is repeated here for convenience:

$$\mathbf{q}(t) = \mathrm{C}_p^{\mathrm{S}} \mathrm{B}_v \mathbf{v}(t) + \mathrm{C}_p^{\mathrm{S}} \Theta' \mathbf{u}(t).$$

As expected, the charge output of the piezoelectric elements is a linear combination of a term due to the input voltage and a term due to the motion of the structure. This, of course, is due to electromechanical coupling in the piezoelectric material. Substituting equation (10.2) into equation (10.50) results in

$$\mathrm{M}_c \ddot{\boldsymbol{\eta}}(t) + \mathrm{D}_c \dot{\boldsymbol{\eta}}(t) + \mathrm{K}_c \boldsymbol{\eta}(t) = \mathrm{G}_c' \mathrm{B}_v' \left[\mathrm{C}_p^{\mathrm{S}} \mathrm{B}_v \mathbf{v}(t) + \mathrm{C}_p^{\mathrm{S}} \Theta' \mathbf{u}(t)\right]. \tag{10.51}$$

Incorporating the feedback control law, equation (10.48), into equation (10.51) and rewriting terms yields the equation

$$\mathrm{M}_c \ddot{\boldsymbol{\eta}}(t) + \mathrm{D}_c \dot{\boldsymbol{\eta}}(t) + \left(\mathrm{K}_c - \mathrm{G}_c' \mathrm{B}_v' \mathrm{C}_p^{\mathrm{S}} \mathrm{B}_v \mathrm{G}_c\right) \boldsymbol{\eta}(t) - \mathrm{G}_c' \mathrm{B}_v' \mathrm{C}_p^{\mathrm{S}} \Theta' \mathbf{u}(t) = 0. \tag{10.52}$$

The matrix expressions for the closed-loop system are written by combining equations (10.49) and (10.52),

$$\begin{bmatrix} \mathrm{M} & 0 \\ 0 & \mathrm{M}_c \end{bmatrix} \begin{Bmatrix} \ddot{\mathbf{u}}(t) \\ \ddot{\boldsymbol{\eta}}(t) \end{Bmatrix} + \begin{bmatrix} 0 & 0 \\ 0 & \mathrm{D}_c \end{bmatrix} \begin{Bmatrix} \dot{\mathbf{u}}(t) \\ \dot{\boldsymbol{\eta}}(t) \end{Bmatrix} + \begin{bmatrix} \mathrm{K}^{\mathrm{E}} & -\Theta C_p^{\mathrm{S}} \mathrm{B}_v \mathrm{G}_c \\ -\mathrm{G}_c' \mathrm{B}_v' C_p^{\mathrm{S}} \Theta' & \mathrm{K}_c - \mathrm{G}_c' \mathrm{B}_v' C_p^{\mathrm{S}} \mathrm{B}_v \mathrm{G}_c \end{bmatrix}$$
$$\times \begin{Bmatrix} \mathbf{u}(t) \\ \boldsymbol{\eta}(t) \end{Bmatrix} = \begin{bmatrix} \Phi_f \\ 0 \end{bmatrix} \mathbf{f}(t). \tag{10.53}$$

Comparing equation (10.53) with equation (10.33), we note that there are similarities in the form of the expression. Both expressions are symmetric second-order systems if the matrices of the control law are chosen to be symmetric. Also, the coupling of the control system to the structure occurs due to the coupling introduced in the stiffness matrix. One difference between the two expressions is the fact that the charge feedback produces an additional term in the lower right-hand partition of the closed-loop stiffness matrix. This term does not appear in the analysis of a structural system with positive-position feedback.

Example 10.3 Using the parameters from Example 10.1, write the equations of motion for the system with a second-order feedback compensator of the form

$$m_c\ddot{\eta}(t) + d_c\dot{\eta}(t) + k_c\eta(t) = g_c q(t).$$

Plot the frequency response of the closed-loop system for the parameters $m_c = 1$, $d_c = 2828$, $k_c = 50 \times 10^6$, and $g_c = 39 \times 10^6$.

Solution The closed-loop equations of motion are expressed in equation (10.53) for the general second-order control system. The open-loop equations of motion are written from Example 10.2:

$$\begin{bmatrix} 0.5 & 0 \\ 0 & 0.5 \end{bmatrix} \begin{pmatrix} \ddot{u}_1 \\ \ddot{u}_2 \end{pmatrix} + \begin{bmatrix} 61 \times 10^6 & -30 \times 10^6 \\ -30 \times 10^6 & 30 \times 10^6 \end{bmatrix} \begin{pmatrix} u_1 \\ u_2 \end{pmatrix} = \begin{bmatrix} 0 \\ 1 \end{bmatrix} f_e + \begin{bmatrix} -0.6240 \\ 0 \end{bmatrix} v.$$

Combining the open-loop equations with the control system yields

$$\begin{bmatrix} 0.5 & 0 & 0 \\ 0 & 0.5 & 0 \\ 0 & 0 & m_c \end{bmatrix} \begin{pmatrix} \ddot{u}_1 \\ \ddot{u}_2 \\ \ddot{\eta} \end{pmatrix} + \begin{bmatrix} 0 & 0 & 0 \\ 0 & 0 & 0 \\ 0 & 0 & d_c \end{bmatrix} \begin{pmatrix} \dot{u}_1 \\ \dot{u}_2 \\ \dot{\eta} \end{pmatrix}$$

$$+ \begin{bmatrix} 61 \times 10^6 & -30 \times 10^6 & 0.6240 g_c \\ -30 \times 10^6 & 30 \times 10^6 & 0 \\ 0.6240 g_c & 0 & k_c - \left(13.9 \times 10^{-9}\right) g_c^2 \end{bmatrix} \begin{pmatrix} u_1 \\ u_2 \\ \eta \end{pmatrix} = \begin{bmatrix} 0 \\ 1 \\ 0 \end{bmatrix} f_e.$$

Substituting the parameters into the closed-loop equations of motion yields

$$\mathbf{M}_{\text{cl}} = \begin{bmatrix} 0.5 & 0 & 0 \\ 0 & 0.5 & 0 \\ 0 & 0 & 1 \end{bmatrix}$$

$$\mathbf{D}_{\text{cl}} = \begin{bmatrix} 0 & 0 & 0 \\ 0 & 0 & 0 \\ 0 & 0 & 2828.4 \end{bmatrix}$$

$$\mathbf{K}_{\text{cl}} = \begin{bmatrix} 61.0000 & -30.0000 & 24.3360 \\ -30.0000 & 30.0000 & 0 \\ 24.3360 & 0 & 28.8642 \end{bmatrix} \times 10^6.$$

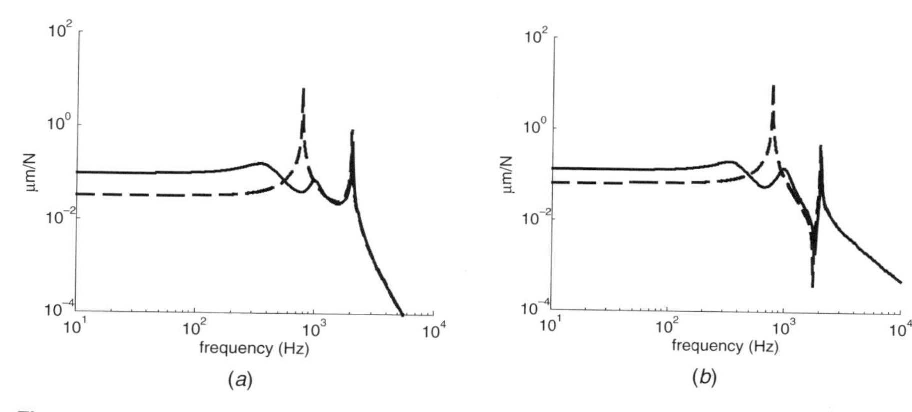

Figure 10.4 Open-loop (dashed) and closed-loop (solid) frequency response for the system studied in Example 10.3. (*a*) u_1/f_e; (*b*) u_2/f_e.

The closed-loop frequency response is computed from the expression

$$\begin{pmatrix} u_1(j\omega) \\ u_2(j\omega) \\ \eta(j\omega) \end{pmatrix} = (\mathrm{K_{cl}} - \omega^2\mathrm{M_{cl}} + j\omega\mathrm{D_{cl}}) \begin{bmatrix} 0 \\ 1 \\ 0 \end{bmatrix} f_e(j\omega).$$

The frequency response computed with this expression is shown in Figure 10.4 along with the open-loop frequency response. The result demonstrates the reduction in peak response due to feedback control. The reduction is similar for the closed-loop frequency response functions at both locations. Also, the frequency response illustrates the similarity between this type of active feedback control and the passive methods studied earlier. The closed-loop frequency response exhibits the double peak that is characteristic of passive methods of vibration suppression such as inductive–resistive shunts. The similarity is not simply coincidence. Comparing the closed-loop equations of motion with the equations of motion for a resistive–inductive shunt, we will see that the equations of motion have the same form.

10.3.2 Self-Sensing Actuation

A clever way to circumvent this difference in the closed-loop equations of motion is to use a concept called *self-sensing actuation*, or, *simultaneous sensing and actuation*, using the piezoelectric material. Self-sensing actuation is a concept by which the same piece of piezoelectric material is used simultaneously as a sensor and an actuator. How is this done? Consider the expression for the charge output, equation (10.2), due to the application of an external voltage and the motion of the structure. Linear piezoelectric theory predicts that the charge output will be a linear combination of two terms, one due to the applied voltage and the second due to the structural motion. Since the applied voltage is a prescribed variable, we can form a measured output that consists of the measured charge, $\mathbf{q}(t)$, and a term that will eliminate the dependence on the applied voltage. Specifically, if we form the measured output

$$\mathbf{q}_f(t) = \mathrm{B}_v'\mathbf{q}(t) - \mathrm{B}_v'\mathrm{C}_p^{\mathrm{S}}\mathrm{B}_v\mathbf{v}(t), \tag{10.54}$$

then combining with equation (10.2) yields

$$\mathbf{q}_f(t) = \mathrm{B}_v'\mathrm{C}_p^{\mathrm{S}}\Theta'\mathbf{u}(t). \tag{10.55}$$

The key difference between the actual charge, $\mathbf{q}(t)$, and the charge used for feedback, $\mathbf{q}_f(t)$, is that the charge used for feedback is only a function of the structural motion. Thus, we have created a signal that is directly correlated with the structural motion and is not a function of the applied voltage. Self-sensing actuation can be implemented by combining the piezoelectric electric material with a bridge circuit that adds the term associated with the capacitance into the measured output. The additional bridge circuit uses a capacitor of the same value as the piezoelectric element to eliminate the term due to the applied voltage in the output signal.

Assuming that we are using self-sensing actuation for feedback, we can form the feedback control law

$$\mathrm{M}_c\ddot{\boldsymbol{\eta}}(t) + \mathrm{D}_c\dot{\boldsymbol{\eta}}(t) + \mathrm{K}_c\boldsymbol{\eta}(t) = \mathrm{G}_c'\mathbf{q}_f(t). \tag{10.56}$$

When this control law is combined with the structural equations of motion, the matrix expressions for the closed-loop system are

$$\begin{bmatrix} \mathrm{M} & 0 \\ 0 & \mathrm{M}_c \end{bmatrix} \begin{Bmatrix} \ddot{\mathbf{u}}(t) \\ \ddot{\boldsymbol{\eta}}(t) \end{Bmatrix} + \begin{bmatrix} 0 & 0 \\ 0 & \mathrm{D}_c \end{bmatrix} \begin{Bmatrix} \dot{\mathbf{u}}(t) \\ \dot{\boldsymbol{\eta}}(t) \end{Bmatrix} + \begin{bmatrix} \mathrm{K}^{\mathrm{E}} & -\Theta\mathrm{C}_p^{\mathrm{S}}\mathrm{B}_v\mathrm{G}_c \\ -\mathrm{G}_c'\mathrm{B}_v'\mathrm{C}_p^{\mathrm{S}}\Theta' & \mathrm{K}_c \end{bmatrix} \begin{Bmatrix} \mathbf{u}(t) \\ \boldsymbol{\eta}(t) \end{Bmatrix}$$
$$= \begin{bmatrix} \Phi_f \\ 0 \end{bmatrix} \mathbf{f}(t). \tag{10.57}$$

Using self-sensing feedback, we see that equation (10.57) is identical to the expressions derived for position feedback in Section 10.3.1.

Self-sensing actuation is a novel method for creating a virtual displacement sensor using a piezoelectric material that is also being used as an actuator. One of the difficulties with the method is that it requires accurate knowledge of the piezoelectric material properties to be effective. For example, let us assume that the term added to the charge output is equal to

$$\mathbf{q}_f(t) = \mathrm{B}_v'\mathbf{q}(t) - (1+\mu)\mathrm{B}_v'\mathrm{C}_p^{\mathrm{S}}\mathrm{B}_v\mathbf{v}(t), \tag{10.58}$$

where μ represents a mistuning parameter due to the uncertainty in knowledge of the piezoelectric properties. A value of $\mu = 0$ represents perfect knowledge of the piezoelectric properties, whereas a nonzero value indicates that there is some mismatch between the actual capacitance of the piezoelectric material and the capacitance used to eliminate the voltage term in the charge output. Combining equation (10.58) with the expression for the charge, equation (10.2), produces the expression

$$\mathbf{q}_f(t) = \mathrm{B}_v' \mathrm{C}_p^S \Theta' \mathbf{u}(t) - \mu \mathrm{B}_v' \mathrm{C}_p^S \mathrm{B}_v \mathbf{v}(t). \tag{10.59}$$

Equation (10.59) illustrates that any mistuning in the capacitance of the self-sensing actuator bridge circuit will reintroduce a dependence on the applied voltage into the charge signal used for feedback. Regrouping equation (10.59) into

$$\mathbf{q}_f(t) = \mathrm{B}_v' \mathrm{C}_p^S [\Theta' \mathbf{u}(t) - \mu \mathrm{B}_v \mathbf{v}(t)], \tag{10.60}$$

we can see that if the mistuning is large enough such that

$$|\Theta' \mathbf{u}(t)| << |\mu \mathrm{B}_v \mathbf{v}(t)|, \tag{10.61}$$

the signal $\mathbf{q}_f(t)$ will be dominated by feedthrough of the applied voltage and will not be correlated with displacement of the structure. In this situation the self-sensing actuation will not be very effective as a method for measuring the displacement of the structure using the same piezoelectric.

Example 10.4 Compute the frequency response between the output variable q_f and the input voltage for a self-sensing feedback loop with $\mu = 0, -1$, and 1 for the representative system studied in this chapter. Compare the result to the the output $u_1(j\omega)/v(j\omega)$.

Solution The equation for the charge is obtained by transforming equation (10.2) into the frequency domain,

$$q(j\omega) = \mathrm{C}_p^S B_v \mathbf{v}(j\omega) + \mathrm{C}_p^S \Theta' \mathbf{u}(j\omega),$$

where the frequency response $u(j\omega)$ was computed in Example 10.2:

$$\mathbf{u}(j\omega) = \frac{1}{0.25\omega^4 - (45.5 \times 10^6)\omega^2 + 930 \times 10^{12}} \begin{bmatrix} (-0.6240)(30 \times 10^6 - 0.5\omega^2) \\ -18.72 \times 10^6 \end{bmatrix} \times v(j\omega).$$

Combining the preceding two expressions yields

$$q(j\omega) = 13.9 \times 10^{-9} \left[1 - \frac{(27.8 \times 10^6)(30 \times 10^6 - 0.5\omega^2)}{0.25\omega^4 - (45.5 \times 10^6)\omega^2 + 930 \times 10^{12}} \right] v(j\omega).$$

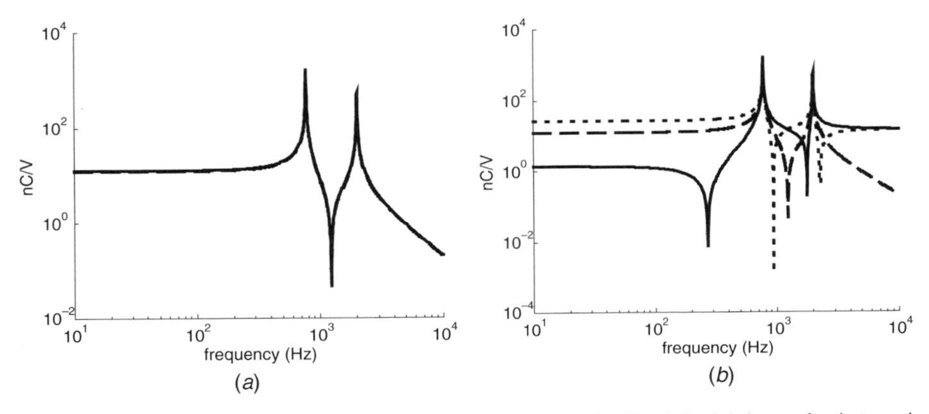

Figure 10.5 Frequency response of the self-sensing charge feedback for (*a*) the perfectly tuned case and (*b*) the case where the bridge circuit is mistuned.

The expression for the self-sensing charge feedback is obtained from equation (10.58):

$$q_f(j\omega) = 13.9 \times 10^{-9}\left[1 - \frac{(27.8 \times 10^6)(30 \times 10^6 - 0.5\omega^2)}{0.25\omega^4 - (45.5 \times 10^6)\omega^2 + 930 \times 10^{12}}\right]$$

$$\times\, v(j\omega) - (1+\mu)13.9 \times 10^{-9} v(j\omega)$$

$$= -13.9 \times 10^{-9}\left[\frac{(27.8 \times 10^6)(30 \times 10^6 - 0.5\omega^2)}{0.25\omega^4 - (45.5 \times 10^6)\omega^2 + 930 \times 10^{12}} + \mu\right] v(j\omega).$$

Plotting the frequency response for $\mu = 0$ and comparing it to the frequency response between u_1 and the input voltage demontrates that the self-sensing charge feedback is correlated with displacement for a perfectly tuned bridge circuit, as shown in Figure 10.5*a*. Mistuning of the circuit produces a variation in the self-sensing circuit output and changes the zeros associated with the transfer function. This is illustrated in Figure 10.5*b*, but can be inferred from the preceding expression through the fact that the parameter μ does not change the poles of the transfer function. It acts only as a direct feedthrough term that changes the transfer function zeros. Physically, this is representative of the fact that mistuning of the circuit produces a feedthrough between the applied voltage and the charge output q_f.

10.4 DISTRIBUTED SENSING

In Chapter 4 we analyzed the concept of shaped sensors and actuators using piezoelectric material. In the analysis we demonstrated that shaping the transducer (or electrode) according to the mode shape of the structure produced a transducer that would measure or actuate only a single structure mode. The discussion in previous sections highlights the importance of this technique and illustrates that a sensor or actuator that is coupled to only a single mode will simplify the design of feedback control systems. In a number of applications, though, it is difficult to shape a sensor or actuator such that it measures only a single mode. For systems that consist of beam and plate elements, it may be possible, and this technique has been shown to be effective in certain applications in noise and vibration control. For more geometrically complex structures, it may be difficult to measure accurately enough the mode shape to enable shaped sensing and actuation.

In the absence of the ability to shape a transducer to measure a single mode, there are additional means of achieving *distributed sensing or actuation*. One of the benefits afforded by smart material systems is the ability to place potentially a large number of sensors or actuators on the structure to achieve a distribution of sensing or control authority. In this section we develop generalized methods for designing control laws that enable distributed sensing and actuation.

Consider the model derived for a piezoelectric element coupled to a vibrating structure shown in equation (10.3). Let us assume that there are n_q piezoelectric elements to be used as sensors, for which the expression for the charge output is shown in equation (10.2). Let us substitute the modal expansion $\mathbf{u}(t) = \mathbf{Pr}(t)$ into the equations of motion and assume that the sensors are in a short-circuit condition, or equivalently, that we are using the concept of self-sensing actuation to eliminate the dependence between the charge output and the applied voltage. Under these assumptions the expression for the charge output is

$$\mathbf{q}(t) = \mathrm{C}_p^{\mathrm{S}}\Theta'\mathbf{Pr}(t). \tag{10.62}$$

Assume that we would like to control a certain number of structural modes that are equal to the number of charge sensors that we have for measurement, n_q. Furthermore, let us separate the modal expansion into those modes that are to be controlled, $\mathbf{r}_c(t)$, and those that are not to be controlled, $\mathbf{r}_u(t)$, where the size of $\mathbf{r}_c(t)$ is $n_q \times 1$ and the size of $\mathbf{r}_u(t)$ is $(n_r - n_q) \times 1$, where n_r is the size of the modal expansion. Denoting $\Phi_q = \mathrm{C}_p^{\mathrm{S}}\Theta'\mathrm{P}$, we can write the expression as

$$\mathbf{q}(t) = \Phi_{qc}\mathbf{r}_c(t) + \Phi_{qu}\mathbf{r}_u(t). \tag{10.63}$$

Assuming that Φ_{qc} is not a singular matrix, we can *prefilter* the charge signal by writing

$$\mathbf{q}_f(t) = \Phi_{qc}^{-1}\mathbf{q}(t), \tag{10.64}$$

resulting in an expression for the feedback signal

$$\mathbf{q}_f(t) = \mathbf{Ir}_c(t) + \Phi_{qc}^{-1}\Phi_{qu}\mathbf{r}_u(t). \tag{10.65}$$

Let us expand equation (10.65) to illustrate the point more strongly. The filtered charge output is expressed as

$$\begin{Bmatrix} q_{f1}(t) \\ \vdots \\ q_{fn_q} \end{Bmatrix} = \begin{bmatrix} 1 & \cdots 0 \cdots & 0 \\ \cdots 0 \cdots & 1 & \cdots 0 \cdots \\ 0 & \cdots 0 \cdots & 1 \end{bmatrix} \begin{Bmatrix} r_{c1}(t) \\ \vdots \\ r_{cn_q} \end{Bmatrix} + \begin{bmatrix} \odot & \odot & \odot \\ \vdots & \ddots & \vdots \\ \odot & \odot & \odot \end{bmatrix} \mathbf{r}_u(t), \tag{10.66}$$

where $\odot$ represents a (potentially) nonzero matrix element.

The derivation of equation (10.66) demonstrates that the use of multiple transducers to mimic a distributed sensor does not produce a perfect modal measurement. We see from the analysis that n_q sensors will enable the ability to measure n_q individual modes. Any modes not in the set of controlled modes will be present in the sensor signal as shown in equation (10.66).

10.5 STATE-SPACE CONTROL METHODOLOGIES

The control methodologies discussed in Section 10.4 are represented by ordinary differential equations that are second order in time. For structural systems this representation is often very convenient because the mass, damping, and stiffness terms can be readily identified from the equations of motion. In the analysis of feedback control systems the second-order form is useful because it provides insight between the type of feedback control and the impact on the physical properties of the structural system. For example, it is clear from the analysis of positive-position feedback that the stiffness matrix of the system is affected by the feedback control for this type of compensator.

Although second-order form has certain advantages for the design of structural control systems, the majority of systematic methods for feedback control design utilize first-order, or state-space, representation of the open-loop system. A full discussion of all of the design methods for state-space analysis is well beyond the scope of this book (and is described in a number of separate textbooks), but in this section we provide an introduction to the use of state-space models for structural control design utilizing the models of piezoelectric material systems derived earlier. Specifically, we focus on methods for transforming second-order models to state-space form and methods for designing full-state feedback control laws with state estimation.

10.5.1 Transformation to First-Order Form

The fundamental properties of a state-space model of a linear time-invariant system were introduced in Section 3.3. The use of state-space models for control design require that we transform the second-order equations of motion for the piezoelectric material system, equation (10.3), into the first-order form of a state-space model. Beginning with equation (10.3), we first solve for the acceleration term as a function of the displacement and input forces:

$$\ddot{\mathbf{u}}(t) = -M_s^{-1}(\mathrm{K}_s + \mathrm{K}^{\mathrm{E}})\mathbf{u}(t) + M_s^{-1}\mathrm{B}_f\mathbf{f}(t) + M_s^{-1}\mathrm{C}_p^{\mathrm{S}}\Theta'\mathrm{B}_v\mathbf{v}(t). \tag{10.67}$$

Making the substitutions

$$\begin{aligned} \mathbf{z}_1(t) &= \mathbf{u}(t) \\ \mathbf{z}_2(t) &= \dot{\mathbf{u}}(t) \end{aligned} \tag{10.68}$$

into equation (10.67) yields

$$\dot{\mathbf{z}}_2(t) = -M_s^{-1}\left(\mathrm{K}_s + \mathrm{K}^{\mathrm{E}}\right)\mathbf{z}_1(t) + M_s^{-1}\mathrm{B}_f\mathbf{f}(t) + M_s^{-1}\mathrm{C}_p^{\mathrm{S}}\Theta'\mathrm{B}_v\mathbf{v}(t). \tag{10.69}$$

Combining with the definition of the state variables

$$\dot{\mathbf{z}}_1(t) = \mathbf{z}_2(t), \tag{10.70}$$

produces a matrix set of equations:

$$\begin{Bmatrix} \dot{\mathbf{z}}_1(t) \\ \dot{\mathbf{z}}_2(t) \end{Bmatrix} = \begin{bmatrix} 0 & \mathrm{I} \\ -M_s^{-1}\left(\mathrm{K}_s + \mathrm{K}^{\mathrm{E}}\right) & 0 \end{bmatrix} \begin{Bmatrix} \mathbf{z}_1(t) \\ \mathbf{z}_2(t) \end{Bmatrix} + \begin{Bmatrix} 0 \\ M_s^{-1}\mathrm{B}_f \end{Bmatrix}\mathbf{f}(t) + \begin{Bmatrix} 0 \\ M_s^{-1}\mathrm{C}_p^{\mathrm{S}}\Theta'\mathrm{B}_v \end{Bmatrix}\mathbf{v}(t). \tag{10.71}$$

Equation (10.71) assumes that the damping of the uncontrolled system is neglegible. If a viscous structural damping term is included in the second-order model, the state equations for the first-order system are

$$\begin{Bmatrix} \dot{\mathbf{z}}_1(t) \\ \dot{\mathbf{z}}_2(t) \end{Bmatrix} = \begin{bmatrix} 0 & \mathrm{I} \\ -M_s^{-1}\left(\mathrm{K}_s + \mathrm{K}^{\mathrm{E}}\right) & -M_s^{-1}\mathrm{D}_s \end{bmatrix} \begin{Bmatrix} \mathbf{z}_1(t) \\ \mathbf{z}_2(t) \end{Bmatrix} + \begin{Bmatrix} 0 \\ M_s^{-1}\mathrm{B}_f \end{Bmatrix}\mathbf{f}(t) + \begin{Bmatrix} 0 \\ M_s^{-1}\mathrm{C}_p^{\mathrm{S}}\Theta'\mathrm{B}_v \end{Bmatrix}\mathbf{v}(t). \tag{10.72}$$

Equations (10.71) and (10.72) represent the relationship between the state vectors and the inputs as represented in first-order form for both undamped and damped structural systems. A state-space model also requires that we define the outputs of the system to completely define the input–output relationships. In general, the outputs of the state model must be represented as a linear combination of the states and, in certain circumstances, the inputs. The exact definition of the outputs is dependent on the specific problem under consideration, but let us define two sets of output expressions: the first, denoted $\mathbf{w}(t)$, which defines the *performance* outputs. The performance outputs are a set of states or input functions that we would like to observe to assess the quality of our control system. For example, it is often the case that we would like to observe a linear combination of displacements, velocities, and the voltage input to the piezoelectric elements. In this case let us define the performance outputs as

$$\mathbf{w}(t) = \begin{bmatrix} H_d\mathbf{u}(t) + H_v\dot{\mathbf{u}}(t) \\ \mathbf{v}(t) \end{bmatrix}. \tag{10.73}$$

Writing the performance outputs in matrix form in terms of the defined states produces

$$\mathbf{w}(t) = \begin{bmatrix} H_d & H_v \\ 0 & 0 \end{bmatrix} \begin{Bmatrix} \mathbf{z}_1(t) \\ \mathbf{z}_2(t) \end{Bmatrix} + \begin{bmatrix} 0 \\ \mathrm{I} \end{bmatrix} \mathbf{v}(t). \tag{10.74}$$

The second set of outputs required to define the state model are the *control* outputs $\mathbf{y}$(t). Once again, the exact definition of the control outputs is problem dependent, but common definitions of the control outputs are a linear combination of the states:

$$\mathbf{y}(t) = \begin{bmatrix} H_d & H_v \\ 0 & 0 \end{bmatrix} \begin{Bmatrix} \mathbf{z}_1(t) \\ \mathbf{z}_2(t) \end{Bmatrix}, \tag{10.75}$$

or it is often the case that control outputs are the charge signals of the piezoelectric elements. In the case of collocated charge feedback, the control output is

$$\mathbf{y}(t) = \mathrm{B}_v'\mathbf{q}(t) = \mathrm{B}_v'\mathrm{C}_p^{\mathrm{S}}\Theta'\mathbf{u}(t) + \mathrm{B}_v'\mathrm{C}_p^{\mathrm{S}}\mathrm{B}_v\mathbf{v}(t). \tag{10.76}$$

Writing equation (10.76) in terms of the state variables and placing into matrix form yields

$$\mathbf{y}(t) = \begin{bmatrix} \mathrm{B}_v'\mathrm{C}_p^{\mathrm{S}}\Theta' & 0 \end{bmatrix} \begin{Bmatrix} \mathbf{z}_1(t) \\ \mathbf{z}_2(t) \end{Bmatrix} + \mathrm{B}_v'\mathrm{C}_p^{\mathrm{S}}\mathrm{B}_v\mathbf{v}(t). \tag{10.77}$$

It is important to emphasize at this point that we have done nothing more than transform the second-order equations of motion into first-order form and define them in terms of the state variable representation for linear time-invariant systems. To complete the transformation, let us define the state vector

$$\mathbf{z}(t) = \begin{Bmatrix} \mathbf{z}_1(t) \\ \mathbf{z}_2(t) \end{Bmatrix} \tag{10.78}$$

and rewrite the state equations as

$$\begin{aligned} \dot{\mathbf{z}}(t) &= \mathrm{A}\mathbf{z} + \mathrm{B}_e\mathbf{f}(t) + \mathrm{B}_c\mathbf{v}(t) \\ \mathbf{w}(t) &= \mathrm{C}_{\mathrm{w}}\mathbf{z}(t) + \mathrm{D}_{\mathrm{we}}\mathbf{f}(t) + \mathrm{D}_{\mathrm{wc}}\mathbf{v}(t) \\ \mathbf{y}(t) &= \mathrm{C}_{\mathrm{y}}\mathbf{z}(t) + \mathrm{D}_{\mathrm{ye}}\mathbf{f}(t) + \mathrm{D}_{\mathrm{yc}}\mathbf{v}(t). \end{aligned} \tag{10.79}$$

Example 10.5 Transform the second-order equations studied in Example 10.1 into first-order form. Write out the state matrices and the input matrices for the system.

Solution Transforming the second-order equations of motion according to equation (10.67), gives us

$$\ddot{\mathbf{u}}(t) = -\begin{bmatrix} 2 & 0 \\ 0 & 2 \end{bmatrix} \begin{bmatrix} 61 \times 10^6 & -30 \times 10^6 \\ -30 \times 10^6 & 30 \times 10^6 \end{bmatrix} \mathbf{u}(t) + \begin{bmatrix} 2 & 0 \\ 0 & 2 \end{bmatrix} \begin{bmatrix} 0 \\ 1 \end{bmatrix} f_e(t) + \begin{bmatrix} 2 & 0 \\ 0 & 2 \end{bmatrix} \begin{bmatrix} -0.6240 \\ 0 \end{bmatrix} v(t).$$

Performing the multiplications results in:

$$\ddot{\mathbf{u}}(t) = \begin{bmatrix} -122 \times 10^6 & 60 \times 10^6 \\ 60 \times 10^6 & -60 \times 10^6 \end{bmatrix} \mathbf{u}(t) + \begin{bmatrix} 0 \\ 2 \end{bmatrix} f_e(t) + \begin{bmatrix} -1.248 \\ 0 \end{bmatrix} v(t).$$

Placing the equations into first-order form yields

$$\mathrm{A} = \begin{bmatrix} 0 & 0 & 1 & 0 \\ 0 & 0 & 0 & 1 \\ -122 \times 10^6 & 60 \times 10^6 & 0 & 0 \\ 60 \times 10^6 & -60 \times 10^6 & 0 & 0 \end{bmatrix}$$

$$\mathrm{B}_{\mathrm{zf}} = \begin{bmatrix} 0 \\ 0 \\ 0 \\ 2 \end{bmatrix}$$

$$\mathrm{B}_{\mathrm{zv}} = \begin{bmatrix} 0 \\ 0 \\ -1.248 \\ 0 \end{bmatrix}.$$

10.5.2 Full-State Feedback

One of the most important reasons for transforming the second-order equations of motion to first-order form is that we can take advantage of a (very) large number of systematic design methodologies. In this chapter we apply design methodologies that enable us systematically to design vibration control systems. The first methodology that we study is *full-state feedback*. As the name implies, full-state feedback is based on the assumption that the control input, $\mathbf{v}(t)$, is related to the states of the system. For the moment, assume that there is only a single control input $v(t)$, and this control input is expressed as

$$v(t) = -\mathbf{g}'\mathbf{z}(t), \tag{10.80}$$

where $\mathbf{g}$ is a $n \times 1$ vector of control gains. The objective of the control design is to choose a set of gains that achieve a desired control objective. Substituting equation (10.80) into the state equations yields

$$\dot{\mathbf{z}}(t) = (\mathrm{A} - \mathrm{B}_c\mathbf{g}')\mathbf{z}(t) + \mathrm{B}_e\mathbf{f}(t). \tag{10.81}$$

Equation (10.81) illustrates that the closed-loop state matrix assuming full-state feedback is equal to $\mathrm{A} - \mathrm{B}_c\mathbf{g}'$. We know that the characteristic equation for the closed-loop system, denoted $\xi_{cl}(\mathsf{s})$, is equivalent to

$$\xi_{cl}(\mathsf{s}) = |\mathsf{s}\mathrm{I} - \mathrm{A} + \mathrm{B}_c\mathbf{g}'|. \tag{10.82}$$

The characteristic equation can be written as an nth-order polynomial of the form

$$\xi_{cl}(\mathsf{s}) = \mathsf{s}^n + a_1\mathsf{s}^{n-1} + \cdots + a_{n-1}\mathsf{s} + a_n. \tag{10.83}$$

The coefficients a_i are computed from the determinant $|\mathsf{s}\mathrm{I} - \mathrm{A} + \mathrm{B}_c\mathbf{g}'|$. In general, each coefficient is a function of the n unknown gains from the gain vector $\mathbf{g}$.

The feedback gains are computed by first assuming that we have a desired characteristic equation, $\xi_d(\mathsf{s})$, which is expressed as

$$\xi_d(\mathsf{s}) = (\mathsf{s} - \lambda_{d1})(\mathsf{s} - \lambda_{d2})\cdots(\mathsf{s} - \lambda_{dn}), \tag{10.84}$$

where λ_{di} are desired closed-loop eigenvalues. Judicious choice of these eigenvalues will be discussed shortly, but for now assume that the desired characteristic equation is expanded to the form

$$\xi_d(\mathsf{s}) = \mathsf{s}^n + b_1\mathsf{s}^{n-1} + \cdots + b_{n-1}\mathsf{s} + b_n. \tag{10.85}$$

To ensure that the closed-loop system has the desired eigenvalues, we must equate the coefficients of the polynomials expressed in equations (10.83) and (10.85). Writing the n equations as

$$\begin{aligned} a_1(g_1, g_2, \ldots, g_n) &= b_1 \\ a_2(g_1, g_2, \ldots, g_n) &= b_2 \\ &\vdots \\ a_n(g_1, g_2, \ldots, g_n) &= b_n, \end{aligned} \tag{10.86}$$

it is clear that solving for the desired control gains is equivalent to solving n equations for the n unknown gains g_i. If the equations can be solved for the control gains, it is *guaranteed* that the closed-loop system will have the desired eigenvalues λ_{di}.

The question of whether or not the n equations in equation (10.86) can be solved for the desired control gains is answered by considering the concept of the controllability of the system. A system is said to be *controllable* if and only if there is a control input that will take the system from any initial condition to any desired position in the state space in a finite amount of time. Derivations of the controllability conditions can be found in several textbooks. For our purposes we introduce the concept of controllability in relation to the solution of equation (10.86). If a system is controllable for the control input vector B_c, the n equations can be solved for the n control gains g_i. The controllability for a linear time invariant system is determined by finding the rank of the matrix

$$\mathcal{C} = [\mathrm{A} \quad \mathrm{AB_c} \quad \mathrm{A^2B_c} \cdots \mathrm{A^{n-1}B_c}]. \tag{10.87}$$

If rank$(\mathcal{C}) = n$, the system is said to be controllable. If rank$(\mathcal{C}) < n$, the system is said to be *uncontrollable*. For systems that are uncontrollable from the input vector B_c, the equations for the control gains will not yield a unique solution and the closed-loop system is not guaranteed to have the eigenvalues desired.

Full-state feedback is a very systematic approach to control design, due to the fact that the *eigenvalues of a controllable system can be placed anywhere* in the s-plane. Thus, the designer has a substantial amount of freedom in shaping the closed-loop response through judicious choice of the closed-loop poles. Since the closed-loop poles are related directly to the system response, the vibration control design can be cast within a very systematic framework.

Algorithms for computing the full-state feedback control gains are very well developed. Computer-aided design packages (e.g., MATLAB) have functions that will compute the full-state feedback control gains using the state representation of the open-loop system. The question of how to choose the gains requires some discussion of the basic properties of structural systems. If the open-loop system is assumed to be undamped, which is not strictly correct but a very good assumption for systems with light damping, recall that the natural frequencies of the system will be real valued and can be ordered in such a way that

$$\omega_{n1}^2 \leq \omega_{n2}^2 \leq \omega_{n3}^3 \cdots. \tag{10.88}$$

Under this assumption, the eigenvalues of the first-order system will be equal to a set of complex-conjugate pairs:

$$\begin{aligned} \lambda_{1,2} &= \pm j\omega_{n1} \\ \lambda_{2,3} &= \pm j\omega_{n2} \\ &\vdots \end{aligned} \tag{10.89}$$

The arrangement of eigenvalues is visualized in the real and imaginary plane in the manner shown in Figure 10.6. Let us assume that each eigenvalue pair is "moved" to a desired location

$$\lambda_d = -\zeta\omega_d \pm j\omega_d\sqrt{1-\zeta^2}, \tag{10.90}$$

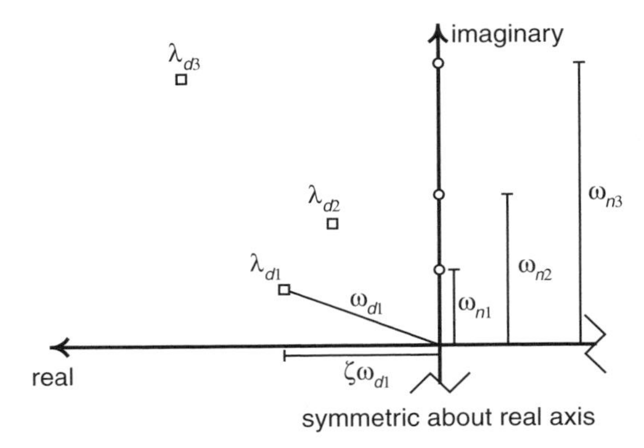

Figure 10.6 Visualization of the open- and closed-loop poles of a structural system.

where $\zeta < 1$. This concept is shown in Figure 10.6. We know that taking the inverse Laplace transform of each individual closed-loop eigenvalue will produce a decaying oscillatory response whose period is related to $\omega_d\sqrt{1-\zeta^2}$ and whose exponential decay is related to $\zeta\omega_d$. Increasing $\zeta\omega_d$ will yield a response that decays more quickly, and increasing the magnitude of $\omega_d\sqrt{1-\zeta^2}$ will decrease the period of the response.

Using this concept, we can envision two straightforward methods for choosing the eigenvalues of the closed-loop system. One approach would be to move the poles along a circle whose radius is equal to the open-loop natural frequency of the system. In this method the period of the open- and closed-loop response remains the same but the decay rate of the response increases (i.e., the response decays faster) as the angle that the eigenvalue makes with the imaginary axis increases. This method for choosing the poles is represented in Figure 10.7.

An alternative approach to choosing the closed-loop poles would be to move the poles farther into the left-half plane by increasing the real component of the eigenvalue

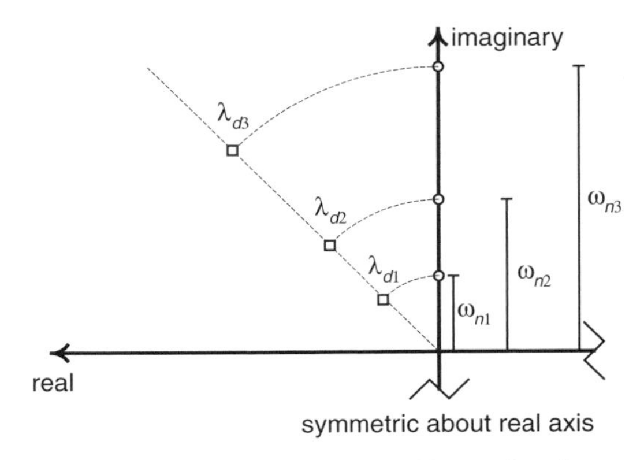

Figure 10.7 Visualization of the open- and closed-loop poles of a structural system.

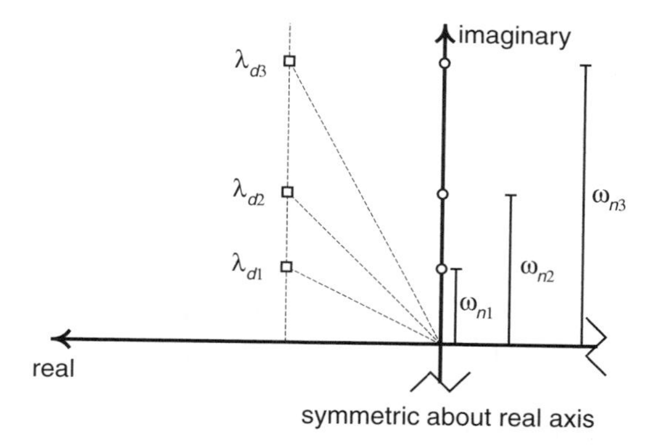

Figure 10.8 An alternative method for choosing closed-loop poles.

while leaving the imaginary component of the eigenvalue constant. This concept is visualized in Figure 10.8. If this approach is applied to multiple modes of a structure in which the real component of the eigenvalues are set to be constant for all the modes, the modal response will exhibit larger damping in the lower-frequency modes than in the higher-frequency modes. This can be visualized by examining the angle that the eigenvalue makes with the imaginary axis. For lower-frequency eigenvalues the angle will be larger than those at higher frequency. Larger angles will produce larger closed-loop damping values.

Example 10.6 Compute the full-state feedback gains so that the closed-loop eigenvalues of the system are

$$\begin{aligned}\lambda_{d1,d2} &= -1.26 \times 10^4 \pm j1.26 \times 10^4 \\ \lambda_{d3,d4} &= -0.484 \times 10^4 \pm j0.484 \times 10^4\end{aligned}$$

for the representative system studied in this chapter.

Solution The state matrices are obtained from Example 10.5:

$$\mathrm{A} = \begin{bmatrix} 0 & 0 & 1 & 0 \\ 0 & 0 & 0 & 1 \\ -122 \times 10^6 & 60 \times 10^6 & 0 & 0 \\ 60 \times 10^6 & -60 \times 10^6 & 0 & 0 \end{bmatrix}$$

$$\mathrm{B}_{\mathrm{zv}} = \begin{bmatrix} 0 \\ 0 \\ -1.248 \\ 0 \end{bmatrix}.$$

The closed-loop state matrix, assuming full-state feedback, is

$$\mathrm{A} - \mathrm{B}_{zf}\mathbf{g}' = \begin{bmatrix} 0 & 0 & 1 & 0 \\ 0 & 0 & 0 & 1 \\ -122 \times 10^6 & 60 \times 10^6 & 0 & 0 \\ 60 \times 10^6 & -60 \times 10^6 & 0 & 0 \end{bmatrix} - \begin{bmatrix} 0 \\ 0 \\ -1.248 \\ 0 \end{bmatrix} \begin{bmatrix} g_1 & g_2 & g_3 & g_4 \end{bmatrix}$$

$$= \begin{bmatrix} 0 & 0 & 1 & 0 \\ 0 & 0 & 0 & 1 \\ -122 \times 10^6 + 1.248 g_1 & 60 \times 10^6 + 1.248 g_2 & 1.248 g_3 & 1.248 g_4 \\ 60 \times 10^6 & -60 \times 10^6 & 0 & 0 \end{bmatrix}.$$

The characteristic equation of the closed-loop system is

$$\begin{aligned} \left|\mathrm{sI} - \mathrm{A} + \mathrm{B}_{zv}\mathbf{g}'\right| = {} & \mathrm{s}^4 - 0.6240 g_3 \mathrm{s}^3 \\ & + (-0.6240 g_1 + 0.1820 \times 10^9)\mathrm{s}^2 + (-0.3744 \times 10^8 g_3 - 0.3744 \times 10^8 g_4)\mathrm{s} \\ & + 0.3720 \times 10^{16} - 0.3744 \times 10^8 g_1 - 0.3744 \times 10^8 g_2. \end{aligned}$$

The characteristic desired equation is obtained from the expression

$$\begin{aligned} \xi_d(\mathrm{s}) &= (\mathrm{s} + 1.26 - j1.26)(\mathrm{s} + 1.26 + j1.26)(\mathrm{s} + 0.484 - j0.484)(\mathrm{s} + 0.484 + j0.484) \times 10^{16} \\ &= \mathrm{s}^4 + 15,520\mathrm{s}^3 + 1.204 \times 10^8 \mathrm{s}^2 - 1.8929 \times 10^{12}\mathrm{s} + 1.4876 \times 10^{16}. \end{aligned}$$

Equating the desired characteristic equation with the characteristic equation of the closed-loop state matrix yields a set of four equations and four unknowns:

$$\begin{aligned} -0.6240 g_3 &= 34{,}880 \\ -0.6240 g_1 + 0.1820 \times 10^9 &= 6.0831 \times 10^8 \\ -0.3744 \times 10^8 g_3 - 0.3744 \times 10^8 g_4 &= 4.2542 \times 10^{12} \\ 0.3720 \times 10^{16} - 0.3744 \times 10^8 g_1 - 0.3744 \times 10^8 g_2 &= 1.4876 \times 10^{16}. \end{aligned}$$

Solving the four equations for the four gains yields

$$\mathbf{g} = \begin{bmatrix} -6.8175 \times 10^8 \\ 3.8378 \times 10^8 \\ -5.5897 \times 10^4 \\ -5.7730 \times 10^4 \end{bmatrix}.$$

10.5.3 Optimal Full-State Feedback: Linear Quadratic Regulator Problem

In Section 10.5.2 the concept of full-state feedback was introduced in the context of designing control systems using a first-order representation of the smart structure. An important aspect of full-state feedback is that the closed-loop eigenvalues of the system can be chosen arbitrarily for controllable systems. This gives the designer substantial freedom in shaping the closed-loop response. Two methods for choosing the closed-loop eigenvalues were discussed in relation to the modal response of a system with underdamped modes.

In certain applications, choice of the closed-loop eigenvalues is complicated by the fact the goals of the control design are specified in terms of requirements related to the size of the closed-loop response. As an example, in many structural vibration suppression applications, the specifications of the closed-loop system are written in terms of the amount of vibration at particular critical points on the structure. Similarly, it is always the case that there are physical limitations to the size of the control voltage; therefore, there are always control specifications that limit the size of the control effort.

Choosing closed-loop eigenvalues that limit the size of the vibration at particular points or limit the control effort required is not always an easy task. The problem is complicated by the fact that there is not typically a clear relationship between the location of the closed-loop poles and structural vibration and control effort. Generally, a number of iterations are required to achieve a satisfactory design.

There are numerous optimal control methods that overcome the challenges associated with full-state feedback designs. An optimal control method is one in which the control gains are chosen to minimize a specified *cost function*. Many optimal control methods have been developed over the years that address various aspects of designing feedback control systems given specifications on the closed-loop system. A large number of these methodologies are derivatives of a technique called *linear quadratic regulator design*, in which the control gains are chosen to minimize the cost function

$$\mathcal{J} = \int_0^\infty [\mathbf{z}'(t)\mathrm{Q}\mathbf{z}(t) + \mathbf{v}'(t)R\mathbf{v}(t)]\, dt. \tag{10.91}$$

The cost function shown in equation (10.91) can be interpreted as the size of the closed-loop response. The first term,

$$\mathcal{J}_z = \int_0^\infty z(t)\mathrm{Q}\mathbf{z}(t)\, dt, \tag{10.92}$$

is interpreted as the size of the state response. More rigorously, equation (10.92) is equal to the mean-square state response. Similarly, the term

$$\mathcal{J}_v = \int_0^\infty \mathbf{v}'(t)R\mathbf{v}(t)\, dt \tag{10.93}$$

is the mean square of the voltage response or control effort. The great advantage of defining this cost function and using it as a means of control design is that there is an analytical solution for the control gains that minimize the cost function $\mathcal{J}$. Thus, in contrast to full-state feedback methods, in which the control gains are chosen based on the choice of the closed-loop eigenvalues, in optimal control methods the control gains are computed by first choosing the matrices Q and R that specify the cost function and then using an analytical solution to compute the control gains.

The solution for the control gains will be discussed shortly. First, though, let us discuss the physical significance of the cost function $\mathcal{J}$. A common type of cost function is to minimize a weighted sum of the system states. Recall from equation (10.68) that the states have been chosen to be a combination of the displacements and velocities of the structural system. As an example, consider a response cost function of the form

$$\mathcal{J}_z = \int_0^\infty \sum_i^N \alpha_i u_i^2(t)\, dt. \tag{10.94}$$

This cost function can be specified by partitioning the matrix Q into four submatrices:

$$\mathrm{Q} = \begin{bmatrix} \mathrm{Q}_{11} & 0 \\ 0 & 0 \end{bmatrix}, \tag{10.95}$$

where

$$\mathrm{Q}_{11} = \begin{bmatrix} \alpha_1 & 0 & \cdots & 0 \\ 0 & \alpha_2 & 0\cdots & 0 \\ \vdots & \vdots & \ddots & \vdots \\ 0 & 0 & \cdots & \alpha_N \end{bmatrix}. \tag{10.96}$$

Similarly, the response cost function

$$\mathcal{J}_z = \int_0^\infty \sum_i^N \alpha_i \dot{u}_i^2(t)\, dt \tag{10.97}$$

could be represented by the partitioned matrix

$$\mathrm{Q} = \begin{bmatrix} 0 & 0 \\ 0 & \mathrm{Q}_{11} \end{bmatrix}. \tag{10.98}$$

These two illustrations highlight the relationship between the choice of the cost function and the physical interpretation of the system response. The size of the control effort can be controlled by choosing the elements of the matrix R. For example, a control effort cost function of the form

$$\mathcal{J}_v = \int_0^\infty \sum_i^{n_v} \beta_i v_i^2(t)\, dt \tag{10.99}$$

will produce a control effort cost that is influenced by the response of each control voltage. The relative importance of the control voltage terms is determined by the choice of the coefficients β_i. The matrix R that reflects this cost function is

$$\mathrm{R} = \begin{bmatrix} \beta_1 & 0 & \cdots & 0 \\ 0 & \beta_2 & 0\cdots & 0 \\ \vdots & \vdots & \ddots & \vdots \\ 0 & 0 & \cdots & \beta_{n_v} \end{bmatrix}. \tag{10.100}$$

With the cost function matrices chosen in this manner, we can study the trade-offs associated with the optimal control design. The control gains computed will be a function of the $N + n_v$ weighting coefficients of the matrices Q and R. Although this could potentially represent a large number of iterations, it is good news that there are well-established numerical algorithms for computing the optimal control gains. Thus, iterations on the choice of the weighting functions can be performed rather quickly.

One of the most compelling reasons to use optimal control methods for the control design is that it enables a clear understanding of one of the fundamental trade-offs in feedback control. It is physically intuitive that there is generally a trade-off on the size of the closed-loop system response and the size of the control effort. Generally speaking, a larger amount of control effort is required to produce a greater reduction in the closed-loop response. This is an example of competing design specifications.

Linear quadratic regulator control theory provides a systematic method of understanding this fundamental trade-off. Continuing with the line of reasoning of the previous discussion, we see that increasing the values of the weighting coefficients α_i will probably produce an increase in the control effort required, while increasing the values of the coefficients β_i will produce a decrease in the control effort at the expense of an increase in the closed-loop system response.

This fundamental trade-off can be studied very concisely by choosing the matrices of the cost function to be

$$\mathrm{Q} = \alpha \mathrm{I} \qquad \mathrm{R} = \beta \mathrm{I}. \tag{10.101}$$

For this choice of the matrices, the cost function for the optimal control system reduces to

$$\mathcal{J} = \int_0^\infty \left\{ \alpha \left[\sum_{i=1}^{N} z_i^2(t) + \dot{z}_i^2(t) \right] + \beta \sum_{j=1}^{n_v} v_j^2(t) \right\} dt. \tag{10.102}$$

Physically, this represents a cost function in which all of the displacements and velocities of the structural system are weighted equally and all of the control voltages are weighted equally. The relative weighting between the system response and the control effort is a function of the ratio α/β. For large values of α/β, the optimal control system will tend to emphasize minimization of the system response at the expense of a large control effort. In contrast, small values of α/β will produce a closed-loop system in which minimizing the control effort is emphasized at the expense of large system response. With this choice of cost function, we see that the trade-off between system response and control effort can be reduced to the choice of a single design parameter.

This discussion highlights one of the main advantages of optimal control methods over pole placement methods for full-state feedback design. Assuming that we have chosen the weighting matrices in the manner described by equation (10.101), computation of the control gains can be reduced to the choice of a single parameter that has physical significance. In contrast, a pole placement approach would require the choice of $2N$ eigenvalues whose relationship to the physical response and control effort is not especially clear. With present-day computer-aided design packages, there is no compelling computational advantage of either approach, so generally, optimal control methods are deemed superior to pole placement methods.

Solutions for the optimal control problem are obtained by solving a set of algebraic equations for the optimal control gains. The control gains are obtained by solving the *algebraic Riccati equation*,

$$\mathrm{A}'\mathrm{X}_{\mathrm{lqr}} + \mathrm{X}_{\mathrm{lqr}}\mathrm{A} - \mathrm{X}_{\mathrm{lqr}}\mathrm{B}_c R^{-1}\mathrm{B}'\mathrm{X}_{\mathrm{lqr}} + \mathrm{Q} = 0. \tag{10.103}$$

The solution of this equation is a function of the state matrix and input vector as well as the choice of the weighting matrices for the LQR cost function. The solution to this equation has been implemented in a number of algorithms and can be solved using computer-aided design packages. The unknown matrix in equation (10.103) is $\mathrm{X}_{\mathrm{lqr}}$. Once $\mathrm{X}_{\mathrm{lqr}}$ has been obtained, the optimal control gains are obtained from the matrix equation

$$\mathrm{G} = R^{-1}\mathrm{B}_c'\mathrm{X}_{\mathrm{lqr}}. \tag{10.104}$$

Example 10.7 (a) Compute the LQR feedback gains using voltage as the control input for the representative system studied in this chapter. Assume the weighting matrices

$$\mathrm{Q} = 10^6\mathrm{I} = \mathrm{R} = 1.$$

(b) Write the state-space representation of the closed-loop system assuming that the three outputs are the displacements of mass 1 and mass 2 and the control voltage.

Solution (a) The state matrix is given by

$$\mathrm{A} = \begin{bmatrix} 0 & 0 & 1 & 0 \\ 0 & 0 & 0 & 1 \\ -122\times 10^6 & 60\times 10^6 & 0 & 0 \\ 60\times 10^6 & -60\times 10^6 & 0 & 0 \end{bmatrix}$$

and the input matrix is

$$B_{zv} = \begin{bmatrix} 0 \\ 0 \\ -1.248 \\ 0 \end{bmatrix}.$$

The LQR gains are computed using a numerical solver (e.g., MATLAB). First compute

$$B_{zv}R^{-1}B_{zv} = \begin{bmatrix} 0 \\ 0 \\ -1.248 \\ 0 \end{bmatrix} (1) \begin{bmatrix} 0 & 0 & -1.248 & 0 \end{bmatrix}$$

$$= \begin{bmatrix} 0 & 0 & 0 & 0 \\ 0 & 0 & 0 & 0 \\ 0 & 0 & 1.5575 & 0 \\ 0 & 0 & 0 & 0 \end{bmatrix}$$

Using a numerical solver to compute X_{lqr} yields

$$X_{lqr} = \begin{bmatrix} 1.2459 \times 10^{11} & -5.3702 \times 10^{10} & 4.5457 \times 10^{5} & 8.9497 \times 10^{5} \\ -5.3883 \times 10^{10} & 6.9741 \times 10^{10} & -4.9805 \times 10^{5} & -3.8856 \times 10^{5} \\ 5.7333 \times 10^{5} & -4.3549 \times 10^{5} & 1.1411 \times 10^{3} & 2.5895 \times 10^{2} \\ 9.5533 \times 10^{5} & -5.0861 \times 10^{5} & 2.5730 \times 10^{2} & 1.4229 \times 10^{3} \end{bmatrix}$$

The LQR control gains are computed using equation (10.104):

$$\mathbf{g}_{lqr} = \begin{bmatrix} 0 & 0 & -1.248 & 0 \end{bmatrix} \begin{bmatrix} 1.2459 \times 10^{11} & -5.3702 \times 10^{10} & 4.5457 \times 10^{5} & 8.9497 \times 10^{5} \\ -5.3883 \times 10^{10} & 6.9741 \times 10^{10} & -4.9805 \times 10^{5} & -3.8856 \times 10^{5} \\ 5.7333 \times 10^{5} & -4.3549 \times 10^{5} & 1.1411 \times 10^{3} & 2.5895 \times 10^{2} \\ 9.5533 \times 10^{5} & -5.0861 \times 10^{5} & 2.5730 \times 10^{2} & 1.4229 \times 10^{3} \end{bmatrix}$$

$$= \begin{bmatrix} -7.1551 \times 10^{5} \\ 5.4349 \times 10^{5} \\ -1.4241 \times 10^{3} \\ -3.2317 \times 10^{2} \end{bmatrix}.$$

(b) The closed-loop state matrix is

$$A - B_{zv}\mathbf{g}'_{lqr} = \begin{bmatrix} 0 & 0 & 1 & 0 \\ 0 & 0 & 0 & 1 \\ -1.2289 \times 10^{8} & 6.0678 \times 10^{7} & -1.7772 \times 10^{3} & -4.0331 \times 10^{2} \\ 6.0000 \times 10^{7} & -6.0000 \times 10^{7} & 0 & 0 \end{bmatrix}.$$

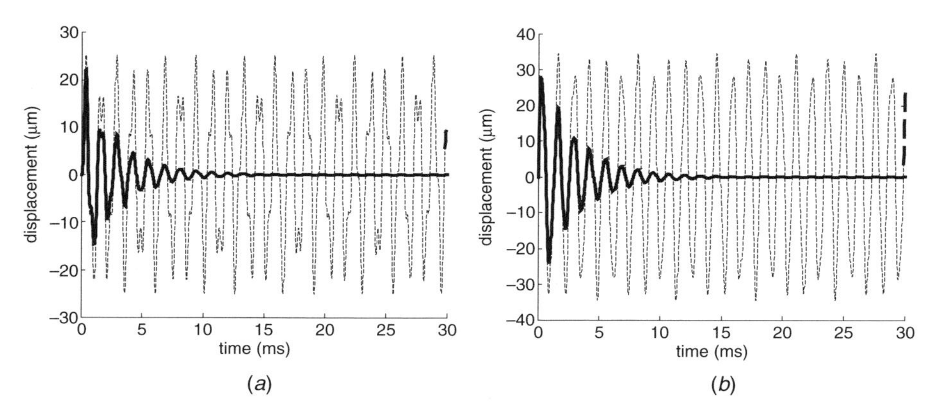

Figure 10.9 Open- (dashed) and closed-loop (solid) impulse response of the system studied in Example 10.7: (*a*) u_1; (*b*) u_2.

The input vector associated with the external force is

$$\begin{bmatrix} 0 \\ 0 \\ 0 \\ 2 \end{bmatrix}.$$

The first output is defined as the displacement of mass 1, the second output is defined as the displacement of mass 2, and the third output is defined as the control voltage. This results in the output matrix

$$\mathrm{C} = \begin{bmatrix} 1 & 0 & 0 & 0 \\ 0 & 1 & 0 & 0 \\ 7.1551 \times 10^5 & -5.4349 \times 10^5 & 1.4241 \times 10^3 & 3.2317 \times 10^2 \end{bmatrix}.$$

The results of Example 10.7 can be used to study the effects of varying the LQR weighting matrices on the response of the closed-loop system. As an example, consider the impulse response of the closed-loop system for the weighting matrices specified in the example, assuming that the magnitude of the impulse is 0.1 N. Figure 10.9 illustrates the open- and closed-loop response of the system to this impulse. The open-loop response exhibits no damping, due to the fact that no viscous energy dissipation has been included in the model. This is an idealization but is a reasonable assumption if the damping of the open-loop system is very small. The closed-loop response exhibits increased damping due to the control action. The decay rate of the response has increased due to the change in eigenvalues of the closed-loop system.

An important design consideration is the voltage required across the piezoelectric stack for closed-loop control. The closed-loop model developed in Example 10.7 can be used to compute the voltage response because the voltage has been included in the last row of the output matrix. Figure 10.10 is a plot of the voltage response for the

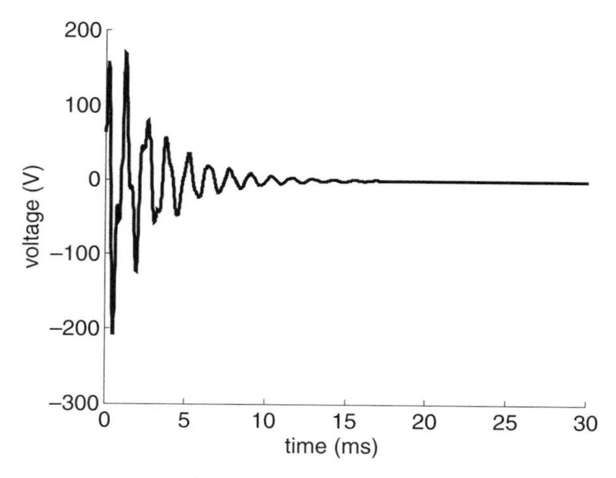

Figure 10.10 Impulse response of the voltage for the control system studied in Example 10.7.

system studied in Example 10.7. An important parameter would be the peak voltage across the stack. For the choice of weighting matrices in Example 10.7 the peak voltage is approximately 200 V. Recalling that the thickness of the piezoelectric layer is 0.25 mm, we can determine that the peak electric field across the stack is on the order of 0.8 MV/m, which is well within the range of typical piezoelectric materials.

Using weighting matrices of the form expressed in equation (10.101) provides a straightforward method of analyzing the variation in the relative importance of reducing the state response and the control response. In Example 10.7 the value of α/β was chosen to be 1×10^6 and the gains were computed to form the closed-loop state equations. Decreasing α/β reduces the relative importance of the state response in the cost function, whereas increasing α/β increases the importance of the state response.

Repeating the computation of the control gains for $\alpha/\beta = 10^5$ produces a change in the closed-loop response. Since the relative importance of the state response in the cost function is decreased, we expect to see an decrease in the decay rate and hence a longer amount of time until the impulse response decays to approximately zero. This situation would correspond to an increase in the mean-square response of the states to the impulse input. Figure 10.11*a* illustrates that this is exactly what we see. Comparing Figure 10.9 with Figure 10.11*a*, we see that the response takes longer to decay to zero when α/β is decreased. Conversely, when α/β is increased to 10^7, the relative importance of the state response in the cost function is increased; therefore, we expect that the mean-square response of the states would decrease. Comparing Figure 10.9 with Figure 10.11*b*, we see that the decay rate of the response has increased and the displacement decays to zero more quickly when α/β is increased.

Varying the weighting matrices to improve the state response comes at a cost in terms of the response of the control voltage. Increasing α/β to improve the state response will produce an increase in the mean-square voltage response, due to the

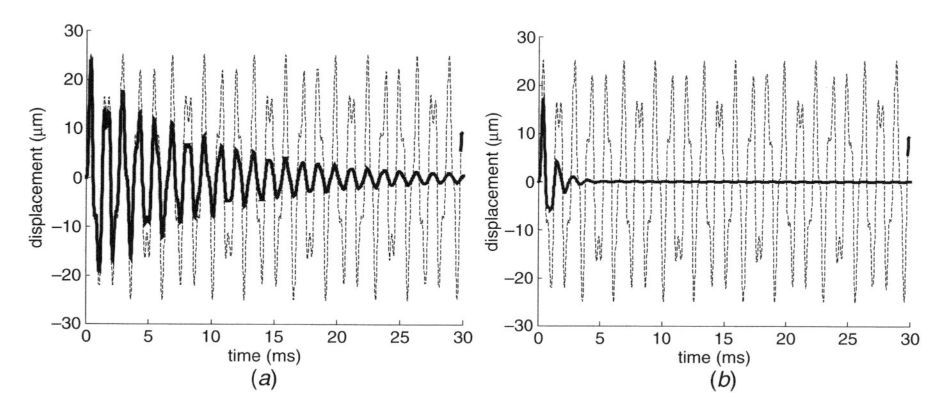

Figure 10.11 Response of u_1 for LQR full-state feedback control with (*a*) $\alpha = 1 \times 10^5$ and (*b*) $\alpha = 1 \times 10^7$.

fact that the relative importance of the control voltage in the cost function has been decreased. Conversely, we would expect that decreasing α/β would also decrease the mean-square response of the control voltage. Figure 10.12 illustrates that these trends do hold for the system studied in Example 10.7. A decrease in α/β to 10^5 produces a decrease in the peak response of the piezoelectric control voltage from approximately 200 V to less than 60 V. Similarly, an increase in α/β produces an increase in peak control voltage to over 450 V, due to the fact that increased voltage is required to reduce the mean-square value of the state response. Comparisons of the state and control costs as a function of the weighting matrices illustrates the basic design trade-off for linear quadratic regulator control. One of the strengths of the technique is that it provides a systematic method for trading off state cost and control cost (Table 10.2). Compare this result to the result obtained with pole placement, where the design

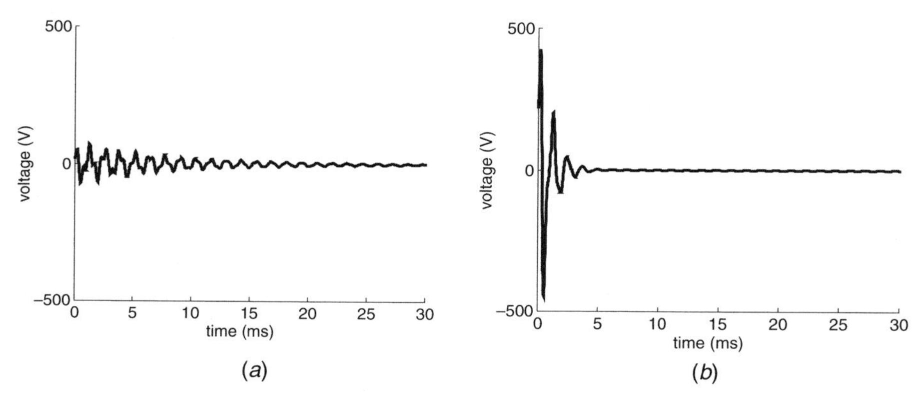

Figure 10.12 Response of the piezoelectric control voltage for LQR full-state feedback control with (*a*) $\alpha = 1 \times 10^5$ and (*b*) $\alpha = 1 \times 10^7$.

Table 10.2 Comparison of state cost and input cost for three values of α/β for the system studied in Example 10.7.

α/β	$\int_0^\infty \mathbf{z}'(t)\mathbf{z}(t)\,dt$	$\int_0^\infty v^2(t)\,dt$
10^5	8.74×10^{-3}	8.64
10^6	2.75×10^{-3}	27.9
10^7	1.08×10^{-3}	89.3

parameters were the closed-loop eigenvalues of the system. Choosing the closed-loop eigenvalues does not guarantee any optimality of the closed-loop performance, whereas the optimality of the closed-loop response is the basis for the computation of the control gains in the LQR design method.

10.5.4 State Estimation

The full-state feedback control laws discussed in Section 10.5.3 all assume that every state in the system is measured directly. Direct measurement of all the states allows the control law stated in equation (10.80) to be implemented for feedback. In a majority of instances, though, direct measurement of all of the system states is not possible; therefore, an important question is how to utilize the concept of full-state feedback for the case in which only a subset of states is measurable directly.

If we assume that only a subset of states is measurable through the system output $\mathbf{y}(t)$, a *state estimator* is required to produce an indirect measurement of the states of the system. A state estimator is implemented in hardware based on a model of the system being controlled. Consider the state estimator representation

$$\dot{\hat{\mathbf{z}}}(t) = \mathrm{A}\hat{\mathbf{z}}(t) + \mathrm{B}_c\mathbf{v}(t) + \mathrm{G}_e\mathbf{r}_e(t), \tag{10.105}$$

where G_e is the state estimator gain and $\mathbf{r}_e(t)$ is the *residual* between the measured output and the estimated output:

$$\mathbf{r}_e(t) = \mathbf{y}(t) - \hat{\mathbf{y}}(t) = \mathrm{C}_y\left(\mathbf{z}(t) - \hat{\mathbf{z}}(t)\right). \tag{10.106}$$

Combining equations (10.105) and (10.106), we can write the estimator state equations as

$$\dot{\hat{\mathbf{z}}}(t) = (\mathrm{A} - \mathrm{G}_e\mathrm{C}_y)\hat{\mathbf{z}}(t) + \mathrm{G}_e\mathrm{C}_y\mathbf{z}(t) + \mathrm{B}_c\mathbf{v}(t). \tag{10.107}$$

The important attribute of equation (10.107) is that the estimator gain matrix is a feedback term that determines the eigenvalues of the estimator state matrix. This occurs due to the presence of the residual feedback term.

The use of an estimator allows us to implement a form of full-state feedback based on the estimated states:

$$\mathbf{v}(t) = -\mathrm{G}\hat{\mathbf{z}}(t). \tag{10.108}$$

Applying this feedback control law to the original system without any forcing input, the estimator yields a set of two equations:

$$\dot{\mathbf{z}}(t) = \mathrm{A}\mathbf{z}(t) - \mathrm{B}_c\mathrm{G}\hat{\mathbf{z}}(t) \tag{10.109}$$

$$\dot{\hat{\mathbf{z}}}(t) = (\mathrm{A} - \mathrm{G}_e\mathrm{C}_y - \mathrm{B}_c\mathrm{G})\hat{\mathbf{z}}(t) + \mathrm{G}_e\mathrm{C}_y\mathbf{z}(t). \tag{10.110}$$

Defining the state error as

$$\mathbf{e}(t) = \mathbf{z}(t) - \hat{\mathbf{z}}(t), \tag{10.111}$$

we can use equation (10.110) to write the state equations for the error as

$$\dot{\mathbf{e}}(t) = (\mathrm{A} - \mathrm{G}_e\mathrm{C}_y)\mathbf{e}(t). \tag{10.112}$$

Equation (10.112) demonstrates that the time evolution of the state error is determined by the eigenvalues of the matrix $\mathrm{A} - \mathrm{G}_e\mathrm{C}_y$. Thus, the choice of the estimator gain matrix G_e plays a critical role in determining how fast the state estimates will converge to the actual states. Rewriting the original state equations in terms of the state error,

$$\dot{\mathbf{z}}(t) = \mathrm{A}\mathbf{z}(t) - \mathrm{B}_c\mathrm{G}\left(\mathbf{z}(t) - \mathbf{e}(t)\right), \tag{10.113}$$

and combining with the state equation for the error yields

$$\begin{Bmatrix} \dot{\mathbf{z}}(t) \\ \dot{\mathbf{e}}(t) \end{Bmatrix} = \begin{bmatrix} \mathrm{A} - \mathrm{B}_c\mathrm{G} & \mathrm{B}_c\mathrm{G} \\ 0 & \mathrm{A} - \mathrm{G}_e\mathrm{C}_y \end{bmatrix} \begin{Bmatrix} \mathbf{z}(t) \\ \mathbf{e}(t) \end{Bmatrix}. \tag{10.114}$$

Equation (10.114) is a representation of the closed-loop system defined by the estimator and the full-state feedback control law that utilizes the state estimates. The closed-loop system was determined by choosing the eigenvalues of the matrix $\mathrm{A} - \mathrm{G}_e\mathrm{C}_y$ and choosing the eigenvalues of the matrix $\mathrm{A} - \mathrm{B}_c\mathrm{G}$. The question now arises: What are the eigenvalues of the *closed-loop system* defined by the estimator and full-state feedback? To answer this question, consider defining the closed-loop state matrix as

$$\mathrm{A}_{cl} = \begin{bmatrix} \mathrm{A} - \mathrm{B}_c\mathrm{G} & \mathrm{B}_c\mathrm{G} \\ 0 & \mathrm{A} - \mathrm{G}_e\mathrm{C}_y \end{bmatrix} \tag{10.115}$$

and noting that the eigenvalues of the closed-loop system are determined from $|\mathrm{sI} - \mathrm{A}_{\mathrm{cl}}|$:

$$|\mathrm{sI} - \mathrm{A}_{\mathrm{cl}}| = \left\| \begin{bmatrix} \mathrm{sI} - \mathrm{A} + \mathrm{B}_c\mathrm{G} & -\mathrm{B}_c\mathrm{G} \\ 0 & \mathrm{sI} - \mathrm{A} + \mathrm{G}_e\mathrm{C}_y \end{bmatrix} \right\|. \tag{10.116}$$

A general expression for the determinant of a matrix that is partitioned into four submatrices is

$$\|\mathrm{M}| = \left|\begin{bmatrix} \mathrm{M}_{11} & \mathrm{M}_{12} \\ \mathrm{M}_{21} & \mathrm{M}_{22} \end{bmatrix}\right| = \|\mathrm{M}_{22}| \left|\mathrm{M}_{11} - \mathrm{M}_{12}\mathrm{M}_{22}^{-1}\mathrm{M}_{21}\right|. \tag{10.117}$$

Applying the definition in equation (10.117) to the partitioned state matrix, we have

$$|\mathrm{sI} - \mathrm{A}_{\mathrm{cl}}| = |\mathrm{sI} - \mathrm{A} + \mathrm{B}_c\mathrm{G}|\ |\mathrm{sI} - \mathrm{A} + \mathrm{G}_e\mathrm{C}_y|. \tag{10.118}$$

The form of the combined state equations for the system and the estimator shown in equation (10.114) illustrates that the eigenvalues of the closed-loop system are the union of the eigenvalues of $\mathrm{A} - \mathrm{B}_c\mathrm{G}$ and $\mathrm{A} - \mathrm{G}_e\mathrm{C}_y$. Thus, the choice of gain matrices for the full-state feedback and the state estimator determine the eigenvalues of the closed-loop system. More important, we are *choosing* the eigenvalues of these two matrices; therefore, the combination of an estimator with full-state feedback allows us to choose the eigenvalues of the closed-loop system directly.

10.5.5 Estimator Design

Combining full-state feedback with state estimation enables a systematic approach to the design of state-space control laws. The estimator design is reduced to the choice of the eigenvalues of $\mathrm{A} - \mathrm{G}_e\mathrm{C}_y$ since this state matrix controls the time evolution of the state error. Another powerful result in state estimation is that the problem of choosing the eigenvalues of $\mathrm{A} - \mathrm{G}_e\mathrm{C}_y$ is analogous to the problem of choosing the eigenvalues of $\mathrm{A} - \mathrm{B}_c\mathrm{G}$. When we have a single output and the estimator gain is represented as the vector $\mathbf{g}_e$, the eigenvalues of the estimator state matrix are obtained from the expression

$$\xi_{\mathrm{est}}(\mathrm{s}) = |\mathrm{sI} - \mathrm{A} + \mathbf{g}'_{\mathrm{e}}\mathrm{C}_{\mathrm{y}}|. \tag{10.119}$$

The characteristic equation can be written as an nth-order polynomial of the form

$$\xi_{\mathrm{est}}(\mathrm{s}) = \mathrm{s}^n + a_1\mathrm{s}^{n-1} + \cdots + a_{n-1}\mathrm{s} + a_n, \tag{10.120}$$

where the coefficients a_i are computed from the determinant $|\mathrm{sI} - \mathrm{A} + \mathbf{g}'_{\mathrm{e}}\mathrm{C}_{\mathrm{y}}|$. In general, each coefficient is a function of the n unknown gains from the gain vector $\mathbf{g}_e$.

The estimator gains are computed by assuming that we have a desired characteristic equation, $\xi_d(\mathrm{s})$, which is expressed as

$$\xi_d(\mathrm{s}) = (\mathrm{s} - \lambda_{d1})(\mathrm{s} - \lambda_{d2})\cdots(\mathrm{s} - \lambda_{dn}), \tag{10.121}$$

where λ_{di} are desired eigenvalues of the estimator state matrix. Expanding equation (10.121), we form the polynomial

$$\xi_d(\mathsf{s}) = \mathsf{s}^n + b_1\mathsf{s}^{n-1} + \cdots + b_{n-1}\mathsf{s} + b_n. \tag{10.122}$$

To ensure that the estimator state matrix has the desired eigenvalues, we must equate the coefficients of the polynomials expressed in equations (10.121) and (10.122). We write the n equations as

$$\begin{aligned} a_1\left(g_{e1}, g_{e2}, \ldots, g_{en}\right) &= b_1 \\ a_2\left(g_{e1}, g_{e2}, \ldots, g_{en}\right) &= b_2 \\ &\vdots \\ a_n\left(g_{e1}, g_{e2}, \ldots, g_{en}\right) &= b_n. \end{aligned} \tag{10.123}$$

Solving for the desired estimator gains is equivalent to solving n equations for the n unknown gains g_{ei}. As with full-state feedback, it is not always the case that the n equations and n unknowns can be solved. The set of equations has a unique solution if and only if the matrix

$$\mathcal{O} = [C_y' \quad \mathsf{A}'C_y' \quad (\mathsf{A}')^2C_y' \cdots \quad (\mathsf{A}')^{n-1}C_y'] \tag{10.124}$$

has rank equal to the number of states in the system. If the rank of $\mathcal{O}$ is equal to n, the system is said to be *observable*.

10.6 CHAPTER SUMMARY

Methods for utilizing piezoelectric material systems for active vibration control were studied in this chapter. The methodologies could be separated into those that used second-order models and those that utilized first-order models. Second-order models have the advantage that they are consistent with the methodologies introduced in earlier chapters for modeling piezoelectric material systems. In addition, the use of displacement, velocity, and acceleration feedback could be related directly to the addition of stiffness, damping, and mass, respectively, using feedback. Dynamic second-order controllers were studied as a means of collocated control. Positive-position feedback was an example of collocated control that is widely used for introduced damping into structural material systems. Self-sensing actuation was derived as a means of utilizing the same piezoelectric material as both a sensor and an actuator. Through examples we illustrated the effects of mistuning of the self-sensing bridge circuit on the feedback measurement.

The second class of vibration control methodologies that we studied were based on first-order forms of the equations of motion. Transforming second-order equations of motion into first-order form enables the application of a large number (almost too large) of control methodologies for active vibration suppression. Pole placement methods and methods based on linear quadratic optimal control theory were introduced and the basic design principles were studied through example. The chapter concluded with a discussion of pole placement techniques for state estimation for cases in which it is not possible to implement full-state feedback.

PROBLEMS

10.1. (**a**) Compute the open- and short-circuit natural frequencies for the system studied in Example 11.1.
(**b**) Use the results of part (a) to compute the generalized coupling coefficients.

10.2. Plot the frequency response between voltage input and charge output for the system studied in Example 10.1.

10.3. Plot the frequency response between the short-circuit charge output and the external forcing input for the system studied in Example 10.1.

10.4. Plot the frequency response between the voltage output and the external force input for the system studied in Example 10.3.

10.5. Redesign the controller studied in Example 10.3 so that damping is added to the second vibration mode. Plot the closed-loop frequency response and compare the result to Figure 10.4.

10.6. Redesign the controller studied in Example 10.3 so that damping is added to both vibration modes. Plot the closed-loop frequency response and compare the result to Figure 10.4.

10.7. Use the results of Example 10.6 and compute the eigenvalues of the closed-loop state matrix $\mathrm{A} - \mathrm{B}_{zv}\mathbf{g}'$. Compare the results to the desired closed-loop pole locations.

10.8. Repeat Example 10.6 using the desired eigenvalues

$$\lambda_{d1,d2} = -0.5 \times 10^4 \pm j1.26 \times 10^4$$
$$\lambda_{d3,d4} = -0.5 \times 10^4 \pm j0.484 \times 10^4.$$

10.9. Confirm that the system studied in Example 10.6 is controllable from the input voltage to the piezoelectric stack.

10.10. Confirm that the system studied in Examaple 10.6 is observable from a measurement of the displacement of mass 1.

10.11. Compute the state estimator gains for the system studied in Example 10.6 using the displacement of mass 2 as the output measurement. Use the desired eigenvalues

$$\lambda_{d1,d2} = -1.26 \times 10^4 \pm j1.26 \times 10^4$$

$$\lambda_{d3,d4} = -0.484 \times 10^4 \pm j0.484 \times 10^4.$$

Confirm that the system is observable from this measurement location.

NOTES

The subject of active vibration control has probably received the most interest in the field of smart material systems. A large number of publications on the topic can be found in the literature from the mid-1980s (starting with the work by Bailey and Hubbard [1]) to the present time. The information in this chapter is based on a variety of articles as well as textbooks on the subject. The matrix theory cited in this chapter may be found in an early textbook by Inman [43] which has recently been republished [114]. Another good reference on active vibration control for structures is a book by Preumont [115].

The discussion of positive-position feedback is based on work of Goh and Caughey [116] and Fanson and Caughey [117]. There are also numerous more recent references on the subject of positive-position feedback. One of the advantages of writing a book is that you can readily cite your own work on a topic [118–120]. A general analysis of using second-order models for control design was presented by Juang and Phan [121]. Articles continue to be published on this technique, due to its robustness and simplicity; see, for example, a work by Rew et al. [122]. The seminal publications on self-sensing using piezoelectric materials are Dosch et al. [123] and Anderson and Hagood [124]. The analysis presented herein is a generalization of the techniques developed in those two papers.

11

POWER ANALYSIS FOR SMART MATERIAL SYSTEMS

Many engineering applications of smart materials and smart material systems require the use of powered electronic systems. For example, using a piezoelectric material as a vibration sensor generally requires powered instrumentation for signal conditioning. Use of a piezoelectric actuator generally requires a power amplifier to transform the input signals to the actuator into a signal at the correct voltage and current. A majority of control systems today are digital control systems that utilize microprocessors and data converters, which, of course, require power.

In certain applications of smart material systems the need for power is an important parameter in the engineering design. For example, the use of a smart material as a vibration damping device often performs the same function in an engineering system as does a passive viscoelastic material. Viscoelastic materials require no external power source to operate; therefore, the power required to implement a solution with smart materials is a valid question for an engineering design. In some instances the need for external power can render a smart material solution infeasible for a particular application. In other applications it requires that the designer clearly specify the benefits associated with the implementation of a smart material over a dumb (or at least coupling-challenged) material.

In this chapter we study techniques for analysis of the power requirements of smart material systems. First, we discuss the concept of electrical power for resistive and capacitive elements and then relate this discussion to the case of general linear impedance that represents a number of smart materials. These analysis techniques will then be applied to understand the concept of system efficiency for a variety of smart materials. The chapter concludes with a discussion of power amplification techniques for smart material systems.

11.1 ELECTRICAL POWER FOR RESISTIVE AND CAPACITIVE ELEMENTS

Our discussion of the constitutive properties of smart materials has introduced us to the electrical behavior of piezoelectric materials, shape memory alloys, and electroactive polymers. For example, we discussed in detail the fact that piezoelectric materials are primarily capacitive devices, due to the fact that they are dielectric materials that contain bound charge. In contrast, shape memory alloys are conductive materials that allow the flow of charge under an applied voltage. For this reason a shape memory alloy is modeled as a resistive electrical element. The electrical properties of electroactive polymers vary. Certain materials, such as dielectric elastomers, are primarily insulators and therefore are primarily capacitive. Other electroactive polymers, such as conducting polymers and ionomeric polymers, exhibit mixed electrical behavior and to first order can be modeled as a network of resistive and capacitive elements.

The predominance of resistive and capacitive behavior in the smart materials studied in this book makes these two electrical elements a good starting point for discussing the power requirements of smart material systems. Consider the standard definition of electrical power in the time domain,

$$P(t) = v(t)i(t). \tag{11.1}$$

Consider a very simple electrical circuit consisting of a voltage source and a resistor in series as shown in Figure 11.1*a*. The current across the resistor is related to the source voltage and resistance through the expression

$$i(t) = \frac{v(t)}{R_L}, \tag{11.2}$$

where R_L represents the load resistance. Substituting equation (11.2) into equation (11.1) yields the expression

$$P(t) = \frac{v^2(t)}{R_L}. \tag{11.3}$$

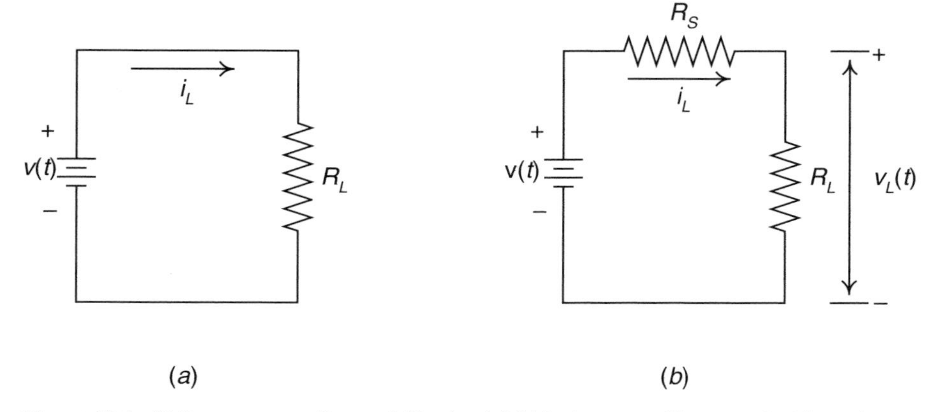

Figure 11.1 Voltage source with a resisitive load: (*a*) ideal source with zero output impedance; (*b*) source with finite output impedance.

Equation (11.3) models the case in which the source is ideal and has zero output impedance. All real sources have a finite output impedance, which can be modeled as a voltage source in series with a source resistance, R_S. This case is shown in Figure 11.1*b*. The source resistance creates a voltage drop between the source and the voltage across the load. We denote the voltage and current across the load as $v_L(t)$ and $i_L(t)$, respectively. Using the rule for voltage dividers, the load voltage is written as

$$v_L(t) = \frac{R_L}{R_S + R_L} v(t). \tag{11.4}$$

The current across the load is

$$i_L(t) = \frac{1}{R_S + R_L} v(t). \tag{11.5}$$

The electrical power across the load is written as the product of equations (11.4) and (11.5):

$$p(t) = v_L(t) i_L(t) = \frac{R_L}{(R_S + R_L)^2} v^2(t). \tag{11.6}$$

Equation (11.6) can be rewritten in terms of the ratio of the load resistance to the source resistance, R_L/R_S:

$$p(t) = \frac{R_L/R_S}{(1 + R_L/R_S)^2} \frac{v^2(t)}{R_S}. \tag{11.7}$$

The term $v^2(t)/R_S$ represents the power output of the source when connected to ground. The first term on the right-hand side becomes close to zero when the ratio of the load resistance to source resistance is approximately zero or as it approaches infinity. The ratio of the power across the load to the short-circuit power is maximized when the ratio of load resistance to source resistance is 1. In this case the ratio of the power across the load to the short-circuit power is $\frac{1}{4}$.

This analysis covers the basic properties of the power across a resistive load. The power is always a positive number, due to the linear proportionality between the voltage and the current. For real sources (i.e., those that have finite output resistance) the power across the load is maximized when the load resistance is equal to the source resistance.

Consider the case when the voltage source is a harmonic function of the form

$$v(t) = V \sin \omega t. \tag{11.8}$$

The power from the source for a short-circuit is now

$$\frac{v^2(t)}{R_S} = \frac{V^2}{R_S} \sin^2 \omega t = \frac{V^2}{2R_S} (1 - \cos 2\omega t). \tag{11.9}$$

The *average power* across a resistive load is equal to $\frac{1}{2}$ the peak power, and the frequency of oscillation of the power is twice that of the original frequency of the source voltage.

Turning our attention to a capacitive load, we note that the relationship between current and voltage across a capacitor is

$$i(t) = C\frac{dv}{dt}, \tag{11.10}$$

where C is the capacitance. Let us return to the case of an ideal voltage source ($R_S = 0$) and assume that the source voltage is a harmonic function as in equation (11.8). Substituting the expression for the voltage into equation (11.10), we have

$$i(t) = CV\omega \cos \omega t. \tag{11.11}$$

One of the central differences between the current induced across a capacitor and the current induced across a resistor is that the capacitive current is proportional to the frequency of the harmonic input. Thus, keeping the amplitude of the voltage constant but increasing the driving frequency increases the output current.

The power across the capacitor is

$$p(t) = i(t)v(t) = CV^2\omega \sin \omega t \cos \omega t. \tag{11.12}$$

Using trigonometric identities, we can write

$$p(t) = \frac{1}{2}CV^2\omega \sin 2\omega t. \tag{11.13}$$

From equation (11.13) we see that the peak power is also proportional to frequency. Once again, this contrasts with the case of a resistor, where the peak power is independent of frequency. Also important is the fact that the *average* power across a capacitor is equal to zero since the mean value of $\sin 2\omega t$ is equal to zero.

An average power of zero is an important concept that produces a substantial contrast with the case of a resistive load. For a resistive load the power (and the average power) is always a positive value, indicating that the *power flow* is always in the direction *from the source to the load*. In the case of a capacitive load, the power oscillates between a positive and a negative value with a zero mean. Physically, this indicates that the power does not only flow from the source to the load, but also from the load to the source.

This result is consistent with our understanding of the physics of capacitive materials. Recall that capacitance is the storage of energy in bound charges within the material. When an electric field is applied to the material, the bound charges respond to the field, resulting in stored energy in the capacitor. These bound charges then release this energy when the electric field is removed, resulting in a flow of power from the capacitor to the source. For an ideal capacitor the amount of energy released is equal to the amount of energy stored, and the average power supplied during a full cycle is equal to zero.

Example 11.1 A multilayer piezoelectric stack consists of 50 layers of 250-μm-thick PZT-5H with side dimensions of 10 mm $\times$ 10 mm. Compute the peak power and average power required to excite this stack with a voltage of 200 V at a frequency of 150 Hz.

Solution To compute the power, we must first compute the capacitance of the stack. The capacitance of the stack is equal to the 50 times the capacitance of a single layer:

$$C = 50\frac{(3800)(8.85 \times 10^{-12}\ \text{F/m})(10 \times 10^{-3}\ \text{m})(10 \times 10^{-3}\ \text{m})}{250 \times 10^{-6}\ \text{m}}$$
$$= 0.672\ \mu\text{F}.$$

The peak power is the amplitude of the expression for power, equation (11.13):

$$P_{\text{pk}} = \frac{1}{2}CV^2\omega = \frac{1}{2}(0.672 \times 10^{-6}\ \text{F})(200\ \text{V})^2(2\pi)(150)$$
$$= 12.7\ \text{W}.$$

Since we are assuming an ideal capacitive load, the average power is equal to zero since the average value of equation (11.13) is equal to zero.

Example 11.2 A shape memory alloy wire with a resistivity of 80 $\mu\Omega \cdot$ cm is 20 cm long and has a circular cross section with a 50-μm radius. Compute the power dissipated in the wire for an applied voltage of 5 V.

Solution The resistance of the wire is computed using the resistivity, length, and cross-sectional area. The cross-sectional area is

$$A = \pi(0.005\ \text{cm})^2 = 7.85 \times 10^{-5}\ \text{cm}^2.$$

The resistance is

$$R = \frac{(80 \times 10^{-6}\ \Omega \cdot \text{cm})(20\ \text{cm})}{7.85 \times 10^{-5}\ \text{cm}^2}$$
$$= 20.4\ \Omega.$$

The power dissipated in the wire is

$$P = \frac{V^2}{R} = \frac{(5\ \text{V})^2}{20.4\ \Omega} = 1.23\ \text{W}.$$

The analysis of purely resistive and purely capacitive loads is a good starting point for power analysis of smart materials. As shown in Examples 11.1 and 11.2, piezoelectric materials and shape memory alloys are examples of smart materials that have primarily resistive or capacitive electric properties. From the analysis we see that the resistive properties of shape memory alloys make them dissipate energy when heated by an electrical input. Of course, this dissipated energy is the mechanism that induces the phase transformations from martensite to austenite. In contrast, piezoelectric materials are capacitive and thus, in the ideal case, they do not dissipate energy when an electric potential is applied. The peak power across the piezoelectric material is proportional to the capacitance and excitation frequency and is proportional to the square of the voltage amplitude.

There are a number of instances, though, in which the electrical load of a smart material system is not a pure capacitor or a pure resistor. For example, piezoelectric materials are not truly ideal capacitors. The loss in a piezoelectric material can be modeled as a resistive element in the electrical capacitance which introduces energy dissipation into the power analysis. Also, we know that coupling a piezoelectric device to a mechanical system introduces a change in the electrical properties of the material. Thus, if a piezoelectric device is coupled to a lossy mechanical system, the loss in the mechanical system would also give rise to a resistive component in the electrical properties. We also learned in Chapter 7 that the electrical properties of certain electroactive polymers is modeled as a combination of resistive and capacitive elements. For example, ionomeric electroactive polymers are resistive at low frequencies, exhibit a capacitive behavior in a midfrequency range, and are then resistive again at high frequencies, due to the conductivity of the polymer matrix.

A more general power analysis of smart material systems is obtained by assuming that the source and load are modeled as a linear impedance function. Referring to Figure 11.2, we see that the system under consideration consists of a voltage source, a source impedance $Z_S(j\omega)$, and a load impedance $Z_L(j\omega)$. The expression for the voltage across the load is obtained by applying the impedance generalization of a voltage divider,

$$v_L(j\omega) = \frac{Z_L(j\omega)}{Z_S(j\omega) + Z_L(j\omega)} v_S(j\omega), \tag{11.14}$$

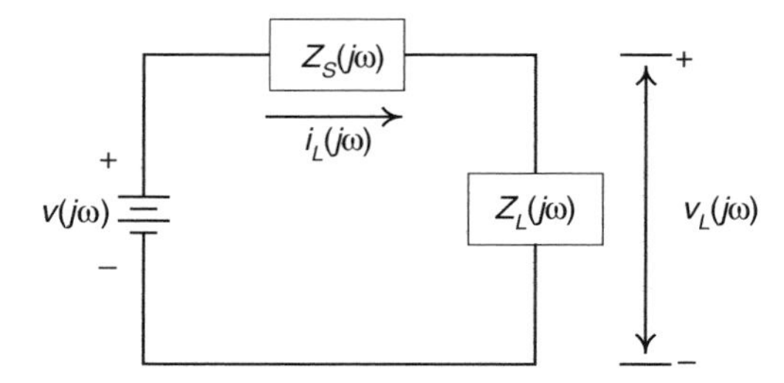

Figure 11.2 System for general power analysis of a smart material system.

where the frequency dependence is implicit in the expression. The expression for the current across the load is

$$i_L(j\omega) = \frac{1}{Z_S(j\omega) + Z_L(j\omega)} v_S(j\omega). \tag{11.15}$$

When the source impedance is much smaller in magnitude than the load impedance, we can approximate the voltage-to-current relationship across the load as

$$v_L(j\omega) = v_S(j\omega) \qquad v_S(j\omega) = Z_L(j\omega) i_L(j\omega). \tag{11.16}$$

Assuming that the current across the load is a harmonic function,

$$i_L(t) = I_L \sin \omega t, \tag{11.17}$$

the voltage across the load can be written as

$$v_L(t) = I_L \, |Z_L(j\omega)| \sin(\omega t + \phi), \tag{11.18}$$

where

$$\tan \phi = \frac{\Im Z_L(j\omega)}{\Re Z_L(j\omega)}. \tag{11.19}$$

The power across the load is

$$p(t) = i_L(t) v_L(t) = I_L^2 \, |Z_L(j\omega)| \sin(\omega t + \phi) \sin \, \omega t. \tag{11.20}$$

Applying some trigonometric identities, we have

$$\begin{aligned} p(t) &= I_L^2 |Z_L(j\omega)| (\sin^2 \omega t \cos \phi + \sin \omega t \cos \omega t \sin \phi) \\ &= \frac{I_L^2 \, |Z_L(j\omega)|}{2} [\cos \phi - \cos(2\omega t + \phi)]. \end{aligned} \tag{11.21}$$

The harmonic term in equation (11.21) has an average value equal to zero, therefore the real power dissipated in a load with a complex impedance is equal to

$$< p(t) > = \frac{I_L^2 \, |Z_L(j\omega)|}{2} \cos \phi. \tag{11.22}$$

We see that the amount of real power dissipated depends on the phase angle ϕ associated with the load impedance. For a phase angle of zero, the real power dissipated is equal to $I_L^2 |Z_L(j\omega)|/2$, which is equivalent to the real power dissipated across a resisitor. In contrast, a phase angle of $\pi/2$ is equivalent to the case of a pure capacitive load, and the average power dissipated across the load is equal to zero. Intermediate values of the phase angle indicate that the load is neither purely resistive nor purely capacitive.

The phase angle of the impedance $Z_L(j\omega)$ tells us directly about the power dissipation characteristics of the load. In electrical circuit theory the term $\cos\phi$ is called the *power factor* of the load. A power factor of 1 corresponds to a purely resistive load while a power factor of zero corresponds to a purely capacitive load. Power factors between 0 and 1 indicate how closely the load models a pure capacitor or a pure resistor.

Example 11.3 The load impedance of an electroactive polymer has been measured to be

$$Z_L(\mathsf{s}) = 1000\frac{\mathsf{s}/20\pi + 1}{\mathsf{s}/0.02\pi + 1} \quad \Omega. \tag{11.23}$$

Determine the power factor at 0.01 and 1 Hz.

Solution The power factor is determined directly from the phase angle of the load impedance. Substituting $\mathsf{s} = j\omega$ into the transfer function of the impedance, we have

$$Z_L(j\omega) = 1000\left(\frac{j\omega/20\pi + 1}{j\omega/0.02\pi + 1}\right) \; \Omega.$$

At a frequency of 0.01 Hz, the load impedance is

$$\begin{aligned} Z_L(0.02\pi j) &= 1000\left(\frac{0.02\pi j/20\pi + 1}{0.02\pi j/0.02\pi + 1}\right) \; \Omega \\ &= 500.5 - 499.5j \;\; \Omega. \end{aligned}$$

The phase angle of the load impedance is

$$\phi = \tan^{-1}\frac{-499.5}{500.5} = -0.784 \text{ rad}.$$

The power factor PF is

$$\text{PF} = \cos(-0.784) = 0.707.$$

Repeating the analysis for a frequency of 1 Hz, we have

$$\begin{aligned} Z_L(2\pi j) &= 1000\left(\frac{2\pi j/20\pi + 1}{2\pi j/0.02\pi + 1}\right) \; \Omega \\ &= 1.1 - 10j \; \Omega. \end{aligned}$$

The phase angle of the load impedance is

$$\phi = \tan^{-1} \frac{-10}{1.1} = -1.461 \text{ rad.}$$

The power factor is

$$\text{PF} = \cos(-1.461) = 0.109.$$

We see from the analysis that the power factor for the electroactive polymer changes as a function of frequency. At lower frequencies the power factor is closer to 1 and therefore the material more closely resembles a resistive load. The decrease in the power factor as the frequency increases to 1 Hz indicates that the material becomes more capacitive. Note that in both cases the power factor is not exactly equal to 0 or 1 as it would be for a purely resistive or capacitive material.

Note that the power factor does not provide any information about the magnitude of the power across the load. This is illustrated in the following example.

Example 11.4 Determine the expression for the power across the electroactive polymer studied in Example 11.3. Assume an applied voltage of 2 V.

Solution The expression for the power is shown in equation (11.21). The first factor on the right in this expression can be substituted for

$$\frac{I_L^2 \, |Z_L(j\omega)|}{2} = \frac{V_L^2}{2 \, |Z_L(j\omega)|}.$$

At a frequency of 0.01 Hz,

$$|Z_L(0.02\pi j)| = |500.5 - 499.5j| = 707 \ \Omega.$$

The expression for the power across the polymer at 0.01 Hz is

$$\begin{aligned} p(t) &= \frac{(2 \text{ V})^2}{(2)(707 \ \Omega)} [0.707 - \cos(0.02\pi t - 0.784)] \\ &= (2.8)[0.707 - \cos(0.02\pi t - 0.784)] \text{ mW.} \end{aligned}$$

At a frequency of 1 Hz the magnitude of the impedance function is

$$|Z_L(2\pi j)| = |1.1 - 10j| = 10.1 \ \Omega$$

and the expression for the power is

$$
\begin{aligned}
p(t) &= \frac{(2\ \text{V})^2}{(2)\,(10.1\ \Omega)}[0.109 - \cos(2\pi t - 1.461)] \\
&= (198)[0.109 - \cos(2\pi t - 1.461)\}\ \text{mW}.
\end{aligned}
$$

The peak-to-peak power at 0.01 is only 2.8 mW, while the peak-to-peak power at 1 Hz is 198 mW. A larger portion of the power at 1 Hz is capacitive due to the smaller power factor. Note that the average power at 1 Hz is greater than the average power dissipated at 0.01 Hz, due to the decrease in impedance function.

11.2 POWER AMPLIFIER ANALYSIS

In Section 11.1 we saw that the electrical characteristics of a smart material system can be modeled as a generalized impedance that consists of resistive and capacitive elements. The power factor at any frequency of operation represents that relative contribution of the resistive and capacitive properties of the smart material. The analysis also demonstrated that the average power dissipated across the load was a function of the power factor. Purely capacitive loads (e.g., ideal piezoelectric materials) did not dissipate any power, due to the storage and release of electrical energy when a potential was applied.

In a system-level design it is often important to know how much power must be supplied to a smart material to perform its function. For example, when using a piezoelectric actuator in a battery-operated system, the amount of energy supplied to the actuator will determine the battery life, and hence the lifetime of the system. Designing for low-power operation will increase the lifetime and utility of the system.

In this section we study how to analyze the electrical power requirements for smart material systems. The analysis focuses on determining the power supplied to the smart material system for different types of power amplifiers.

11.2.1 Linear Power Amplifiers

A common method of supplying power to a smart material system is a linear power amplifier. A linear power amplifier consists of an electronic component known as an *operational amplifier*, or *op-amp* that supplies the necessary voltage and current to the material. An operational amplifier has two inputs, a positive terminal and a negative terminal, and a single output. The symbol for an operational amplifier is a rotated triangle with a positive and a negative sign on the left side as shown in Figure 11.3. The two additional inputs to the op-amp are the *supply rails*. These two inputs supply the power for the op-amp output.

There are two "golden rules" for an operational amplifier:

1. The op-amp will do whatever is necessary to make the difference between the voltages at the input terminals' zero.
2. The inputs draw no current.

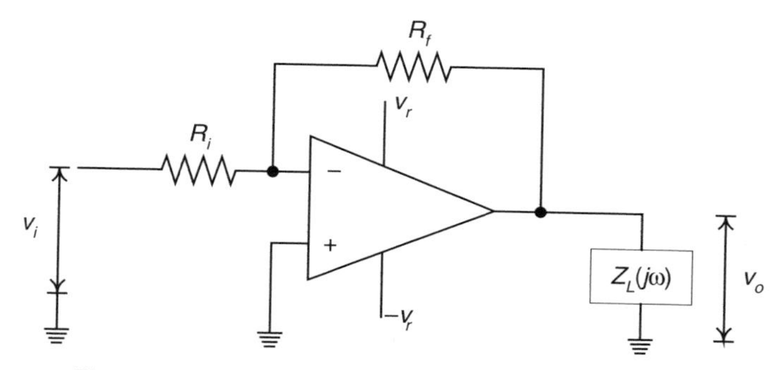

Figure 11.3 Linear power amplifier with a generalized load.

So far in our discussion of inputs to smart material systems, we generally have modeled this as a prescribed voltage or current input. To understand how this input voltage is prescribed, consider an operational amplifier with an input voltage, v_i, an output voltage, v_o, and two terminal voltages, v_- and v_+. The purpose of the operational amplifier is to supply a voltage v_o to the load, which is the smart material input, with the prescribed voltage and current. Generally, it is assumed that whatever device is supplying the input voltage to the operational amplifier does not have the ability to produce the necessary current for the load on the amplifier.

One of the most common implementations of the op-amp is a negative feedback amplifier, as shown in Figure 11.3. The negative feedback amplifier consists of an op-amp whose positive terminal is connected to ground and whose negative input terminal is connected to the electronic circuit shown in the figure. The input voltage v_i is connected to the negative terminal through a resistor, R_i, and the output terminal is connected to the negative terminal through the feedback resistor, R_f. Our analysis will determine how the output voltage v_o depends on the input voltage v_i. To do so, first apply rule 1 for an ideal op-amp. Applying this rule we see that

$$v_+ = v_- = 0 \tag{11.24}$$

since the positive terminal is connected to ground. Examining the circuit node at the negative terminal, we can write the expression

$$i_{\text{in}} + i_f - i_- = 0. \tag{11.25}$$

The currents due to the input and feedback are

$$i_{\text{in}} = \frac{v_i - v_-}{R_i} \qquad i_f = \frac{v_o - v_-}{R_f}. \tag{11.26}$$

Combining the expressions and recalling that $i_- = 0$ by the assumption of an ideal op-amp, we can write

$$\frac{v_i - 0}{R_i} + \frac{v_o - 0}{R_f} + 0 = 0. \tag{11.27}$$

Solving equation (11.27) for v_o, we have

$$v_o = -\frac{R_f}{R_i} v_i. \tag{11.28}$$

An ideal operational amplifier can produce an output voltage that is proportional to the input voltage. The constant of proportionality is set by the choice of the feedback resistor and input resistor. In the case in which $R_f = R_i$, the constant of proportionality is equal to 1.

One may wonder why an amplifier is necessary if the output voltage is proportional to the input voltage and in certain cases may be made equal to the input voltage. Recall that the purpose of the amplifier is to supply a voltage and current to the load. It is assumed that the component that is producing v_i cannot supply the necessary current and the current must be supplied by the operational amplifier. The question then becomes: How does the operational amplifier supply current to the load?

A conventional linear amplifier has a *push–pull output stage*, which can be modeled as a transistor that is connected to an input voltage and two supply rails. The voltage at one supply rail is assumed to be $-v_r$ and the other is assumed to be $+v_r$, which is called a *symmetric supply voltage*. The amplifier is designed such that the output voltage can never be above or below the voltage of the supply rails; therefore, $|v_o| < |v_r|$. If this condition is satisfied, the power supplied by the operational amplifier is a function of the sign of the output voltage.

A push-pull amplifier has the characteristic that the current output $i_o(t)$ is supplied by the voltage rails. When the current output of amplifier is positive, current flows from the positive supply rail to the output stage; when the current output of the amplifier is negative, current flows from the negative supply rail to the output stage. The voltage drop between the supply rail to the output stage determines the power dissipation in the amplifier according to the expressions

$$p(t) = \begin{cases} [v_r - v_o(t)]i_o(t) & i_o(t) > 0 \\ -[v_o(t) + v_r]i_o(t) & i_o(t) < 0. \end{cases} \tag{11.29}$$

Combining these two expressions, we can write the power dissipation in the linear amplifier as

$$p_{\text{diss}}(t) = v_r \, |i_o(t)| - v_o(t)i_o(t). \tag{11.30}$$

The power dissipation in the amplifier is a function of the voltage and current across the load as well as the supply rail voltage. Under the assumption that the load has an impedance $Z_L(j\omega)$ and that the current across the load is a harmonic function $i_o(t) = I_o \sin \omega t$, the power dissipation in the amplifier is written as

$$p_{\text{diss}}(t) = v_r I_o |\sin \omega t| - \frac{I_o^2 \, |Z_L(j\omega)|}{2}[\cos\phi - \cos(2\omega t + \phi)]. \tag{11.31}$$

The average power dissipation is computed by integrating over a single period T:

$$< p_{\text{diss}} > = \frac{1}{T}\int_0^T p_{\text{diss}}(t)\,dt = \frac{2v_r I_o}{\pi} - \frac{I_o^2|Z_L(j\omega)|}{2}\cos\phi. \tag{11.32}$$

The average power dissipation is a function of the power factor of the load. The two special cases for a resistive and capacitive load are

$$< p_{\text{diss}} > = \begin{cases} \dfrac{2v_r I_o}{\pi} - \dfrac{I_o^2 R}{2} & \text{resistive load} \\ \dfrac{2v_r I_o}{\pi} & \text{capacitive load} \end{cases} \tag{11.33}$$

Example 11.5 A piezoelectric actuator is being excited at a frequency of 100 Hz with a voltage amplitude of 110 V. The linear amplifier supplying the voltage and current operates on supply rails of ±150 V. Compute the average power dissipation in the amplifier assuming a purely capacitive load of 1.3 μF.

Solution The average power dissipation is obtained from equation (11.33) for a purely capacitive load. The supply voltage $v_r = 150$ V. The current amplitude across the load is computed from

$$i(t) = C\frac{dv(t)}{dt}.$$

Assuming that $v(t) = 110\sin(200\pi t)$ V, the current is

$$\begin{aligned} i(t) &= (1.3\times10^{-6}\ \text{F})(110\ \text{V})(200\pi\ \text{rad/s})\cos(200\pi t)\ \text{A} \\ &= 89.8\ \text{mA}. \end{aligned}$$

Thus, $I_o = 89.8$ mA. Substituting this result into equation (11.33), we have

$$\begin{aligned} < p_{\text{diss}} > &= \frac{(2)(150\ \text{V})(89.8\times10^{-3}\ \text{A})}{\pi} \\ &= 8.58\ \text{W}. \end{aligned}$$

Example 11.6 Compute the average power dissipation at 1 Hz for the electroactive polymer studied in Example 11.4 assuming that the linear amplifier exciting the actuator has a supply rail of ±15 V.

Solution The power factor of the polymer actuator at 1 Hz is neither purely real nor purely capactive; therefore, the average power dissipation is given by equation (11.32).

The term $I_o^2 R/2$ was computed in Example 11.4:

$$\frac{I_o^2 R}{2} = 198 \text{ mW}$$

$$I_o = \frac{V}{|Z_L(j\omega)|} = \frac{2 \text{ V}}{10.1\ \Omega} = 198 \text{ mA}.$$

The power factor of the electroactive polymer at 1 Hz is 0.109. Substituting these results into equation (11.32), we have

$$\begin{aligned} < p_{\text{diss}} > &= \frac{(2)(15 \text{ V})(198 \times 10^{-3} \text{ A})}{\pi} - (198 \times 10^{-3} \text{ A})(0.109) \\ &= 1.86 \text{ W}. \end{aligned}$$

Note the large discrepancy between the average power dissipation in the actuator and the average power dissipated in the linear amplifier. The average power dissipated across the actuator is only approximately 200 mW, but the average power dissipated in the linear amplifier is close to 2 W. If we define the *efficiency* as the ratio of these two numbers, we realize that the power efficiency of the linear amplifier is only about 10%.

11.2.2 Design of Linear Power Amplifiers

In Section 11.2.1 we analyzed the power flow in linear amplifiers as a function of the electrical behavior of smart materials. An additional consideration in the design of power amplifiers are properties such as the relationship between amplifier gain and the speed of response, or *bandwidth*, of the amplifier. For linear amplifiers, these properties are related to the characteristics of the amplifier as well as the electrical behavior of the smart material.

Consider isolating the open-loop characteristics of the operational amplifier. For an operational amplifier without any feedback paths or any load, we can write the input–output relationship as

$$V_a(\mathsf{s}) = G_a(\mathsf{s})[V_+(\mathsf{s}) - V_-(\mathsf{s})]. \tag{11.34}$$

Note that this expression is valid only when the amplifier is not saturated and is operating in its linear regime. The term V_a represents the amplifier output. All amplifiers have a finite output impedance which we denote R_o. The output voltage of the amplifier, V_o, is the voltage that is measured at the output terminal of the amplifier. Without any load, the output voltage is equal to the amplifier voltage. With a load, though, these voltages can differ. In the case in which the amplifier has a resistive output load, R_L, the output voltage and amplifier voltage are related through the expression

$$V_o(\mathsf{s}) = \frac{R_L}{R_o + R_L} V_a(\mathsf{s}), \tag{11.35}$$

which is simply the expression for a voltage divider. In the case in which $R_o \ll R_L$, the output voltage is approximately equal to the output voltage, and the output impedance can generally be ignored in the amplifier analysis. This may be the case if the amplifier is being used to drive a shape memory alloy wire which is primarily a resistive load.

A case that requires more careful consideration is the case in which the load is capacitive, such as the case of a piezoelectric material or electroactive polymer. In this case the load impedance is frequency dependent, and the relationship between the output voltage and the amplifier voltage is

$$V_o(\mathrm{s}) = \frac{1}{\tau_L \mathrm{s} + 1} V_a(\mathrm{s}) = G_L(\mathrm{s}) V_a(\mathrm{s}), \tag{11.36}$$

where

$$\tau_L = R_o C_L \tag{11.37}$$

is a time constant that is associated with the output impedance and load capacitance.

The open-loop characteristics of the operational amplifier are modeled as a first-order system of the form

$$V_a(\mathrm{s}) = \frac{g_a}{\tau_a \mathrm{s} + 1} [V_+(\mathrm{s}) - V_-(\mathrm{s})] = G_a(\mathrm{s})[V_+(\mathrm{s}) - V_-(\mathrm{s})], \tag{11.38}$$

where τ_a is the amplifier time constant and g_a represents the dc gain of the operational amplifier. The operational amplifiers are designed so that the dc gain is very large, often on the order of 10^5 to 10^8. Combining equations (11.36) and (11.38), we have

$$V_o(\mathrm{s}) = G_L(\mathrm{s}) G_a(\mathrm{s})[V_+(\mathrm{s}) - V_-(\mathrm{s})]. \tag{11.39}$$

Consider once again the case in which we are using a resistor network on the input and output to control the amplifier feedback. Assuming that the positive terminal is connected to ground, we can write

$$\begin{aligned} V_+(\mathrm{s}) &= 0 \\ \frac{V_i(\mathrm{s}) - V_-(\mathrm{s})}{R_i} + \frac{V_0(\mathrm{s}) - V_-(\mathrm{s})}{R_f} &= 0. \end{aligned} \tag{11.40}$$

The purpose of this analysis is to determine an expression between the input and outputs voltage of the amplifier. To do this, solve equation (11.40) for V_-:

$$V_-(\mathrm{s}) = \frac{R_i}{R_i + R_f} V_o(\mathrm{s}) + \frac{R_f}{R_i + R_f} V_i(\mathrm{s}) \tag{11.41}$$

and substitute into equation (11.39). The result is

$$\frac{V_o(\mathsf{s})}{V_i(\mathsf{s})} = \frac{-[R_f/(R_i + R_f)]G_L(\mathsf{s})G_a(\mathsf{s})}{1 + [R_i/(R_i + R_f)]G_L(\mathsf{s})G_a(\mathsf{s})}. \tag{11.42}$$

This expression represents the closed-loop equation for the input–output amplifier response, once again assuming that the amplifier is responding in its linear regime. The equation illustrates the importance of the high input impedance of the amplifier. At frequencies in which $[R_i/(R_i + R_f)]G_L(j\omega)G_a(j\omega)| \gg 1$, the input–output response is approximately $-R_f/R_i$, which is identical to the analysis performed earlier in the chapter for an ideal amplifier. The additional dynamics associated with the capacitive load produce stability considerations for the closed-loop system. Substituting in the expressions for G_a and G_L, we can write the characteristics equation of the closed-loop system as

$$1 + G_{\text{LOOP}}(\mathsf{s}) = 0, \tag{11.43}$$

where

$$G_{\text{LOOP}}(\mathsf{s}) = \frac{g_a}{1 + R_f/R_i} \frac{1}{(\tau_a \mathsf{s} + 1)(\tau_L \mathsf{s} + 1)} \tag{11.44}$$

is the *loop transfer function*. From the expression for the characteristic equation it is clear that the capacitive load produces an additional pole into the loop transfer function. The frequency of this pole is $1/\tau_L$, which, for a majority of systems, is at a frequency that is greater than $1/\tau_a$. Equation (11.44) also illustrates that the ratio of the feedback resistor to the input resistor determines the gain of the loop transfer function. Basic stability considerations demonstrate that the ratio R_f/R_i must be set *high enough* to reduce the loop gain such that the phase at gain crossover—defined as the frequency at which the loop gain becomes 1—does not approach $-180°$.

The stability considerations for an operational amplifier with a capacitive load are considered by analyzing the general characteristics of the loop transfer function. A typical amplifier transfer function has a high dc gain, g_a, and a time constant that produces a first-order roll-off in the magnitude. The dc gain of the amplifier is often on the order of 10^5 or 10^8 and the roll-off is typically in the frequency range 10 to 1000 Hz. The frequency of the second pole is a function of the output impedance and load capacitance. This is generally at a frequency that is significantly higher than the roll-off of the amplifier. A typical frequency response of the amplifier with a capacitive load is shown in Figure 11.4. Increasing the output resistance or the load capacitance will *decrease* the frequency at which the roll-off occurs in the amplifier frequency response. The roll-off in the magnitude is accompanied by an additional 90° of phase lag, which begins at a frequency approximately 1/10 the frequency associated with τ_L.

One might question why *increasing* the feedback resistance *increases* the stability of the closed-loop amplifier. Recall that the feedback resistor controls the amount of current that is fed back from the amplifier output to the negative terminal. Increasing

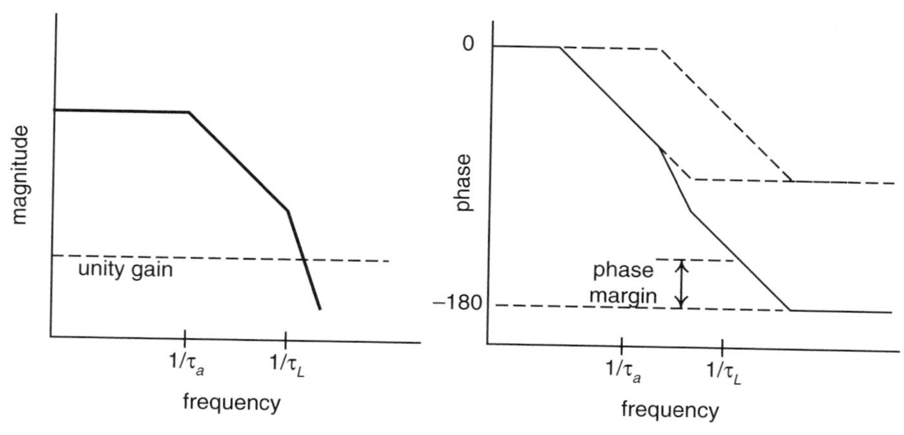

Figure 11.4 Representative frequency response plots for a power amplifier with a capacitive load.

the feedback resistor reduces the feedback from the output to the negative input and therefore increases the closed-loop stability.

Decreasing the feedback resistance does benefit the closed-loop performance in that it increases the closed-loop bandwidth of the amplifier. The fundamental compromise in the amplifier design is the trade-off between speed of response, or bandwidth, and fidelity of the amplifier output in representing the amplifier input. Decreasing the bandwidth of the amplifier produces an amplifier step response that exhibits a first-order rise to the steady-state value. Increasing the bandwidth increases the oscillations in the amplifier output, and increasing the bandwidth too much (by reducing the value of R_f) produces a purely oscillatory response that is associated with marginal closed-loop stability.

The fidelity of the output to changes in the input can be designed by choosing the phase margin of loop transfer function. The phase margin is determined by the phase of G_{LOOP} when the magnitude is equal to 1. Two typical designs are of interest. Choosing the phase margin to be 90° will produce a first-order response to the steady-state value, while choosing a phase margin of 45° will yield small oscillations but a higher bandwidth. Both designs can be studied by first expressing

$$G_{\mathrm{LOOP}}(j\omega) = \frac{g_a}{1 + R_f/R_i} \frac{1 - \tau_a\tau_L\omega^2 - j\omega(\tau_a + \tau_L)}{\left(1 - \tau_a\tau_L\omega^2\right)^2 + \omega^2(\tau_a + \tau_L)^2}. \tag{11.45}$$

The frequency at which the loop transfer function has a phase of −90° (which is equivalent to a phase margin of 90°) can be determined by solving for the frequency at which the real part of the loop transfer function becomes zero. This is

$$\omega^{90} = \frac{1}{\sqrt{\tau_a\tau_L}}. \tag{11.46}$$

To find the value of the feedback resistance that achieves a phase margin of 90°, we solve for the gain that produces $|G_{\text{LOOP}}(j\omega^{90})| = 1$. The result is

$$\left(\frac{R_f}{R_i}\right)^{90} = \frac{g_a\sqrt{\tau_a\tau_L}}{\tau_a + \tau_L} - 1. \tag{11.47}$$

Choosing the gain according to equation (11.47) produces a closed-loop bandwidth of ω^{90} rad/s.

The frequency and gain at which the phase margin is 45° can be found by solving for the frequency at which

$$\frac{-\omega(\tau_a + \tau_L)}{1 - \tau_a\tau_L\omega^2} = 1. \tag{11.48}$$

There are two solutions to this expression. Taking the one at the higher frequency produces the result

$$\omega^{45} = \frac{1}{2}\left(\frac{1}{\tau_a} + \frac{1}{\tau_L} + \sqrt{\frac{1}{\tau_a^2} + \frac{6}{\tau_a\tau_L} + \frac{1}{\tau_L^2}}\right). \tag{11.49}$$

Substituting equation (11.49) into equation (11.45) and simplifying yields

$$G_{\text{LOOP}}(j\omega^{45}) = \frac{g_a}{1 + R_f/R_i}\frac{1}{1 - \tau_a\tau_L(\omega^{45})^2}\frac{1+j}{2}. \tag{11.50}$$

The value of the feedback resistance that produces $|G_{\text{LOOP}}(j\omega^{45})| = 1$ is

$$\left(\frac{R_f}{R_i}\right)^{45} = \frac{g_a\sqrt{2}}{2}\left|\frac{1}{1 - \tau_a\tau_L(\omega^{45})^2}\right| - 1. \tag{11.51}$$

A simple rule of thumb can be derived if we assume that $\tau_L \ll \tau_a$. Under this assumption, equation (11.49) reduces to the approximation

$$\omega^{45} \approx \frac{1}{\tau_L}, \tag{11.52}$$

which indicates that if the time constant of the load in series with the amplifier is much faster than the time constant of the amplifier, the closed-loop time constant of the system for a 45° phase margin is approximately equal to the time constant τ_L.

Example 11.7 A power amplifier is designed for a piezoelectric actuator that has a short-circuit capacitance of 1.2 μF. Determine the bandwidth of the actuator for a design with phase margins of 90° and 45°. The output impedance of the power amplifier is 50 Ω and the time constant of the amplifier is 50 ms.

Solution The variable τ_a can be determined from the time constant,

$$\tau_a = 0.05 \text{ s}.$$

The time constant of the amplifier in series with the load is determined from equation (11.37),

$$\tau_L = (50\ \Omega)(1.2\ \mu\text{F}) = 60\ \mu\text{s}.$$

Equation (11.46) is used to compute the closed-loop bandwidth for a phase margin of 90°,

$$\begin{aligned}\omega^{90} &= \frac{1}{\sqrt{(0.05\text{ s})(60\times 10^{-6}\text{ s})}}\\ &= 408 \text{ rad/s}.\end{aligned} \tag{11.53}$$

The closed-loop time constant assuming a phase margin of 90° is approximately 2.4 ms.

The closed-loop bandwidth for a phase margin of 45° is determined from equation (11.49). Substituting in the values of the time constants yields

$$\begin{aligned}\omega^{45} &= \frac{1}{2}\left[\frac{1}{0.05\text{ s}} + \frac{1}{60\times 10^{-6}\text{ s}}\right.\\ &\quad \left. + \sqrt{\frac{1}{(0.05\text{ s})^2} + \frac{6}{(0.05\text{ s})\left(60\times 10^{-6}\text{ s}\right)} + \frac{1}{\left(60\times 10^{-6}\text{ s}\right)^2}}\right]\\ &= 16{,}687 \text{ rad/s}.\end{aligned}$$

The closed-loop time constant assuming a 45° phase margin is approximately 60 μs. The compromise for the increased speed of response is the fact that additional oscillation will occur in the amplifier output due to the decrease in phase margin.

The preceding analysis assumes that the electrical load is purely capacitive. In certain instances the load is not a pure capacitor or a pure resistor but can be modeled as a combination of resistors and capacitors. This is the case when dissipative mechanisms are added to a piezoelectric model, or we are modeling the electrical load of an electroactive polymer. A general analysis for power amplifier design is obtained if we assume that the load is modeled as a linear impedance function $Z_L(\mathsf{s})$. In this case the load transfer function is

$$G_L(\mathsf{s}) = \frac{Z_L(\mathsf{s})}{R_o + Z_L(\mathsf{s})}. \tag{11.54}$$

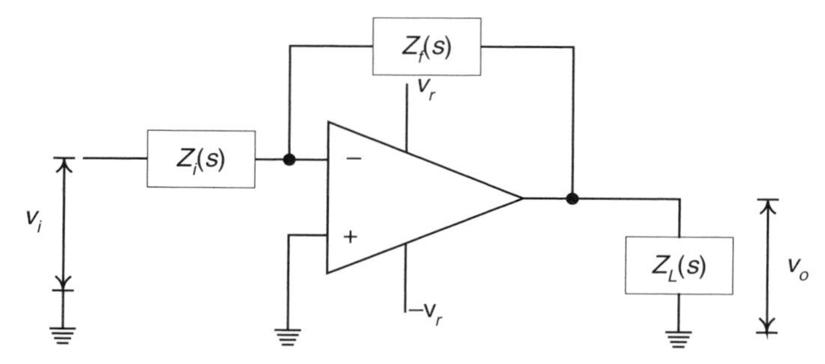

Figure 11.5 Block diagram of a power amplifier design for a generalized load impedance and generalized compensation networks.

The feedback analysis of the power amplifier can be generalized by assuming that the feedback network consists of a feedback impedance $Z_f(\mathrm{s})$ and an input impedance $Z_i(\mathrm{s})$. This is shown in Figure 11.5.

The block diagram of the system can be reduced to an input–output relationship:

$$\frac{V_o(\mathrm{s})}{V_i(\mathrm{s})} = \frac{[Z_f(\mathrm{s})/(Z_i(\mathrm{s}) + Z_f(\mathrm{s}))]G_a(\mathrm{s})G_L(\mathrm{s})}{1 + [Z_i(\mathrm{s})/(Z_i(\mathrm{s}) + Z_f(\mathrm{s}))]G_a(\mathrm{s})G_L(\mathrm{s})}. \tag{11.55}$$

We can see that in the generalized case the feedback and input networks can be chosen to vary the loop transfer function

$$G_{\mathrm{LOOP}}(\mathrm{s}) = \frac{Z_i(\mathrm{s})}{Z_i(\mathrm{s}) + Z_f(\mathrm{s})} G_a(\mathrm{s})G_L(\mathrm{s}). \tag{11.56}$$

The choice of input and feedback networks allow for feedback compensation of the loop transfer function. The compensation network can be chosen to enable better compromise between closed-loop bandwidth and overshoot in the amplifier response. Many of the techniques for proportional–derivative control can be employed to design the compensation networks $Z_i(\mathrm{s})$ and $Z_f(\mathrm{s})$.

11.2.3 Switching and Regenerative Power Amplifiers

Power analysis for smart material systems illustrates that energy dissipation increases and bandwidth decreases when the load is capacitive. This occurs, of course, when the material is piezoelectric or an electroactive polymer. In a number of engineering systems the power dissipation associated with a capacitive load can be a severe limitation to the application of smart material systems. This challenge has motivated the development of novel ways of amplifying the signal while minimizing the energy dissipation.

One of the important attributes of capacitive loads is that the power flow between the source and the load is bidirectional. As discussed in Section 11.1, an ideal capacitive load will have zero average power and a power flow that will oscillate between positive

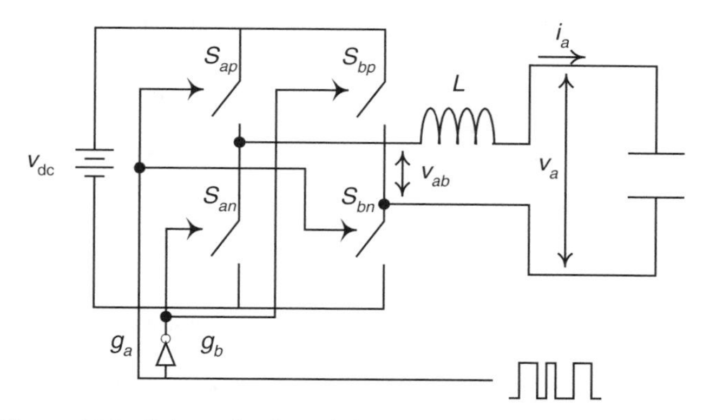

Figure 11.6 Schematic of a switching amplifier for a capacitive load.

and negative. The primary problem of a linear amplifier is that the recirculation of power between the source and the load will cause additional energy dissipation. Essentially, there is no way for the amplifier to accept the recirculated power, and it is therefore dissipated.

A more efficient amplifier is one that is able to recycle the recirculating power associated with a capacitive load. How is this accomplished? Consider a set of switches that are connected to a dc power supply with a maximum potential of V_{dc} (Figure 11.6). These switches have only two states, on and off, and when they are on they have very low resistance and hence dissipate very little energy. The switches are labeled S_{ap}, S_{an}, S_{bp}, and S_{bn}. When switches S_{ap} and S_{bn} are activated, the potential v_{ab}, is equal to v_{dc}. Closing these switches and opening S_{an} and S_{bp} produces a potential difference v_{ab} which is $-v_{dc}$. Alternating the opening and closing of the switch pairs can produce a signal v_{ab} that modulates controllably between $\pm v_{dc}$.

Placing the switching voltage v_{ab} across a piezoelectric material would produce undesirable current pulses in a capacitive load due the fact that $i = C\,dv/dt$. This problem is eliminated by placing an inductor at the output of the switching amplifier. The relationship between current and voltage in an inductor is

$$v(t) = L\frac{di}{dt}, \tag{11.57}$$

where L is the inductance. Equation (11.57) can be rewritten as

$$i(t) = \frac{1}{L}\int v(t)\,dt. \tag{11.58}$$

With an inductor at the output of the amplifier, the current will be transformed from a series of pulses to a triangular wave.

Without any additional electronics, the triangular wave output current will be a signal whose fundamental frequency is equal to the switching frequency of the amplifier. One final set of components is required to control the switching behavior of the amplifier. Consider the case in which the switches are on the same amount of time that they are off. If we define the *duty cycle* of the switches as the percentage of time they are on versus the total time, a duty cycle of 50% is the case in which the on time is equal to the off time. An extreme case is when the duty cycle of the switches is equal to 0% or 100%. In this case the output current would be a straight line with a slope of v_{ab}/L, due to the voltage–current relationship of the inductor. From this analysis we see that if we control the duty cycle of the switches, we can moduluate the frequency of the output current.

The duty cycle of the switches is controlled by introducing a *pulse width modulator* into the amplifier circuit. A pulse width modulator consists of a circuit component known as a *comparator*. The comparator has two inputs and the output of the comparator is a signal that is +1 when the signal at terminal 1 is greater than the signal at terminal 2, and the output is 0 when the signal at terminal 2 is greater than the signal at terminal 1. The comparator is a basic logic circuit. If we apply a triangular signal known as a *carrier signal* to terminal 2 and a *reference signal* to terminal 1, the duty cycle of the comparator output will be modulated by the relative size of the two signals. It will be 1 when the reference is greater than the carrier signal and 0 when the reference is less than the carrier signal.

The output of the comparator becomes the signal that opens and closes the switches of the amplifier. If we connect the output of the comparator to a gate that inverts the logic, we now have two *gating signals* that control the opening and closing of the switches. Furthermore, the duty cycle of the gating signals is controlled by the reference signal. Controlling the duty cycle is tantamount to controlling the current flow across the inductor and the load. Figure 11.7 is an illustration of the gating signals and the resulting current across the piezoelectric actuator (assuming a purely capacitive load). Note that the variation in the duty cycle produces a current waveform that consists of a low-frequency component with a small ripple signal superimposed on it. The ripple is due to the switching of the gating signals and represents an additional source of noise for the piezoelectric actuator. Typically, the switching frequency is much higher than the frequency of operation of the piezoelectric device to minimize the effects of ripple on the output of the piezoelectric material.

The efficiency of a switching amplifier comes from the low on resistance of the electronic switches. Unlike a linear amplifier, which does not allow for recirculation of the current, a switching amplifier can recirculate the current with only a small loss. The loss in the amplifier comes from any resistance in the switches when they are closed. This resistance can be minimized, thus increasing the overall efficiency of the switching amplifier. Switching amplifiers have been built with efficiencies on the order of 80% to even greater than 95%.

A switching amplifier does have some disadvantages compared to a linear amplifier. The need for an inductor to smooth the output often increases the size of a switching amplifier. Depending on the amount of power dissipation, it is possible for a switching amplifier to be larger than an equivalent linear amplifier since increasing the size of the inductor will decrease the ripple in the switching output. Ripple is also another disadvantage of a switching amplifier compared to a linear amplifier. As we know

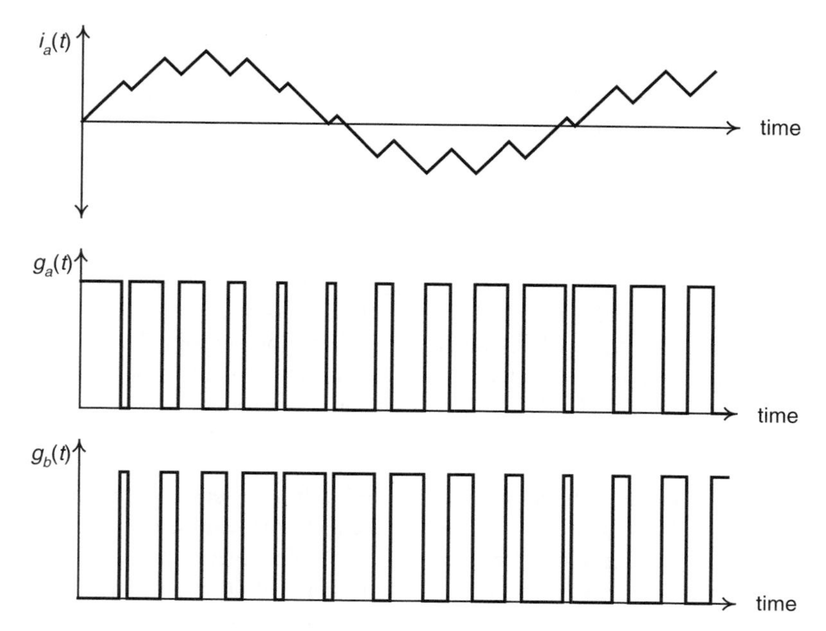

Figure 11.7 Gating signals and resulting current across a capacitive load for a switching amplifier.

from our analysis of smart material systems, it is possible for resonant frequencies to occur at frequencies much higher than the operating range of an actuator. If the switching frequency of an amplifier coincides with a resonant mode, the small ripple in the amplifier can be amplified by the mechanical resonance and cause undesirable vibration in the system.

11.3 ENERGY HARVESTING

Most of our discussion of power has centered on the use of smart materials as actuators. We have found that the electrical properties of the smart material determine the energy dissipation properties of the power amplifier.

A novel use of certain smart materials is to use the reciprocal property of the energy conversion to create *energy-harvesting* devices that convert mechanical energy into stored electrical energy. The concept of energy harvesting is gaining interest due to the explosion of battery-powered devices. Although great strides have been made in the development of low-power electronic devices, which, in turn, increases the lifetime of devices that must carry their own energy source with them, battery-powered devices still have a finite lifetime before they have to be disposed of or recharged. This may be unacceptable in certain applications, such as the use of a device in a remote location, which prohibits recharging or changing of a battery.

In this section we concentrate on the analysis of energy-harvesting devices that convert ambient mechanical energy to stored electrical energy. Of the materials that we have studied in this book, the most appropriate material for this application is piezoelectric polymers and ceramics. Although it has been shown that electroactive polymers can be used as energy-harvesting devices, the low conversion efficiency reduces their utility in small-strain applications. Probably the most often used materials are piezoelectric materials, due to their relatively high energy conversion efficiency and their ability to be integrated into structural components. For this reason we focus on analyzing the mechanical-to-electrical energy conversion of piezoelectric materials.

We have already studied the mechanical-to-electrical energy conversion of piezoelectric materials for the analysis of mechanical motion sensors. The primary difference in the sensor analysis is that a piezoelectric motion sensor does not produce any appreciable *power*. The function of a sensor is to produce a voltage or current that is strongly correlated (hopefully, through a linear relationship) with the physical quantity of interest.

A piezoelectric energy-harvesting device differs from a sensor in that its function is to convert mechanical energy into electrical energy that can be *stored* in a device, such as a battery or capacitor, for future use. Thus, the energy-harvesting material must extract useful energy from the ambient mechanical vibration.

The basic components of a piezoelectric energy harvester are the piezoelectric material, a rectifier that transforms the oscillatory output of the piezoelectric to a unipolar signal, and a storage device such as a battery or capacitor. The primary components of an energy-harvesting system are shown in Figure 11.8. For simplicity we assume that the energy-harvesting device is oscillating at a single frequency, although this is not a constraint, and that the output of the piezoelectric material is a single-frequency harmonic. This oscillatory signal is connected to a rectifier circuit that produces a unipolar signal, as shown in the figure. The unipolar voltage signal also produces a unipolar current signal. The output of the rectifier circuit is smoothed so that it is a dc signal with a small ripple at the oscillation frequency of the piezoelectric elements. This is the same type of ac-to-dc conversion that is performed in a transformer for household appliances. The dc signal (with its small ripple) is then connected to an energy storage device such as a capacitor. The capacitor then stores the electrical energy as charge. An ideal capacitor will hold the charge when

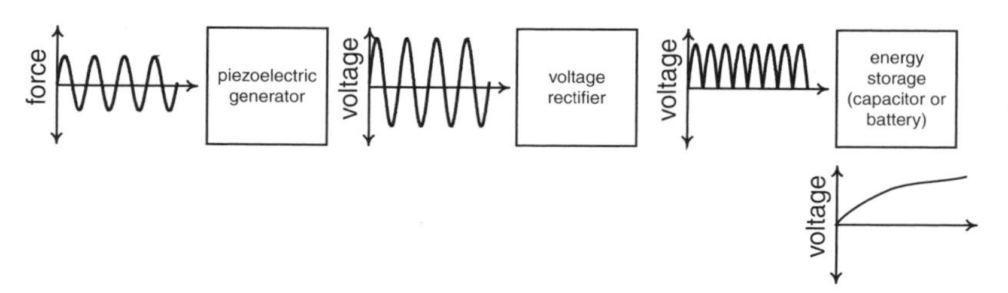

Figure 11.8 Components of a piezoelectric energy-harvesting system.

disconnected from the energy-harvesting device and then discharge the energy when connected to a load.

Our analysis of piezoelectric energy-harvesting circuits is based on the analysis that we developed for active–passive dampers in Chapter 9. Consider the equation for a piezoelectric element subjected to an external force

$$\begin{aligned} m\ddot{u}(t) + ku(t) + k_a^{\mathrm{D}}u(t) - g_{33}q(t) &= f(t) \\ -g_{33}u(t) + \frac{1}{C^{\mathrm{S}}}q(t) &= v(t). \end{aligned} \tag{11.59}$$

The energy-harvesting analysis motivates us to compute the output voltage as a function of the input force. Transforming both equations to the Laplace domain yields

$$\begin{aligned} \left(m\mathrm{s}^2 + k + k_a^{\mathrm{D}}\right)U(\mathrm{s}) - g_{33}Q(\mathrm{s}) &= F(\mathrm{s}) \\ -g_{33}U(\mathrm{s}) + \frac{1}{C^{\mathrm{S}}}Q(\mathrm{s}) &= V(\mathrm{s}). \end{aligned} \tag{11.60}$$

The equations can be written with displacement and voltage as the dependent variables

$$\begin{aligned} \left(m\mathrm{s}^2 + k + k_a^{\mathrm{E}}\right)U(\mathrm{s}) - g_{33}C^{\mathrm{S}}V(\mathrm{s}) &= F(\mathrm{s}) \\ g_{33}C^{\mathrm{S}}U(\mathrm{s}) + C^{\mathrm{S}}V(\mathrm{s}) &= Q(\mathrm{s}). \end{aligned} \tag{11.61}$$

From these expressions we can derive a relationship between the applied force and the output voltage of the piezoelectric. Before doing so, though, we have to consider the properties of the rectifier circuit shown in Figure 11.8. The purpose of the rectifier circuit is to transform the alternative voltage of the piezoelectric output to a unipolar voltage source that will charge the energy storage device. There are a number of ways to do this, the most straightforward method being to implement a *full-wave rectifier* (Figure 11.9*a*). Understanding the full-wave rectifier requires a basic understanding

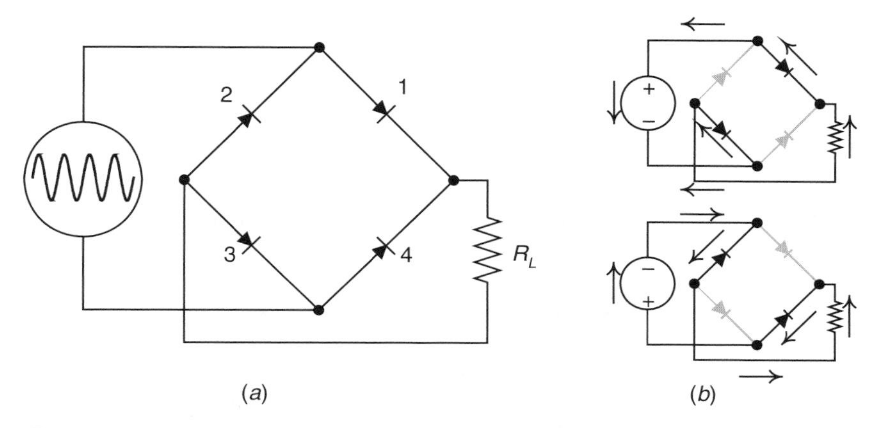

Figure 11.9 (*a*) Full-wave rectifier. (*b*) Current flow in a rectifier for an oscillatory input.

of the diode. The diode is a nonlinear circuit element that acts as a voltage-dependent switch for current flow. The fundamental property of a diode is that it will conduct current in only one direction—until a large negative voltage called the *breakdown voltage* is reached—and current will conduct only when the forward voltage becomes greater than a threshold value.

The full-wave rectifier works in the following manner. Assume that the output of the piezoelectric material is a sinusoidal voltage. When the piezoelectric output is positive, diodes 1 and 3 will conduct and current will flow as shown at the top of Figure 11.9*b*. When the voltage across the piezoelectric is negative, diodes 2 and 4 will conduct and voltage will flow in the circuit as shown at the bottom of Figure 11.9*b*. The important fact is that *current is always flowing in the same direction across the load.*

The energy-harvesting circuit is analyzed by considering the expression for the voltage and current across the rectifier load. Consider modeling the energy storage element as an ideal resistor with resistance R_L, which incorporates the small on-resistance of the diodes; then the relationship between voltage and current across the load when the diodes are conducting is

$$i_L(t) = -\frac{v_L}{R_L}. \tag{11.62}$$

Transforming equation (11.62) into the Laplace domain (and ignoring initial conditions) yields

$$I_L(\mathsf{s}) = -\frac{1}{R_L} V_L(\mathsf{s}). \tag{11.63}$$

To analyze the energy-harvesting circuit, we recognize that when the diodes are conducting, $v = v_L$ and $i_L = dq(t)/dt$. Under this assumption we can substitute equation (11.63) into equation (11.61) to compute the expression for the voltage as a function of the input force. Writing $Q(\mathsf{s}) = I_L(\mathsf{s})/\mathsf{s}$ and substituting into equation (11.61), we have

$$\begin{aligned} \left(m\mathsf{s}^2 + d_v\mathsf{s} + k + k_a^{\mathrm{E}}\right) U(\mathsf{s}) - g_{33}C^{\mathrm{S}} V(\mathsf{s}) &= F(\mathsf{s}) \\ g_{33}C^{\mathrm{S}} U(\mathsf{s}) + \frac{\mathsf{s}R_L C^{\mathrm{S}} + 1}{\mathsf{s}R_L} V(\mathsf{s}) &= 0. \end{aligned} \tag{11.64}$$

The voltage-to-force relationship is solved by matrix manipulation

$$\frac{V(\mathsf{s})}{F(\mathsf{s})} = \frac{-g_{33}(R_L C^{\mathrm{S}}\mathsf{s})}{\left(m\mathsf{s}^2 + k + k_a^{\mathrm{E}}\right)\left(\mathsf{s}R_L C^{\mathrm{S}} + 1\right) + \left(g_{33}^2 C^{\mathrm{S}}\right)\left(R_L C^{\mathrm{S}}\mathsf{s}\right)}. \tag{11.65}$$

Several basic properties of the energy-harvesting system can be obtained from equation (11.65). First, we note that the voltage is zero if the coupling parameter g_{33} is equal to zero. This is sensible since no electromechanical coupling exists if this term is zero. Second, we also notice that the voltage is zero if the forward resistance of the diodes R_L is equal to zero. This occurs because zero forward resistance would produce a short-circuit condition once the diodes are conducting, thus producing zero voltage drop across the piezoelectric material.

It is important to note that this analysis ignores three important properties of the rectifier circuit. This linear analysis ignores any transients associated with the switching of the rectifier circuit. Second, the analysis does not account for the nonzero voltage drop that occurs across the diodes when they transition from nonconducting to conducting. Ignoring the voltage drop across the diodes is appropriate if the voltage output of the piezoelectric is much greater than the voltage drop across the diodes, which is generally on the order of 0.3 to 0.7 V, depending on the diode material. Finally, our analysis has ignored the existence of the reverse breakdown voltage by assuming that the diodes can conduct in only one direction. This aspect of the analysis is reasonable as long as the voltage of the piezoelectric is kept lower than the reverse breakdown voltage of the diodes.

If these assumptions are valid, we can write an expression for the current output of the energy harvester by combining equations (11.63) and (11.65):

$$\frac{I(\mathsf{s})}{F(\mathsf{s})} = \frac{-g_{33}C^{\mathrm{S}}\mathsf{s}}{\left(m\mathsf{s}^2 + k + k_a^{\mathrm{E}}\right)\left(\mathsf{s}R_L C^{\mathrm{S}} + 1\right) + \left(g_{33}^2 C^{\mathrm{S}}\right)\left(R_L C^{\mathrm{S}}\mathsf{s}\right)}. \tag{11.66}$$

Derivation of equations (11.65) and (11.66) allow us to perform a nondimensional analysis of the voltage and current relationships. Making the definitions

$$\begin{aligned} \omega_n^{\mathrm{E}} &= \sqrt{\frac{k + k_a^{\mathrm{E}}}{m}} \\ K^2 &= \frac{g_{33}^2 C^{\mathrm{S}}}{k + k_a^{\mathrm{E}}} \\ \tilde{\mathsf{s}} &= \frac{\mathsf{s}}{\omega_n^{\mathrm{E}}}, \end{aligned} \tag{11.67}$$

we can substitute the previous three expressions into equation (11.66) and rewrite the expression as

$$\frac{I(\tilde{\mathsf{s}})}{F(\tilde{\mathsf{s}})} = \frac{-g_{33}C^{\mathrm{S}}\omega_n^{\mathrm{E}}}{k + k_a^{\mathrm{E}}} \frac{\tilde{\mathsf{s}}}{(\tilde{\mathsf{s}}^2 + 1)\left(\tilde{\mathsf{s}}R_L C^{\mathrm{S}}\omega_n^{\mathrm{E}} + 1\right) + K^2\left(R_L C^{\mathrm{S}}\omega_n^{\mathrm{E}}\tilde{\mathsf{s}}\right)}. \tag{11.68}$$

Now that the expression is separated into a coefficient in parentheses that has the units of current per force, and the remaining terms are nondimensional. The nondimensional component of the equation is in terms of two parameters. The term $R_L C^{\mathrm{S}}\omega_n^{\mathrm{E}}$ is a ratio of time constants. The second term, K, is a modified coupling coefficient that reflects the additional stiffness associated with the passive spring in the model. When $k = 0$, the modified coupling coefficient reduces to the coupling coefficient of the material.

The voltage-to-force expression is nondimensionalized in the same manner. The result is

$$\frac{V(\tilde{s})}{F(\tilde{s})} = \frac{-g_{33}}{k + k_a^{\mathrm{E}}} \frac{R_L C^{\mathrm{S}} \omega_n^{\mathrm{E}} \tilde{s}}{(\tilde{s}^2 + 1)\left(\tilde{s} R_L C^{\mathrm{S}} \omega_n^{\mathrm{E}} + 1\right) + K^2 \left(R_L C^{\mathrm{S}} \omega_n^{\mathrm{E}} \tilde{s}\right)}. \tag{11.69}$$

We can plot the frequency dependent term in equations (11.68) and (11.69) to understand how the voltage and current vary as a function of the design parameters. Let us define

$$\begin{aligned} \frac{\tilde{I}(\tilde{s})}{\tilde{F}(\tilde{s})} &= \frac{\tilde{s}}{(\tilde{s}^2 + 1)\left(\tilde{s} R_L C^{\mathrm{S}} \omega_n^{\mathrm{E}} + 1\right) + K^2 \left(R_L C^{\mathrm{S}} \omega_n^{\mathrm{E}} \tilde{s}\right)} \\ \frac{\tilde{V}(\tilde{s})}{\tilde{F}(\tilde{s})} &= \frac{R_L C^{\mathrm{S}} \omega_n^{\mathrm{E}} \tilde{s}}{(\tilde{s}^2 + 1)\left(\tilde{s} R_L C^{\mathrm{S}} \omega_n^{\mathrm{E}} + 1\right) + K^2 \left(R_L C^{\mathrm{S}} \omega_n^{\mathrm{E}} \tilde{s}\right)} = \left(R_L C^{\mathrm{S}} \omega_n^{\mathrm{E}}\right) \frac{\tilde{I}(\tilde{s})}{\tilde{F}(\tilde{s})}. \end{aligned} \tag{11.70}$$

as the nondimensional functions that relate force to current and voltage.

Assuming an ideal rectifier circuit, we can write the steady-state current across the load as

$$i_{\mathrm{ss}}(t) = \frac{-g_{33} C^{\mathrm{S}} \omega_n^{\mathrm{E}}}{k + k_a^{\mathrm{E}}} \left| \frac{\tilde{I}(\tilde{s})}{\tilde{F}(\tilde{s})} \right| \left| \sin\left(\omega t + \angle \frac{\tilde{I}(\tilde{s})}{\tilde{F}(\tilde{s})}\right) \right| F_o, \tag{11.71}$$

where F_o is the amplitude of the input force. The steady-state voltage is

$$v_{\mathrm{ss}}(t) = \frac{-g_{33}}{k + k_a^{\mathrm{E}}} R_L C^{\mathrm{S}} \omega_n^{\mathrm{E}} \left| \frac{\tilde{I}(\tilde{s})}{\tilde{F}(\tilde{s})} \right| \left| \sin\left(\omega t + \angle \frac{\tilde{I}(\tilde{s})}{\tilde{F}(\tilde{s})}\right) \right| F_o. \tag{11.72}$$

The expression for the steady-state power is the product of equations (11.71) and (11.72):

$$P_{\mathrm{ss}}(t) = \frac{g_{33}^2 C^{\mathrm{S}} \omega_n^{\mathrm{E}}}{\left(k + k_a^{\mathrm{E}}\right)^2} R_L C^{\mathrm{S}} \omega_n^{\mathrm{E}} \left| \frac{\tilde{I}(\tilde{s})}{\tilde{F}(\tilde{s})} \right|^2 \left| \sin\left(\omega t + \angle \frac{\tilde{I}(\tilde{s})}{\tilde{F}(\tilde{s})}\right) \right|^2 F_o^2. \tag{11.73}$$

The first term on the right-hand side

$$\frac{g_{33}^2 C^{\mathrm{S}} \omega_n^{\mathrm{E}}}{\left(k + k_a^{\mathrm{E}}\right)^2} = \frac{K^2}{\sqrt{m\left(k + k_a^{\mathrm{E}}\right)}}, \tag{11.74}$$

which allows us to write equation (11.73) as

$$P_{ss}(t) = \frac{F_o^2}{\sqrt{m\left(k + k_a^{\mathrm{E}}\right)}} K^2 \left(R_L C^{\mathrm{S}} \omega_n^{\mathrm{E}}\right) \left|\frac{\tilde{I}(\tilde{\mathrm{s}})}{\tilde{F}(\tilde{\mathrm{s}})}\right|^2 \left|\sin\left(\omega t + \angle \frac{\tilde{I}(\tilde{\mathrm{s}})}{\tilde{F}(\tilde{\mathrm{s}})}\right)\right|^2. \tag{11.75}$$

The average power over a single period can be found by integrating the harmonic function as a function of time and dividing by the period. The result is

$$< P_{ss}(t) > = \frac{F_o^2}{\sqrt{m\left(k + k_a^{\mathrm{E}}\right)}} \frac{K^2 \left(R_L C^{\mathrm{S}} \omega_n^{\mathrm{E}}\right)}{2} \left|\frac{\tilde{I}(\tilde{\mathrm{s}})}{\tilde{F}(\tilde{\mathrm{s}})}\right|^2. \tag{11.76}$$

Let us take a moment and analyze equation (11.76) to highlight some important results of the analysis. Equation (11.76) illustrates that the average steady-state power dissipated across the load for the energy-harvesting circuit is proportional to the ratio of time constants, $R_L C^{\mathrm{S}} \omega_n^{\mathrm{E}}$, as well as the square of the generalized coupling coefficient, K^2. It is reasonable that the average power should be proportional to the ratio of time constants. In the limit of zero load resistance this parameter will go to zero and the average power will also go to zero. This is sensible. Also sensible is the fact that the average steady-state power dissipated across the load will approach zero as the generalized coupling coefficient approaches zero. A small coupling coefficient implies that the energy conversion in the system is low, implying that only a small amount of energy can be harvested from the mechanical vibration source. Note that in this analysis the generalized coupling coefficient is a function of the material coupling coefficient and the stiffness of the system. A material with a high coupling coefficient can still have a small generalized coupling parameter if it is placed at a position on the structure where the passive stiffness is much greater than the short-circuit stiffness of the material.

Returning to the nondimensional analysis, we see from equation (11.76) that we can plot the the nondimensional function $< P_{ss}(t) > \sqrt{m(k + k_a^{\mathrm{E}})}/F_o^2$ to obtain an understanding of how the average steady-state power varies with frequency, the ratio of time constants, and the effective coupling coefficient. Figure 11.10*a* is a plot of this function for $K = 0.5$ and five different values of $R_L C^{\mathrm{S}} \omega_n^{\mathrm{E}}$. A viscous damping ratio of 0.01 was also incorporated into the analysis to represent a reasonable amount of structural damping. We note from the plot that values of $R_L C^{\mathrm{S}} \omega_n^{\mathrm{E}}$ that are too small (1/1000) or too large (1000) yield substantially smaller average steady-state power output from the energy-harvester. Choosing values in the range of $\frac{1}{10}$ to 10 produces a larger amount of steady-state power, but only near the system resonance, $\omega/\omega_n^{\mathrm{E}} \approx 1$. Off resonance the average steady-state power produced by the energy-harvesting system drops off substantially. The relationship between the resonance and the average steady-state power becomes even more pronounced when the effective coupling coefficient is reduced. For values of $K = 0.15$, we see from Figure 11.10*b* that to achieve maximum steady-state power we must tune the resonance of the energy-harvesting circuit to the excitation frequency of the input mechanical force, $\omega/\omega_n^{\mathrm{E}} = 1$; otherwise, the steady-state average power is reduced significantly. As the figure shows,

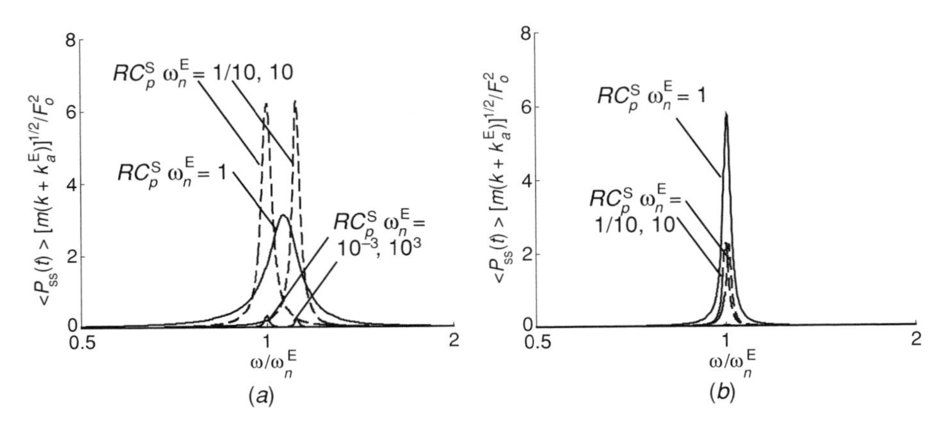

Figure 11.10 Plot of $< P_{ss}(t) > \sqrt{m(k+k_a^E)}/F_o^2$ for (a) $K = 0.5$ and (b) $K = 0.15$.

though, tuning the circuit properly allows us to achieve steady-state power output that is equal to the power output at larger values of the effective coupling coefficient.

Taken together, these results demonstrate that mechanical and electrical tuning of the circuit are critical to optimizing the steady-state power of the energy-harvesting circuit. For low values of the critical coupling coefficient it is important to tune the circuit so that both ω/ω_n^E and $R_L C^S \omega_n^E$ are approximately 1. For larger values of the coupling coefficient we see that the ratio of time constants $R_L C^S \omega_n^E$ can be varied around 1 to achieve maximum energy harvesting.

Example 11.8 An energy-harvesting circuit is being designed for a mechanical system with a stiffness of 6300 N/m. The piezoelectric material used for the energy-harvesting circuit has a short-circuit capacitance of 150 nF, a stiffness of 5000 N/m, and a strain coefficient of g_{33} = 45,000 N/C. The input vibration frequency is 40 Hz and the amplitude of the input force is 100 mN. Assume that the mechanical system has a damping ratio of 0.01. Compute (a) the value of the moving mass and the resistance that will approximately maximize the steady-state average power, and (b) the steady-state average power for the parameters chosen.

Solution (a) The average steady-state power can be obtained from equation (11.76), which requires that we know the ratio of the resonance frequency to the excitation frequency, the effective coupling coefficient, and the load resistance. The effective coupling coefficient is computed from

$$
\begin{aligned}
K &= \sqrt{\frac{g_{33}^2 C_p^S}{k + k_a^E}} \\
&= \sqrt{\frac{(45{,}000\ \text{N/C})^2\,(150 \times 10^{-9}\ \text{F})}{6300\ \text{N/m} + 5000\ \text{N/m}}} \\
&= 0.16.
\end{aligned}
$$

We see from Figure 11.10*b* that for a value of K in this range we want to choose to match the natural frequency of the system to the excitation frequency. Thus, we want $\omega/\omega_n^{\mathrm{E}} = 1$ to optimize the average steady-state power. Solving for the moving mass, we obtain the expression

$$\begin{aligned} m &= \frac{k + k_a^{\mathrm{E}}}{\omega^2} \\ &= \frac{6300\ \mathrm{N/m} + 5000\ \mathrm{N/m}}{(80\pi)^2} \\ &= 170\ \mathrm{g}. \end{aligned}$$

The load resistance is also computed by assuming that we want to set R_L such that $R_L C^{\mathrm{S}} \omega_n^{\mathrm{E}} = 1$. The result is

$$\begin{aligned} R_L &= \frac{1}{C_p^{\mathrm{S}} \omega_n^{\mathrm{E}}} \\ &= \frac{1}{\left(150 \times 10^{-9}\ \mathrm{F}\right)(80\pi\ \mathrm{rad/s})} \\ &= 26.53\ \mathrm{k\Omega}. \end{aligned}$$

(b) From the plot in Figure 11.10 we see that the function $< P_{\mathrm{ss}}(t) > \sqrt{m(k + k_a^{\mathrm{E}})}/F_o^2 \approx 6$ when $K = 0.15$ with a viscous damping ratio of 0.01. From this result we can estimate the average steady-state power as

$$\begin{aligned} < P_{\mathrm{ss}} > &\approx 6 \frac{F_o^2}{\sqrt{m\left(k + k_a^{\mathrm{E}}\right)}} \\ &\approx 6 \frac{\left(100 \times 10^{-3}\ \mathrm{N}\right)^2}{\sqrt{(0.170\ \mathrm{kg})(6300\ \mathrm{N/m} + 5000\ \mathrm{N/m})}} \\ &\approx 1.4\ \mathrm{mW}. \end{aligned}$$

A more accurate computation is obtained by realizing that $\tilde{s} = j$ and substituting the result into equation (11.76). The nondimensional expression for the current-to-force relationship is

$$\begin{aligned} \frac{\tilde{I}(\tilde{\mathrm{s}})}{\tilde{F}(\tilde{\mathrm{s}})} &= \frac{\tilde{\mathrm{s}}}{(\tilde{\mathrm{s}}^2 + 2\zeta\tilde{s} + 1)\left(\tilde{\mathrm{s}} R_L C^{\mathrm{S}} \omega_n^{\mathrm{E}} + 1\right) + K^2\left(R_L C^{\mathrm{S}} \omega_n^{\mathrm{E}} \tilde{\mathrm{s}}\right)} \\ \frac{\tilde{I}(j)}{\tilde{F}(j)} &= \frac{j}{(j0.02)(j+1) + (0.16)^2 j} \\ &= 18.39 - 8.07j. \end{aligned}$$

The magnitude of this function is 20.08. Substituting the result into equation (11.76) yields

$$\left[\frac{K^2\left(R_L C^S \omega_n^E\right)}{2}\right]\left|\frac{\tilde{I}(\tilde{s})}{\tilde{F}(\tilde{s})}\right|^2 = \frac{(0.16)^2\,(1)\,(20.08)^2}{2} = 5.16.$$

Recomputing the average steady-state power with this result yields

$$\begin{aligned} < P_{ss} > &\approx 5.16\frac{F_o^2}{\sqrt{m\left(k + k_a^E\right)}} \\ &\approx 5.16\frac{\left(100 \times 10^{-3}\ \text{N}\right)^2}{\sqrt{(0.170\ \text{kg})\,(6300\ \text{N/m} + 5000\ \text{N/m})}} \\ &\approx 1.2\ \text{mW}. \end{aligned}$$

The steady-state average power obtained from the circuit is reasonable for typical values of piezoelectric material parameters. Note that the steady-state power harvested by the circuit is proportional to the square of the input force; therefore, a tenfold increase in the force (to 1 N) would yield a 100-fold increase in the steady-state average power (to over 100 mW). As Figure 11.10 reveals, small changes in the resonant frequency or the excitation frequency can have a drastic impact on the steady-state average power when the effective coupling coefficient is low. This emphasizes the need to tune the circuit parameters properly to optimize the power obtained by the energy-harvesting circuit.

11.4 CHAPTER SUMMARY

Power analysis can play an important role in the implementation of smart material systems for many of the applications discussed in this book. In this chapter the fundamental properties of powering smart materials were analyzed to quantify the power necessary for actuation using piezoelectric, shape memory alloy, and electroactive polymer materials. The analysis was discussed in terms of the electrical properties of the various smart materials. Capacitive materials (e.g., piezoelectric ceramics) are an important subclass of smart materials because they present particular difficulties in the design of power amplifiers; that is, the additional capacitance of the material produces phase lag in the amplifier and can lead to amplifier instability. In addition, capacitive loads produce increased power dissipation in the amplifier, due to the fact that all of the power delivered to the material is returned and dissipated within the amplifier. The topic of switching amplifiers was then introduced as a means of overcoming the problem of driving capacitive loads with a linear amplifier. Switching amplifiers dissipate much less energy when driving a capacitive load but introduce additional distortion into the drive signal due to the on–off behavior of the switches.

The topic of energy harvesting with piezoelectric materials was analyzed with a single-mode model of a piezoelectric structure. Energy harvesting takes advantage of the generator properties of piezoelectric material to transform applied mechanical energy into stored electrical energy. An energy-harvesting device that incorporated a perfect rectifier was analyzed to highlight the importance of the electromechanical coupling and resonance behavior in maximizing the stored energy in the harvesting device. The expressions derived at the close of the chapter allow a quantification of the power and energy stored due to harmonic excitation of the piezoelectric material.

PROBLEMS

11.1. A 10-cm-long shape memory alloy wire with a circular cross section of 2 mm is being heated with a 3 V dc voltage. Assuming that the resistivity of the wire is 80 $\mu\Omega \cdot$ cm, compute the power across the wire.

11.2. A pieozoelectric stack actuator with a strain-free capacitance of 1.2 μF is being excited with a 100 V signal at 40 Hz.

(a) Compute the real power dissipated in the piezoelectric and the peak power.

(b) Repeat part (a) for a 200-V signal.

(c) Repeat part (a) for a signal frequency of 80 Hz.

11.3. The electrical impedance of an electroactive polymer actuator has been measured to be

$$Z(\mathsf{s}) = 1000 + \frac{10}{\mathsf{s}+2}.$$

(a) Compute the average power and peak power to a 2-V sine wave at 0.01 Hz.

(b) Repeat part (a) for a 2-V sine wave at 1 Hz.

(c) Repeat part (a) for a 2-V sine wave at 100 Hz.

11.4. Compute the power dissipated in a linear amplifier with supply voltage of 15 V for the shape memory alloy actuator analyzed in Problem 11.1.

11.5. Compute the power dissipated in a linear amplifier with suppy voltage of 150 V for the piezoelectric actuator analyzed in Problem 11.2a.

11.6. A linear amplifier with an open-loop gain of 10^7 and $\tau_a = 10$ rad/s is being used to excite a piezoelectric stack with a capacitance of 6 μF. The output resistance of the amplifier is 50 Ω.

(a) Plot the frequency response of the loop transfer function and identify the gain crossover frequency.

(b) Choose an input resistance and feedback resistance of the amplifier that makes the phase margin of the amplifier 45°.

(c) Compute the closed-loop bandwidth of the actuator with a phase margin of 45°.

11.7. Repeat Problem 11.6 for a piezoelectric bimorph actuator consisting of two 0.25-mm-thick plates of dimensions 20 by 50 mm. Assume that the actuators are connected in parallel and that the relative dielectric constant of the piezoelectric material is 4500.

11.8. Replot Figure 11.10 for $K = 0.3$ and $K = 0.7$.

11.9. A piezoelectric stack with stiffness 25 N/μm, strain-free capacitance 1.2 μF, and $g_{33} = 45{,}000$ N/C is being used in an energy-harvesting application. The attachment point of the structure has a stiffness of 10 N/μm and an effective mass of 0.8 kg.

(a) Compute the frequency at which the power harvested will be maximized.

(b) Compute the resistance required to maximize the frequency.

(c) Compute the power harvested at the frequency and resistance computed in parts (a) and (b).

11.10. Repeat Problem 11.9 when the frequency is $\frac{1}{10}$ the frequency that maximizes the power harvested.

NOTES

Additional information on basic electric analysis can be found in any number of introductory texts on the topic (e.g., Tse [125]). The linear amplifier analysis discussed in this chapter was also based on design information found in the Applications Notes for Apex Microtechnology. Additional considerations for piezoelectric drive electronics may be found in Main et al. [126] and Leo [127]. The discussion of switching power amplifiers was based on the work by Chandrekekaran et al. [128]. The concept of energy harvesting with a variety of materials has been in the literature for a number of years. Early patents on the concept can be found; note that these patents were approved as early as the late 1970s. Recently, there has been a renewed interest in the topic due to the development of ultralow-power sensor technologies. Recent work on the use of piezoelectric energy-harvesting devices for storing electrical energy is described in articles by Sodano et al. [129, 130]. Development of efficient power electronics for energy-harvesting technologies is discussed by Ottman et al. [131, 132]. Recently, the *Journal of Intelligent Material Systems and Structures* ran a special issue on energy harvesting. Details may be found in Clark [133].

REFERENCES

[1] Bailey, T., and Hubbard, J., 1985, "Distributed Piezoelectric-Polymer Active Vibration Control of a Cantilever Beam," *Journal of Guidance, Control, and Dynamics*, **8**(5), pp. 605–611.

[2] Crawley, E. F., and de Luis, J., 1987, "Use of Piezoelectric Actuators as Elements of Intelligent Structures," *AIAA Journal*, **25**(10), pp. 1373–1385.

[3] Fujishima, S., 2000, "The History of Ceramic Filters," *IEEE Transactions on Ultrasonics, Ferroelectrics, and Frequency Control*, **47**(1), pp. 1–7.

[4] Ikeda, T., 1996, *Fundamentals of Piezoelectricity*, Oxford University Press, New York.

[5] Jaffe, B., Jaffe, H., and Cook, W., 1971, *Piezoelectric Ceramics*, Academic Press, New York.

[6] IEEE, 1988, *IEEE Standard on Piezoelectricity*, Std. 176-1987, IEEE, Piscataway, NJ.

[7] Culshaw, B., 1996, *Smart Structures and Materials*, Artech House, Norwood, MA.

[8] Gandhi, M., and Thompson, B., 1992, *Smart Materials and Structures*, Chapman & Hall, London.

[9] Srinivasan, A., and McFarland, D., 2000, *Smart Structures*, Cambridge University Press, New York.

[10] Clark, R. L., Saunders, W., and Gibbs, G., 1998, *Adaptive Structures: Dynamics and Control*, Wiley, New York.

[11] Smith, R. C., 2005, *Smart Material Systems: Model Development*, SIAM, Philadelphia, PA.

[12] Gere, J. M., and Timoshenko, S. P., 1991, *Mechanics of Materials*, Chapman & Hall, London.

[13] Allen, D., and Haisler, W., 1985, *Introduction to Aerospace Structural Analysis*, Wiley, New York.

[14] Pilkey, W. D., and Wunderlich, W., 1994, *Mechanics of Structures: Variational and Computational Methods*, CRC Press, Boca Raton, FL.

[15] Reddy, J., 1984, *Energy and Variational Methods in Applied Mechanics*, Wiley, New York.

[16] Crandall, S., Karnopp, D., Kurtz, E., and Pridmore, D., 1968, *Dynamics of Mechanical and Electromechanical Systems*, McGraw-Hill, New York.

[17] Inman, D. J., 1994, *Engineering Vibration*, Prentice Hall, Upper Saddle River, NJ.

[18] Meirovitch, L., 1986, *Elements of Vibration Analysis*, McGraw-Hill, New York.

[19] Meirovitch, L., 2002, *Fundamentals of Vibrations*, McGraw-Hill, New York.

[20] Chen, C.-T., 1998, *Linear System Theory and Design*, Oxford University Press, New York.

[21] Friedland, B., 1986, *Control System Design: An Introduction to State-Space Methods*, McGraw-Hill, New York.

[22] Franklin, G., Powell, J., and Emami-Naeini, A., 2002, *Feedback Control of Dynamic Systems*, Prentice Hall, Upper Saddle River, NJ.

[23] Beranek, L. L., 1954, *Acoustics*, McGraw-Hill, New York.

[24] Giurgiutiu, V., and Rogers, C. A., 1997, "Power and Energy Characteristics of Solid-State Induced-Strain Actuators for Static and Dynamic Applications," *Journal of Intelligent Material Systems and Structures*, **8**(9), pp. 738–750.

[25] Crawley, E. F., and Anderson, E. H., 1990, "Detailed Models of Piezoceramic Actuation of Beams," *Journal of Intelligent Material Systems and Structures*, **1**(1), pp. 4–25.

[26] Crawley, E. F., and Lazarus, K. B., 1991, "Induced Strain Actuation of Isotropic and Anisotropic Plates," *AIAA Journal*, **29**(6), pp. 945–951.

[27] Leo, D. J., Kothera, C., and Farinholt, K., 2003, "Constitutive Equations for an Induced-Strain Bending Actuator with a Variable Substrate," *Journal of Intelligent Material Systems and Structures*, **14**(11), pp. 707–718.

[28] Zhou, S.-W., Liang, C., and Rogers, C., 1996, "An Impedance-Based System Modeling Approach for Induced Strain Actuator-Driven Structures," *Journal of Vibration and Acoustics*, **118**(3), pp. 323–331.

[29] Liang, C., Sun, F., and Rogers, C., 1996, "Electro-mechanical Impedance Modeling of Active Material Systems," *Smart Materials and Structures*, **5**, pp. 171–186.

[30] Liang, C., Sun, F. P., and Rogers, C. A., 1997, "An Impedance Method for Dynamic Analysis of Active Material Systems," *Journal of Intelligent Material Systems and Structures*, **8**(4), pp. 323–334.

[31] Giurgiutiu, V., Rogers, C. A., and Chaudhry, Z., 1996, "Energy-Based Comparison of Solid-State Induced-Strain Actuators," *Journal of Intelligent Material Systems and Structures*, **7**(1), pp. 4–14.

[32] Damjanovic, D., and Newnham, R., 1992, "Electrostrictive and Piezoelectric Materials for Actuator Applications," *Journal of Intelligent Material Systems and Structures*, **3**(2), pp. 190–208.

[33] Hom, C. L., and Shankar, N., 1994, "A Fully Coupled Constitutive Model for Electrostrictive Ceramic Materials," *Journal of Intelligent Material Systems and Structures*, **5**(6), pp. 795–801.

[34] Pablo, F., and Petitjean, B., 2000, "Characterization of 0.9PMN–0.1PT Patches for Active Vibration Control of Plate Host Structures," *Journal of Intelligent Material Systems and Structures*, **11**(11), pp. 857–867.

[35] Hagood, N., Chung, W., and von Flotow, A., 1990, "Modeling of Piezoelectric Actuator Dynamics for Active Structural Control," *Journal of Intelligent Material Systems and Structures*, **1**(3), pp. 327–354.

[36] Burke, S. E., and Hubbard, J. E., 1991, "Distributed Transducer Vibration Control of Thin Plates," *Journal of the Acoustical Society of America*, **90**, pp. 937–944.

[37] Burke, S. E., and Sullivan, J. M., 1995, "Distributed Transducer Shading via Spatial Gradient Electrodes," in *Proceedings of the SPIE*, **2443**, pp. 716–726.

[38] Lee, C., and Moon, F., 1990, "Modal Sensors/Actuators," *Journal of Applied Mechanics*, **57**, pp. 434–441.

[39] Lee, C., 1990, "Theory of Laminated Piezoelectric Plates for the Design of Distributed Sensors/Actuators, Part I: Governing Equations and Reciprocal Relationships," *Journal of the Acoustical Society of America*, **87**, pp. 1144–1158.

[40] Lee, C., Chiang, W.-W., and O'Sullivan, T., 1991, "Piezoelectric Modal Sensor/Actuator Pairs for Critical Active Damping Vibration Control," *Journal of the Acoustical Society of America*, **90**, pp. 374–384.

[41] Sullivan, J. M., Hubbard, J. E., and Burke, S. E., 1996, "Modeling Approach for Two-Dimensional Distributed Transducers of Arbitrary Spatial Distribution," *Journal of the Acoustical Society of America*, **99**, pp. 2965–2974.

[42] Clark, R. L., Burdisso, R. A., and Fuller, C. R., 1993, "Design Approaches for Shaping Polyvinylidene Fluoride Sensors in Active Structural-Acoustic Control," *Journal of Intelligent Material Systems and Structures*, **4**, pp. 354–365.

[43] Inman, D. J., 1989, *Vibration—with Control, Measurement, and Stability*, Prentice Hall, Englewood Cliffs, NJ.

[44] Blevins, R., 1984, *Formula for Natural Frequency and Mode Shape*, R.E. Kreiger, Melbourne, FL.

[45] Liang, C., and Rogers, C. A., 1990, "One-Dimensional Thermomechanical Constitutive Relations for Shape Memory Materials," *Journal of Intelligent Material Systems and Structures*, **1**(2), pp. 207–234.

[46] Liang, C., and Rogers, C. A., 1997, "Design of Shape Memory Alloy Actuators," *Journal of Intelligent Material Systems and Structures*, **8**(4), pp. 303–313.

[47] Zhang, X. D., Rogers, C. A., and Liang, C., 1997, "Modelling of the Two-Way Shape Memory Effect," *Journal of Intelligent Material Systems and Structures*, **8**(4), pp. 353–362.

[48] Brinson, L. C., 1993, "One-Dimensional Constitutive Behavior of Shape Memory Alloys: Thermomechanical Derivation with Non-constant Material Functions and Redefined Martensite Internal Variable," *Journal of Intelligent Material Systems and Structures*, **4**(2), pp. 229–242.

[49] Brinson, L. C., and Huang, M. S., 1996, "Simplifications and Comparisons of Shape Memory Alloy Constitutive Models," *Journal of Intelligent Material Systems and Structures*, **7**(1), pp. 108–114.

[50] Boyd, J., and Lagoudas, D., 1996, "A Thermodynamical Constitutive Model for Shape Memory Materials, Part I: The Monolithic Shape Memory Alloy," *International Journal of Plasticity*, **12**(6), pp. 805–842.

[51] Boyd, J., and Lagoudas, D., 1996, "A Thermodynamical Constitutive Model for Shape Memory Materials, Part II: The SMA Composite Material," *International Journal of Plasticity*, **12**(7), pp. 843–873.

[52] Zhang, Q. M., Bharti, V., and Zhao, X., 1998, "Giant Electrostriction and Relaxor Ferroelectric Behavior in Electron-Irradiated Poly(Vinylidene Fluoride-Trifluoroethylene) Copolymer," *Science*, **280**, pp. 2101–2104.

[53] Bar-Cohen, Y., ed., 2004, *Electroactive Polymer Actuators as Artificial Muscles*, SPIE Press, Bellingham, WA.

[54] Pelrine, R., Kornbluh, R., Pei, Q., and Joseph, J., 2000, "High-Speed Electrically Actuated Elastomers with Strain Greater than 100%," *Science*, **287**, pp. 836–839.

[55] Pelrine, R. E., Kornbluh, R. D., and Joseph, J. P., 1998, "Electrostriction of Polymer Dielectrics with Compliant Electrodes as a Means of Actuation," *Sensors and Actuators A*, **64**, pp. 77–85.

[56] Goulbourne, N., Mockensturm, E., and Frecker, M., 2005, "A Nonlinear Model for Dielectric Elastomer Membranes," *Journal of Applied Mechanics*, **72**(6), pp. 899–906.

[57] Baughman, R. H., 1996, "Conducting Polymer Artificial Muscles," *Synthetic Metals*, **78**, pp. 339–353.

[58] Santa, A. D., Rossi, D. D., and Mazzoldi, A., 1997, "Characterization and Modeling of a Conducting Polymer Muscle-like Linear Actuator," *Smart Materials and Structures*, **5**, pp. 23–34.

[59] Madden, J., Cush, R., Kanigan, T., and Hunter, I., 2000, "Fast Contracting Polypyrrole Actuators," *Synthetic Metals*, **113**, pp. 185–192.

[60] Spinks, G., Zhou, D., Liu, L., and Wallace, G., 2003, "The Amounts per Cycle of Polypyrrole Electromechanical Actuators," *Smart Materials and Structures*, **12**, pp. 468–472.

[61] Spinks, G., Campbell, T., and Wallace, G., 2005, "Force Generation from Polypyrrole Actuators," *Smart Materials and Structures*, **14**, pp. 406–412.

[62] Oguro, K., Kawami, Y., and Takenaka, H., 1992, "Bending of an Ion-Conducting Polymer Film–Electrode Composite by an Electric Stimulus at Low Voltage," *Journal of the Micromachine Society*, **5**, pp. 27–30.

[63] Kanno, R., Tadokoro, S., Takamori, T., Hattori, M., and Oguro, K., 1995, "Modeling of ICPF Actuator: Modeling of Electrical Characteristics," in *Proceedings of the International Conference on Industrial Electronics, Control, and Instrumentation*, pp. 913–918.

[64] Kanno, R., Tadokoro, S., Takamori, T., and Hattori, M., 1996, "Linear Approximate Dynamic Model of ICPF Actuator," in *Proceedings of the IEEE International Conference on Robotics and Automation*, pp. 219–225.

[65] Shahinpoor, M., Bar-Cohen, Y., Simpson, J., and Smith, J., 1998, "Ionic Polymer–Metal Composites (IPMCs) as Biomimetic Sensors, Actuators and Artificial Muscles: a Review," *Smart Materials and Structures*, **7**(6), pp. R15–R30.

[66] Shahinpoor, M., and Kim, K., 2002, "Solid-State Soft Actuator Exhibiting Large Electromechanical Effect," *Applied Physics Letters*, **80**(18), pp. 3445–3447.

[67] Bennett, M., and Leo, D. J., 2004, "Ionic Liquids as Stable Solvents for Ionic Polymer Transducers," *Sensors and Actuators A: Physical*, **115**, pp. 79–90.

[68] Akle, B., Bennett, M., and Leo, D., 2006, "High-Strain Ionic Liquid-Ionomeric Electroactive Actuators," *Sensors and Actuators A: Physical*, **126**(1), pp. 173–181.

[69] Li, J., and Nemat-Nasser, S., 2000, "Micromechanical Analysis of Ionic Clustering in Nafion Perfluourinated Membrane," in *Proceedings of the SPIE*, **3987**, pp. 103–109.

[70] Nemat-Nasser, S., and Thomas, C. W., 2001, "Ionomeric Polymer–Metal Composites," in *Electroactive Polymer Actuators as Artificial Muscles*, Chapter 6, pp. 139–191, SPIE Press, Bellingham, WA.

[71] Nemat-Nasser, S., 2002, "Micro-mechanics of Actuation of Ionic Polymer–Metal Composites," *Journal of Applied Physics*, **92**(5), pp. 2899–2915.

[72] Newbury, K. M., and Leo, D. J., 2002, "Electromechanical Modeling and Characterization of Ionic Polymer Benders," *Journal of Intelligent Material Systems and Structures*, **13**(1), pp. 51–60.

[73] Newbury, K. M., and Leo, D. J., 2003, "Linear Electromechanical Model of Ionic Polymer Transducers, Part I: Model Development," *Journal of Intelligent Material Systems and Structures*, **14**(6), pp. 343–342.

[74] Newbury, K. M., and Leo, D. J., 2003, "Linear Electromechanical Model of Ionic Polymer Transducers, Part II: Experimental Validation," *Journal of Intelligent Material Systems and Structures*, **14**(6), pp. 343–358.

[75] Royster, L., 1970, "The Flextensional Concept: A New Approach to the Design of Underwater Transducers," *Applied Acoustics*, **3**, pp. 117–126.

[76] Sugawara, Y., Onitsuka, K., Yoshikawa, S., Xu, Q., and Newnham, R., 1992, "Metal–Ceramic Composite Actuators," *Journal of the American Ceramic Society*, **75**, pp. 996–998.

[77] Dogan, A., Uchino, K., and Newnham, R. E., 1997, "Composite Piezoelectric Transducer with Truncated Conical Endcaps 'Cymbal,'" *IEEE Transactions on Ultrasonics, Ferroelectrics, and Frequency Control*, **44**, pp. 597–605.

[78] Haertling, G., 1994, "Chemically Reduced PLZT Ceramics for Ultra High Displacement Actuators," *Ferroelectrics*, **154**, pp. 101–106.

[79] Barron, B., Li, G., and Haertling, G., 1996, "Temperature Dependent Characteristics of Cerambow Actuators," in *IEEE International Symposium on Applications of Ferroelectrics*, **96CH35948**, pp. 305–308.

[80] Chandran, S., Kugel, V., and Cross, L., 1997, "Crescent: A Novel Piezoelectric Bending Actuator," in *Proceedings of the SPIE*, **3041**, pp. 461–469.

[81] Mossi, K., Selby, G. V., and Bryand, R. G., ????, "Thin-Layer Composite Unimorph Ferroelectric Driver and Sensor Properties," *Materials Letters*, **35**, pp. 39–49.

[82] Moskalik, A. J., and Brei, D., 1997, "Deflection-Voltage Model and Experimental Results for Polymeric Piezoelectric C-Block Actuators," *AIAA Journal*, **35**(9), pp. 1556–1558.

[83] Moskalik, A. J., and Brei, D., 1999, "Analytical Dynamic Performance Modeling for Individual C-Block Actuators," *Journal of Vibration and Acoustics*, **121**(2), pp. 221–229.

[84] Ervin, J., and Brei, D. E., 1998, "Recurve Piezoelectric-Strain-Amplifying Actuator Architecture," *IEEE/ASME Transactions on Mechatronics*, **3**, pp. 293–301.

[85] Ervin, J., and Brei, D. E., 2004, "Dynamic Behavior of Piezoelectric Recurve Actuation Architectures," *Journal of Vibration and Acoustics*, **126**, pp. 37–46.

[86] Nasser, K. M., and Leo, D. J., 2000, "Efficiency of Frequency-Rectified Piezohydraulic and Piezopneumatic Actuation," *Journal of Intelligent Material Systems and Structures*, **11**(10), pp. 798–810.

[87] Mauck, L., and Lynch, C. S., 2000, "Piezoelectric Hydraulic Pump Development," *Journal of Intelligent Material Systems and Structures*, **11**(10), pp. 758–764.

[88] Near, C. D., 1996, "Piezoelectric Actuator Technology," in *Proceedings of the Smart Structures and Materials Conference*, **2717**, pp. 246–258.

[89] Niezrecki, C., Brei, D., Balakrishnan, S., and Moskalik, A., 2001, "Piezoelectric Actuation: State of the Art," *Shock and Vibration Digest*, **33**(4), pp. 269–280.

[90] Moskalik, A., and Brei, D., 2001, "Dynamic Performance of C-Block Array Architectures," *Journal of Sound and Vibration*, **243**(2), pp. 317–346.

[91] Mauck, L., and Lynch, C. S., 1999, "Piezoelectric Hydraulic Pump," in *Proceedings of the SPIE*, **3668**, pp. 844–852.

[92] Sirohi, J., and Chopra, I., 2003, "Design and Development of a High Pumping Frequency Piezoelectric–Hydraulic Hybrid Actuator," *Journal of Intelligent Material Systems and Structures*, **14**(3), pp. 135–147.

[93] Tan, H., Hurst, W., and Leo, D., 2005, "Performance Modelling of a Piezohydraulic Actuation System with Active Valves," *Smart Materials and Structures*, **14**, pp. 91–110.

[94] Mallavarapu, K., and Leo, D. J., 2001, "Feedback Control of the Bending Response of Ionic Polymer Actuators," *Journal of Intelligent Material Systems and Structures*, **12**(3), pp. 143–155.

[95] Kothera, C. S., and Leo, D. J., 2005, "Bandwidth Characterization in the Micropositioning of Ionic Polymer Actuators," *Journal of Intelligent Material Systems and Structures*, **16**(1), pp. 3–13.

[96] Hagood, N. W., and von Flotow, A., 1991, "Damping of Structural Vibrations with Piezoelectric Materials and Passive Electrical Networks," *Journal of Sound and Vibration*, **146**(2), pp. 243–268.

[97] Beranek, L., and Ver, I., 1992, *Noise and Vibration Control Principles and Applications*, Wiley, New York.

[98] Harris, C. M., ed., 1988, *Shock and Vibration Handbook*, McGraw-Hill, New York.

[99] Johnson, C. D., and Kienholz, D. A., 1981, "Finite Element Prediction of Damping in Structures with Constrained Viscoelastic Layers," *AIAA Journal*, **20**(9), pp. 1284–1290.

[100] Johnson, C. D., 1995, "Design of Passive Damping Systems," *Journal of Mechanical Design*, **117**, p. 171.

[101] Lesieutre, G. A., 1998, "Vibration Damping and Control Using Shunted Piezoelectric Materials," *Shock and Vibration Digest*, **30**(3), p. 187.

[102] Tsai, M., and Wang, K., 1999, "On the Structural Damping Characteristics of Active Piezoelectric Actuators with Passive Shunt," *Journal of Sound and Vibration*, **221**(1), pp. 1–22.

[103] Park, C. H., and Inman, D. J., 2003, "Enhanced Piezoelectric Shunt Design," *Shock and Vibration*, **10**(2), pp. 127–133.

[104] Park, J.-S., Kim, H. S., and Choi, S.-B., 2006, "Design and Experimental Validation of Piezoelectric Shunt Structures Using Admittance Analysis," *Smart Materials and Structures*, **15**(1), pp. 93–103.

[105] Hollkamp, J. J., Starchville, J., and Thomas, F., 1994, "A Self-Tuning Piezoelectric Vibration Absorber," *Journal of Intelligent Material Systems and Structures*, **5**(4), pp. 559–566.

[106] Morgan, R. A., and Wang, K.-W., 1998, "An Integrated Active-Parametric Control Approach for Active-Passive Hybrid Piezoelectric Network with Variable Resistance," *Journal of Intelligent Material Systems and Structures*, **9**(7), pp. 564–573.

[107] Morgan, R. A., and Wang, K.-W., 2002, "Active-Passive Piezoelectric Absorbers for Systems Under Multiple Non-Sationary Harmonic Excitationse," *Journal of Sound and Vibration*, **255**(4), pp. 685–700.

[108] Hollkamp, J. J., 1994, "Multimodal Passive Vibration Suppression with Piezoelectric Materials and Resonant Shunts," *Journal of Intelligent Material Systems and Structures*, **5**(1), pp. 49–57.

[109] Cunefare, K. A., de Rosa, S., Sadegh, N., and Larson, G., 2000, "State-Switched Absorber for Semi-active Structural Control," *Journal of Intelligent Material Systems and Structures*, **11**(4), pp. 300–310.

[110] Cunefare, K. A., 2002, "State-Switched Absorber for Vibration Control of Point-Excited Beams," *Journal of Intelligent Material Systems and Structures*, **13**(2–3), pp. 97–105.

[111] Holdhusen, M. H., and Cunefare, K. A., 2003, "Damping Effects on the State-Switched Absorber Used for Vibration Suppression," *Journal of Intelligent Material Systems and Structures*, **14**(9), pp. 551–561.

[112] Clark, W. W., 2000, "Vibration Control with State-Switched Piezoelectric Materials," *Journal of Intelligent Material Systems and Structures*, **11**(4), pp. 263–271.

[113] Corr, L, R., and Clark, W. W., 2002, "Comparison of Low-Frequency Piezoelectric Switching Shunt Techniques for Structural Damping," *Smart Materials and Structures*, **11**, pp. 370–376.

[114] Inman, D. J., 2006, *Vibration with Control*, Wiley, Hoboken, NJ.

[115] Preumont, A., 2002, *Vibration Control of Active Structures*, Springer-Verlag, New York.

[116] Goh, C., and Caughey, T., 1985, "On the Stability Problem Caused by Finite Actuator Dynamics in the Collocated Control of Large Space Structures," *International Journal of Control*, **41**(3), pp. 787–802.

[117] Fanson, J., and Caughey, T., 1990, "Positive Position Feedback Control for Large Space Structures," *AIAA Journal*, **28**(4), pp. 717–724.

[118] Leo, D. J., and Inman, D. J., 1993, "Pointing Control and Vibration Suppression of a Slewing Flexible Frame," *Journal of Guidance, Control, and Dynamics*, **17**, pp. 529–536.

[119] Leo, D. J., and Inman, D. J., 1993, "Modeling and Control Simulations of a Slewing Frame Containing Active Members," *Smart Materials and Structures Journal*, **2**, pp. 82–95.

[120] McEver, M., and Leo, D. J., 2000, "Autonomous Vibration Suppression Using On-Line Pole-Zero Identification," *Journal of Vibration and Acoustics*, **123**(4), pp. 487–495.

[121] Juang, J.-N., and Phan, M., 1992, "Robust Controller Design for Second-Order Dynamic Systems: A Virtual Passive Approach," *Journal of Guidance, Control, and Dynamics*, **15**(5), pp. 1192–1198.

[122] Rew, K.-H., Han, J.-H., and Lee, I., 2002, "Multi-modal Vibration Control Using Adaptive Positive Position Feedback," *Journal of Intelligent Material Systems and Structures*, **13**(1), pp. 13–22.

[123] Dosch, J., Inman, D. J., and Garcia, E., 1992, "A Self-Sensing Piezoelectric Actuator for Collocated Control," *Journal of Intelligen Material Systems and Structures*, **3**(1), pp. 166–185.

[124] Anderson, E., and Hagood, N., 1994, "Simultaneous Piezoelectric Sensing-Actuation: Analysis and Application to Controlled Structures," *Journal of Sound and Vibration*, **174**(5), pp. 617–639.

[125] Tse, C. K., 1998, *Linear Circuit Analysis*, Addison-Wesley, Reading, MA.

[126] Main, J., Newton, D., Massengil, L., and Garcia, E., 1996, "Efficient Power Amplifiers for Piezoelectric Applications," *Smart Materials and Structures*, **5**(6), pp. 766–775.

[127] Leo, D. J., 2000, "Energy Analysis of Piezoelectric-Actuated Structures," *Journal of Intelligent Material Systems and Structures*, **10**(1), pp. 36–45.

[128] Chandrekekaran, S., Lindner, D., and Smith, R., 2000, "Optimized Design of Switching Power Amplifiers for Piezoelectric Actuators," *Journal of Intelligent Material Systems and Structures*, **11**, pp. 887–901.

[129] Sodano, H. A., Inman, D. J., and Park, G., 2005, "Comparison of Piezoelectric Energy Harvesting Devices for Recharging Batteries," *Journal of Intelligent Material Systems and Structures*, **16**(10), pp. 799–807.

[130] Sodano, H. A., Inman, D. J., and Park, G., 2005, "Generation and Storage of Electricity from Power Harvesting Devices," *Journal of Intelligent Material Systems and Structures*, **16**(1), pp. 67–75.

[131] Ottman, G., Hofmann, H., Bhatt, A., and Lesieutre, G., 2002, "Adaptive Piezoelectric Energy Harvesting Circuit for Wireless Remote Power Supply," *IEEE Transactions on Power Electronics*, **17**(5), pp. 669–676.

[132] Ottman, G., Hofmann, H., and Lesieutre, G., 2003, "Optimized Piezoelectric Energy Harvesting Circuit Using Step-down Converter in Discontinuous Conduction Mode," *IEEE Transactions on Power Electronics*, **18**(2), pp. 696–703.

[133] Clark, W. W., 2005, Preface: Special Issue on Energy Harvesting, *Journal of Intelligent Material Systems and Structures*, **16**(10), p. 783.

INDEX

S

T

V

W

Z